Women, Science, and Technology

Second Edition

Women, Science, and Technology is an ideal reader for courses in feminist science studies, science studies more generally, women's studies, and gender and education. The second edition updates the first edition with readings that:

- extend content coverage into areas not originally included, such as reproductive, agricultural, medical, and imaging technologies;
- reflect new feminist theory and research on biology, language, the global economy, and the intersection of race and class with gender;
- provide current statistical information about the representation of women and people of color in science, technology, engineering, and mathematics; and
- are more accessible for students.

Section introductions are also fully updated to cover the latest controversies, such as Harvard president Lawrence Summers's widely debated discussion about women and science and the current debates around reports on the low number of women engineers.

Mary Wyer is Assistant Dean of Interdisciplinary Studies at North Carolina State University, Raleigh.

Mary Barbercheck is Professor of Entomology at Pennsylvania State University, University Park.

Donna Giesman is Research Integrity Officer in the School of Medicine, Duke University, Durham.

Hatice Örün Öztürk is Teaching Associate Professor at the Department of Electrical and Computer Engineering at North Carolina State University, Raleigh.

Marta Wayne is Associate Professor in the Department of Zoology at the University of Florida, Gainesville.

Women, Science, and Technology

A Reader in Feminist Science Studies

Second Edition

Mary Wyer, Mary Barbercheck,
Donna Giesman, Hatice Örün Öztürk,
and Marta Wayne

EDITORS

NEW YORK AND LONDON

First edition published 2001
by Routledge

This edition first published 2009
by Routledge
270 Madison Ave, New York, NY 10016

Simultaneously published in the UK
by Routledge
2 Park Square, Milton Park, Abingdon, Oxon OX14 4RN

Routledge is an imprint of the Taylor & Francis Group, an informa business

Typeset in Minion by RefineCatch Limited, Bungay, Suffolk
Printed and bound in the United States of America on acid-free paper by
Edwards Brothers, Inc.

Library of Congress Cataloging in Publication Data
Women, science and technology : a reader in feminist science studies / edited by
Mary Wyer . . . [et al.]. – 2nd ed.
p. cm.
ISBN 978–0–415–96039–7 (alk. paper) – ISBN 978–0–415–96040–3 (alk. paper) – ISBN 0–203–89565–8 1. Women in science. 2. Women in technology. 3. Feminism and science. I. Wyer, Mary.
Q130.W672 2008
500.82—dc22 2007048799

ISBN10: 0–415–96039–8 (hbk)
ISBN10: 0–415–96040–1 (pbk)
ISBN10: 0–203–89565–7 (ebk)

ISBN13: 978–0–415–96039–7 (hbk)
ISBN13: 978–0–415–96040–3 (pbk)
ISBN13: 978–0–203–89565–8 (ebk)

We renew our dedication of this book to all those women in science and engineering who have been denied the good friendship and encouragement that have sustained us. Know that we are cheering for you all.

Contents

Acknowledgments for the Second Edition

Though many of those mentioned in the acknowledgments in the first edition have continued to be sturdy supporters of our collaboration, we have also discovered new allies in recent years. At NC State, these include Dean Louis Martin-Vega in the College of Engineering, Dean Linda Brady and Dean Toby Parcel in the College of Humanities and Social Sciences, and Dean Dan Solomon in the College of Physical and Mathematical Sciences. At University of Florida-Gainesville, these include the Director for the Center for Women's Studies and Gender Research, Milagros Peña, and the Interim Associate Provost, Angel Kwolak-Folland, as well as Jane Brockmann in the Department of Zoology. At Duke University, this includes Jody Power, Executive Director of the Institutional Review Board.

A series of course development and educational research activities also uncovered friends of this work in departments of plant biology, zoology, psychology, biomedical engineering, entomology, women's and gender studies, and technology transfer at our institutions. The list of names of those to whom we owe thanks is very long but should begin with Jennifer Schneider and Elizabeth Adams, psychology doctoral students *par excellence*, who were determined, organized, and good-humored junior colleagues who assembled the permissions, article excerpts, and authors' information for this edition, as well as making substantive contributions throughout the writing and editing processes. We have included both as contributing authors for the Section V introduction by way of recognizing their efforts. Stephanie Wenzel provided excellent editorial support. Among others who should be mentioned are the many faculty who participated in an NSF ADVANCE project (SBE-0123604) at NC State (you know who you are), as well as Laura Severin, Chris Pierce, Marcia Gumpertz, Margo Daub, Jane Lubischer, Maria Correa, Ellen Damschen, Kristen Rosenfeld, Tom Wentworth, and Linda Smith, to name but a few. Of course, none of the above are implicated in our errors.

Acknowledgments for the Second Edition

Preface to the Second Edition

In the first edition of this book, we began the preface with the warning that this book may be unlike anything you have previously encountered, because the work as a whole does not fit neatly into conventional academic categories. As we revisited our material, we found this still to be accurate. Though the volume of work that can be categorized as feminist science studies has blossomed, academic institutions continue to honor the traditions of disciplinary boundaries. It was somewhat of a novelty when we first began to work together, to be a group of five people from such a range of backgrounds. This continues to be true as well. The five of us—an entomologist, a molecular biologist, a geneticist, an engineer, and a social scientist—are no longer at the same institution, but we have sustained our working relationships by meeting at conferences, teaching courses and working on committees together, developing grant proposals, and consulting with one another about our professional dilemmas and dramas. It is fair to say that, despite various forms of discouragement from our departments and colleagues over the years, we have all benefited from this common project that we simply call The Book. Each of us has advanced professionally, so that we have acquired the trappings of "insider" status, with tenure, promotions, administrative responsibility, and even executive influence within our institutions. Some of us are as well known for this book as we are for our research. We can hardly complain. It is a success story to be still working together at all. We never imagined a second edition in the beginning, but our continuing conversations opened the door to rethinking our collection of essays in the field of feminist science studies.

In this new edition, we have dropped essays that were old friends to us but that we discovered, with feedback from colleagues and students, were too advanced for the book's mission as an introduction to the field. We have again featured work that we consider to be "classic" in the sense that it continues to offer an original and groundbreaking perspective. We have expanded the contents to include topics that were not well represented in the first edition, in particular a new emphasis on technologies as implicated in sustaining (or challenging) the oppressive contexts of women's lives. We have also added readings that reflect the growing literature on women internationally, with a twin emphasis on how women impact and are impacted by scientific and technological innovations driven by capitalist, rather than humanitarian, imperatives. There are still gaps in our coverage, including articles in the important areas of communication technologies, biomedical engineering, genetic technologies, cyberspace, and sustainable agriculture.

The 2008 edition of *Women, Science, and Technology* marks the tenth year since we began teaching our course, titled "Women and Gender in Science and Technology," at NC State, out of which the book emerged. The first year we offered the course, there were just five students enrolled, with all five of us teaching it. The course now routinely attracts over sixty students a year and satisfies a university general education requirement at our own university, and it has been adopted widely in the United States. Our decisions about content for the new edition are in no small part due to the evolving interests and enthusiasms of students. This is true for most of the new articles, which were tested as classroom readings at our university. What has become apparent to us in the last decade is that our students are more activist than they were when we began our work, that the topic of "feminist science studies" is not as scary to them as it once may have been, and that the topics that resonate most profoundly for them are related to the body. By "the body" we refer to women and men as embodied, laboring, thinking, breathing humans whose individual desires, dreams, and choices are contained by barely visible social, institutional, and economic boundaries. And so our new edition reflects this development, with several articles that focus literally on the body as an object and subject of scientific and technological innovation.

In our hunt for new material and our review of the earlier introductions, we discovered several arenas in which significant change has taken place, and we want to mention these here—not to lay out a claim of discovering that all is well with the world, but rather to mark the moment and honor the change that has taken place. The arenas in which we note improvements, to name a few, are: the increasing representation of women as undergraduate and graduate degree earners in science, mathematics, and engineering; the increasing visibility of women's health care in public policy discussions; the increasing recognition that women scientists and engineers bring useful and (perhaps) distinct experiences to the table in the development, implementation, and adaptation of new discoveries; the increasingly institutionalized commitments of colleges and universities to denounce gender bias in education, employment, and training in science and engineering fields; and the decreasing representation of science and scientists as necessarily masculine (never mind engineering or computer science for the moment).

There is no shortage of new concerns, or persisting concerns, however; most obvious is the need to enliven and expand undergraduate students' (and faculty members') exposure to cross-talk between feminist and scientific/technological perspectives. We leave to you the task of considering the possibilities of this dialog in light of your conversations around essays and themes herein. This book represents our effort to make a small wave in a sea of change.

General Introduction

Science, Technology, and Feminism

Since the emergence of women's studies initiatives in academe in the 1980s and 1990s, there has been a steady stream of challenges to the received wisdom of the humanities and social sciences. By placing women at the center of analysis, by asking the simple question "What about women?" researchers revealed the ways in which the grounding assumptions in a variety of disciplines excluded women. In literary studies, history, psychology, philosophy, sociology, and anthropology—in all of these fields, the exclusion of women as subjects of study mattered. It mattered because the interesting questions, grounding assumptions, and accepted answers all changed once women were brought into the picture. The fields grew, in short, as a result of including women as subjects of study.[1]

Many science and engineering fields have been untouched by these developments. There are several reasons for this. (1) Where the subject matter does not include any people, the absence of women as subjects of study may seem irrelevant. (2) In disciplines where there are few or no women in a position to promote change, there are few who have a vested interest in challenging the status quo. (3) Scientists and engineers do not usually have the training to consider the social dynamics that have shaped their fields, and so most assume that scientific perspectives are necessarily "objective" and outside the influence of these dynamics. (4) Most women's studies scholars come from the humanities and social sciences, so their work does not address or challenge issues in the physical, biological, or engineering sciences. (5) Whatever their discipline, faculty evaluations of their peers tend to reinforce disciplinary boundaries by demanding that research and teaching address discipline-specific questions. Those who focus on women's lives must cross these institutionally determined boundaries, leaving themselves vulnerable to criticism from colleagues and the institution.

This book is our attempt to begin breaching those boundaries in order to help ensure that readers have the knowledge and perspectives they need to participate fully in the research, teaching, and public policy decision-making of the twenty-first century. To the extent that women continue to be second-class citizens socially, economically, and politically, the issue of our access to training and expertise in scientific and technological fields has renewed urgency in the face of growing centrality of these fields in a global economy. The belief that scientific and feminist perspectives are incompatible contributes to a persisting ignorance about the importance of scientific and technological innovations in shaping the conditions of women's lives, the limitations of research and teaching that exclude women, and the talents and abilities of women in science and engineering. It is our goal to confront this ignorance.

THE SCIENTIFIC METHOD

We generally think of scientific perspectives as true and factual. Because these perspectives are presumed to be untainted by political considerations, we assume that they produce reliable and complete knowledge that we can use with predictable results. In a world full of uncertainties, scientific facts and truth lead us to clarity and understanding. In contrast, we often think of feminist perspectives in different terms. Feminist perspectives create knowledge that challenges what we take for granted to be true and factual. Critical analysis from a feminist perspective provokes change. It encourages us to ask new questions about how to understand the world around us. Feminists are often accused of being politically biased, of creating knowledge that is not purely objective. What possibly could these two perspectives—feminist and scientific—have in common? We argue that they share many commitments, including an emphasis on understanding the social and physical worlds in which we live, and on striving toward more complete accounts of those worlds.

What sets science apart from any other way to gather information or explain how something works? How do we differentiate between scientific knowledge and nonscientific knowledge? Everyone has some notion of what science is and what scientists do. If we were to watch a group of people working in a laboratory while they were dissecting animals and arguing about their data, and if we knew nothing of science, we might describe them naively like this:

> Perhaps these animals are being processed for eating. Maybe we are witnessing oracular prophecy through the inspection of rat entrails. Perhaps the individuals spending hours discussing scribbled notes and figures are lawyers. Are the heated debates in front of the blackboard part of a gambling contest? Perhaps the occupants of the laboratory are hunters of some kind, who, after patiently lying in wait by a spectrograph for several hours, suddenly freeze like a gun dog fixed on a scent.[2]

However, most of us have had some exposure to images of real or imagined scientists in television, movies, and advertising, and we have at least some notion of what scientists do. But what do ordinary scientists do that defines their activity as science? Most scientists would define science as an approach called the scientific method. The scientific method encompasses the procedures and principles regarded as necessary for scientific investigation and the production of scientific knowledge. The scientific method is actually quite simple in concept and is made up of the following "steps":

1. Observations are made of objects or events, either natural or produced by experimentation. From these observations a falsifiable hypothesis is developed.
2. The hypothesis is tested by conducting repeatable experiments.
3. If the experiments refute the hypothesis, it is rejected, and new hypotheses are formulated.
4. Hypotheses that survive are condensed into generalized empirical relationships, or laws.
5. Laws are synthesized into larger theories that explain the nature of the empirical relationships and provide a conceptual framework for understanding how the natural world operates.

The facts of nature are uncovered by a step-wise winnowing out of inadequate hypotheses and refinement of the remaining hypotheses. The explanations or facts that remain are considered true until someone is able to falsify them using the scientific method. Therefore, ideally, scientists must be prepared to abandon or alter hypotheses, laws, and theories when new knowledge contradicts them. However, science does not always proceed in this idealized way. The major questions in feminist analyses of science concern the degree to which the scientific method ensures that the knowledge produced is unbiased and untouched by social and cultural norms or the conscious or unconscious political commitments of scientists.

SCIENTIFIC BEHAVIOR AND THE SCIENTIFIC METHOD

In using the scientific method, it is assumed that scientists will adhere to a number of behavioral norms.[3] These are truisms or prescriptions for the behavior expected to arrive at scientific knowledge. They contain the essence, or spirit, of scientific inquiry. The scientific norms describe the ethics to which the community of science subscribes and form an informal code by which the behavior of most scientists is judged by their peers. The scientific norms include (1) originality (scientific results should be original); (2) detachment (the motive of scientists should be the advancement of science, and the scientist should have no psychological commitment to a particular point of view); (3) universality (claims and arguments should be judged according to their intrinsic merits; there are no privileged sources of scientific knowledge); (4) skepticism (all claims should be scrutinized for error); and (5) public accessibility (all scientific knowledge should be freely available to anyone).

The characteristics listed above are ideals, and many working scientists would admit that in doing science it can be difficult to live up to them. We can get a better understanding of why these qualities are accepted as norms by examining the philosophical and historical roots of Western science.

A BRIEF HISTORY OF THE SCIENTIFIC METHOD: THE STANDARD STORY

To better understand how science as a method of knowing was invented, we will briefly review its long history associated with and informed by European-Greek philosophy. Science as a practice has been promoted for at least 400 years as a unique form of knowledge production whose power transcends the peculiarities of time and place. A strong association with Western philosophy has influenced how scientists and society think about science. Some influential philosophers who are credited with formalizing our ideas about science as a way of knowing are Aristotle, Galileo, Sir Isaac Newton, René Descartes, Sir Francis Bacon, Sir Karl Popper, and Thomas Kuhn.

From Aristotle we inherited a method of reasoning known as logical deduction. In this method one uncovers the truth by postulating premises, then uses deduction to derive the consequences implicit in the premises. For example, "All men are mortal. Socrates is a man, therefore, Socrates is mortal" demonstrates deductive logic. So does the example, "All women are irrational. Marie Curie is a woman, therefore, Marie Curie is irrational." The method of logical deduction can be used whether or not the premises are true. The method merely requires that the conclusion follows from the premises. The conclusion is considered valid (even if, as in the example with Marie Curie, it is not accurate) so long as the correct method for deducing the conclusion is used.[4]

Galileo is credited with the notion of conducting experiments under controlled conditions. If the scientist has an hypothesis about how something works, she or he must construct an experiment in which all the variables except the one of interest are held constant. The scientist can then measure the variable of interest. Until Galileo, most important discussions about the natural world were conducted as theoretical debates rather than through practice. Newton is credited with the formalization of the idea of the description of nature in mathematical terms and invented the idea and means of using mathematics to analyze experimental data. From Descartes we inherited the dominant mode of analysis in science, which is Cartesian reductionism. Some assumptions of this mode are that the whole is equal to the sum of the parts, that the parts can exist in isolation, and that causes are unambiguously separate from effects.

Sir Francis Bacon developed the idea of a laboratory and codified the procedure for research now called the scientific method. He is considered by many scientists to be the "father" of the scientific method. According to Bacon, scientific inquiry begins with the careful recording of observations in a disinterested, impartial, and totally objective way. The observations should not be influenced by any prior prejudice or theoretical preconception. When enough observations are accumulated, the investigator generalizes from these via the process of induction to some hypothesis describing a general pattern present in the observations.[5] By the inductive approach to the scientific method, one makes a series of observations and forms a universal generalization.

Karl Popper was probably one of the most influential philosophers of science in modern times. Popper recognized that there was a problem with Baconian induction for producing scientific knowledge. The problem with induction is that you cannot be certain that what you observed in the past will be true in the future. Also, it is not possible to be totally objective because the decisions about what is a relevant observation will inevitably be influenced heavily by background assumptions which are often highly theoretical in character. For example, in his book *Paradigms Lost* mathematician John Casti (1989) offers the following illustration of the problem of induction:

> You may remember on IQ or college board tests where a sequence of numbers is given and you are supposed to pick the right continuation of the sequence as a demonstration of your intelligence. Suppose the initial sequence is {1, 2, 4, 8} and you are asked to choose the natural continuation of the sequence. Some possible answers would be {16, 32, 64, 128} (doubling) or {9, 11, 15} (differences in original sequence) or {who do we appreciate} (high school football stadium). In the absence of context, that is, additional information, there is no such thing as a "natural" continuation of the sequence. The "right" answer is dictated by sociological considerations rather than any kind of objective reality for number sequences.

Because of the problem of induction, most scientists consider the only way to legitimate scientific knowledge is through the scientific method based on testing of falsifiable hypotheses, rather than by amassing observations. The hypothesis must be falsifiable because one can show conclusively that an hypothesis is false but not that it is true. To refute or falsify a hypothesis, only one piece of evidence is needed. For example, consider the hypothesis that all ravens are black. We can make observations of ravens in many locations. Even if all the ravens that we see are black, we cannot verify the hypothesis, because we cannot observe every raven in the world. If we observe even one white raven, then the hypothesis is falsified and this leads to more refined hypotheses, perhaps about factors that influence the distribution or occurrence of black as opposed to white ravens. In Popper's use of the scientific method, learning about nature is carried out by a process of elimination of falsified hypotheses. In the pursuit of scientific knowledge scientists come closer and closer to the truth through a series of falsified observations or experiments, but they can never know with 100% certainty if they have reached their goal of scientific fact.[6]

Thomas Kuhn, a physicist who published *The Structure of Scientific Revolutions* in 1962, challenged Popper's philosophy of science. Many would argue that Kuhn's book launched a field of research into the practices of science, sometimes called science studies, whose project has been to show how social phenomena deflect science practice from the presumed ideal. Kuhn challenged the fundamental assumptions that scientists are objective about and receptive to new ideas. He argued that every scientist works within a distinctive paradigm, which is an intellectual framework that influences the way that nature is perceived. Scientists, just like the rest of humanity, carry out their day-to-day business within a framework of presuppositions about what constitutes a problem, a solution, and a method. This background of shared assumptions makes up a paradigm, and at any given time a particular scientific community will have a prevailing paradigm that shapes and directs research in a field of science. Paradigms have great utility and practical value, because without them no one would know how to set up an experiment and collect data.

A paradigm is like a filter through which scientists see and construct research problems, and a paradigm shift (or in Kuhn's terms, a scientific revolution) takes place when a new filter transforms the scientist's perspective. Once this shift takes place, the next generation of scientists is trained according to the new vision of truth. According to Kuhn, there is no such thing as an empirical observation or fact; we always see by interpretation and the interpretation we use is filtered through the prevailing paradigm. Because paradigms are influenced by the social and economic contexts in which scientists live and work, science is not independent of the broader culture, but is itself an integral part of culture.[7]

Kuhn described two phases of science, normal and revolutionary. During a normal phase in science, the level of consensus is high, and most scientists working in a field accept the validity of the current paradigm. During a revolutionary phase, consensus breaks down but in time will again consolidate around the new paradigm. During periods of normal science observations are influenced by theory, as opposed to theory being influenced by observations, and, therefore, the data that scientists collect are theory-laden. Rather than experimental and observational data being the determinant of the course of science, theories determine what evidence is looked for and what evidence is taken seriously. During normal science, scientists accept almost without question the dominant scientific theories in their research areas even if there are observations that the theory is unable to explain or which suggest that the theory is wrong. Failure of an experiment to confirm a theory or achieve results expected by a widely accepted theory casts doubts not on the theory but upon the ability of the experimenter.[8] Kuhn contends that the observer, the theory, and the equipment are all an expression of that point of view as well. If Kuhn's ideas about science are accurate, then several of the scientific norms and an essential foundation of the scientific method, objectivity (that the observer can be essentially separate from the experimental apparatus used to test the theory and the subject being studied), are open to challenge.

WOMEN AND THE DEVELOPMENT OF WESTERN SCIENCE

Very little is known of women's involvement in science and technology in the ancient world. In ancient Egypt, women were able to attend medical school, mainly to specialize in midwifery.[9] Western science is considered largely a European-Greek invention, and we have inherited the tradition of Western philosophy and the ideals of reason and objectivity from white males, many of whom have been demonstrably sexist or misogynist.[10] For example, Aristotle, who rejected his teacher Plato's moderate views on women, promoted the traditional Greek view that women were inferior to men and incapable of rational thought. During the Middle Ages in Europe, education centered on the cathedral schools for boys and men, which later developed into universities. Independent women could join convents, which expanded their function to become schools for upper-class girls and women. In this way medieval women could participate in science-related activities. Two such women were the Abbess of Hohenburg, Harrad, who wrote the encyclopedia *Hortus deliciarum* to teach convent students, and Hildegard of Bingen, who wrote the book *On Nature and Man, the Moral World and the Material Universe, the Spheres, the Winds and the Humours, Birth and Death, on the Soul, the Resurrection of the Dead, and the Nature of God.* In Italy, opportunities for women were much broader, and upper-class women were able to participate in university life both as students and as teachers, especially in medical schools.

The fifteenth through eighteenth centuries are considered an especially important era in the development of modern science. Women's involvement in science at this time seems to have been restricted by the persistence of women's subordinate social and presumed biological status. Francis Bacon, the "father" of the scientific method, saw women as an impediment to the achievements of men and believed that women should be avoided altogether. His dismissal of women is somewhat ironic given that Bacon's writings on the philosophy of science are threaded with sexual imagery and metaphor. Bacon's vision of science was articulated as mastery of man over nature, with nature characterized as a bride and science as the means to bind her to the will of men.[11]

It has been argued that the underlying metaphors of the identification of nature, subjectivity, and emotion with women, on the one hand, and culture, objectivity, and reason with men, on the other hand, have provided a rationale for the exploitation and control of both women and nature. Ideologies of gender, nature, and science that arose in the seventeenth century supported the increasing split between men's and women's worlds, between the public and private spheres, and supported the exclusion of women from intellectual pursuits such as science.[12]

Although formal educational opportunities for women existed mainly in the highest socioeconomic classes, we know that women during this time were interested in science and corresponded and studied with their contemporary male scientists.[13] For example, Descartes corresponded on scientific topics with several women, including Princess Elizabeth of Bohemia and Queen Christina of Sweden, among others. Margaret Cavendish (1617–1673), the Duchess of Newcastle, produced fourteen books on subjects from natural history to atomic physics. Emilie du Châtelet (1706–1749) exchanged ideas with the foremost mathematicians and scientists of Paris and earned a reputation as a physicist and interpreter of the theories of Leibniz and Newton. She duplicated Newton's experiments in the great hall of her chateau.

During the seventeenth and eighteenth centuries, upper-class women in England and France participated in scientific discussion in salons. The grand salons of Paris were run almost exclusively by women. Influential women, effectively acting as intellectual power brokers, were patrons of young men hoping to make careers in science.[14] Unfortunately, the great majority of men who were aided by these unconventional women produced writings that either ignored women entirely or upheld the most traditional views of them. The salons and the women who ran them were ridiculed even by the influential men who benefited from the interactions there. For example, Jean-Jacques Rousseau, in his novel *Emile*, wrote:

> I would a thousand times rather have a homely girl . . . than a learned lady and a wit who would make a literary circle of my house. . . . A female wit is a scourge to her husband, her children, her friends, her servants, to everybody. From the lofty height of her genius, she scorns every womanly duty, and she is always trying to make a man of herself.

German philosopher Immanuel Kant wrote that women like the Marquise du Châtelet who "carries on fundamental controversies about mechanics . . . might as well even have a beard."[15]

And yet, women developed at least three science-related fields: midwifery, nursing, and home economics.[16] These fields have generally been ignored by historians of science, perhaps because most historians of science have been men who focused on the achievements of men. Nonaristocratic European women participated in scientific activities through the craft tradition as daughters, wives, and apprentices; independent artisans; or widows who inherited the family business. The participation of these women in fields such as astronomy, physics, and the natural sciences in the seventeenth and eighteenth centuries has been largely unacknowledged, presaging the "invisible assistants"—the many women who worked in laboratories but seldom received credit for their contributions—of the nineteenth and twentieth centuries.

During the eighteenth century, treatises on physics, chemistry, and natural history became popular, and lectures in these areas were attended by women. Popular scientific newspapers and journals arose to satisfy women's interest in these topics. This accepted interest prepared society to at least acknowledge the possibility of a woman scientist. By the nineteenth century, women participated in some aspects of almost all science, mainly in data-gathering, but also in idea creation. Educational reforms, the Industrial Revolution, and political changes increased the ability of women to be active in science, which has resulted in increasing the contributions of women to science practice and theory up to the present.

BECOMING A SCIENTIST

It is one thing to talk about the scientific method and how it developed historically but quite another to consider how research actually is done. In Western society, rarely can someone who is not formally trained produce information that will be accepted as scientific fact. As with many professions, in science there is a long accreditation/training period in which the aspirant gains not only technical knowledge but is also introduced to the paradigms, culture, and norms of the field. In the U.S. the training period usually starts in high school with coursework focused on mathematics and science, continues through undergraduate education, and includes more specialized training at the graduate level. Research scientists usually work in academic settings, in government research organizations, or in industry.[17] Usually scientists are aided in their work by support staff who have varying levels of academic and scientific training, who may or may not be given credit by the scientific community for their work. The acceptance of scientific information is dependent upon consensus by other scientists in the field, and therefore scientists act as gatekeepers on the production of scientific facts.

BUILDING CONSENSUS IN SCIENCE

Information collected through the scientific method is not automatically accepted as scientific fact by the scientific community. The scientific community accepts scientific knowledge through a process of consensus building. Theoretically, consensus is based on the unbiased evaluation of the empirical evidence.

There are many necessary steps after experimentation in order to convince other scientists that a new piece of information is a "fact." For instance, Dr. Susan A. Scientist at Prestigious University has conducted experiments to test the hypothesis that the moon actually consists of blue rather than green cheese. After her brilliant discovery, she writes up her experiments in manuscripts that detail the methods used, the results, and significance of her results relative to other knowledge in her field and submits them for publication to a scientific journal. The editor of the journal sends her manuscript to two or more of her peers for anonymous review. If the reviewers and the editor of the journal agree that

Dr. Scientist's research is repeatable and the results are valid and significant, her manuscript will be published. At the same time, Dr. Scientist will attend scientific conferences and present her results formally as lectures and informally in discussion with other scientists to make them aware of her results and to convince them of their validity.

When the manuscript is published it becomes available for the scrutiny of other scientists, who may or may not challenge any aspect of Dr. Scientist's research: her methods, the results, or her interpretation of the results. Sometimes, in the case of especially significant research, results will be announced in the popular media at the same time as publication in the scientific literature.

Unfortunately, sometimes a scientist's research is challenged not only on the merits of the research but on the personal characteristics of the scientist herself. For example, James Watson, a recipient of the Nobel Prize for his development of the model of the helical structure of DNA, describes Rosalind Franklin, another scientist investigating the structure of DNA in this way:

> I suspect that in the beginning Maurice hoped that Rosy would calm down. By choice she did not emphasize her feminine qualities. Though her features were strong, she was not unattractive and might have been quite stunning had she taken even a mild interest in clothes. This she did not. There was never lipstick to contrast with her straight black hair. Clearly, Rosy had to go or be put in her place. The real problem, then was Rosy. The thought could not be avoided that the best home for a feminist was in another person's lab.[18]

Rosalind Franklin died of cancer at the age of 37 in 1958, just four years before the Nobel Prize was awarded to Watson, Francis Crick, and Maurice Wilkins. Until recently, the critical importance of Franklin's research for the description of the structure of DNA was not generally acknowledged. The unauthorized use of Franklin's crystallographic data allowed Watson and Crick to develop the structural model for DNA.[19]

Sometimes the results of scientific research take many years to be accepted, especially if they challenge the validity of currently held beliefs of what has been previously accepted as a scientific fact. The research of Barbara McClintock, who received the Nobel Prize in 1988 for her research in corn genetics, is a good example of this. Her results, which challenged then-current beliefs about the static behavior of genes on chromosomes, were not fully appreciated until 40 years after she had completed her experiments.

WOMEN AND TECHNOLOGY

How is technology different from science? How are they related? Like science, technology typically relies on empirical tests (e.g., experiments) to gain evidence for claims about its products. However, science is most widely considered a process for understanding and creating knowledge about nature that may or may not address a practical need, whereas technology is often described as a process of creating artifacts and systems to meet a need. Many consider that technology is the practical application of science. In this sense, technology and science are wrapped together into a single conceptual package known simply as "science" or "science and technology," with technology as a dependent entity. Others consider a different relationship between science and technology in which science and technology are two ends of a continuum. On one end is "pure science" with no readily apparent application and on the other is "pure technology" that can be directly used to address a need or solve a problem. Along the continuum there is a broad overlap that could be referred to as "applied science." Some consider that science and technology are not necessarily related.

Change in the material environment is the explicit purpose of technology, and not, as with science, the understanding of nature; accordingly its solutions are not right or wrong, verifiable or falsifiable, but more or less effective from different points of view. In reality, sharp, neat distinctions between science and technology do not exist.

Regardless of how one considers the relationship between science and technology, few historians of science and technology have considered the role that women have always and continue to play in the development of technologies. In her book, *Mothers and Daughters of Invention: Notes for a Revised History of Technology*, Autumn Stanley suggests that inventive ability is independent of biological gender.[20] Throughout prehistory and history, both sexes have devised, made, or improved tools and procedures necessary for the tasks in which they were engaged. She posits that the absence of women in the story of technology has to do with the traditional identification of invention with men and machines, and that technology has been defined to exclude women's work. "Real" technology is related to weapons, machines, and chemical compounds created in laboratories. Accounts by anthropologists, especially before the 1980s, have been androcentric in their focus on hunting tools and weapons and whatever males do, while women's activities are trivialized, underreported, or not considered but are merely an "invisible" part of the landscape. However, even work traditionally identified with women is dependent on technology. Consider,

> . . . if women were in charge of butchering the animals, or slicing and drying the meat, or of debittering the acorns or detoxifying the cassava and processing its starch, preparing the animal skins or whatever cooking was done, then they most likely invented and improved the processes and compounds and tools used in those activities: knives and scrapers, tanning methods and chemistry; food-preserving and—processing technology—including refinements in the exploitation of fire, storage containers, preservative agents, steatite cooking surfaces or vessels, cooking stones, ovens, mortars, pestles, grain-grinders, etc.

In industrialized societies, evidence of technological creativity or invention is often provided by patents. Instituted in most countries in the seventeenth or eighteenth century, the patent process is a relatively modern record of human technological activity. Because women throughout history have faced social, educational, and legal limitations to applying for and receiving patents, women's role in the development of technologies may be biased. Until fairly recently, married women lacked the economic power or legal right to produce, market, or profit from an invention in their own name. So, as with scientific knowledge created by women, many products and processes developed by women were credited to a husband, a father, a brother, or a male partner, rendering women's contributions invisible. Additionally, the U.S. Patent and Trademark Office does not require gender, racial, or ethnic identification in patent or trademark applications, further obscuring women's contributions to patented inventions. The number of patents issued to women increased with changes in property-rights laws and educational opportunities for women in the mid-nineteenth century.[21] Women's inventions were exhibited at the 1876 Centennial Exposition in Philadelphia and the 1893 Columbian Exposition in Chicago.

Women's experience has always been a strong motivation for invention.[22] Early women inventors registered from urban areas frequently patented articles pertaining to dress and household furnishing, whereas rural women often patented agricultural and household labor-saving devices. Patents by women diversified when women were largely responsible for traditionally male tasks during wartime. For example, in the period around the U.S. Civil War, women received an annual average of 10.1 patents during 1855–1865 and 67.3 during 1865–1875.[23] Today, hundreds of thousands of women apply for and receive a patent every year, and an estimated 20% of all inventors are female. It is probably safe to predict that the number of patents applied for and awarded to women should continue to rise through the benefits of their increasing education and employment in technical and scientific fields and access to laboratories or workshops, equipment, assistants, and financial support.

A MULTIDISCIPLINARY PERSPECTIVE ON SCIENCE AND TECHNOLOGY

Understanding that scientists are people who do research within particular social and historical contexts requires looking at science as a social activity. As we have pointed out, the scientific method is an approach that philosophers and scientists developed and formalized over time. This method allows scientists to collect and evaluate data systematically and then develop generalizations that provide a shared conceptual framework for information exchange and interpretations of natural phenomena. From a social constructivist perspective, the method and the knowledge that results from it are socially constructed. This means that human activity has produced the knowledge. Scientists (or anyone who does research) can only represent their understandings of the natural world according to the conventions of their disciplinary theories, vocabulary, and professional practices. These conventions constrain how and what they can credibly claim about the natural world. Within the social constructivist view, language mediates our understanding of nature because it is through language that all humans learn about and represent the world. In contrast, the positivist perspective sees knowledge as facts that humans uncover rather than produce. Knowledge about the natural world is something that is revealed to the scientist by careful hypothesis development and testing. Language is a transparent medium through which humans relay their insights. Within the positivist view, nothing of importance rests on language.[24]

In this book, we borrow from each of these perspectives. We argue that social processes leading to scientific knowledge and technologies can produce adequate and useful explanations of the natural world and a means to address needs with technology. At the same time, they can also produce inadequate and regressive explanations, theories, and technologies. Feminist science studies seek to sort out which is which in relation to women. *Women, Science, and Technology* is meant to give readers tools for doing the same. Many of the original works and classic texts in feminist scholarship on science and technology are too difficult in their original and unabridged form for most newcomers to the field. Nonetheless, the analyses detailed in the texts provide valid and critically useful insights about scientific research and technology. To make these readings more accessible and available, we have excerpted from them and included section introductions that provide background information on the salient points of key arguments.

When deciding how to organize this book given our diverse interests and expertise, we agreed on a list of six major questions: Who does science? Who develops technology? How does culture shape science and technology? How do science and technology shape culture? Can we redefine and reform science and technology to include feminist perspectives? How can feminist perspectives on science and technology improve the day-to-day lives of women (and men)? These questions are a distillation of the issues raised in the five sections. There is interplay between the sections, in itself an example of the complexity of feminist critiques of science and technology. For example, the answer to the question of who does science and technology is inseparable from the question of how culture shapes science and technology. The questions we ask are thus an organizing device that focuses readers' attention around core issues in order to survey the material. The overlaps among sections are meant to open debate and foster discussions about their inseparability.

In our first section, entitled "Educating Women for Scientific Careers," we include as a backdrop a brief overview of the history of women's exclusion from—and struggles to obtain training in—the sciences. The book's sequence of readings begins with accounts of women's experiences in science as outsiders in their training programs. Other essays in section one review the qualitative and quantitative data on why women decide to avoid or pursue scientific careers and on how they fare in those professions relative to men. Even with the great strides that have been made, only eleven women have been awarded the Nobel Prize in scientific fields. Marie Curie was awarded the Nobel Prize two times, in 1903 (Physics) and in 1911 (Chemistry).

We take the theme of exclusion into our second section, "Stereotypes, Rationality, and Masculinity in Science and Engineering," which explores how our cultural images of men and women shape our beliefs about who can and cannot be scientists and engineers. We examine images of scientists drawn from the culture of science as well as images drawn from the culture at large, including in particular advertising images from scientific publications. We have also included readings on the demographic trends, culture, symbolism, and language of science and how they affect the climate for women and relationships between men and women in science and engineering.

The continuity between, and the complex intertwining of, science, technology, medicine, culture, and the physical body is the theme organizing the third section, "Technologies Born of Difference: How Ideas about Women and Men Shape Science and Technology." We explore why it matters that women generally have been excluded from the culture of science and technology, and why it matters that where women have been able to make contributions these have been lost, ignored, or appropriated. Some of the most persuasive examples of the effects of culture on science come from work used expressly to reinscribe dominant social paradigms, such as studies documenting the biological differences between men and women. In these cases, it is clear (at least in hindsight) that investigators were predisposed to finding results which were consistent with their beliefs about the social and political status of people of different sex and race and that the physical body can be understood separately from the sociocultural milieu that it inhabits. Cultural influences on interpretations of observed phenomena are not limited to the early history of science and technology—they also occur in contemporary theoretical assumptions, interpretation of data, experimental design, language use, and technological applications. So, for instance, the exclusive use of male subjects in health research until the early 1990s was assumed to be "scientific," even though it leads to a nonrepresentative sample (excluding 51% of the population), because the male body was accepted as the norm. It is acceptable to use terms such as "sexy," "penetrating," and "seminal" to describe exciting research because male sexuality is considered the norm. A reductive scientific approach obscures the full understanding of the definition of disease and approaches to medical treatment and technology.

In the fourth section, "The Next Generation: Bringing Feminist Perspectives into Science and Technology Studies," we consider the question "Can we redefine and reform science and technology to include feminist perspectives?" Feminist science and technology is not simply or necessarily science and technology by women. It is science and technology that considers topics considered irrelevant or obvious by men but nonetheless important to improving women's lives. It is science and technology that recognizes the place of scientists and engineers in a social system. Striving for objectivity, for complete detachment of observer from observed, may seem irreconcilable with many feminist philosophies. However, the critical issue is not whether or not objectivity can be accomplished, but rather whether or not it is a useful tool. If it is useful, and we think it is, then there clearly is a place for feminist science and engineering—science done outside of and analyzing the traditional male-centered viewpoint—because it enhances scientific knowledge and technology by enlarging the circle of legitimate questions, topics, and needs, and because feminist perspectives can help in the detection and analysis of distorting biases. Paradoxically, acknowledging bias is often a crucial step toward more complete, meaning more objective, descriptions of the world and technologies that address the needs of many kinds of people. Where social and cultural diversity is lacking, however, it is hard to imagine a vigorous discussion of such biases taking place in the sciences and in the development of useful and appropriate technologies.

In "Reproducible Insights: Women Creating Knowledge, Social Policy, and Change," we close the book with a section that explores how feminist perspectives on science and technology can improve the day-to-day lives of women (and men). This involves looking at what happens when scientific knowledge is translated into use in public policy, corporate decision-making, or technological innovation. The topics included in section five range from an analysis of environmental policy to an exploration of social, legal, economic, and political tensions in feminist commitments to reproductive choice.

Current economic and social trends mean that the more scientifically and technologically literate we *all* are, the better able we will be to make informed decisions about everything from the most personal issue involving medical care to the most public issues involving emerging local, state, and federal legislation about the environment. In addition, the United States is becoming a country in which an increasingly diverse majority of the educated will need to have training in scientific or engineering fields in order to reach their career potential.[25] Feminist perspectives on science and technology provide an intellectual map that can help readers find their way through the morass of available information. Understanding how social dynamics shape our educational experiences and options, workplace experiences and job opportunities, and the directions and content of scientific knowledge and technological innovation reduces the distance between the "experts" and the "citizens." Similarly, scientific and technological knowledge can ground feminist teaching and learning in contemporary issues, inviting feminist scholars to consider if (or how) our theories and research are adequately informed by those disciplines in which women are just beginning to make our presence felt.

Notes

1 There are many good survey texts in women's and gender studies. For an early compendium reviewing the development of women's studies across the disciplines, including in particular the emergence of research on women of color and on masculinity, see Cheris Kramarae and Dale Spender, *The Knowledge Explosion: Generations of Feminist Scholarship*, New York: Teachers College Press, 1992. For recent reviews of major themes and topics, see E. L. Kennedy and A. Beins, eds., *Women's Studies for the Future: Foundations, Interrogations, Politics*, New Brunswick, NJ: Rutgers University Press, 2005; and K. Davis, M. Evans, and J. Lorber, eds., *Handbook of Gender and Women's Studies*, Thousand Oaks, CA: Sage, 2006.

2 B. Latour and S. Woolgar, *Laboratory Life: The Social Construction of Scientific Facts*. Princeton, NJ: Princeton University Press, 1986.

3 R. Merton, *The Sociology of Science*. Chicago: University of Chicago Press, 1973.

4 For a critique of the use of the generic male in the standard example, see J. Moulton, "The Myth of the Neutral 'Man,' " in *Feminism and Philosophy*, ed. M. Vetterling-Brraggin, F. Elliston, and J. English, pp. 124–37, Totowa, NJ: Littlefield, Adams, 1977.

5 M. Peltonen, ed., *The Cambridge Companion to Bacon*. New York: Cambridge University Press, 1996.

6 K. Popper, *The Logic of Scientific Discovery*. London: Hutchinson, 1959.

7 T. Kuhn, *The Structure of Scientific Revolutions*. Chicago: University of Chicago Press, 1970.

8 Barbara McClintock is one such scientist. Her scientific insights in genetics were so advanced in comparison to her colleagues' that they were not readily accepted as she was publishing them, though now her work is recognized as important and valid by the scientific community. See E. F. Keller, *A Feeling for the Organism: The Life and Work of Barbara McClintock*, New York: Freeman, 1983.

9 M. B. Ogilvie, *Women in Science: Antiquity through the Nineteenth Century.* Cambridge, MA: MIT Press, 1983.
10 L. M. Antony and C. Witt, eds., *A Mind of One's Own: Feminist Essays on Reason and Objectivity,* Boulder, CO: Westview Press, 1993; R. Agonito, *History of Ideas on Women,* New York: Perigee, 1977.
11 E. F. Keller, *Reflections on Gender and Science.* New Haven, CN: Yale University Press, 1985.
12 C. Merchant, *The Death of Nature,* London: Wildwood House, 1980; and V. Plumwood, *Feminism and the Mastery of Nature,* New York: Routledge, 1993.
13 B. S. Anderson and J. P. Zinsser, *A History of Their Own: Women in Europe from Prehistory to the Present,* Vol. II. New York: Harper and Row, 1988.
14 L. Schiebinger, *The Mind Has No Sex? Women in the Origins of Modern Science.* Cambridge, MA: Harvard University Press, 1989.
15 Anderson and Zinsser, *A History of Their Own.*
16 Ogilvie, *Women in Science.*
17 For a discussion on women doing science in nontraditional places, see M. A. Eisenhart and E. Finkel, *Women's Science: Learning and Succeeding from the Margins,* Chicago: University of Chicago Press, 1998.
18 J. Watson, *The Double Helix: A Personal Account of the Discovery of the Structure of DNA.* New York: Atheneum, 1968.
19 For a view of Rosalind Franklin that differs greatly from that of Watson's, see A. Sayre, *Rosalind Franklin and DNA,* New York: Norton, 1975; and B. Maddox, *Rosalind Franklin: The Dark Lady of DNA,* New York: HarperCollins, 2002.
20 A. Stanley, *Mothers and Daughters of Invention: Notes for a Revised History of Technology.* New Brunswick, NJ: Rutgers University Press, 1995.
21 D. J. Warner, "Women Inventors at the Centennial," in *Dynamos and Virgins Revisited: Women and Technological Change in History,* ed. Martha Moore Trescott, p. 104. Metuchen, NJ: Scarecrow Press, Inc., 1979.
22 E. H. Showell and F. M. B. Amram, *From Indian Corn to Outer Space: Women Invent in America.* Peterborough, NH: Cobblestone Publishing, 1995.
23 Carroll Pursell, "Women Inventors in America," *Technology and Culture* 22, no. 3 (July 1981): 546.
24 Popper, *The Logic of Scientific Discovery.*
25 Sigma Xi, *Entry-level Undergraduate Courses in Science, Mathematics and Engineering: An Investment in Human Resources,* Research Triangle Park, North Carolina, 1990; W. Greider, *One World, Ready or Not: The Manic Logic of Global Capitalism,* New York: Simon and Schuster, 1997.

Section I
Educating Women for Scientific Careers

In the United States, the history of women's participation in science is entangled with debates about women's intellectual capacities and our roles and responsibilities in relation to men and children. Until the mid-1800s, women were expressly and specifically excluded from all but basic literacy education, since it was thought that educated women would develop deviant social and political behavior. It was said that women would refuse to do housework and they would disobey their husbands if their education were too advanced. They would become masculinized and expect to be included in men's activities. They would try to take over men's jobs. If women knew too much, their intellects would undermine their health and that of their children. The education of women was against the natural order; God would not approve. By the mid-1800s, advocates for women's education argued that women could not fulfill their God-given duties to their husbands without a complete education, but the depth and range of subjects deemed appropriate for women's minds was a matter of controversy.[1] By the late 1800s, scientists had published a number of studies that purported to document a decrease in the health of upper-middle-class white women due to the strain education was putting on women's limited intellectual capacities. (The physiological impact of education on working-class white women and African Americans was dismissed as an issue since it was assumed that they would not seek higher education.) These studies argued that the strain of learning Latin, Greek, or advanced mathematics drew blood to the brain and away from women's reproductive organs. Education threatened white female fertility and racial superiority, and it upset the progress of human evolution. Indeed, the uterus was considered to be the "controlling organ" in the female body, "as if the Almighty, in creating the female sex, *had taken the uterus and built up a woman around it*" [emphasis in original].[2]

Though elementary education was available to women, these attitudes were prevalent enough so that the literacy rate for women (defined as the ability to write one's own name) was half that of men. The general literacy rate during the eighteenth century is estimated at 40% for women and 80% for men. By the middle of the nineteenth century, historians estimate that literacy rates for white women and men became about equal, but for African American women literacy rates remained at about 50%. Even with these restricted educational opportunities, there were some early American white women who contributed to scientific knowledge through research and writing. In the 1750s, for instance, Jane Colden, taught by her botanist father, identified and classified over 300 species of plants in the Hudson River Valley. She is best known for her work with the gardenia, which she identified and described.[3]

A factor contributing to white women's increased literacy was the emergence of female academies, which were private schools for upper-class white women. These schools created audiences for general texts on science topics. So, for instance, in 1785 Rousseau's posthumously published *Letters on the Elements of Botany Addressed to a Lady* appeared. In addition, women wrote popular science texts, participating in the process of educating the reading public about basic science. In 1796, Priscilla Bell Wakefield published her *Introduction to Botany in a Series of Familiar Letters*, which was in print in England and America for fifty years. A few years later, Jane Marcet published her *Conversations on Chemistry,* of which there were fifteen editions before 1860.[4]

The belief that it was unnatural for women to be educated began to give way to the idea that education improved the ability of women to do what was most natural to them: to be mothers and wives. The popularity of science texts directed at women legitimated the idea that women were interested in and could understand scientific information, leading the way to the introduction of science topics into the curriculum for women and girls who were less privileged. Emma Hart Willard argued that it was the responsibility of the state to fund the advanced education of all women because women were responsible for the character of future American citizens. The moral fiber of the country depended on the education of women. In 1819 Willard founded the Troy Female Seminary in New York and secured public funding of secondary education for women. She offered the first classes available to women in mathematics, physics, physiology, and natural history. Her sister, Almira Hart Lincoln, published *Familiar Lectures on Botany* in 1829, which sold 275,000 copies in the next forty years, appearing in seventeen editions. Almira became wealthy writing and publishing popular texts in botany, chemistry, and natural philosophy. Ironically, she did not believe that women should become scientists, only that science would make women better wives and mothers, enriching their domestic lives.[5]

Changes in attitudes toward women, in particular the idea that women were entitled to a publicly funded education, just as were men, provoked dramatic changes in the educational opportunities available to women. New colleges sprung up that were dedicated to the education of women. Public, land-grant institutions opened across the country as new states entered the Union, and some of these opened their doors to women. In the meantime, some women were being educated in science by their fathers or husbands, as they had been for centuries. These two processes contributed to a slow and determined growth in the numbers of women who gained access to scientific knowledge. In 1833, Oberlin College opened its doors and announced that it would be coeducational and that it would accept African American students. Private women's colleges opened later in the century, including Vassar College (1865), Smith College (1871), Wellesley College (1875), and Barnard College (1889).

Maria Mitchell is perhaps the earliest woman who was a professional scientist in the United States. She was educated by her father and worked part-time as a librarian on Nantucket while she did research in her father's laboratory. In 1847, at the age of twenty-eight, she discovered a new comet and became something of a celebrity as the first woman elected to the American Academy of Arts and Sciences (1848), as well as one of the first women in the American Philosophical Society of Philadelphia (1869). She was hired at the newly opened Vassar College (1865) as a professor of science, the first in the United States. Mitchell was always aware of her unique presence in science and was an advocate for including women in science. She was the founding president of the Association for the Advancement of Women and reported routinely to members on women's advances and setbacks in science. In her 1875 presidential address she said: "In my younger days, when I was pained by the half-educated, loose, and inaccurate ways which we [women] all had, I used to say, 'how much women need exact science,' but since I have known some workers in science who were not always true to the teachings of nature, who have loved self more than science, I have now said, 'how much science needs women.' "[6]

All women in the United States confronted resistance to their participation in science, no matter how talented, but those who had money or whose fathers were scientists could secure some training despite the obstacles. African American women in general did not have these advantages, but there were important exceptions. Josephine Silone Yates, for instance, became the first African American woman who was a professor of science when she was appointed head of the Natural Sciences Department at Lincoln University in Missouri. She was born in 1859 to a well-respected, established, and educated family in Mattituck, New York. She was taught reading, writing, and arithmetic at home before entering school. Her parents saw that she had the best available educational opportunities, sending her to both black private schools and Newport, Rhode Island, public schools (in which she was the only black student). She finished a four-year high school course of study in just three years and went on to receive the highest score ever recorded on the Rhode Island state teachers' examination. In 1879, she became head of the department of natural sciences at Lincoln University. Because state law prohibited married women from holding teaching positions, she resigned her position in 1889 in order to marry. At this point, she, like Maria Mitchell, became an activist on behalf of women, and from 1901 to 1906 she was president of the National Association of Colored Women. Despite Yates's considerable talents and interests in science, prevailing beliefs about women and African Americans severely restricted her access to the career opportunities and research environments necessary for scientific achievement.[7]

In general, the first generation of women scientists who were employed as faculty at women's colleges and historically black colleges and universities published little because they carried heavy teaching burdens and they had uneven research backgrounds given their exclusion from most graduate programs in the United States. Because of laws prohibiting married women from teaching, many women literally dedicated their lives to the education of a new generation of women scientists. They remained unmarried and stayed on faculties for thirty to forty years, retiring to a cottage near campus with a sister or retired colleague. When they died, the college would honor them with an endowed chair or a building named after them. Students would write obituaries defending their lack of publications and celebrating their commitment to educating students. Many bequeathed their estates to the colleges that had been the center of their adult lives.[8]

Their efforts and dedication were having an effect, however slow and dispersed it must have seemed at the time. By 1889 a total of 25 doctorates had been awarded to women in the United States, six of them in science. By 1900 another 204 doctorates were awarded to women, with 36 of them in science. These increases, as small as they seem, mark the end of men's exclusive claim to scientific expertise and the beginning of women's march toward entry into scientific professions.[9]

This march has not been an even and easy one, however. The increasing numbers of women who had advanced training led to discussions about the need for expanded job opportunities for women. Science was no longer considered, at least among women who had doctorates in science, simply as a way to improve their domestic skills as mothers and wives. Still, the old attitudes did not completely give way. Expansions in scientific research directions and practices led to larger research staffs, and the expansion of land-grant institutions led to more research staffs in science, all of which opened opportunities for women. Still limited by stereotypes about women, however, employers defined these new jobs as "women's work." So, for instance, in astronomy the use of new technologies, including spectroscopes and cameras, meant that there was a need for fewer "observers" (i.e., men) and more assistants (i.e., women). Women were thought to be especially good at detailed and methodical work, and they would work for less money. So they were hired to examine mountains of photographs of stellar activities, and though many made important contributions to astrophysics, their work did not lead to wages that matched men's or to advancement and recognition on the job.[10]

For this next generation, there is story after story of highly qualified and talented women, dedicated to their research, who were employed in low-paying, sometimes nonpaying, jobs. They were exploited for their skills and abilities, but they were nonetheless willing to do difficult and tedious work because it was the only work available to them in science. One researcher, psychologist Leta Stetter Hollingworth, used her scientific expertise to question the assumptions that grounded these employment practices. Hollingworth decided to challenge the notion that women and men had different physical, motor, and intellectual abilities, since women's workplace opportunities were severely constrained by beliefs about sex differences in these abilities. A student at Barnard, she was a self-proclaimed feminist who was sparked to undertake this research when she had to resign her teaching position upon marrying. Between 1913 and 1916, she published a series of academic studies on sex differences in work performance that concluded such differences did not exist.[11]

There were others, particularly in the suffrage movement, who, like Hollingworth, argued that women were not different from or inferior to men. But their voices were overwhelmed by public acceptance of the idea that fundamental differences between the sexes were a natural fact informing social arrangements. Two world wars opened doors for women's participation in science as part of the war effort, even while after the wars women were required to leave their positions. Still, optimism ran high. In 1921 the president of Bryn Mawr declared that "the doors of science have been thrown wildly open to women," and in that decade an annual average of 50 doctorates in the sciences were awarded to women. By the end of the 1930s, 165 doctorates in the sciences annually were awarded to women. Barriers to African American women's participation also began to fall: in 1933 Ruth Moore became the first African American to be awarded a doctorate in bacteriology (Ohio State University); in 1935 Jessie Mark became the first African American to be awarded a doctorate in botany (Iowa State University); in 1940 Roger Arliner Young became the first African American woman to be awarded a doctorate in zoology (University of Pennsylvania).[12]

World War II, in particular, had a dramatic impact on employment opportunities for women who had scientific training as the federal government's investment in science grew exponentially. Women scientists were in demand as the scientific labor force expanded. Women were allowed entry into a variety of settings where they had previously been excluded, particularly in academia. Women were appointed temporary department chairs, were hired as faculty, and were enrolled as graduate students. They were recruited into careers in science and engineering with a barrage of books and articles that painted a cheery picture of the prosperous and fulfilling futures that awaited them. Unfortunately, conservative social attitudes prevailed in the postwar period, as women were displaced by returning veterans taking advantage of the generous terms of the GI Bill and the Serviceman's Readjustment Act of 1944, which provided veterans with up to five years of full tuition plus living expenses.[13] As campus enrolments bulged, quotas were set on the number of women students admitted, so that male students could be housed in women's dorms. Despite admission caps for graduate women, the number of science doctorates awarded to women continued to increase steadily from 120 in 1940 to 290 in 1954. But because the numbers of men in science programs were skyrocketing, the percentage of doctorates awarded to women shrank to only 6% of the total.[14]

A few extraordinary women continued to work productively in science, however, and international recognition for women as scientists in the United States became a reality when in 1947 Gerta Cori, a biochemist (and a wife and mother), became the first American woman to win a Nobel Prize for Medicine. Resistance to women in science continued, however. One California housewife and educator, Olive Lewis, in 1948 published an article in the American Medical Association journal, *Hygeia*, arguing that scientists who were mothers were always so preoccupied with their own domestic problems that they did not do much good work of their own and even often disrupted the experiments of their colleagues. But such attitudes did not discourage an increasing number of women from becoming scientists. Between 1947 and 1963, a total of 782 doctorates were awarded to women in chemistry, 1,107 in the biosciences, 236 in mathematics, and 40 in engineering.[15]

By the second half of the twentieth century, women had established their interest in scientific careers and their ability to compete with men for the privileges of practicing science. However, new barriers in hiring and promotion emerged, particularly for careers in academe where the most prestigious positions were to be filled. In 1958 there were 138 doctorates awarded to women in the biosciences, but only 26 women doctorates in this pool were hired into jobs in colleges and universities. In the physical sciences, 74 women earned doctorates that year, but only 26 were hired into colleges and universities.[16]

Women in science organized to confront this trend. A series of national initiatives to eliminate discrimination in U.S. higher education sprung up as a response to the modern women's movement in the 1960s. Academic women in sociology, anthropology, and psychology researched and published a series of articles and books that documented gender inequalities in the employment status of women in their fields. These efforts prompted changes in discriminatory policies on the enrolment, funding, and housing of women on college campuses. As increasing numbers of women were accepted into majors in the sciences, and as increasing numbers of women were accepted and graduated from doctoral programs, the number of women scientists on college campuses grew.

Despite these increases in the overall numbers of women doctorates, individuals were often the only woman in a graduate program. Sarah Hrdy, an anthropologist, offers this account in the "Acknowledgements" section of one of her essays to describe her experiences in the 1970s:

> *Acknowledgements*—In the preface to her recent book "Mother Care," my colleague in behavioral biology Sandra Scarr (1984, p. ix) notes, "I wish I could thank all the wonderful graduate school professors who helped me to realize the joys of combining profession and motherhood; unfortunately there weren't any at Harvard in the early 1960s." A decade later, things at Harvard—at least in the biologically oriented part of Harvard that I encountered—had changed remarkably little. As I think back on those postgraduate years (my undergraduate experience at Harvard was a wonderful and very different story), I cannot recall a single moment's fear of success, but what I do distinctly recall was the painful perception that there were professors and fellow students (no women in those years) who acted as if they *feared* that I might succeed. Intellectually, it was a tremendously exciting environment, filled with stimulation and occasionally inspirational teachers and coworkers. It was also an environment that was socially and psychologically hostile to the professional aspirations of women. But there were exceptions, exceptions made all the more significant because they were rare.[17]

Promoting access to education and careers as scientists and engineers is only part of the story, however.[18] A handful of women scientists began to examine how sexism and male-centered perspectives could influence scientific theories and empirical research. Among the earliest was biologist Donna Haraway, who in 1978 published a groundbreaking feminist theoretical critique of primate behavior studies.[19] Another pioneer was Harvard biologist Ruth Hubbard, who published, along with colleagues, one of the earliest collections of essays examining the ways in which scientific understandings of women's biology were gender-biased, in *Women Look at Biology Looking at Women* (1979). Sociobiologist Sarah Hrdy was one of the first feminists to challenge sociobiological arguments about male superiority, in *The Woman that Never Evolved* (1981). Perhaps influenced by the early work of the pioneering Ruth Herschberger in *Adam's Rib* (1948), neurobiologist Ruth Bleier began publishing articles on social influences on biology in the 1970s and published a major study, *Science and Gender*, in 1984. At the same time, biological physicist Evelyn Fox Keller began publishing articles exploring objectivity, gender, and feminism, followed by a major theoretical essay that appeared in *Signs: Journal of Women in Culture and Society* in 1982. Similarly, biologist Anne Fausto-Sterling's *Myths of Gender: Biological Theories about Women and Men* appeared in 1985. Sue Rosser, also a biologist, took this work in new directions when in the mid-1980s she began to write on integrating feminist perspectives into science education, publishing *Female-Friendly Science: Applying Women's Studies Methods and Theories to Attract Students to Science* in 1990. Since then, a wide variety of articles and books have elaborated the theoretical groundwork laid by these scientists, enriching our understanding of feminist perspectives in/on the sciences with philosophy, literature, sociology, anthropology, psychology, history, and cultural and technology studies.

The growth of work by and about women in science is testimony to the intellectual richness that feminist perspectives can bring to science and science studies. Yet, despite two decades of research, and despite a wealth of new information and perspectives on the causes of women's underrepresentation in science, women remain outsiders in scientific culture. The essays in this section focus on the experiences of women in a variety of fields, the processes that exclude them, the ways that they have responded to these exclusionary practices, and analyses of the connections between women's exclusion and key ideologies in the scientific community. We have divided the section into two parts, "Education: Out of the Frying Pan" and "Careers: And into the Fire," in order to emphasize that women face both persisting and distinct challenges throughout both their education and their careers.

Evelyn Fox Keller's "The Anomaly of a Woman in Physics" and Aimée Sands' interview with Evelynn Hammonds, "Never Meant to Survive, a Black Woman's Journey," illustrate by firsthand accounts the personal and social impediments to women's success in science. Fox Keller recalls her painful experiences in the Physics Department at Harvard in the late 1950s and relates how she transformed her feelings of rage about the blatant biases and differential treatment of men and women students into political consciousness. Hammonds describes the special problems brought about by the interplay of gender and race. Both articles describe the alienation and self-doubt engendered by being part of an extreme minority. Both Fox Keller and Hammonds doubted their own abilities before questioning the culture of science that worked against their survival. The broader focus helped them to understand that the difficulties they confronted were not of their own making. Their stories are also about strength, self-realization, and eventual success—Fox Keller as a molecular biologist and philosopher of science, and Hammonds as a physicist and historian of science. Fox Keller is a MacArthur "Genius Award" winner and a professor emeritus at MIT in the science, technology, and society program. Hammonds is professor of the history of science and African and African American studies and the inaugural senior vice provost for diversity and faculty development at Harvard University.

Hammonds' and Fox Keller's accounts are compelling examples of how individual women experience discrimination. But what are the social processes within higher education that lead women toward or away from careers in science or engineering? The third reading in this section, by Banu Subramaniam, is written in the form of a fairytale that tells the story of a young woman from India who travels to the United States to become a scientist. The hero, Sneha, is troubled by her experiences in graduate school and seeks advice from two authorities: the Senior White Patriarch and the Wise Matriarch. Her dreams about becoming a scientist fade, however, as her confidence erodes in the face of the racism and sexism that she encounters. Though "Snow Brown and the Seven Detergents" is a fictional account, many have read and appreciated it because it resonates deeply with their experiences. The author's story, unlike that of the heroine in her fairytale, is a story of survival, reinvention, and success. Subramaniam obtained her Ph.D. in ecology and evolution from Duke University and has gone on to become a tenured faculty member at the University of Massachusetts at Amherst in the women's studies program, doing research and teaching at the interface of science, gender, and colonialism.

A short essay by Dara Horn addresses the ways in which women's contributions to science have been lost, indicating specific cases in which scientific knowledge generated by women has been devalued. Horn further discusses how these losses, omissions, and marginalizations challenge science's privileged position as a source of objective, unbiased knowledge. When she wrote this article, Horn was an undergraduate student in comparative literature at Harvard University; she is now regarded as one of America's best young novelists (*Granta* magazine, 2007). Horn's viewpoint is quite different from that of the other writers in this section: she is an "outsider" in science as a nonscientist but an "insider" with respect to her status as a woman in academia. Horn challenges scientists to turn the lens of objectivity and detachment on themselves, gently reminding scientists of the special position scientific information occupies in our society and demanding that scientists hold themselves to the ideals outsiders believe they aspire to, particularly with respect to sexism.

The articles in the previous group, with the subtitle of "Out of the Frying Pan," document the ongoing and systematic, if perhaps unconscious, discouragement of women and people of color in the sciences during graduate school. Fox Keller, Hammonds, and Subramaniam have used these challenges to their advantage, to create unique, valued intellectual niches for themselves at the intersections of the natural and social sciences. What then of those who persist in the mainstream natural sciences, perhaps despite their graduate school experiences? Is graduate school a uniquely discriminatory experience, beyond which the meritocracy and objectivity of the sciences triumph? Unfortunately, the data suggest otherwise. The next section, "And into the Fire," explores issues that confront those who have succeeded in becoming professional scientists. The underlying dynamics are similar across the different perspectives and topics, however the specific experiences differ.

After completion of an advanced degree, success in science requires the publication of peer-reviewed papers, which depends on acquiring resources in order to test theories by collecting and analyzing data to ground the research, writing, and publication process. These publications make up the "shoulders of the giants" on which new scientists stand (see Dara Horn's article, this volume), permitting the thoughtful testing of existing hypotheses and the background reasoning for the deductive development of new ones. Excellence in science is measured primarily by the quantity of such publications, their quality, and the reputations of the journals in which they are published. These easily quantifiable variables serve as a proxy for the more elusive and more important variables: uniqueness, importance, and innovation. Objectivity, that crown jewel of scientific thought, should result in the best papers appearing in the best journals and the appropriate distribution of resources to the best scientists, regardless of the gender or color of the authors. Wenneras and Wold, in their study, "Nepotism and Sexism in Peer Review," produce a detailed quantitative analysis of the effect of gender on the peer review process and demonstrate compellingly that the peer review process is not gender-blind.

In "Nine Decades, Nine Women, Ten Nobel Prizes: Gender Politics at the Apex of Science" Hilary Rose reviews the obstacles that even the most remarkable women scientists have faced on their way to the highest recognition of achievement in science—the Nobel Prize. Rose also tells the stories of women who were overlooked—whose achievement might have been recognized in a more equitable world. Rose points out historical and social factors that played into recognition (or nonrecognition) of the women included in the chapter. She describes how the late timing of the award in the lives of women Nobel Prize winners differs from awards made to men, rendering award-winning women less influential to young scientists and less able to broker the power that accompanies the prize. At the time of this writing, two additional women (and 88 men) have won Nobel Prizes in the sciences: Linda Buck and Christiane Nüsslein-Volhard, both in medicine and physiology. The age trend, though difficult to extrapolate from only two women awardees, seems to be shifting: the women were in their fifties, while the mean age of their accompanying cohort of men was 63, though the youngest man was 40.

Jaekyung Lee extends our analysis of those excluded from participating in the creation of scientific knowledge to ethnicity and the intersections of ethinicity and gender. Lee documents the underrepresentation of women within each ethnic group, providing data that suggest the "double jeopardy" experience of Evelynn Hammonds is neither unique nor a thing of the past. Lee uses the especially dramatic drop in participation of Asian American women between completion of the Ph.D. and entering the professoriate to document the more general pattern of the increasing underrepresentation of women with each rung of the academic ladder.

Moving from descriptions of achievements by groups to achievements of individuals, Wayne documents intellectual and personal growth as a process of recognizing that the same social forces at play in creating a hostile environment for women and people of color also create a scientific body of knowledge which is informed by pernicious assumptions about gender, race, agency, and ability. While Horn's piece in the previous half of this section alluded to the knowledge "lost" by restriction of scientific careers to a small subset of our population, Wayne's suggests that the partial scientific knowledge we do have is skewed by the socialization of its discoverers. Like the other autobiographers included in this book, Wayne has persisted in academic science: she wrote this piece during her first year as an assistant professor of zoology at the University of Florida; she has since earned tenure, and is now the inaugural director of the UF Graduate Program in Genetics and Genomics as well as affiliated faculty with the Center for Women's Studies and Gender Research.

Notes

1 See Janice Law Trecker, "Sex, Science, and Education," in *Women, Science, and Technology*, ed. M. Wyer *et al.*, pp. 88–98, New York: Routledge, 2001. See also Glenda Riley, "Origins of the Argument for Improved Female Education," *History of Education Quarterly* 9, no. 4 (Winter 1969): 455–70.
2 Margaret Rossiter, *Women Scientists in America: Struggles and Strategies to 1940*, Baltimore: Johns Hopkins University Press, 1982, pp. 2–3; Barbara Ehrenreich and Deirdre English, *For Her Own Good: 150 Years of the Experts' Advice to Women*, New York: Doubleday, 1978, p. 108.
3 S. Schwager, "Educating Women in America," *Signs: Journal of Women in Culture and Society* 12, no. 2 (1987): 333–72, esp. pp. 339–40; Rossiter, *Women Scientists to 1940*, pp. 2–3.
4 Ibid. See also H. J. Mozans, *Woman in Science*. Notre Dame, IN: University of Notre Dame Press, 1991 [1913].
5 Rossiter, *Women Scientists to 1940*, pp. 4–7; Schwager, "Educating Women in America," pp. 340–46.
6 P. Mack, "Straying from Their Orbits: Women in Astronomy in America," in *Women of Science: Righting the Record*, ed. G. Kass-Simon and P. Farnes, Bloomington: Indiana University Press, 1990; Rossiter, *Women Scientists to 1940*, pp. 14–15.
7 D. C. Hine *et al.*, *Black Women in America*. New York: Carlson Publishing, 1993.
8 Rossiter, *Women Scientists to 1940*, pp. 19–21.
9 Ibid., p. 35.
10 Mack, "Straying from Their Orbits," pp. 81–91; Rossiter, *Women Scientists to 1940*, pp. 55–57.
11 S. Shields, "Ms. Pilgrim's Progress: The Contributions of Leta Stetter Hollingworth to the Psychology of Women," *American Psychologist* 30 (1975): 852–57.
12 Rossiter, *Women Scientists to 1940*, p. 127; Hine *et al.*, *Black Women in America*, pp. 1298–99.
13 Women veterans were technically also eligible (there were some 400,000), but this was not widely known or applied. M. Rossiter, *Women Scientists in America: Before Affirmative Action, 1940–1972*, Baltimore: Johns Hopkins University Press, 1995, pp. 30–31.
14 Ibid., pp. 30–33.
15 Ibid., p. 41; National Science Foundation, *Women, Minorities, and Persons with Disabilities in Science and Engineering*, Washington DC: U.S. Government Printing Office, 1994.
16 Rossiter, *Women Scientists 1940–1972*, pp. 195–96.
17 S. B. Hrdy, "Empathy, Polyandry, and the Myth of the Coy Female," in *Feminist Approaches to Science*, ed. Ruth Bleier, p. 141. New York: Teachers College Press, 1991.
18 The full integration of women into science and engineering education challenges assumptions about the professionalization process itself. Even when male scientists are supportive of women students, gender dynamics can shape the mentor/student relationship in unproductive ways. See B. Subramaniam and M. Wyer, "Assimilating the 'Culture of No Culture' in Science: Feminist Interventions in (De)Mentoring Graduate Women," *Feminist Teacher* 12, no. 1 (June 1998): 12–28.
19 D. Haraway, "Animal Sociology and a Natural Economy of the Body Politic, Part I. A Political Physiology of Dominance," and "Part II. The Past Is the Contested Zone: Human Nature and Theories of Production and Reproduction in Primate Behavior Studies," *Signs: Journal of Women in Culture and Society* 4, no. 1 (1978): 21–36; 37–60.

1
The Anomaly of a Woman in Physics

Evelyn Fox Keller

A couple of months ago I was invited to give a series of lectures at a major university as one of a "series of distinguished guest lecturers" on mathematical aspects of biology. Having just finished teaching a course on women at my own college, I somehow felt obliged to violate the implicit protocol and address the anomalous fact of my being an apparently successful woman scientist. Though I had experienced similar vague impulses before, for a variety of reasons arising from a mix of anger, confusion, and timidity, it had never seemed to me either appropriate or possible to yield to such an impulse. Now, however, it seemed decidedly inappropriate, somewhat dishonest, and perhaps even politically unconscionable to deliver five lectures on my work without once making reference to the multitude of contradictions and conflicts I had experienced in arriving at the professional position presumed on this occasion. Therefore, in a gesture that felt wonderfully bold and unprofessional, I devoted the last lecture to a discussion of the various reasons for the relative absence of women in science, particularly in the higher ranks. The talk formed itself—with an ease, clarity, and lack of rancor that amazed me. I felt an enormous sense of personal triumph. Somehow, in the transformation of what had always appeared to me an essentially personal problem into a political problem, my anger had become depersonalized, even defused, and a remarkable sense of clarity emerged. It suggested to me that I might, now, be able to write about my own rather painful and chaotic history as a woman in science.

Origins are difficult to determine and obscure in their relation to final consequences. Suffice it to say that in my senior year of college I decided I would be a scientist. After several years of essentially undirected intellectual ambition, I majored in physics partly for the sake of discipline and partly out of the absence of any clear sense of vocation; and in my last year I fell in love with theoretical physics.

I invoke the romantic image not as a metaphor, but as an authentic, literal description of my experience. I fell in love, simultaneously and inextricably, with my professors, with a discipline of pure, precise, definitive thought, and with what I conceived of as its ambitions. I fell in love with the life of the mind. I also fell in love, I might add, with the image of myself striving and succeeding in an area where women had rarely ventured. It was a heady experience. In my adviser's fantasies, I was to rise, unhampered, right to the top. In my private fantasies, I was to be heralded all the way.

It was 1957. Politics conspired with our fantasies. Graduate schools, newly wealthy with National Science Foundation money, competed vigorously for promising students, and a promising female student was a phenomenon sufficiently unique to engage the interest and curiosity of recruiters from Stanford to Harvard. Only Cal Tech and Princeton were closed to me—they were not yet admitting women—and I felt buoyant enough to challenge them. I particularly wanted to go to Cal Tech to study with Richard Feynman—a guru of theoretical physics—on whose work I had done my senior thesis. In lieu of my being accepted at Cal Tech, an influential friend of mine volunteered to offer Feynman a university chair at MIT, where I would be admitted. Heady indeed.

Even then I was aware that the extreme intoxication of that time was transitory—that it had primarily to do with feeling "on the brink." Everything that excited me lay ahead. I had fantasies of graduate school and becoming a physicist; what awaited me, I thought, was the fulfillment of those fantasies. Even the idea of "doing physics" was fantasylike. I could form no clear picture of myself in that role, had no clear idea of what it involved. My conception of a community of scholars had the airiness of a dream. I was intoxicated by a vision that existed primarily in my head.

Well, Feynman was not interested in leaving Cal Tech, and so I went to Harvard. More accurately, I was pressured, and eventually persuaded, by both a would-be mentor at Harvard and my adviser, to go to Harvard. At Harvard I was promised the moon and the sun—I could do anything I wanted. Why I was given this extraordinary sales pitch seems, in retrospect, all but inexplicable. At the time, it seemed quite natural. I dwell on the headiness of this period in order to convey the severity of the blow that graduate school at Harvard actually was.

The story of my graduate school experience is a difficult one to tell. It is difficult in part because it is a story of behavior so crude and so extreme as to seem implausible.

Moreover, it is difficult to tell because it is painful. In the past, the telling of this story always left me so badly shaken, feeling so exposed, that I became reluctant to tell it. Many years have passed, and I might well bury those painful recollections. I do not because they represent a piece of reality—an ongoing reality that affects others, particularly women. Even though my experiences may have been unique—no one else will share exactly these experiences—the motives underlying the behavior I am going to describe are, I believe, much more prevalent than one might think, and detectable in fact in behavior much less extreme.

I tell the story now, therefore, because it may somehow be useful to others. I *can* tell the story now because it no longer leaves me feeling quite so exposed. Let me try to explain this sense of exposure.

Once, several months into my first year in graduate school, a postdoctoral student in an unusual gesture of friendliness offered me a ride home from a seminar and asked how I was doing. Moved by his gesture, I started to tell him. As I verged on tears, I noticed the look of acute discomfort on his face. Somehow, I had committed a serious indiscretion. It was as if I had publicly disrobed. Whatever I said, then and always after, it somehow seemed I had said too much. Some of this feeling remains with me even now as I write this article. It is a consequence of the assumption in the minds of others that what I am describing must have been a very personal, private experience—that is, that it was produced somehow by forces within myself. It was not. Although I clearly participated in and necessarily contributed to these events, they were *essentially* external in origin. That vital recognition has taken a long time. With it, my shame began to dissolve, to be replaced by a sense of personal rage and, finally, a transformation of that rage into something less personal—something akin to a political conscience.

That transformation, crucial in permitting me to write this, has not, however, entirely removed the pain from the process of recollecting a story that retains for me considerable horror. If I falter at this point, it is because I realize that in order for this story to be meaningful, even credible, to others, I must tell it objectively—I must somehow remove myself from the pain of which I write. The actual events were complex. Many strands weave in and out. I will describe them, one by one, as simply and as fairly as I can.

My first day at Harvard I was informed, by the very man who had urged me to come, that my expectations were unrealistic. For example, I could not take the course with Schwinger (Harvard's answer to Feynman) that had lured me to Harvard, and I ought not concern myself with the foundations of quantum mechanics (the only thing that did concern me) because, very simply, I was not, could not be, good enough. Surely my ambition was based on delusion—it referred to a pinnacle only the very few, and certainly not I, could achieve. Brandeis, I was told bluntly, was not Harvard, and although my training there might have earned me a place at Harvard, distinction at Brandeis had no meaning here. Both I and they had better assume I knew nothing. Hence I ought to start at the beginning. The students they really worried about, I was informed, were those who were so ignorant and naive that they could not apprehend the supreme difficulty of success at Harvard.

These remarks were notable for their blatant class bias and arrogance, as well as for their insistent definition of me on the basis of that bias—a gratuitous dismissal of my own account that I experienced recurrently throughout graduate school. The professor's remarks were all the more remarkable in that I had expressed exactly the same intentions in our conversation the previous spring and had then been encouraged. What could account for this extraordinary reversal? There had been no intervening assessment of my qualifications. Perhaps it can be explained simply by the fact that the earlier response was one of someone in the position of selling Harvard, while now it seemed there was an obligation to defend her. (It is ironic that universities should be associated with the feminine gender.) Nor was it coincidental, I suspect, that this man was shortly to assign to one of the senior graduate students (male, of course) the task of teaching me how to dress.[1]

Thus began two years of almost unmitigated provocation, insult, and denial. Lacking any adequate framework—political or psychological—for comprehending what was happening to me, I could only respond with personal rage: I felt increasingly provoked, insulted, and denied. Where political rage would have been constructive, personal rage served only to increase my vulnerability. Having come to Harvard expecting to be petted and fussed over (as I had been before) and expecting, most of all, validation and approval, I was entirely unprepared for the treatment I received. I could neither account for nor respond appropriately to the enormous discrepancy between what I expected and what I found. I had so successfully internalized the cultural identification between male and intellect that I was totally dependent on my (male) teachers for affirmation—a dependency made treacherous by the chronic confusion of sexuality and intellect in relationships between male teachers and female students. In seeking intellectual affirmation, I sought male affirmation, and thereby became exquisitely vulnerable to the male aggression surrounding me.

I had in fact been warned about the extreme alienation of the first year as a graduate student at Harvard, but both my vanity and my naiveté permitted me to ignore these warnings. I was confident that things would be different for me. That confidence did not last long. Coming from everywhere, from students and faculty alike, were three messages. First, physics at Harvard was the most difficult enterprise in the world; second, I could not possibly understand the things I thought I understood; and third, my lack of fear was proof of my ignorance. At first, I adopted a wait-and-see attitude and agreed to take the conventional curriculum, though I privately resolved to audit Schwinger's course. Doing so, as it turned out, seemed such an act of bravado that, daily, all eyes turned on me as I entered the class and, daily, I was asked by half a dozen people with amusement if I still thought I understood. Mysteriously, my regular courses seemed manageable, even easy, and as I became increasingly nervous about my failure to fear properly, I spent more and more evenings at the movies. In time, the frequent and widespread iteration of the message that I could not understand what I thought I understood began to take its toll. As part of a general retreat, I stopped attending Schwinger's course. I had begun to lose all sense of what

I did or did not understand, there and elsewhere. That I did well in my exams at the end of the semester seemed to make no difference whatever.

Meanwhile, it was clear that I was becoming the subject—or object—of a good deal of attention in the Physics Department. My seriousness, intensity, and ambition seemed to cause my elders considerable amusement, and a certain amount of curiosity as well. I was watched constantly, and occasionally addressed. Sometimes I was queried about my peculiar ambition to be a theoretical physicist—didn't I know that no woman at Harvard had ever so succeeded (at least not in becoming a *pure* theoretical physicist)? When would I too despair, fail, or go elsewhere (the equivalent of failing)? The possibility that I might succeed seemed to be a source of titillation; I was leered at by some, invited now and then to a faculty party by others. The open and unbelievably rude laughter with which I was often received at such events was only one of many indications that I was on display—for purposes I could either not perceive or not believe. My fantasy was turning into nightmare. . . .

It is sometimes hard to separate affront to oneself as a person from affront to one's sensibilities. Not only do they tend to generate the same response—one feels simply affronted—but it is also possible (as I believe was true here) that the motives for both affronts are not unrelated. I went to graduate school with a vision of theoretical physics as a vehicle for the deepest inquiry into nature—a vision perhaps best personified, in recent times, by Einstein. The use of mathematics to further one's understanding of the nature of space, time, and matter represented a pinnacle of human endeavor. I went to graduate school to learn about foundations. I was taught, instead, how to do physics. In place of wisdom, I was offered skills. Furthermore, this substitution was made with moralistic fervor. It was wrong, foolhardy, indeed foolish, to squander precious time asking why. Proper humility was to bend to the grindstone and learn techniques. Contemporary physics, under the sway of operationalism, had, it seemed, dispensed with the tradition of Einstein—almost, indeed, with Einstein himself. General relativity, the most intellectually ambitious venture of the century, seemed then (wrongly) a dead subject. Philosophical considerations of any sort in the physical sciences were at an all-time low. Instead, techniques designed to calculate nth-order corrections to a theory grievously flawed at its base were the order of the day. . . .

My naiveté and idealism were perfect targets. Not only did I not know my place in the scheme of things as a woman, but by a curious coincidence, I was apparently equally ingenuous concerning my place as a thinker. I needed to be humbled. Though I writhed over the banality of the assignments I was given, I did them, acknowledging that I needed in any case to learn the skills. I made frequent arithmetic errors—reflecting a tension that endures within me even today between the expansiveness of conception and the precision of execution, my personal variation perhaps of the more general polar tension in physics as a whole. When my papers were returned with the accuracy of the conception ignored and the arithmetic errors streaked with red—as if with a vengeance—I wondered whether I was studying physics or plumbing. Who has not experienced such a wrenching conflict between idealism and reality? Yet my fellow students seemed oddly untroubled. From the nature of their responses when I tried to press them for deeper understanding of the subject, I thought perhaps I had come from Mars. Why, they wondered, did I want to know? That they were evidently content with the operational success of the formulas mystified me. Even more mystifying was the absence of any appearance of the humility of demeanor that one would expect to accompany the acceptance of more limited goals. I didn't fully understand then that in addition to the techniques of physics, they were also studying the techniques of arrogance. This peculiar inversion in the meaning of humility was simply part of the process of learning how to be a physicist. It was intrinsic to the professionalization, and what I might even call the masculinization, of an intellectual discipline.

To some extent the things I describe here are in the nature of the academic subculture. They reflect the perversion of academic style—familiar in universities everywhere—a perversion that has become more extensive as graduate schools have tended to become increasingly preoccupied with professional training. My experiences resemble those of many graduate students—male and female alike. What I experienced as a rather brutal assault on my intellectual interests and abilities was I think no accident, but rather the inevitable result of the pervasive attempt of a profession to make itself more powerful by weeding out those sensibilities, emotional and intellectual, that it considers inappropriate. Not unrelated is a similar attempt to maintain the standards and image of a discipline by discouraging the participation of women—a strategy experienced and recounted by many other women. Viewed in this way, it is perhaps not surprising that the assault would be most blatant in a subject as successful as contemporary physics, and in a school as prestigious as Harvard.

Perhaps the most curious, undoubtedly the most painful, part of my experience was the total isolation in which I found myself. In retrospect, I am certain that there must have been like-minded souls somewhere who shared at least some of my disappointments. But if there were, I did not know them. In part, I attribute this to the general atmosphere of fear that permeated the graduate student body. One did not voice misgivings because they were invariably interpreted to mean that one must not be doing well.[2] The primary goal was to survive, and, better yet, to *appear* to be surviving, even prospering. So few complaints were heard from anyone. Furthermore, determined not to expose the slightest shred of ignorance, few students were willing to discuss their work with any but (possibly) their closest friends. I was, clearly, a serious threat to my fellow students' conception of physics as not only a male stronghold but a male *retreat*, and so I was least likely to be sought out as a colleague. I must admit that my own arrogance and ambition did little to allay their anxieties or temper their resistances. To make matters even worse, I shared with my fellow classmates the idea that a social or sexual relationship could only exist between male and female students if the man was "better" or "smarter" than the woman—or at the very least, comparable. Since both my self-definition and my performance labeled me as a superior student, the field of sociability and companionship was considerably narrowed.

There was one quite small group of students whom I did view as like-minded and longed to be part of. They too were concerned with foundations; they too wanted to know why. One of them (the only one in my class) had in fact become a close friend during my first semester. Though he preached to me about the necessity of humility, the importance of learning through the tips of one's fingers, the virtue of precision—he also listened with some sympathy. Formerly a Harvard undergraduate, he explained to me the workings of Harvard and I explained to him how to do the problems. With his assistance, I acquired the patience to carry out the calculations. We worked together, talked together, frequently ate together. Unfortunately, as the relationship threatened to become more intimate, it also became more difficult—in ways that are all too familiar—until, finally, he decided that he could no longer afford the risk of a close association with me. Out of sympathy for his feelings, I respected his request that I steer clear of him and his friends—with the consequence that I was, thereafter, totally alone. The extent of my isolation was almost as difficult for *me* to believe as for those to whom I've attempted to describe it since. Only once, years later in a conversation with another woman physicist, did I find any recognition. She called it the "sea of seats": you walk into a classroom early, and the classroom fills up, leaving a sea of empty seats around you.

Were there no other women students? There were two, who shared neither my ambition, my conception of physics, nor my interests. For these reasons, I am ashamed to say, I had no interest in them. I am even more ashamed to admit that out of my desire to be taken seriously as a physicist I was eager to avoid identification with other women students who I felt could not be taken seriously. Like most women with so-called male aspirations, I had very little sense of sisterhood.

Why did I stay? The Harvard Physics Department is not the world. Surely my tenacity appears as the least comprehensible component of my situation. At the very least, I had an extraordinary tolerance for pain. Indeed, one of my lifelong failings has been my inability to know when to give up. The very passion of my investment ruled out alternatives.

I had, however, made some effort to leave. At the very beginning, a deep sense of panic led me to ask to be taken back at Brandeis. Partly out of disbelief, partly out of the conviction that success at Harvard was an invaluable career asset, not to be abandoned, I was refused, and persuaded to continue. Although I had the vivid perception that rather than succeed I would be undone by Harvard, I submitted to the convention that others know better; I agreed to suspend judgment and to persevere through this stinging "initiation rite." In part, then, I believed that I was undergoing some sort of trial that would terminate when I had proven myself, certainly by the time I completed my orals. I need be stoic only for one year. Unfortunately, that hope turned out to be futile. The courses were not hard, never became hard in spite of the warnings, and I generally got As. But so did many other students. Exams in fact were extremely easy.

When I turned in particularly good work, it was suspected, indeed sometimes assumed, that I had plagiarized it. On one such occasion, I had written a paper the thesis of which had provoked much argument and contention in the department. This I learned, by chance, several weeks after the debate was well under way. In an effort to resolve the paradox created by my results, I went to see the professor for whom I had written the paper. After an interesting discussion, which incidentally resolved the difficulty, I was asked, innocently and kindly, from what article(s) I had copied my argument.

The oral exams, which I had viewed as a forbidding milestone, proved to be a debacle. My committee chairman simply failed to appear. The result was that I was examined by an impromptu committee of experimentalists on mathematical physics. Months later, I was offered the following explanation: "Oh, Evelyn, I guess I owe you an apology. You see, I had just taken two sleeping pills and overslept." The exam was at 2:00 P.M. Nevertheless, I passed. Finally, I could begin serious work. I chose as a thesis adviser the sanest and kindliest member of the department. I knocked on his door daily for a month, only to be told to come back another time. Finally I gained admittance, to be advised that I'd better go home and learn to calculate.

My second year was even more harrowing than the first. I had few courses and a great deal of time that I could not use without guidance. I had no community of scholars. Completing the orals had not served in any way to alleviate my isolation. I was more alone than ever. The community outside the Physics Department, at least that part to which I had access, offered neither solace nor support. The late fifties were the peak of what might be called home-brewed psychoanalysis. I was unhappy, single, and stubbornly pursuing an obviously male discipline. What was wrong with me? In one way or another, this question was put to me at virtually every party I attended. I was becoming quite desperate with loneliness. And as I became increasingly lonely, I am sure I became increasingly defensive, making it even more difficult for those who might have been sympathetic to me or my plight to approach me to commiserate. Such support might have made a big difference. As it was, I had neither colleagues nor lovers, and not very many friends. The few friends I did have viewed my situation as totally alien. They gave sympathy out of love, though without belief. And I wept because I had no friend whose ambition I could identify with. Was there no woman who was doing, had done, what I was trying to do? I knew of none. My position was becoming increasingly untenable. . . .

I recognize that this account reads in so many ways like that of a bad marriage—the passionate intensity of the initial commitment, the fantasies on which such a commitment (in part) is based, the exclusivity of the attachment, the apparent disappearance of alternative options, the unwillingness and inability to let go, and finally, the inclination to blame oneself for all difficulties. Although I can now tell this story as a series of concrete, objective events that involved and affected me, at the time I eventually came to accept the prevalent view that what happened to me at Harvard simply manifested my own confusion, failure, neurosis—in short that *I* had somehow "made" it happen. The implications of such internalization were—as they always are—very serious.

Now I had to ask *how* I had "made" it happen—what in me required purging? It seemed that my very ambition and seriousness were at fault, and that these qualities—qualities I had always admired in others—had to be given up. Giving up physics, then, seemed to mean giving up parts of me so central to my sense of myself that a meaningful extrication was next to impossible. I stayed on at Harvard, allowing myself to be convinced once again that I must finish my degree, and sought a dissertation project outside the Physics Department.

After drifting for a year, I took advantage of an opportunity to do a thesis in molecular biology while still nominally remaining in the Physics Department. That this rather unusual course was permitted indicated at least a recognition, on the part of the then-chairman, of some of the difficulties I faced in physics. Molecular biology was a field in which I could find respect, and even more important, congeniality. I completed my degree, came to New York to teach (physics!), married, bore children, and ultimately began to work in theoretical biology, where I could make use of my training and talents. This proved to be a rewarding professional area that sustained me for a number of critical years. If my work now begins to take me outside this professional sphere, into more political and philosophical concerns, this reflects the growing confidence and freedom I have felt in recent years.

Inner conflict, however, was not to disappear with a shift in scientific specialization. While it is true that I was never again to suffer the same acute—perhaps bizarre—discomfort that I did as a graduate student in physics, much of the underlying conflict was to surface in other forms as I assumed the more conventional roles of wife, mother, and teacher. The fundamental conflict—between my sense of myself as a woman and my identity as a scientist—could only be resolved by transcending all stereotypical definitions of self and success. This took a long time, a personal analysis, and the women's movement. It meant establishing a personal identity secure enough to allow me to begin to liberate myself from everyone's labels—including my own. The tension between "woman" and "scientist" is not now so much a source of personal struggle as a profound concern.

After many years, I have carved out a professional identity very different from the one I had originally envisioned, but one that I cherish dearly. It is, in many important ways, extraprofessional. It has led me to teach in a small liberal arts college that grants me the leeway to pursue my interests on my own terms and to combine the teaching I have come to love with those interests, and that respects me for doing so. It has meant acquiring the courage to seek both the motives and rewards for my intellectual efforts more within myself. Which is not to say that I no longer need affirmation from others; but I find that I am now willing to seek and accept support from different sources—from friends rather than from institutions, from a community defined by common interests rather than by status.

As I finished writing this essay, I came across an issue of the annals of the *New York Academy of Sciences* (March 15, 1973) devoted to "Successful Women in the Sciences." The volume included brief autobiographical accounts of a dozen or so women, two of whom were trained in physics and one in mathematics. Because material of this kind is almost nonexistent, these first-person reports are an important contribution "to the literature." I read them avidly. More than avidly, for the remarks of these women, in their directness and honesty, represent virtually the only instance of professional circumstances with which to compare my own experience.

It may be difficult for those removed from the mores of the scientific community to understand the enormous reticence with which anyone, especially a woman, would make public his or her personal impressions and experiences, particularly if they reflect negatively on the community. To do so is not only considered unprofessional; it jeopardizes one's professional image of disinterest and objectivity. Women, who must work so hard to establish that image, are not likely to take such risks. Furthermore, our membership in this community has inculcated in us the strict habit of minimizing any differences due to our sex. I wish therefore to congratulate women in the mainstreams of science who demonstrate such courage.

Their stories, however, are very different from mine. Although a few of these women describe discrete experiences similar to some of mine, they were generally able to transcend their isolation and discomfort, and in their perseverance and success, to vindicate their sex. I am in awe of such fortitude. In their stories I am confirmed in my sense that with more inner strength I would have responded very differently to the experiences I've recorded here. The difficulty, however, with success stories is that they tend to obscure the impact of oppression, while focusing on individual strengths. It used to be said by most of the successful women that women have no complaint precisely because it has been demonstrated that with sufficient determination, anything can be accomplished. If the women's movement has achieved anything, it has taught us the folly of such a view. If I was demolished by my graduate school experiences, it was primarily because I failed to define myself as a rebel against norms in which society has heavily invested. In the late fifties, "rebel" was not a meaningful word. Conflicts and obstacles were seen to be internal. My insistence on maintaining a romantic image of myself in physics, on holding to the view that I would be rewarded and blessed for doing what others had failed to do, presupposed a sense of myself as special, and therefore left me particularly vulnerable. An awareness of the political and social realities might have saved me from persisting in a search for affirmation where it could not and would not be given. Such a political consciousness would have been a source of great strength. I hope that the political awareness generated by the women's movement can and will support young women who today attempt to challenge the dogma, still very much alive, that certain kinds of thought are the prerogative of men.

Notes

1 My attire, I should perhaps say, was respectable. It consisted mainly of skirts and sweaters, selected casually, with what might have been called a bohemian edge. I wore little or no makeup.

2 Indeed, most people then and later assumed I had done badly—particularly after hearing my story. Any claims I made to the contrary met with disbelief.

2

Never Meant to Survive, a Black Woman's Journey*

An Interview with Evelynn Hammonds

Aimée Sands

Aimée. What was it that sparked becoming a scientist in your mind?

Evelynn. I thought I'd like to be a scientist when at nine I had my first chemistry set. I had such a good time with all the experiments. I wanted to know more, and I wanted to get the advanced Gilbert chemistry set so I could do more interesting experiments.

A. Who gave you the chemistry set?

E. My father. And he gave me a microscope a year later. I always had sets like that. I had chemistry sets or microscopes or building sets or race car sets or different kinds of project-kit things to build stuff. My father and I always spent some time together working on them, and he was always interested in what I was finding out . . . figuring out. . . .

A. When did you start doing science in school?

E. We always had science in elementary school. The, you know, "go out and look at the plants," and the general basic (I guess in elementary school) science curriculum that I took along with everything else. I didn't think of taking more science courses than just the requirements until I was in high school. But I really liked science, I always did. But my basic interest was that I wanted to go to a good college. So I wanted to have a good background to do that. And I felt that the more science and math I could take the better off I would be.

So I started seriously . . . I guess in my high school we had to take up through chemistry, but then I went on and took physics. We only had math up through trigonometry, but I begged my math teacher to let us have a pre-calculus class because I wanted to go on. And that pre-calculus class came about because in my junior year in high school I was accepted into a National Science Foundation summer program for high achievers in mathematics for high school students. So I spent the summer in Emory University studying math.

There were three Black students in the program, and we were all just totally baffled by what was going on. We were taking a course in analytical geometry when we didn't know what analytical geometry was. We were taking an introductory course in group theory, and I can't remember the third course, but, some of

* The phrase "never meant to survive" is from a line in the poem "A Litany for Survival" by Audre Lorde, published in *A Black Unicorn* (New York: Norton, 1978).

the concepts it seemed all the other students had studied before and we hadn't studied at all, 'cause all three of us had gone to segregated high schools or recently integrated high schools. And it was a very painful experience because I felt that I was as smart as the other kids, the white kids in the class, but I had this gap in my background. I didn't know what to do about it, how to go and find the information I didn't have, and I didn't know how to prove I was still good, even though I didn't understand what was going on in class.

A. Did you know what the gap was even called? What you were missing?

E. No, I didn't have any words for it. It was just very painful. The three of us sort of haunted the libraries trying to find the books that would help us understand what was going on. It was supposed to be a summer program, so we were supposed to have fun, but the three of us weren't having fun at all. We were miserable and scared, and wondered if we were going to make it. And I was also completely angry at my parents and at my teachers that I'd had at my high school, who I felt hadn't pushed me and hadn't given me the right preparation. And that was the beginning for me to begin to understand that I'd had a deficient education ... because I'd gone to predominantly Black schools, that that deficiency showed up most strongly in math and science. So it made me angry and made me start looking over what had happened to me.

A. What did you see?

E. I felt I'd been cheated ... I felt I'd been denied that opportunity to have a good education because I was a Black person and I lived in the South. So I went back to my high school, and I took another year of science when I didn't have to. And I took another year of math and asked my teacher for the pre-calculus course.

A. Did that turn out to be what you were missing? Pre-calculus?

E. In part.

A. I want to go back a little before the Emory experience, when you were in grade school and high school. Were there teachers that encouraged your interest in science? I mean you said you took what everyone else took, and you implied by that you didn't have any special interest in school and it sounds like your special interest was more outside of school with the chemistry stuff. Is that right?

E. Yeah, uh ... in elementary school and probably in junior high and the early years of high school I pursued my interest outside. I'd read books, science books and books about science and ideas just on my own, and I would talk to the few friends that I had who were interested in those things.

A. Were there teachers who encouraged or discouraged you or did they just not know anything about that side of your life?

E. Most of my teachers didn't know that I was doing it. It would show up when I'd come back to school that I had read all this interesting stuff, and I'd talk about it in class. They were very encouraging but they didn't push or anything. I always had—particularly math and science—teachers who took an interest in me most of the way—*except*, I have to say, for a couple of the times when I was being bused. I had two teachers, both math, one science, who were just outright racist and ... one math teacher, who would, if I raised my hand for a question (I was the only Black student in the class) she would stop, call the roll, ask everybody if they had a question, skip my name and *then* ask me what my question was at the end. So I had those kinds of experiences. Or I never quite had enough points to make an A on a test or ... I always seemed to get an A–. It always seemed there were points to be taken off for something—you know that kind of stuff—and I noticed. Those were the kinds of things I didn't know how to fight, at that time.

A. Why do you think it was that those kinds of experiences didn't discourage you from pursuing science?

E. Because I was angry, and I wasn't going to let that stop me. And . . . because my parents wouldn't have let me stop, to a certain extent. If I had given that kind of reason they wouldn't have—they would have thought I was unacceptable—especially my mother.

A. What would she have said?

E. She would have said that I could stop it if I didn't *like* it, but I couldn't stop because someone was discriminating against me—or making it difficult for me—because I had to understand people were going to make it difficult for me in the world because I was Black. . . .

A. Describe the remainder of your college years and what happened?

E. I entered Spelman College in the Dual Degree Program, which was a program between the five Black colleges and the Atlanta University Center and Georgia Tech where students would spend 2½ years at one of the Black colleges and then 2½ years at Georgia Tech and at the end of that time have bachelor's degrees from both schools.

It was important to me to have the experience of being at Spelman. Even though I rebelled at first, I began to like being there. At the end of my junior year at Spelman, I was about to begin my time at Georgia Tech. I had declared at Spelman that I was a physics major, so I was predominantly taking most of my physics classes at Morehouse because Spelman College actually didn't have a Physics Department. So there I was again—there were about four women in my class, at Morehouse.

It was in the spring of that year that I really came to terms with what it was going to mean to be a female and be a serious scientist, because at that time we had a speaker who was Shirley Jackson, who had just gotten her Ph.D. in physics from MIT and was the first Black woman to do so. She came and spoke, and it created quite a furor in our department and a whole conversation engendered about whether or not women could be women and scientists at the same time.

It was really an ugly way that it all came about. In choosing officers for a society of physics students organization, the men students in the class fought really seriously against any women being officers of the organization. In the midst of that election all the faculty members, who were male, voted for the male students. Afterwards they apologized to me for doing so. It just started coming out more and more that you couldn't be a serious scientist and a woman. That was a prevailing attitude in the department. I was startled, I was completely shocked. I had never had anybody, the Black students I had gone to school with, question whether or not I could do what I wanted to do. I expected opposition from white people, and I expected that to be because I was a Black person, but I never expected opposition because I was a woman. In my usual fashion I went to the library to find out about Black women in science and women scientists . . . and there was *nothing!* Then I started getting worried.

At the end of that school term I went for a summer at Bell Laboratories to the Summer Research Program for Minorities and Women. It was a great program! We would go up to the labs, and we were given scientists to work with. We had a project for the summer, and we could report on it at a big presentation we had at the end of the summer, or lots of us were able to get our names on the work published in scientific journals. So it was a really good program. That also really honed my interest in being a scientist because I really like the projects that I worked on there, and I did well. I got a paper published with my name on it as one of the authors. It was exciting to be around famous people. The labs were well equipped so I got to see I . . . I remember saying to the person who was my advisor for the summer, that I was really interested in lasers, and he pointed to a laser sitting on the table, and I didn't know that was what a laser looked like [laugh]. I had only seen really small ones because we didn't have that kind of equipment at Morehouse. We had minimal equipment. So here I am saying, "I love lasers" and "I'm really interested in them," and I don't know what they look like! I was really embarrassed.

But I didn't encounter any opposition at Bell Labs. All of us were there because we were bright, and we were encouraged to do well and to take the opportunity we were being given there seriously. Though among the students themselves there was still *a lot* of talk about women not being serious scientists, and that was difficult, and I began to see it as more and more of a problem. I was *very* angry about it, and I thought of myself as a feminist for the first time as a result of that experience, both in the spring at school and the summer at Bell Labs.

A. What year was this?

E. This was 1974. . . .

A. What about your family, were they of any help at this point?

E. Let me clarify at this point. I wasn't getting discouraged by teachers or other scientists at Bell Labs at all. I was getting a lot of encouragement, also from my professors at Morehouse. I wasn't getting encouraged by my *peers*, though. That's where it was coming from. . . . All the social pressure. To give up going out with someone because I wanted to stay home and work was seen as weird, *I* was seen as weird and different, and *wrong*!—somehow not being a "right kind of woman." And *that* was what was disturbing me a lot.

And what I got by reading about the women's movement and reading all those books was that I wasn't the only woman in this world that was having this problem. That helped me tremendously, even though I was one of the few . . . I didn't know any other feminists! I was a—you know—bookstore feminist. Certainly there were women around me beginning to call into question men making outrageous sexist remarks. So reading about the women's movement helped me a lot, and the women that I met in the summer program at Bell Labs were also beginning to think of themselves seriously as having careers in science—going to graduate school, getting a Ph.D., and being serious scholars. So we were beginning to talk about it, beginning to see what was happening in terms of our relationships in the social world that we lived in.

So I left Bell Labs at the end of that summer and came back to face Georgia Tech. And that was something. What I faced there was that I was the only woman in my engineering class and one of the three Black students. The racism was unbelievable! . . .

A. So you moved from Spelman and Morehouse which were sister/brother Black colleges over to a predominantly white college?

E. Predominantly male. And the students in the Dual Degree Program were viewed by many people as only there because of affirmative action programs. It was felt if we didn't have good enough grades to come to Georgia Tech from the beginning, then we weren't as strong as other students, and we didn't deserve to be there. We were only there because the government was forcing them to let us in—that was the prevailing view of us.

A. So, on the whole, your experience at Georgia Tech—how would you sum it up insofar as your experience as a Black woman in science?

E. I think it was an extremely *difficult* period for me. If I hadn't had the support to pursue physics, if I hadn't had (after my first year at Georgia Tech) another summer at Bell Labs where I had that same nurturing, encouraging environment, I would not have gone on. At Georgia Tech the people there were *not* interested in Black students' development at all. So just basic things—like going into somebody to ask, "what's going to be on a test"—you could *never* get that information from people. We were being *denied* that kind of information; we were being left out of the environment there in really serious ways. It was as if you—and there were Black students who did *well* and they were *really bright*—they could really get the information on their own and didn't need to have other students to bounce ideas off and help them understand. They really basically did it on their own, and they were real bright—the students who did it. Those who weren't struggled *a lot*, with no encouragement whatsoever, I think. The people in the Dual Degree Program administration basically told us we were going to have a tough time, and nobody was going to help us out. Being a woman—nobody wanted to address that at all.

A. Wasn't that an issue there?

E. For me it was. I was going to lab and having a male lab partner who would set up the experiment. What we would usually do is set the experiment up, run the experiment, and get our data. They'd usually come in and set it up and want me to take notes. If I got in and *I* set it up, usually they would take it apart, and I would have to fight with some guy about "Don't take it apart." They just assumed I couldn't set it up correctly!

A. These were white *and* Black guys?

E. Yes. So I had to deal with that all the time, and it was very hard. But I wasn't the only Black woman in that program who started out; I was the only Black woman in my class, in the Dual Degree Program, who finished. And I was not the brightest woman. I know that. The other three women were stronger in math than me, and two of them were certainly stronger in chemistry. I was definitely the strongest in physics, but they didn't finish because of the lack of encouragement. I feel that very strongly. I don't think it was anything intellectual. I think it was the lack of encouragement. One woman—the other woman who was a physics major—when it was time to transfer to Georgia Tech, she was so terrified of what we heard about how hard it was—about the pressures—that she refused to go. She stayed at Spelman where it was more comfortable for her.

A. It sounds as if the fact that you majored in physics was ultimately what saved you. You also mentioned a professor who helped you along at Georgia Tech, and the other woman, I guess, didn't have that person?

E. No, they didn't and they didn't finish. I think it was the lack of encouragement from professors, what people felt was going to be a hostile atmosphere at Georgia Tech that they didn't want to face. *And* the fact that again, what was happening in our personal lives was that more and more we were experiencing "you can't be a woman and do this." That pull from boyfriends, from male friends, was causing a lot of conflict. And there was nobody to talk to about that who wasn't just saying "you have to be tough," and that was the only solace that we had. So it was very hard . . . and I'd always look around to see who was there and wonder why people weren't with me now that I'd started out with. And I know that for those women, that's why they weren't there.

A. It seems like you got to be tough in terms of racism but not in terms of sexism, how come? I mean it worked as a way to keep going for you when you were facing racism, not when you were facing sexism, is that right?

E. Ah, both. I guess I was prepared to face the fact that I was going to be having difficulties because I was Black. I wasn't *prepared* to face difficulties because I was female. And there were just too few people around to even acknowledge it and help me understand it. I think that there were lots of people around to talk to about how tough it was to be Black and do this. There were just too few people around, I think, until I came to MIT.

A. And it also sounds like, starting with your family, there was a real strong identification of racism as white people's problem, but sexism was *not* strongly identified as men's problem so that you would more easily internalize that.

E. I did internalize a lot of that and see it as my own deficiency. And that created a lot of doubt for me and was just hard for me to deal with. But I was still excited enough about science and physics to keep going. I had *no* intention of being an electrical engineer, at all. So I applied to MIT and was accepted, and I received a fellowship from Xerox. So, I was ready to leave Atlanta and turn my back on that and start. I really saw myself as a serious science student at that time, and I really saw coming to MIT as a big adventure.

And being at MIT was very difficult; again, I faced the racism and the sexism. And even in some ways it was as overt as at Georgia Tech, and in other ways that I felt in the end was more damaging . . . it was very subtle. I think the sexual, the male/female issues were probably stronger then. And my growing, *growing* consciousness as a feminist was almost like—at first I had no words and no one to talk to, and then, when I found them, my whole way of looking at the world changed. I think it's difficult for women students to continue in graduate school, to go on to the Ph.D. without being very single-minded and not let themselves be distracted . . . it's real important to do that. But the nature of the system . . . I have to say sometimes that the feminism distracted me. It made it hard for me to tolerate what I saw around me. It made it hard for me to tolerate the way I saw women treated when they came to speak. And I'd sit there in the back of the room and hear professors and fellow students talk about how the woman was dressed, and not ever talk about the content of her presentation. And there were very few women who came to talk at MIT in the Physics Department. So that was real disheartening. As my feminist consciousness grew, I was in great conflict about this. Did I really want to be involved with these people? The *culture* of physics was beginning to bother me a lot when I saw what was happening. And the work was hard, and I was beginning to have a lot of doubts about whether I could do the work.

A. But you were really a success as a young physicist.

E. I was. My research was successful. As a physics student, I was not successful. I was having a hard time with my courses, for the most part. I was having a hard time with my exams, but I was doing good research.

A. Did your problems in the courses and exams have to do with the atmosphere that you were experiencing there in terms of the sexism you described before?

E. I think a lot of it, again, is not being prepared for MIT in certain kinds of ways. . . .

A. In the educational holes again?

E. Yeah, the educational holes. My first week at MIT, the first person I had as a sort of advisor on which courses to take suggested that I start with freshman physics . . . essentially saying I should start all over!

A. This is graduate school?

E. Graduate school. And I was angry. I was very angry. I was insulted. What that meant was that instead of taking . . . you know looking at the educational holes and saying where do I really need to build up . . . I was mad! I was determined that I was going to take what every other graduate student was going to take. And the hell with them! Which was a mistake. Because there were some things I shouldn't have taken. I remember there were times when people said I had a chip on my shoulder. But, I remember that all of us Black students that came in that year, there were four of us, felt the same way; we felt we had been insulted. We felt we were being . . . um . . . we couldn't ask anybody and get a *reasonable* answer. It was an *unreasonable* answer to say to me I had to take freshman physics. We didn't want to be seen as deficient! We didn't want to be seen as, you know, not supposed to be there once again! This whole thing I think, really results from affirmative action; there are a lot of people who are resentful about affirmative action and they tended to label us. We didn't want to be labeled as deficient students. We wanted to be students and pursue our work like everybody else. There was lots of students . . . in fact, there was a student I had gone to Georgia Tech with, who came in and had (you know) educational holes in the same areas because . . .

A. Was he a white student?

E. Yeah, a white student. Because if you go to a . . . anybody who went to, say, a small college coming to MIT may not have the whole range of courses, or the background that lots of students that have gone to larger research institutions as undergraduates might have had. That's true across the board. And white students are given more reasonable, and in this case, the white student that I knew, was given more reasonable advice—"You should take junior-level this, or the senior undergraduate course in this." Not that he should take freshman physics. If someone had said to me, "Take junior level quantum mechanics or junior or senior level this," I would have said, "fine." You know, uh, but we didn't get that kind of advice. Because I knew it was racist, it angered me. And I went with the anger—I went with the emotional response. And I didn't have any way to distance myself and say, "Wait a minute, what do I really need to do?"

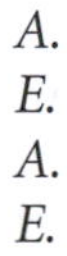

A. And once again, there was nobody to help you distance yourself?

E. Yeah, exactly.

A. You were again on your own?

E. Yeah. We were on our own. We were the other Black students that we had for support. There were very few people around that had just been through the department. Cheryl Jackson was the only Black woman who had ever been through the department and managed to survive, through to the Ph.D., and she had studied under the only *Black* professor in the department. There was nobody who had that step back and was there to say, "Take this approach to being successful."

A. Why were the Black students isolated? How did that come about?

E. I don't know how it comes about anywhere, but we were. In certain ways, we banded together. We all knew each other—except for one. Three of us had been at Bell Labs together—the summer program. So we stuck together because we knew each other. And we were isolated because it took us a long time to be willing to reach out of our group to other people and nobody reached towards us. So I don't see on white campuses, it's simply that Black students isolate themselves. The few Black students there who are in a class stick together; the other white students don't feel like they can reach across into that group and make alliances and work with people. So I think it's on both sides, because I think when *I* reached outside of that group there were people willing to be friends with me and study with me and stuff. But they didn't reach out.

A. What do you mean "they didn't reach out?" I missed what you meant.

E. When I was ready to reach outside of my group of Black students, there were people who were receptive to that. But there weren't white students who were coming around who were trying to reach into *our* group.

A. What was the role of your advisor in encouraging you to get through or not encouraging you to get through this time at MIT?

E. My advisor was very encouraging in a sense. He wanted me to get through, but I thought he was real tough on me, and that was hard. And I don't know if I can say that it's a lot more complicated than that. I think in part it was difficult for him to know *how* to advise a student like me.

A. Like *you* meaning what?

E. Black and female. And not just advise, I mean to move to the next step. An *advisor* just imparts information to you. A *mentor* really prepares you in a larger way to become a part of the profession in the same way that they are. I think that was difficult for him.

A. How long did you stay at MIT?

E. 3½ years.

A. And did you in fact get a Ph.D.?

E. No.

A. What happened?

E. I chose to leave.

A. Why?

E. I chose to leave because I finished my master's degree work, and I had to prepare for my Ph.D. exams, and I really came to a crisis. I didn't know if I really wanted to go on. I questioned whether I really wanted to be a physicist.

A. Is that because you questioned that you wanted to do physics?

E. I didn't question . . . I don't think I really questioned whether or not I wanted to do physics. I questioned whether or not I could really *be* a physicist. I questioned whether or not I had the skills. I questioned whether or not I was going to make it through my exams. I had a lot of doubts, and it was a real crisis for me. So I decided to leave.

A. What about the factor that you mentioned before, the social milieu of doing science? Did that play a role in this decision?

E. Oh yeah. I mean it was clear to me that I was going to be the only Black woman. You know, the social experience of going to an international conference and being the only Black woman there was difficult. That's the kind of isolation that was beginning to bother me tremendously. People were very nice to me, but I didn't have any friends. I didn't have anybody that I was close to that I could share my work with. And I knew that it wasn't going to get any better; it was going to continue, and I was going to continue to be isolated. And that isolation . . . that's what I mean by the culture of it was bothering me. It was the isolation. It was the fact that Black scientists are questioned more severely. Our work is held up to greater scrutiny; we have a difficult time getting research, getting university positions. All of that, and I didn't really want to fight that. One of my friends, the other Black woman who finished MIT, was supposed to come to a particular international conference, so I was feeling ok. "Well, she was going to be there, I'll be fine." And she decided not to come, because a piece of work that she was doing wasn't finished, and she knew that she was going to be given a really difficult time and asked very pointed questions about her work that she was not prepared to face. So she chose not to come. But that meant I was there alone. Ok, so what does that mean? That the three Black women in physics that I knew in the entire country at that time if we weren't all at the same conference we were going to be alone. And as I said before, my consciousness as a feminist was growing and growing. I wanted to become more active. Raising the issues of racism and sexism and trying to get my degree out of that department seemed to be at odds and more and more difficult for me. I would spend more time on those kinds of issues than I would on my science sometimes. So I was in a lot of conflict and that's why I chose to leave.

A. You said that you experienced sexism as an overriding problem at MIT when we first started talking about it. But the way you talked about it since then it really sounds like it was racism that really affected you the most.

E. They are *not* separate. Because they aren't separate in me. I am always Black and female. I can't say "well, that was just a sexist remark" without wondering would he have made the same sexist remark to a white woman. So, does that make it a racist, sexist remark? You know, I don't know. And that takes a lot of energy to be constantly trying to figure out which one it is. I don't do that anymore, I just take it as, you know, somebody has some issues about *me* and who *I am* in the world. *Me* being Black, female, and wanting to do science and be taken seriously. That's it.

3

Snow Brown and the Seven Detergents

A Metanarrative on Science and the Scientific Method

Banu Subramaniam

Once upon a time, deep within a city in the Orient, lived a young girl called Snehalatha Bhrijbhushan. She spent her childhood merrily playing in the streets with her friends while her family and the neighbors looked on indulgently. "That girl, Sneha [as they called her], is going to become someone famous someday," they would all say. Sneha soon became fascinated with the world of science. One day she announced, "I am going to sail across the blue oceans to become a scientist!"

There was silence in the room. "You can be a scientist here, you know."

"But I want to explore the world," said Sneha. "There is so much to see and learn."

"Where is this place?" they asked.

"It's called the Land of the Blue Devils."

"But that is dangerous country," they cried. "No one has ever been there and come back alive."

"Yes, I know," said Sneha. "But I have been reading about it. It is in the Land of the Kind and Gentle People. In any case, I can handle it."

Her friends and family watched her animated face and knew that if anyone could do it, it would be brave Sneha, and so they relented. The city watched her set out and wished her a tearful farewell. She promised to return soon and bring back tales from lands afar. For forty-two days and nights Sneha sailed the oceans. Her face was aglow with excitement, and her eyes were filled the stars. "It's going to be wonderful," she told herself.

And so one fine day she arrived in the Land of the Blue Devils. She went in search of the Building of Scientific Truth. When she saw it, she held her breath. There it stood, tall and slender, almost touching the skies. Sneha shivered. "Don't be silly," she told herself. She entered the building. The floors were polished and gleaming white. It all looked so grand and yet so formidable. She was led into the office of the Supreme White Patriarch. The room was full. "Welcome, budding Patriarchs," he said, "from those of us in the Department of the Pursuit of Scientific Truth. But let me be perfectly frank. These are going to be difficult years ahead. This is no place for the weak or the emotional or the fickle. You have to put in long, hard hours. If you think you cannot cut it, you should leave now. Let me introduce you to our evaluation system. Come with me."

He led them across the hall into a huge room. At the end of the room stood a mirror, long and erect and oh so white. "This is the Room of Judgment," he continued. "The mirror will tell you how you're doing. Let me show you." He went to the mirror and said, "Mirror, mirror on the wall, who is the fairest scientist of them all?"

"You are, O Supreme White Patriarch!" said the mirror.

The Patriarch laughed. "That is what all of you should aspire to. And one day when it calls out your name, you will take my place. But until then, you will all seek Truth and aspire to be number one. We want fighters here, Patriarchs with initiative and genius. And as for those who are consistently last in the class for six months . . . well, we believe they just do not have the ability to pursue Scientific Truth, and they will be expelled. Go forth, all ye budding Patriarchs, and find Scientific Truth."

Everyone went their way. Sneha found herself in the middle of the hallway all alone. "Go find Truth?" she said to herself. Was this a treasure hunt? Did Truth fall from the sky? She was very confused. This was not what she had thought it would be like. She went looking for her older colleagues. "Where does one find Scientific Truth?" she asked.

"Well," said he, "first you have to find the patronage of an Associate Patriarch or an Assistant Patriarch. You will have a year to do that. Until then, you take courses they teach you and you learn about Truths already known and how to find new Truths. During this time you have to learn how to be a scientist. That is very important, but don't worry, the mirror will assist you."

"How does the mirror work?" asked Sneha.

"Well, the mirror is the collective consciousness of all the Supreme White Patriarchs across the Land of the Kind and Gentle People. They have decided what it takes to be the ideal scientist, and it is what we all must dream of and aspire and work toward if we want to find Scientific Truth. You must check with the mirror as often as you can to monitor your progress."

Sneha tiptoed to the Room of Judgment, stood in front of the mirror, and said, "Mirror, mirror on the wall, who is the fairest scientist of them all?"

The mirror replied, "Not you, you're losing this game, you with the unprounceable name!"

Sneha was very depressed. Things were not going as she had expected. "Oh, mirror," she cried, "everything has gone wrong. What do I do?"

"More than anything," said the mirror, "you have to learn to act like a scientist. That's your first task. Deep within the forests lives the Wise Matriarch in the House of the Seven Detergents. Go see her, she will help you."

Sneha set out to meet the Wise Matriarch. "Come in, child," she said. "What seems to be the problem?" She appeared to be a very kind woman, and Sneha poured out her misery.

"I know this is a very difficult time for you, but it is also a very important one," the Matriarch said.

"Why do they call you the Wise Matriarch?" Sneha inquired.

"I joined the Department of the Pursuit of Scientific Truth some twenty years ago," the Matriarch replied. "That is why I understand what you're going through. I was expelled. When the department offered me this position, I felt I could begin changing things. Over the years I have advised many budding Patriarchs. You could say I've earned my reputation.

"My child," she went on, "this is where the department sends its scientific misfits. Let me show you what they would like me to have you do." She led Sneha to a room, and in it stood seven jars. "These are the seven detergents," she said. "With them you can wash away any part of yourself you don't want. But the catch is that once you wash it away, you have lost it forever."

Sneha was excited. "First I'd like to get rid of my name and my accent. The mirror told me that."

The Wise Matriarch shook her head, "My child, do not give away your identity, your culture—they are part of you, of who you are," she cried.

"But," said Sneha, "I've always dreamed of being a scientist. I spent all my savings coming here, and I cannot go back a failure. This is truly what I want." Sneha got into the

Great Washing Machine with the first detergent. *Rub-a-dub-a-dub, rub-a-dub-a-dub*, went the detergent.

"You may come out now, Snow Brown. Good luck."

Snow Brown went back amazed at how differently her tongue moved. For the next week she met the other budding Patriarchs, decided on her courses, and went out socializing with her colleagues. But everything was new in this land: how people ate and drank, even what people ate and drank. She felt stupid and ignorant. And just as she expected, when she went to the mirror, it told her that such behavior was quite unscientific and that she had to learn the right etiquette. Off she went again to the House of the Seven Detergents and used two other detergents that worked their miracles in the Great Washing Machine.

"Now I act like everyone else," she said, satisfied.

Snow Brown went to her classes. She thought them quite interesting. But the professors never looked her in the eye and never asked for her opinions. "Maybe they think I'm stupid," she said to herself. In class discussions everyone spoke up. Some of the things they said were pretty stupid, she thought. And so she would gather up her courage and contribute. She was met with stony silence. On some occasions others would make the same point, and the professor would acknowledge it and build on it.

She knew the mirror would be unhappy with her, and sure enough, she was right. "You have to be more aggressive," it said. "It doesn't matter so much what you say as how you say it."

"But that's ridiculous," she said. "Most of what is said is just plain dumb. Have you listened to some of them? They like the sound of their voices so much."

"That may be true, but that is the way. You have to make an impression, and sitting and listening like a lump of clay is not the way. And another thing—why did you let the others operate the machine in the lab? You have to take initiative."

"That was a ten-thousand-dollar machine. What if I broke it? I've never used it before."

"Leave your Third World mentality behind. The Patriarchs see it as a lack of initiative. They think you are not interested. You have to shoot for number one, be the very best. You have to act like a scientist, like a winner. Girl, what you need is a good dose of arrogance and ego."

Snow Brown was a little perturbed. She was disturbed by what she saw around her. Did she really want to act like some of the people she had met? What had happened to kindness, a little humility, helping each other? Just how badly did she want this, anyway? Her family was going to hate her when she went back. They would not recognize her. She thought long and hard and finally decided to go ahead.

She went back to the House of the Seven Detergents and used the anti-Third World detergent. When the Great Washing Machine was done, she came striding out, pride oozing out of every pore. The next day the Supreme White Patriarch called for her. "So, what kind of progress are you making in your search for Scientific Truth?" he asked.

"Well," she said, "the mirror has kept me occupied with learning to act like a scientist. Surely you can't expect me to make as much progress as the others, considering."

"We don't like students making excuses, Snow Brown. You had better make some progress, and real soon. There is no place for laziness here."

Snow Brown started developing some of her ideas. She went to the mirror to talk them over.

"I'm thinking of working with mutualisms," she said. "Organisms associate with each others in numerous ways ecologically. They can both compete for the same resources as in competition. Some live off other organisms, and that's called parasitism. When organisms get into ecological relationships with each other that are mutually beneficial, it's called mutualism."

"To be frank, Snow Brown, I would recommend studying competition or parasitism."

"But most of the studies of ecological interactions have focused on them," Snow Brown said. "I am amazed that there has been so little study of mutualisms. We know of some examples, but just how prevalent mutualisms are is still up in the air. For all we know, they may be a fundamental principle that describes demographic patterns of organisms on our planet."

"Whoa! Whoa!" cried the mirror. "You're getting carried away with your emotions. We would all like a and-they-lived-happily-ever-after kind of fairy tale. You are violating one of the fundamentals of doing science—objectivity. You don't pursue a study because you think it would be 'nice.' You base it on concrete facts, data. Then you apply the scientific method and investigate the problem."

"I do agree that the scientific method may have merit," she said. "I will use it to study mutualisms. But tell me, why do you think competition has been so well studied?"

"That's because competition is so important. Just look around you," the mirror replied. "Are the Patriarchs working with each other for their mutual benefit, or are they competing? This is what I do—promote competition. It is nature's way."

"Aha!" cried Snow Brown triumphantly. "You throw emotionalism and subjectivity at me. Listen to yourself. You are reading into nature what you see in yourself. I happen to believe that mutualisms are very important in the world. The Patriarchs have decided to work with a particular model. It doesn't mean that it's the only way."

"Get realistic," said the mirror, laughing. "You need the patronage of an Associate or Assistant Patriarch. You need to get money from the Supreme White Patriarch to do the research. Don't forget you need to please the Patriarch to get ahead. And you are still way behind in the game. This is not the time to get radical, and you are not the person to do it."

Convinced that pragmatism was the best course, the overconfident Snow Brown developed her ideas, talked in classes, and aggressively engaged the Patriarchs in dialogue. She was supremely happy. Things were finally going her way. She went to the mirror and said, "Mirror, mirror on the wall, who is the fairest scientist of them all?"

And the mirror replied, "It sure ain't you, Snow Brown. You're still the last one in town."

Snow Brown could not believe her ears. "I act and think like everyone around me. I am even obnoxious at times. What could I possibly still be doing wrong?"

"You're overdoing it," said the mirror. "You don't know everything. You should be a little more humble and subservient."

"Am I hearing things? I don't see anyone else doing that. This place does not validate that. You told me that yourself. What is really going on here?"

"When I advised you last," answered the mirror, "I advised you the way I would advise anyone, but I've been watching how the other Patriarchs interact with you. Apparently their expectations of you are different. You're brown, remember?"

Snow Brown was furious. She stormed out and went to the House of the Seven Detergents, and the sixth detergent washed her brownness away. She was now Snow White. She marched back to the Department of Scientific Truth. All the Patriarchs stared at her. They suddenly realized that what stood before them was a woman, and a beautiful one at that.

"Well, am I white enough for the lot of you now?" she demanded.

"Oh, but you're too pretty to be a scientist," cried the Supreme Patriarch.

"You can be a technician in my lab," said another.

"No, in mine!" urged yet another.

The Wise Matriarch had been right. Sneha had now lost her whole identity, and for what? Why had she not seen this coming? she asked herself. How could she ever have been the fairest scientist? How could she have been anything but last when judged by a mirror that wanted to produce clones of the Supreme White Patriarch? She went to the House of the Seven Detergents.

"It's too late, my child," said the Wise Matriarch. "You cannot go back now. I warned you about it. I wish I had more resources to support you and others like you. I have seen this happen far too often. It is important for you to communicate this to others. You must write down what has happened to you for future generations."

Two days later they discovered Sneha's cold body on the floor of her room. Her face looked tortured. In her sunken eyes was the resigned look of someone who had nothing more to lose to the world she had come to live in. On the nightstand by her body rested a tale entitled "Snow Brown and the Seven Detergents."

ENDING 1: AND INJUSTICE PREVAILS

The Patriarchs stood around the body. "It is so sad," said one. "But she was too emotional, a very fuzzy thinker. Some people are just not meant to pursue Scientific Truth. I wish they would accept it and leave instead of creating all this melodrama."

The other Patriarchs nodded in agreement at the unfortunate event.

"There is no reason for anyone to see this story, is there?" said the Patriarch who had initially spoken.

The others concurred, and they poured the last detergent on her. When they were done, there was nothing left. No pathetic face, no ugly reminders, no evidence.

ENDING 2: INTO EMPIRICISM

Snow Brown in her subversive wisdom sent copies of her story and insights to all in the department. There were some who kept the tale alive. It soon became apparent that there were dissenters within the Patriarchy. They broke their silence, and the movement slowly grew. Scientists began forming coalitions, talking and supporting each other in forming pockets of resistance. They questioned the power inequities. Why are most Patriarchs white? Why are most of them men? Over many decades the negotiations continued. Women scientists and scientists of color rose in the power structure. The collective consciousness was now male, female, and multicolored. But it was still supreme. It was privileged. The Pursuit for Truth continued, although new Truths emerged—Truths from the perspective of women, from the black, brown, yellow, red, and the white. The world had become a better place.

ENDING 3: A POSTMODERN FANTASY

The story of Snow Brown spread like wildfire. The Land of the Blue Devils was ablaze with anger and rage. The Wise Matriarch and a number of budding Patriarchs stormed the Department of the Pursuit of Scientific Truth and took it over. The mirror was brought down. The Room of Judgment was transformed into the Room of Negotiation. In their first meeting after all this occurred, the scientists sat together. "We need a different model," they said. They dismantled the positions of the Supreme White Patriarch, the Emeritus Patriarch, the Associate Patriarch, the Assistant Patriarch, and the Young Patriarch. "We will be self-governing," they decided. They debunked the myth that truth was a monolithic entity. "Truth is a myth," they said. "One person's truth is often privileged over someone else's. This is dangerous. The Patriarchs privileged their worldview over all others. This distorts knowledge and makes an accurate description of the world impossible." Together they decided they could help each other in reconstructing science and rewriting scientific knowledge. The House of the Seven Detergents was dismantled, and the detergents were rendered invisible. The new Department of Scientific Endeavor was very productive. Its faculty and students solved many problems that had eluded the world for years. They became world renowned, and their model was adopted far and wide.

If you are ever in the forests in the Land of the Blue Devils and come across the voice of an old-school scientist arguing vociferously, you know you have stumbled across the ghosts of Snow Brown and the Seven Detergents.

4
The Shoulders of Giants

Dara Horn

I have always been fascinated by scientists, because they appear to be the only people in the world who are immune to personal pettiness. Most people's careers are based on getting ahead, perhaps even at the expense of others. Scientists' goals, on the other hand, are not personal but collective. We imagine that their work is exclusively dedicated to the betterment of humankind.

As a student of literature, I am often asked to consider the life stories, motives, and intentions of the authors whose work I examine. If I were studying politics or history, I would concern myself even more with the personal conduct of the people I studied. But while a juicy biography of Darwin or Einstein would certainly make a good read, there exists a widespread belief among nonscientists that the motivations of researchers are secondary to their discoveries. Scientists are somehow outside of society, freed from its concerns in order to pursue knowledge for us all—or so those of us who are not scientists like to believe.

I imagine that most essays in this section of the book will address the effects of science on society, whether good or bad. But the story I am about to tell demonstrates the effects of society on science—effects that have the potential to be very damaging. In 1925 a twenty-five-year-old graduate student at Harvard discovered what the universe is made of. It was one of the most astonishing discoveries in the history of astronomic research. The problem was that no one believed her.

You have probably never heard of British-born Cecilia H. Payne (later Cecilia Payne-Gaposchkin), who in 1923 came to the United States to study stellar spectra at the Harvard College Observatory. In a remarkably short time Payne managed to quantify and classify the stellar spectra in the plate collection at the observatory, arriving at the startling conclusion that stars are "amazingly uniform" in their composition and that hydrogen is millions of times more abundant than any other element in the universe. Her doctoral dissertation, "Stellar Atmospheres" (1925), demonstrated her theory concerning the chemical composition of stars and earned her the first doctoral degree ever offered to either man or woman by Harvard's astronomy department. A few years later Otto Struve, an eminent astronomer, called it "the most brilliant Ph.D. thesis ever written" (p. 20).[1]

But in 1925 other scholars in the field were less impressed—or perhaps less courageous. Most astronomers at the time believed that stars were made of heavy elements. When her manuscript was presented to Henry Norris Russell, the leading contemporary astronomer dealing with stellar spectra, he wrote that her ideas concerning hydrogen's prevalence were "impossible" (p. 19). The director of Harvard's Observatory, Harlow Shapley, trusted

Russell and convinced Payne to dilute her conclusion substantially. By the end of these machinations Payne, despite the data in her thesis, asserted in writing that the abundance of hydrogen that she had detected was "almost certainly not real" (p. 20). Later, the same scholars who had led her to weaken her thesis steered her away from continuing her work on the observatory's spectra, the area where she had demonstrated both promise and brilliance. At the observatory she was pitted against one of Russell's students, thereby impeding the progress of both, and her research was redirected toward photometry and variable stars, which she studied for the rest of her career. Four years later Russell published a paper of his own announcing that the sun is made mostly of hydrogen.

Payne-Gaposchkin eventually became Harvard's first female tenured professor and later the first female department chair, but her "promotion" did not come until 1956, when a new observatory director finally conceded that she deserved the position and a new university president finally permitted it. She had been passed over for positions several times; once, when the observatory sought to fill a professorship, Shapley, unable to acknowledge the fact that an excellent candidate was standing in front of him, said to her, "What this Observatory needs is a spectroscopist" (p. 223). But by then, at Russell's suggestion, she had already been "pushed against my will into photometry" (p. 223).

Since her death, in 1979, the woman who discovered what the universe is made of has not received so much as a memorial plaque. Her newspaper obituaries did not mention her greatest discovery. Even today, when it has become fashionable for historians to highlight the accomplishments of great female scientists, other astronomers are given precedence, or her name is listed as merely one of many. But there is no need to visit an astronomy hall of fame to see how faint the memory of Payne-Gaposchkin has become. A glance at any elementary physical science textbook will do the trick. Every high-school student knows that Isaac Newton discovered gravity, that Charles Darwin discovered evolution, and that Albert Einstein discovered the relativity of time. But when it comes to the composition of our universe, the textbooks simply say that the most abundant atom in the universe is hydrogen. And no one ever wonders how we came to know this.

I believe that Payne-Gaposchkin's work on stellar spectra was stopped in its tracks by three factors that had absolutely nothing to do with astronomy: she was a woman, she was young, and she was outstanding. The first and second of these factors led other people to underestimate her, either by mistaking her genius for foolishness or by assuming (and perhaps even hoping) that she could not possibly be capable of doing what she did. The third, the brilliance that placed her research beyond the understanding of those who were supposedly older and wiser, ultimately made her underestimate herself—a fact that she acknowledged later in life. Long after the 1920s, when Otto Struve began working on a history of astrophysics, he offered to include her prior discovery of a particular effect in stellar spectra. But Payne-Gaposchkin was too angry with herself to accept. "I was to blame for not having pressed my point," she insisted. "I had given in to authority when I believed I was right. That is another example of How Not To Do Research" (p. 169). Her marriage to astronomer Sergei Gaposchkin seems to have made her even more vulnerable. His work was in variable stars, and Payne-Gaposchkin soon found herself devoting almost all of her research to that field. This, in addition to the challenge of raising their three children, caused her to abandon spectroscopy altogether. In her autobiography, however, she rarely expresses frustration with anyone other than herself.

But more than underestimation and disbelief were working against her. If Payne had merely been misunderstood, her colleagues would have surely encouraged her to continue working on stellar spectra once they realized that she was right. But they did not. Instead, even after the importance of her work had become obvious, Payne was still cajoled into abandoning her specialty. I believe this stemmed not from scientific concerns about the merit of her research, but from something simpler and more universal, an emotion that every scientist and nonscientist can understand. When dressed in the guise of science, jealousy becomes much more destructive than usual, for it can curtail our knowledge of the world. We will probably never be able to confirm why Russell and Shapley made the decisions that they made. Yet it is clear that discrimination as well as personal bitterness precluded scientific progress at many levels throughout Payne-Gaposchkin's career. In her case, one might argue that the public was lucky. Her revelation is ours, even if we do not know her name. But what of the discoveries that might have been made if she had continued working on stellar spectra for another twenty years? Can we even begin to estimate the magnitude of the loss?

Like most people, I have almost no scientific training. What I know about scientific research comes from newspapers, magazines, television programs, and a few ill-remembered high-school chemistry classes. But like most people, I have been taught to see science as an entirely pure and objective pursuit of knowledge, embarked upon for the benefit of people like me. This assumption may be ridiculous. Yet as knowledge expands beyond my grasp it is an assumption that I have to make in order to avoid living in a state of perpetual and paralyzing doubt.

So if I read in the newspaper that a fat substitute is safe for consumption, I do not question it. If a television program tells me that there is no cure for a particular disease, I believe it. If my college textbook explains to me that the universe is made of hydrogen but does not tell me who discovered it, I trust that this fact was so obvious that it did not even need to be discovered. Along with millions of others, I have placed my faith in scientists—not because I am dull-witted, but because their pursuit is reputed to be noble and disinterested, unmarred by the jealousies and desires that motivate most of us. Perhaps I am naive, but then so are many others. If scientists let us down, we will not know it.

The greatest loss to scientific research comes not from anything inherent in science, but rather from something inherent in society: our love of stars, particularly metaphoric ones. As students, we learn to associate the phenomena of our world with the names of the people who discovered them—never with their personalities, or with their networks of teachers and fellow researchers, or with the bibliographies of works upon which they built their own. On the elementary level, evolution is taught not as evolution, but as Darwinian evolution. We study not relativity, but Einstein's theory of relativity. Our textbooks supply us with Planck's constant, Avogadro's number, and Newton's laws. Scarcely a theorem exists without someone's name attached to it, regardless of how many people may have contributed to it.

After spending so many years listening to the great geniuses' names repeated again and again, a young student entering the sciences might understandably believe that the supreme goal of the scientist is not to reach for the stars, but rather to become one. After all, among the constellations of scientific giants, do we ever see the light of their instructors, their colleagues, or those who were their inspirations? Isaac Newton once said of himself, "If I have seen further than other men, it is because I have stood upon the shoulders of giants." But what happens when no one is content to offer his shoulders?

I am not in a position to judge how typical or unusual Payne-Gaposchkin's experience might be today. Nevertheless, I urge scientists to aspire to that which the rest of us already assume is taking place: the effort to ensure that research is not just a solitary effort geared toward individual reward, but a joint effort to push back the boundaries of knowledge. That should be the highest and most impassioned goal. As the sciences become more specialized, stardom will become more elusive. Scientists will then be faced with a choice: to become more competitive in their quest for glory or to become more sincere in their quest for truth. The most crucial contributions to knowledge come not only from those who make revolutionary revelations, but also from those who know how to appreciate and nurture the talents of others.

Cecilia Payne-Gaposchkin wrote in her autobiography that she hoped to be remembered for what she considers her greatest discovery: "I have come to know that a problem does not belong to me, or to my team, or to my Observatory, or to my country; it belongs to the world" (p. 162). The shoulders of that discovery are the only ones strong enough to support us.

Note

1. All quotations are from Cecilia Payne-Gaposchkin, *Cecilia Payne-Gaposchkin: An Autobiography and Other Writings*, 2nd ed., ed. K. Haramundanis (Cambridge: Cambridge University Press, 1986).

Careers: . . . And into the Fire

5 Nepotism and Sexism in Peer-Review

Christine Wennerås and Agnes Wold

Throughout the world, women leave their academic careers to a far greater extent than their male colleagues.[1] In Sweden, for example, women are awarded 44 per cent of biomedical PhDs but hold a mere 25 per cent of the postdoctoral positions and only 7 per cent of professorial positions. It used to be thought that once there were enough entry-level female scientists, the male domination of the upper echelons of academic research would automatically diminish. But this has not happened in the biomedical field, where disproportionate numbers of men still hold higher academic positions, despite the significant numbers of women who have entered this research field since the 1970s.

REASONS FOR LACK OF SUCCESS

Why do women face these difficulties? One view is that women tend to be less motivated and career-oriented than men, and therefore are not as assiduous in applying for positions and grants. Another is that women are less productive than men, and consequently their work has less scientific merit. Yet another is that women suffer discrimination due to gender. We decided to investigate whether the peer-review system of the Swedish Medical Research Council (MRC), one of the main funding agencies for biomedical research in Sweden, evaluates women and men on an equal basis. Our investigation was prompted by the fact that the success rate of female scientists applying for postdoctoral fellowships at the MRC during the 1990s has been less than half that of male applicants.

Our study strongly suggests that peer reviewers cannot judge scientific merit independent of gender. The peer reviewers overestimated male achievements and/or underestimated female performance, as shown by multiple-regression analyses of the relation between defined parameters of scientific productivity and competence scores.

In the peer-review system of the Swedish MRC, each applicant submits a curriculum vitae, a bibliography, and a research proposal. The application is reviewed by one of 11 evaluation committees, each covering a specified research field. The individual applicant is rated by the five reviewers of the committee to which he or she has been assigned. Each reviewer gives the applicant a score between 0 and 4 for the following three parameters: scientific competence, relevance of the research proposal, and the quality of the proposed methodology. The three scores given by each reviewer are then multiplied with one another to yield a product score that can vary between 0 and 64. Finally, the average of the five product scores an applicant has received is computed, yielding a final score that is the basis on which the applicants to each committee are ranked.

The MRC board, which includes the chairmen of the 11 committees, ultimately decides to whom the fellowships will be awarded. Usually each committee chooses between one and three of the top-ranked applicants. Of the 114 applicants for the 20 postdoctoral fellowships offered in 1995, there were 62 men and 52 women, with a mean age of 36 years, all of whom had received a PhD degree within the past five years. Most of the female applicants had basic degrees in science (62 per cent), and the rest had medical (27 per cent) or nursing (12 per cent) degrees; the corresponding figures for the male applicants were 38, 59 and 3 per cent.

Traditionally, peer-review scores are not made public, and indeed the MRC officials initially refused us access to the documents dealing with evaluation of the applicants. In Sweden, however, the Freedom of the Press Act grants individuals access to all documents held by state or municipal authorities. Only documents defined as secret by the Secrecy Act are exempt, for example, those that may endanger the security of the state, foreign relations or citizens' personal integrity. Accordingly, we appealed against the refusal of the MRC to release the scores.

In 1995, the Administrative Court of Appeal judged the evaluation scores of the MRC to be official documents. Hence, to our knowledge, this is the first time that genuine peer-reviewer evaluation sheets concerning a large cohort of applicants has become available for scientific study.

We found that the MRC reviewers gave female applicants lower average scores than male applicants on all three evaluation parameters: 0.25 fewer points for scientific competence (2.21 versus 2.46 points); 0.17 fewer points for quality of the proposed methodology (2.37 versus 2.54); and 0.13 fewer points for relevance of the research proposal (2.49 versus 2.62). Because these scores are multiplied with each other, female applicants received substantially lower final scores compared with male applicants (13.8 versus 17.0 points on average). That year, four women and 16 men were awarded postdoctoral fellowships.

As shown by these figures, the peer reviewers deemed women applicants to be particularly deficient in scientific competence. As it is generally regarded that this parameter is related to the number and quality of scientific publications,[2–5] it seemed reasonable to assume that women earned lower scores on this parameter than men because they were less productive. We explored this hypothesis by determining the scientific productivity of all 114 applicants and then comparing the peer-reviewer ratings of groups of male and female applicants with similar scientific productivity.

PRODUCTIVITY VARIABLES

We measured the scientific productivity of each applicant in six different ways. First, we determined the applicant's total number of original scientific publications, and second, the number of publications on which the applicant was first author. Both figures were taken from the applicant's bibliography, which we doublechecked in the Medline database. (We call these measures "total number of publications" and "total number of first-author publications.")

To take into account the fact that the prestige of biomedical journals varies widely, we constructed measures based on journals' impact factors. The impact factor of a scientific journal is listed in the independent Institute of Scientific Information's *Journal Citation Reports* and describes the number of times an average paper published in a particular journal is cited during one year. Our third measure was to add together the impact factors of each of the journals in which the applicant's papers were published, generating the "total impact measure" of the applicant's total number of publications.

Fourth, we generated the "first-author impact measure" by adding together the impact factors of the journals in which the applicant's first-author papers appeared. The unit of measure for both total impact and first-author impact is "impact points," with one impact point equalling one paper published in a journal with an impact factor of 1.

Fifth, using the science citation database, we identified the number of times the applicant's scientific papers were cited during 1994, which yielded the measure "total citations." And sixth, we repeated this procedure for papers on which the applicant was first author, giving the measure "first-author citations." Did men and women with equal scientific productivity receive the same competence rating by the MRC reviewers? No! As shown in Fig. 5.1 for the productivity variable "total impact," the peer reviewers gave female applicants lower scores than male applicants who displayed the same level of scientific productivity. In fact, the most productive group of female applicants, containing those with 100 total impact points or more, was the only group of women judged to be as competent as men, although only as competent as the least productive group of male applicants (the one whose members had fewer than 20 total impact points).

WHY WOMEN SCORE LOW

Although the difference in scoring of male and female applicants of equal scientific productivity suggested that there was indeed discrimination against women researchers, factors other than the applicant's gender could, in principle, have been responsible for the low scores awarded to women. If, for example, women were mainly to conduct research in

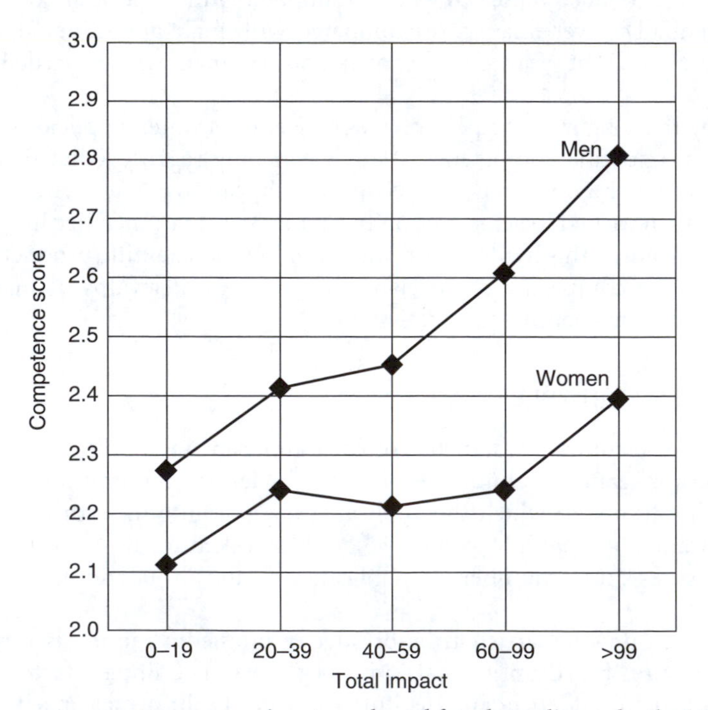

Figure 5.1 The mean competence score given to male and female applicants by the MRC reviewers as a function of their scientific productivity, measured as total impact. One impact point equals one paper published in a journal with an impact factor of 1. (See text for further explanation.)

areas given low priority by the MRC, come from less-renowned universities, or have less collaboration with academic decision-makers, their lower scores could depend on such factors, rather than on their gender *per se*.

To determine the cause of women's lower scores, we performed a multiple-regression analysis, which reveals the factors that exert a primary influence on a certain outcome (for example, competence scores) and the size of such an influence. Multiple regression permits the elimination of factors whose influence on a certain outcome merely reflects their dependence on other factors.

In the multiple-regression analysis, we assumed that the competence scores given to applicants are linearly related to their scientific productivity. We constructed six different multiple-regression models, one for each of the productivity variables outlined above. In each of these models, we determined the influence of the following factors on the competence scores: the applicant's gender; nationality (Swedish/non-Swedish); basic education (medical, science or nursing school); scientific field; university affiliation; the evaluation committee to which the applicant was assigned; whether the applicant had had postdoctoral experience abroad; whether a letter of recommendation accompanied the application; and whether the applicant was affiliated with any of the members of the evaluation committee. The last piece of information is noted on the MRC evaluation protocols, in which case the reviewer in question is not allowed to participate in the scoring of that applicant. It was as frequent for female (12 per cent) as for male (13 per cent) applicants to be associated with a committee member.

The outcome of the regression analysis is shown in Table 5.1. Three out of the six productivity variables generated statistically significant models capable of predicting the competence scores the applicants were awarded: total impact, first-author impact and first-author citations. The model that provided the highest explanatory power was the one based on total impact ($r^2 = 0.47$). In all three models, we found two factors as well as scientific productivity that had a significant influence on competence scores: the gender of the applicant and the affiliation of the applicant with a committee member.

According to the multiple-regression model based on total impact, female applicants started from a basic competence level of 2.09 competence points (the intercept of the multiple, regression curve) and were given an extra 0.0033 competence points by the reviewers for every impact point they had accumulated. Independent of scientific productivity, however, male applicants received an extra 0.21 points for competence. So, for a female scientist to be awarded the same competence score as a male colleague, she needed to exceed his scientific productivity by 64 impact points (95 per cent confidence interval: 35 to 93 impact points).

This represents approximately three extra papers in *Nature* or *Science* (impact factors 25 and 22, respectively), or 20 extra papers in a journal with an impact factor of around 3, which would be an excellent specialist journal such as *Atherosclerosis, Gut, Infection and Immunity, Neuroscience* or *Radiology*. Considering that the mean total impact of this cohort of applicants was 40 points, a female applicant had to be 2.5 times more productive than the average male applicant to receive the same competence score as he ((40 + 64)/40 = 2.6).

FRIENDSHIP BONUS

According to the same multiple-regression model, applicants who were affiliated with a committee member received competence scores 0.22 points higher than applicants of the same gender and scientific productivity who lacked such ties (Table 5.1). This "affiliation bonus" was worth 67 impact points (confidence interval: 29 to 105 impact points). Hence, an applicant lacking personal ties with the reviewers needed to have 67 more impact points

Table 5.1 Factors that Significantly Influenced Peer Reviewers' Rating of Scientific Competence, According to Three Multiple Regression Models.

	Scientific Productivity			*Additional Points Given by the Reviewers for the Following Factors*			*Size of the Influence of the Non-Scientific Factors in Productivity Equivalents*		
Multiple regression model based on:	r^2	Intercept	Competence points per productivity unit	Male gender	Reviewer affiliation	Recommendation letter	Male gender	Reviewer affiliation	Unit of measure
Total impact	0.47	2.09	0.0033 *<0.00005**	0.21 *<0.00005*	0.22 *0.0008*	0.10 *0.04*	64 (35–93)†	67 (29–105)	Impact points
First-author impact	0.44	2.13	0.0094 *<0.0001*	0.24 *<0.00005*	0.20 *0.005*	NS	25 (14–36)	21 (6–36)	Impact points
First-author citations	0.41	2.17	0.0054 *0.001*	0.23 *<0.00005*	0.23 *0.001*	NS	42 (23–61)	42 (17–67)	Citations during 1994

* Italicized numbers indicate *P*-values for the variable in question.
† Numbers in parentheses indicate 95% confidence interval.
NS, not statistically significant, *P*-value > 0.05.

than an applicant of the same sex who was associated with one of the reviewers, to be perceived as equally competent. So, although MRC policy does not allow "biased" reviewers to participate in the scoring of applicants they are associated with, this rule was insufficient, as the "neutral" committee members compensated by raising their scores when judging applicants affiliated with one of their peers.

Because the affiliation bonus was of the same magnitude as the "male gender" bonus, a woman applicant could make up for her gender (–0.21 competence points) by being affiliated with one of the reviewers (+0.22 competence points). On the other hand, a female (–0.21 competence points) lacking personal connections in the committee (–0.22 competence points) had to present an additional 131 impact points to the MRC reviewers to receive the same competence score as a male applicant affiliated with one of the reviewers.

Such a level of productivity was attained by only three of the 114 applicants, one male and two female. Hence, being of the female gender and lacking personal connections was a double handicap of such severity that it could hardly be compensated for by scientific productivity alone.

The two other regression models, based on first-author impact and first-author citations, yielded almost identical results to the first with regard to the effect of gender and affiliation (Table 5.1). This congruity was not a statistical artefact due to a high degree of interrelation between the three productivity variables, as the total impact and first-author impact of the applicants were only moderately correlated ($r = 0.63$), as were total impact and first-author citations ($r = 0.62$). We therefore believe that male gender and reviewer affiliation were real determinants of scientific competence in the eyes of the MRC reviewers.

The applicant's nationality, education, field of research or postdoctoral experience did not influence competence scores in any of the models. A letter of recommendation had a positive effect on the competence score in the model based on total impact, but not in the two others (Table 5.1). By contrast, the evaluation committee that rated individual applicants did influence competence scores, as some committees were "sterner" in their evaluation of competence than the rest (data not shown). However, an applicant who was assigned to a "tough" committee had the same chance of being awarded a fellowship as other applicants, as fellowships were distributed based on the rank the applicant acquired within his or her committee and not on absolute score values.

CHANGING THE SYSTEM

The peer-review system, characterized as "the centerpiece of the modern scientific review process,"[6] has been criticized on many grounds, including poor inter-reviewer reliability[2] and because reviewers may favour projects confirming their own views.[7] Our study is the first analysis based on actual peer-reviewer scores and provides direct evidence that the peer-review system is subject to nepotism, as has already been suggested anecdotally.[8–10]

One might argue that young researchers affiliated with peer reviewers are part of a scientific élite that has received superior training and are therefore more competent than average applicants. Indeed, applicants with such ties had higher total impact levels on average than applicants without such connections (data not shown). Hence, applicants with personal alliances justly benefited from higher competence scores because of their higher scientific productivity. However, on top of that, they were given extra competence points not warranted by scientific productivity. We see no reason why an applicant who manages to produce research of high quality despite not being affiliated with a prestigious research group should not be similarly rewarded.

Several studies have shown that both women and men rate the quality of men's work higher than that of women when they are aware of the sex of the person to be evaluated, but not when the same person's gender is unknown.[11–13] It is somewhat surprising that the results of these studies have not discouraged the scientific community from relying on evaluation systems that are vulnerable to reviewer prejudice.

An interesting question that we could not address here is whether the harsher evaluation of female researchers was due to the paucity of women among the peer reviewers. The small number of women reviewers (5 out of 55) and their uneven distribution among the MRC's committees made a statistical analysis of their scoring behaviour impossible. However, a few studies have indicated that female evaluators may be more objective in assessing the achievement of women than their male counterparts.[14] Nevertheless, we are not confident that a simple increase in the percentage of women reviewers would solve the problem of gender-based discrimination.

If gender discrimination of the magnitude we have observed is operative in the peer-review systems of other research councils and grant-awarding organizations, and in countries other than Sweden, this could entirely account for the lower success rate of female as compared with male researchers in attaining high academic rank. The United Nations has recently named Sweden as the leading country in the world with respect to equal opportunities for men and women, so it is not too far-fetched to assume that gender-based discrimination may occur elsewhere. It is therefore essential that more studies such as ours are conducted in different countries and in different areas of scientific research.

An in-depth analysis of other peer-review systems can be achieved only if the policy of secrecy is abandoned. We could perform our study only because of the Swedish Freedom of the Press Act. It is often claimed that secrecy in scoring will protect reviewers from improper influences. But our results cast doubt on these claims. It has also been suggested that the recruitment of peer reviewers of high quality would be impeded if reviewers were not granted anonymity. Such fears seem to be exaggerated because, although reviewer evaluation scores have been accessible to everyone in Sweden since the court ruling of 1995, there have been no large-scale defections of peer reviewers from the evaluation committees.

Most important, the credibility of the academic system will be undermined in the eyes of the public if it does not allow a scientific evaluation of its own scientific evaluation system. It is our firm belief that scientists are the most suited to evaluate research performance. One must recognize, however, that scientists are no less immune than other human beings to the effects of prejudice and comradeship. The development of peer-review systems with some built-in resistance to the weaknesses of human nature is therefore of high priority. If this is not done, a large pool of promising talent will be wasted.

Notes

1 Widnall, S.E. *Science* **241**, 1740–1745 (1988).
2 Cole, S., Cole, J.R. & Simon, G.A. *Science* **214**, 881–886 (1981).
3 Long, J.S. *Social Forces* **71**, 159–178 (1992).
4 Sonnert, G. *Social Stud. Sci.* **25**, 35–55 (1995).
5 Sonnert, G. & Holton, G. *Am. Sci.* **84**, 63–71 (1996).
6 Glantz, S.A. & Bero, L.A. *J. Am. Med. Assoc.* **272**, 114–116 (1994).
7 Ernst, E., Resch, K.L. & Uher, E.M. *Ann. Intern. Med.* **116**, 958 (1992).
8 Forsdyke, D.R. *FASEB J.* **7**, 619–621 (1993).
9 Calza, L. & Gerbisa, S. *Nature* **374**, 492 (1995).
10 Perez-Enciso, M. *Nature* **378**, 760 (1995).
11 Goldberg, P. *Trans-Action* **5**, 28–30 (1968).
12 Nieva, V.E & Gutek, B.A. *Acad. Manag. Rev.* **5**, 267–276 (1980).
13 O'Leary, V.E. & Wallston, B.S. *Rev. Pers. Soc. Psychol.* **2**, 9–43 (1982).
14 Frieze, I.H. in *Women and Achievement: Social and Motivational Analyses* (eds. Mednick, M.T., Tangri, S.S. & Hoffman, L.W.), 158–171 (Hemisphere, Washington DC, 1975).

6

Nine Decades, Nine Women, Ten Nobel Prizes

Gender Politics at the Apex of Science

Hilary Rose

Gertrude Elion, Rita Levi-Montalcini, Barbara McClintock, Rosalyn Yalow, Dorothy Crowfoot Hodgkin, Maria Goeppert Mayer, Gerty Cori, Irène Joliot Curie and Marie Curie: nine women, ten Nobel Prizes for science (Marie Curie was awarded prizes both in physics—1903—and in chemistry—1911), distributed over the eighty-five years between 1903 and 1988. They range in age, from Curie receiving her first prize at 36, to the three most recent, Elion, Levi-Montalcini and McClintock, being 71, 77, and 81, respectively. Apart from Irene Joliot Curie, who, at 38, emulated her mother in her youthfulness as well as her scientific talent, and was awarded a prize in 1935, the intermediate group of postwar prizewinners were all in their fifties: Gerty Cori, 53 (1947); Maria Goeppert Mayer, 57 (1963); Dorothy Crowfoot Hodgkin, 54 (1964); and Rosalyn Yalow, 56 (1977). These nine women constitute some 2 per cent of the scientific Nobel laureates.

Since Nobel Prizes are not awarded posthumously a number of commentators, including but not only feminists, viewing the three women honoured in the 1980s, have suggested that longevity is increasingly an additional criterion for women scientists to meet. It seems that the Nobel committee, in responding to the new pressure on it to recognize women scientists, feels safer in going back in history, to acknowledge those whose scientific eminence is unquestionable but who have been previously passed over. Perhaps men with the power to give public recognition suffer from an inability to recognize scientific merit in peer-group women, whereas they have no such problem with peer-aged or even younger men. However, in that a central rationale for awarding the cash-rich Nobel Prize was to free creative scientists from concerns about resources, then these most recently honoured women would seem to be ineligible, and certainly other older men scientists have been explicitly excluded on precisely these grounds.[1]

This anomaly, which in its repetition suggests a response to the increasing claims of gender justice and those of scientific merit, while possibly not at the level of conscious intentionality, is demonstrably effective as a means of constraining reform. The overdue recognition of these distinguished but now older women scientists limits the possibility of their exercising the usual powers of a Nobel laureate. Their age means that, however brilliant, they are manifestly less likely to be in touch with younger up-and-coming scientists in their own field and less likely to be able to campaign for them. The move also diminshes the pressure to recognize those others, in their forties or fifties, who would be in a phase of their life and career cycle where they might best utilize the reward and the status. Even before the most recent awards, the time gap between their work and its formal recognition was already more strongly marked for women than men.[2] Nor is this unrecognized by the women scientists themselves, though perhaps it takes someone of Rita Levi-Montalcini's social and scientific confidence to reveal publicly her anger at the lapse of time and of the different treatment accorded to those she sees as in every way her peers. She notes that "two of my university colleagues and close friends, Salvador Luria and Renato Dulbecco, were to receive the Nobel Prize in Physiology and Medicine, respectively seventeen and eleven years before I would receive the same most prestigious award."[3]

THE PRIZE AS CULTURAL CAPITAL

While scientific excellence has, with very rare exceptions, been successfully acknowledged by the Nobel Science committees (the Literature and Peace prizes have long had a more contentious record), the institutional and social origins of the laureates have played a significant part. Just ten colleges, for example, produced 55 per cent of the 71 US laureates studied by Harriet Zuckerman.[4] In a similar way the history of the nine women laureates is in a number of ways a microcosm of the history of gender politics in science this century. The Nobel Prize sits at the apex of the status system of science. The laureates are icons of the fusion of scientific knowledge and cultural power, so that where they are not already members of their national elite groupings of scientists, such as the Royal Society of London or the French or US National Academies, then it is customary that they are rather swiftly elected. Membership of institutionalized national and international scientific elites, as well as confirming such cultural and political power, also offers its bearer the prospect of participation in these institutionalized forms, and hence a close and uncritical relationship with the state. Members of this ultra-elite within science are invited to walk the corridors of power. Governments seeking scientific advice of a politically strategic nature frequently turn to their national academies or to specific disciplinary groups within them. The shadowy JASON group of leading Nobelist and near-Nobelist US physicists advising on US military strategy came into notoriety during the Vietnam war, but has continued ever since, today advising the US government on what is euphemistically termed national security and defence. Nor is the desire for scientific advice limited to powers temporal in the late twentieth century; the pope, wishing to develop his thoughts on the environment, turned to the collective wisdom of Nobel laureates, via the Vatican Academy, for advice.

This is a paradox at the heart of the Nobel system: scientific eminence is achieved through a small but innovatory piece of knowledge concerning a specific aspect of chemistry, physics, physiology or medicine, but winning the prize gives its bearers the ability to advise on global sociopolitical issues far outside their range of expertise. Feminists, in order to explain the systematic undervaluing of women within the labour market, have described women as "inferior bearers of labour"; by contrast Nobel Prize winners become "superior bearers of thought," acquiring the power to speak and be listened to on topics where their competence is either at the same level as that of their fellow citizens, or even demonstrably less. Because this cultural power is rather concrete, few people are entirely consistent in their attitudes to its manifestation. Individual scientists have used their cultural capital to support their ideological and political commitments. Thus I have to admit that, like many antiracists, I tend to point out that Nobel laureate William Shockley received his prize for work on transistors and that he had no special competence to support his unquestionably hereditarian views on intelligence/IQ, but that when anxious to see nuclear power controlled I welcome seeing George Wald throw his political and scientific weight onto the socially critical side of the debate, and am less anxious to point out the modest connection between his Nobel Prize-winning work on receptor pigments in the eye and the scientific debate at issue. Many times have I welcomed the signatures of what seems to be a shrinking handful of anti-militarist British laureates and FRSs, not least Dorothy Hodgkin and Maurice Wilkins, in protests against military aggression, without dwelling on the cultural power that their welcome presence reinforces.

It would also be ungenerous, particularly in periods when nonconformity with the state carries significant penalties, not to acknowledge the personal courage sometimes entailed. For leading non-Jewish German scientists to oppose the Nazis required an act of courage, as it did for leading US scientists to use their cultural capital to protest against witch-hunting during the height of the McCarthy era; it was much easier to deplore the excesses privately and subscribe to the politics of prudential acquiescence. The physicist Sakharov was rightly admired for his courage in using his cultural capital as the father of the Soviet H-bomb to play a leading role in the human rights movement. And although political persecution has not been a significant issue for British scientists, her anti-militarist activities and marriage to a communist meant that Nobel laureate crystallographer Dorothy Hodgkin was proscribed from admission to the US except by a special CIA waiver until she was in her eighties. (Presumably the combination of her age and the collapse of the former Soviet Union led the CIA to think that she was not imminently about to engage in the violent overthrow of US liberal democracy.)

Thus I want both to salute individual laureates and other eminent scientists for their sometimes quite concrete personal courage in the use of their cultural capital, in the face of sanctions which have ranged from exclusion and even death to various levels of social opprobrium, yet also to criticize a system which has amplified the cultural power of science, not least because of the extent to which, during the twentieth century, science has become incorporated and plays a predominantly socially conservative role.[5]

CEREMONY AND SECRECY

. . . The archives of the Royal Society, itself one of the oldest scientific institutions, are accessible after forty years; but those of the Nobel Institute, created at the beginning of this century, are accessible only after fifty years; in consequence proportionately more of the Nobel's iceberg of secrecy is hidden. Such intense and prolonged secrecy about the affairs of the scientific elites, considerably longer even than that of the notoriously secretive British governmental tradition and an anachronism in Sweden, where any citizen may have access to letters written by a minister, is in itself a matter of curiosity.[6] When it is remembered that these elites are choosing to honor creators of "public knowledge" in science, not trade, military or diplomatic secrets, such secrecy speaks of the sense of cultural and political mystery with which Bacon's masculine knowledge has endowed itself.

The archives made it possible to go behind the public face of the Royal Society, with its discourse of the president being above the election of members, and scientific merit being the only criterion for election, and contrast this with the very particular ways, documented by committee minutes and correspondence, through which actual elite men scientists treated the claims of women scientists and finally came to the understanding that they could no longer exclude women. No such possibility, for other than the early years, exists in the case of the Nobel Prize archives. Lacking this account from the perspective of the powerful who manage such events, the story of the election of the women laureates has for the greater part to be built from more outsider sources, including biographies, the rather rare autobiography, the occasional interview, and, as an important set of resources which have remained constant over time, the laureate's Stockholm speech of acceptance, together with the biographical note and the photographic portrait which accompanies its publication.[7]

The occasion of the prize-giving is highly formal, and takes place in the presence of the monarch. It is the Swedish king himself, that symbol of a past military system of power, who awards the medals. The men attending the ceremony are required to wear white tie and tails. While for a number of recent men laureates, perhaps particularly those from the US, who have rarely been known to wear anything except jeans and checked shirts, such dressing up is something of a novelty, it is also—as Virginia Woolf reminded us for the thirties—still very much part of the life of educated men. The academy has a passion not only for secrecy but for distinctive attire, a surrogate uniform on which medals signifying heroic performance on the field of truth may be displayed. The sharing of the military code and its honours is made all the easier in the Nobel ceremony because it is carried out at such a symbolic level; the constitutional monarch of a neutralist country is at once remote from the military and also the descendant of Gustavus Adolphus, the last Swedish king to die leading his troops in battle.[8]

Novelty and innovation are always central within the award of a Nobel Prize, even though the language has shifted over the decades from the "land ho" quality of scientific "discovery" in which the newly recognized phenomenon is equated with finding a new land (or at least new to the discoverers) to what the users doubtless see as a rather more nuanced language of a "seminal contribution." Women laureates have to be innovators in an additional sense. Like the women who were first admitted to the Royal Society, they are likely to be entrants to new and therefore initially low-status areas of science where the discipline has not been fully formed, where there is no clear structure of employment and career, and hence where there is room for unpaid or badly paid pioneers whose passion is knowledge of the natural world. . . .

A DANGEROUS COMBINATION OF LOVE AND SCIENCE: MILEVA EINSTEIN MARÍC

The recently recovered biography of Mileva Einstein Maríc[9] documents the dangerous combination of love and science for women, and its power to render women and their science invisible. After a painful beginning where she conceived a child by Albert Einstein out of wedlock and had the baby adopted, the marriage was initially happy and mutually appreciative. Einstein, for example, explained to a group of Zagreb intellectuals that he needed his wife as "she solves all the mathematical problems for me." Two key episodes document the process by which her work, if not actively appropriated, was certainly lost by her to him. In one episode Mileva, through the collaboration with a mutual friend, Paul Habicht, constructed an innovatory device for measuring electrical currents. Having built the device the two inventors left it to Einstein to describe and patent, as he was at that time working in the patent office. He alone signed the publication and patented the device under the name Einstein-Habicht. When asked why she had not given her own name of Einstein Maríc she asked, "What for, we are both only 'one stone' [*Ein stein*]?" Later when the marriage had collapsed she found that the price of her selfless love and affectionate joke was that her work had become his. She also lost her personal health through trying to do the mathematical work to support his theorizing and simultanously take care of their children. One son suffered from schizophrenia, and after the divorce Einstein was mean about keeping up with the alimony.

Troemel-Ploetz[10] points to the even more disturbing episode of the articles published in 1905 in the Leipzig *Annalen der Physik*. Of the five key papers, two of the originally submitted manuscripts were signed also by Mileva, but by the time of their publication, her name had been removed. These two articles, written in what was widely understood as Einstein's golden age, included the theory of special relativity which was to change the nature of physics, and for which he alone received the Nobel Prize. Thus although the purpose of the biography was to restore Mileva's name as a distinguished and creative scientist, and not to denigrate Einstein, it inevitably raised the issue of his withholding recognition of Mileva's contribution to the achievement. A number of observers have also commented on the puzzle of Einstein's gift of the prize money to Mileva Marić even though they were by then separated. This gift-giving was later emulated by George Hoyt Whipple, a Nobel Prize winner in 1934. Although Whipple had the reputation of being very careful financially, he shared his prize money with Frieda Robsheit Robbins, his co-worker for many years, and with two other women colleagues. In Einstein's and Hoyt Whipple's circumstances, was the money meant to compensate for the system's, and perhaps their own, appropriation of their collaborators' work?

While Mileva's biographer is careful to indicate that Einstein was the creative thinker, she suggests that he could not have realized his theoretical insights without Mileva's mathematics. Between men scientists such a collaboration between theory and technique is rather difficult to ignore; between husband and wife scientists it was—and according to the context still is—rather easy. It was especially so at the turn of the century when bourgeois women, as wives, were only permitted to work as unpaid workers and when scientific work like housework and child care could be constructed—as they were by Mileva—as part of the labor of love. While Trbuhovic Gjuric's biography (not least because it was originally published in Serbian in 1969) has not had the impact of Ann Sayre's study of Rosalind Franklin, it has raised doubts in the physics community;[11] meanwhile feminists will recognize the pattern as characteristic, made possible by that early-twentieth-century scientific labor market in all its unbridled patriarchal power of appropriation. . . .

APPROPRIATION AND ERASURE: ROSALIND FRANKLIN

Between the awards to the biochemist Gerty Cori and the physicist Maria Goeppert Mayer was the triumphalist story of DNA and its soon-to-be revealed subtext of the appalling treatment of the X-ray crystallographer Rosalind Franklin. The account of the erasure of this outstanding woman scientist and the appropriation of her work was told to a wider audience in 1975 in the biography by Anne Sayre, who along with her crystallographer husband was a personal friend of Franklin's. Sayre's book made public the grave disquiet felt among the crystallographic community[12] and was received within a political climate newly sensitized by an increasingly powerful women's movement. The story is brief, as was the life of this scientist who died at 37 of cancer. Born into a well-off North London Jewish family, Rosalind Franklin was sent to St Paul's, a fee-paying girls' school which prided itself on the educational performance of its pupils. She went to Cambridge to read science, did postgraduate work on the physical chemistry of coal, worked with the crystallographer Marcel Mathieu in Paris and then accepted a postdoctoral fellowship in the department of biophysics in King's College, London. The laboratory was one of a number interested in the structure of the giant molecule of DNA, which was already thought to be associated in some way with the genetic mechanisms of heredity, and both Franklin and another scientist, Maurice Wilkins, were engaged in making X-ray diffraction photographs of the rather intractable DNA crystals.

The relationship between the two was far from cordial, a matter not made easier by the anti-woman atmosphere at King's, which in the 1950s still excluded women from the common rooms as a matter of course; by the failure of John Randall as the head of department to clarify the lines of authority between the two researchers; and by the assumption of Maurice Wilkins that the woman scientist, who had more technical experience, was in some automatic sense his junior.[13] Lastly, Rosalind Franklin was regarded by a number of her contemporaries as a "difficult" woman.

While feminism has commented with some sophistication on the construction of "difficult women," not least in the context of independent and creative women such as Franklin, there has been little discussion in this otherwise much examined story concerning the extent of anti-semitism in educational institutions during the immediate postwar period, and what this meant to any Jewish person with a sense of cultural identity. We know that Rosalind Franklin and her family had such a sense. During the war her father worked with the Jewish Board of Deputies to help refugees, and she helped too during school holidays. At Cambridge she had become friends with the metallurgist and French Jewish refugee Adrienne Weill, who was responsible for Franklin working in Paris with Mathieu, who as a communist had egalitarian attitudes to women scientists and was a committed anti-fascist.

Coming to King's must have been something of a shock, not least after Mathieu's laboratory, for not only was King's very much a male bastion; it was also a bastion of the Church of England. The origins of King's were as a Church of England college established in direct opposition to University College, which had been founded by the Utilitarians to provide university education to Unitarians, free thinkers and Jews. Androcentricity and Christian ethnocentricity were thus the twin hallmarks of institutions such as King's. But Christian ethnocentricity in the forties and fifties was not simply a matter of exclusionary or even hostile speech practices; there was also institutionalized anti-Semitism, not least in education. A number of direct-grant schools, particularly those in areas where there was a considerable Jewish community, had a Jewish quota to prevent the stereotypically clever Jews flooding out the Christians. In the discourse of the time, Christians as the privileged group were unmarked; marking was reserved for the Jewish others. Nor was anti-Semitism limited to negative speech and institutionalized exclusion; it also took violent forms, particularly in areas where the poorer sections of the Jewish community lived. Despite the death camps and the war, anti-Semitism was still a virulent force on the streets and a taken-for-granted aspect of everyday British life.

Most of these cruder forms of anti-Semitism faded as the objects of racist abuse were changed. The advent of the Caribbean and Asian migration into Britain resulted in Jews being replaced for some years as the scapegoats of racist fears. Because replacement rather than resistance weakened it, the phenomenon of anti-Semitism within cultural life remains underexplored, but it was there for Jewish men and women who found a number of elite educational institutions difficult places to study and to work at. To be a woman scientist and Jewish during the immediate postwar period in any laboratory where there was no counter-ideology was to carry a double burden, none the less real for not yet being fully named. It is doubtful if it is even healthy not to be "difficult" in such a situation.

In the context of the DNA project, success required the collaboration between theoreticians, or model builders, and experimentalists, who would take the X-ray diffraction photographs to provide the empirical evidence to sustain the models. The former, Francis Crick and James Watson, were based in Cambridge and the latter, Maurice Wilkins and Rosalind Franklin, in London. The crux of what was increasingly seen within crystallography as a shabby affair was that Franklin had made the key photographs which clearly indicated the helical form, but that these had been taken, without her permission, by Wilkins to show to the two Cambridge men with whom he was collaborating. In addition a Cambridge colleague, Max Perutz,[14] who was on the Medical Research Council committee which had received Franklin's research report, also showed this privately to Crick and Watson. Although the crucial papers published in *Nature* included one by Franklin and her colleagues, she did not know just how important her photograph had been to Crick and Watson.

For this and other reasons the situation at King's became intolerable and was resolved in the usual way; the woman, not the man, moved. Franklin went to work at Birkbeck with the crystallographer Desmond Bernal. Bernal's communism, like that of Mathieu, meant that his laboratory was a more congenial environment in which to work. She stayed there until her death, with Bernal writing her obituary memoir.

Thus Franklin was already dead when the Nobel Prize in Medicine and Physiology for the DNA work was awarded to Crick, Watson and Wilkins in 1962. Despite the centrality of her contribution, none of the laureates made a reference in his Stockholm address to her published papers, and Wilkins only spoke of her in very general terms. In Jim Watson's best seller *The Double Helix*, written several years after both Franklin's death and the award of the prize, Rosalind Franklin appears as a bad fairy in the Watson fantasy of himself as artless young man stumbling on the double helix. Despite the enthusiasm shown by a number of men scientists for the "Jack the Giantkiller" quality of Watson's book,[15] Crick considered suing him. Wilkins would have gone along with the action, but the matter was dropped. Similarly the London Science Museum's construction of the DNA story erased Franklin's contribution until her crystallographer friends and colleagues protested and ensured that her work was acknowledged. However, it was not until Sayre questioned Wilkins directly in 1970 as to the probity of taking the photographs to Cambridge that the masculinist appropriation of the work and the erasure of the woman scientist came into full view.

The interesting and unanswerable speculation must be what would have happened if Franklin had not died, given that the prize can by tradition only be shared between three, and that it was her photograph which provided the critical empirical support to the double helix model. For Franklin herself, gender, "race" and cancer colluded to diminish her contribution, yet the combination of personal and scientific friends speaking out in the context of a rising women's moment has meant that her name has become a warning beacon for any who contemplate the erasure of women scientists. . . .

OVERDUE RECOGNITION AND ITS SOCIAL AND SCIENTIFIC IMPLICATIONS

The next three women laureates were awarded prizes for work which they had done between forty and thirty-five years earlier. Zuckerman's general point that women scientists are recognized later for their work is now made almost grotesque. Very few men, other than the ethologists Konrad Lorenz and Niko Tinbergen, have received prizes in their seventies; and their late recognition was intended to flag the new field of ethology, which was seen as of great scientific interest but which the rules had hitherto precluded. In fact little was said about the new field at their prize-giving ceremony. Tinbergen used the occasion to ramble on about the Alexander method while Lorenz chose to explain/explain away his erstwhile support for the Nazis. (Actually he was rather more active than he indicated in his speech, as he was a member of the Nazi Party, a detail which the revisionist history of science omitted in his obituaries.) By contrast the three women prize-winners neither wandered into therapeutic enthusiasms nor used the occasion to explain away unfortunate political associations. They are intensely professional, each speaking technically and elegantly about her science. Only Rita Levi-Montalcini directly confronted the time gap between her science and its recognition, but not even she, on the occasion of the prize or in her subsequent biography, chose to examine the social and scientific meaning of her late recognition.

BARBARA McCLINTOCK

The first of the three, Barbara McClintock, presents her biography in a highly detached manner, only touching the events which were "by far the most influential in my scientific life" in an enigmatic text. None of the social or economic sensitivities which scatter both Hodgkin's and Yalow's biographies appear in this intensely impersonal account. McClintock comes into the world as a student, attending the only course in genetics open to undergraduates at Cornell. For a more intimate account of her childhood and young womanhood we have to read Evelyn Fox Keller's widely read and highly sympathetic biography, which was published just before the prize was awarded.[16]

In both her autobiography and Keller's study we are given a picture of an unusually independent and intellectually purposeful young woman; thus by the time she graduates her research direction is set. Whatever problems there were for some in the economic climate of the 1920s, her self-account gives away nothing. Unlike the earlier generation of US Nobel laureates, McClintock was American-born. Unlike Yalow and Elion she came from a privileged background and despite the harsh times was essentially naive socially and politically, so that her research fellowship to Germany in 1933, where she encountered Nazism and Aryan genetics, was traumatic, and she fled back to Cornell.

Scientifically the biography is a story of a coherent intellectual and academic trajectory, unusual among women and only achievable where women are either without children or have such resources that others take adequate care of them. She reports that she completed her PhD and began a collaborative study locating maize genes to the appropriate one of the ten maize chromosomes. It is as if it is at this point that her history as a scientist begins, and it is the only moment where the dry impersonal prose becomes suffused with the warmth of remembered friendship: "a sequence of events occurred of great significance to me. It began with the appearance in the fall of 1927 of George W. Beadle (a Nobel laureate) . . . to start studies for his PhD degree with Professor Rollins A. Emerson." She then goes on to describe the close-knit group which grew up and which drew in any interested graduate students. "For each of us this was an extraordinary period. . . . Over the years members of this group have retained the warm personal relationship that our early association generated. The communal experience profoundly affected each one of us." We are, from very early on in the autobiography, flagged that this scientist is working as an accepted group member within an elite setting.

Despite a widespread reading of Keller's biographical study as implying that in some way McClintock was not adequately recognized in science, there is little solid evidence of this, except that she did not receive the accolade of a Nobel prize until she was 81 (perhaps not insignificantly, shortly after the publication of the acclaimed Keller biography). Yet McClintock had long been an acknowledged member of the scientific elite, and she was, as Keller points out, early spoken of as a "genius"—a compliment which is more rarely made by one scientist about another than by the media. She was the third woman to be admitted to the National Academy in 1944, when she was 42, for the work for which the Nobel committee honored her almost forty years later in 1983. At the time of her election to the National Academy there were, despite the scale of the US scientific community, rather under 1,000 members; thus the distinction of recognition is considerable. An early recipient of the Association of American University Women's prestigious prize, she had no less than twelve honorary doctorates, from Rochester in 1947 to three in 1983, the year she won the Nobel Prize. Such a biography speaks of McClintock's extraordinary self-sufficiency as part of the small ultra-elite within science. Such people are rare—perhaps particularly so among women, for whom having sufficient privacy in which to be creative is more commonly a problem.[17] McClintock's isolation was not entirely self-chosen, for her work was not easy to communicate and her ideas on the mobility of genes within each chromosome ("transposition") commanded little support. Despite her early recognition, for many years she was relatively isolated in her Cold Spring Harbor laboratory, but—and it is an important but—never without research resources.

Keller interprets this isolation as a problem of language, of the difficulty that McClintock experienced in trying to communicate what she "saw." Keller argues that "seeing" is crucial to many intensely creative scientists; the problem is that appealing to the "seen" when there is no pre-existing understanding about what is out there to be "seen" cannot provide empirical support. In this situation, when the geneticist Joshua Lederberg observed that "the woman is mad or a genius," he was only articulating publicly what many geneticists more privately thought.

But while Keller lets the reader share the scientist's self-doubt at her failure to communicate her theories to her satisfaction, the outside world had the strong suspicion that she was a genius, and scientific honours continued to be bestowed on her, from a non-residential chair at Cornell in 1965 to the Kimber Genetics Medal of the National Academy in 1967 and, in 1970, the National Medal of Science. By the mid-seventies her ideas about transposition, which potentially challenge the central dogma of the fixity of the genome, began to be understood more widely and became influential in shaping the directions of new work. By this time, conventional molecular biological wisdom had already begun to question the earlier seemingly inviolable concept of the stability of the genome, not in the sense that there was a challenge to the understanding of genetic reproduction, but that the genome itself can, under a number of conditions, undergo rearrangement. There was considerable excitement about such "jumping genes" as the flexibility they gave was seen to endow their bearer, whether the salmonella bacteria or maize, with a distinct evolutionary advantage. By transposition McClintock wished to draw attention to the general occurrence of cellular mechanisms which restructure the genome, mechanisms which are called into action by external or internal stress. DNA, far from being the stable macho molecule of the 1962 Watson-Crick prize story, becomes a structure of complex dynamic equilibrium. Such a complex dynamic structure has echoes of Laura Balbo's quilt-making metaphor to describe women's work in maintaining everyday life.[18]

A number of critics have suggested that Keller's account excessively celebrates McClintock's mysticism as if this was some undeclared dimension of femininity, or essentialist feminism, yet such criticism diminishes the very real difficulties in talking about the creative process, of understanding how an alternative vision is developed, how it is possible to "see" something not seen before in nature. The brave attempt by Koestler with his book *The Sleepwalkers*, and the autobiographical accounts of scientists from Einstein to Richard Feynman, go some way towards discussing this process, but Keller attempts to make clearer what Dorothy Hodgkin spoke of when she thanked her colleagues for their "eyes, brains and hands." However, part of the charge of mysticism lies in McClintock's distinctive relationship with nature itself, for her conception constitutes a return to an earlier tradition when nature was seen as active, not passive. Vitalism, however discussed, in the context of the macho reductionist language of contemporary molecular biology, with a nature drained of all subjectivity, would be all too likely to sound like mysticism.

The detached style of McClintock's Nobel lecture makes no genuflections to the occasion, expressing neither pleasure nor gratitude; she notes neither the delay between the date of her work nor its subsequent recognition—yet this could be read as a matter of forty years. The nearest she gets to Yalow-like celebration of the certainty of her vision is when she reports offering "my suggestion to the geneticists at Berkeley who then sent me an amused reply. My suggestion," she says rather mildly, "however, was not without logical support." The lecture, essentially an overview of her work in genetics, describes the crucial experiments, almost entirely during the 1940s, showing how "a genome may react to conditions for which it is unprepared, but to which it responds in a totally unexpected manner." Her view of future research is that "attention will undoubtedly be centred on the genome, and with greater appreciation of its significance as a highly sensitive organ of the cell, monitoring genomic activities and correcting common errors, sensing the unusual and unexpected events, and responding to them, often by restructuring the genome. We know nothing, however," she concludes, "about how the cell senses danger and initiates responses to it that are often truly remarkable."[19] This activist conception of the cell is kin to Lovelock's Gaia, but where he seeks a popular audience, she is primarily concerned with her invisible college.[20]

McClintock's photographic portrait is of a piece with her prose. Despite the grandeur of the occasion her portrait shows her wearing the uniform of East Coast women intellectuals, a shirt collar over a woollen jersey making no concessions. She looks away from the camera as if she is really looking at something else; her lined face has a slight, detached smile.

RITA LEVI-MONTALCINI

Nothing could be more marked than the contrast with Rita Levi-Montalcini's portrait, which speaks of an agreement between photographer and subject—that this is to be an exceptional statement. And indeed she does cut an exceptional figure amongst women scientists. Dressed with silken elegance, she poses with her hand to her chin. On her wrist is a rich bracelet which acts, and has been chosen to act, as a foil to the eyes. Everything about her conveys a theatrical consciousness of her beauty and her presence. While the autobiography she writes does not begin, as did a rather earlier one, "I was crawling out over the palace roof to rescue my kitten,"[21] the world of high culture and wealth is evident in every aspect of her presentation of self and work.

She describes an intellectual dynasty of mutually admiring and affectionate people. Her father is described as a "gifted mathematician," her mother "a talented painter and an exquisite human being."[22] Her three siblings are all named and praised either for their achievements or, if these are not particularly evident, for their good taste. While she describes a domestic world governed by the father, not least in terms of secondary education, where he held strong views about the suitable subjects for girls, the larger context of the Italian university system had different and more liberal traditions of bourgeois women studying and researching from those of the Anglo-Saxon one. As a teenager Rita describes herself as isolated, directionless, uninterested in young men, and spending her time reading Selma Lagerlöf. From a very early age her construction of her own femininity excluded wifehood and motherhood: "My experience in childhood and adolescence had convinced me that I was not cut out to be a wife. Babies did not attract me and I was altogether without the maternal sense so highly developed in small and adolescent girls."

The death of a loved governess turned her towards medicine, and together with her cousin Eugenia she set about preparing herself for university admission. She gives a graphic account, not unlike a story from the eighteenth-century Edinburgh medical grave-snatchers Burke and Hare, of the means by which research students of the brain gained access to human material.[23] She describes travelling on a Rome bus with the corpse of a two-day old baby wrapped inadequately in newspaper. Her reflections as she sees that a small foot is sticking out are solely those of embarrassment from the construction that might be placed on the sight of a young woman carrying a dead baby. The lesson that she derives is not to carry such experimental material on public transport, but what is interesting is the confidence—not to say arrogance—that permits the retelling of this story, without any reflections on either the nature of the material or how it had been secured.[24]

During the 1930s class privilege was only a partial protection from Italian fascism. Her increasingly tenuous place as a Jewish woman scientist in a developing fascistic context became non-viable once Mussolini's 1936 manifesto against Jewish scientists and professionals had been declared. Now the family was left with "two alternatives . . . to emigrate to the States, or to pursue some activity which needed neither support nor connection with the outside Aryan world where we lived. My family chose this second alternative. I then decided to build a small research unit at home and installed it in my bedroom." She then describes how the Jewish biologist Giuseppe Levi, who at university had taught both her (and also Salvador Luria and Renato Dulbecco, both Nobel Prize-winners and her lifelong friends), came to work with her as the universities gradually expelled the Jews.

As the situation became more stringent, even this existence, a scientific Garden of the Finzi Continis, could not be continued. After 1943, Italy was occupied by the German army, and the family went underground; where Italian political culture did not take anti-Semitism entirely seriously, the German did. In 1945 Rita Levi-Montalcini and her family returned to Turin where she was restored to her university post. By 1947 she was involved in collaborative work with the St. Louis-based Viktor Hamburger, a collaboration which lasted thirty years. During this period she held a professorship at the University of Washington from 1956 to her retirement in 1977. With the enthusiastic support of the Italian Science Research Council she established a research unit in Rome in 1962 and divided her time between the two continents. This engagement in the science of both countries may have cost her something at the US end, and certainly her biography, unlike McClintock's, lists few scientific honors, but it did ensure a strong Italian lobby for her Nobel award, and there was long and open discussion about how significant this would be for the morale of Italian science.[25] Undoubtedly *Unita*, the Italian Communist Party newspaper, long anticipated her laureateship, referring to her as "our Nobelist."

The title of her Nobel lecture, "The Nerve Growth Factor: Thirty Five Years Later," makes her scientific claim and political point rally. She then provides a historical perspective so that the audience may share the frustrations experienced by experimental embryologists of the 1940s, despite an earlier period, during the 1920s and 1930s, which had seemed to promise the early resolution of the paradoxes of development. Her work with Hamburger built from her earlier work with Levi, although she continued to suffer technical problems in resolving these immensely complex neurogenetic systems. She ushers in the next phase of the work with the subhead: "The unexpected break: a gift from malignant tissues." But as we read on we learn that the gift came from the imaginative experimental work of one of Hamburger's students; thus it was the created luck of science rather than the accident of fortune, but this is a scientific voice that enjoys storytelling.

She explains how the development of the research work was initially blocked because the group lacked the expertise with tissue culture. This was, however, being developed in Brazil by Hertha Levi, working at the Rio de Janeiro institute directed by Carlos Chagas. In a passage which fuses images of science and femininity, Levi-Montalcini explains that she was invited by Chagas and so "boarded a plane for Rio de Janeiro, carrying in my handbag two mice bearing transplants of mouse sarcomas."[26]

Despite her vivid reporting that it was in Rio de Janeiro that the nerve growth factor "revealed itself . . . in a grand and theatrical way," Levi-Montalcini and her colleagues had difficulty in convincing others. She does not report a story of results refused publication in prestige scientific journals, but infers this failure to convince from the evidence that few followed her down what she saw as an exciting path. Her problem in this respect was similar to that of McClintock during the fifties and sixties. Yet where McClintock had a second immensely creative and communicating period in the 1970s, Levi-Montalcini's significant work was concentrated thirty-five years ago; the gap between achievement and recognition is in her case even harder to explain within the terms of the institution of the prize.

GERTRUDE ELION

The most recent woman to be honoured as a Nobel laureate was Gertrude Elion[27] in 1988. Her parents were both first-generation US immigrants from Europe. Her father had qualified as a dentist but was bankrupted through the stock-market crash in 1929. Nonetheless her parents were able to help her financially within four years of the bankruptcy. She recalls a lost world of the Bronx as a good environment for childrearing, with good public schools and unrivalled opportunities for free tertiary education. She describes herself as "a child with an insatiable thirst for knowledge and remember[s] enjoying all of [her] courses almost equally."[28] She speculates that her affection for her grandfather, who died of cancer when she was 15, motivated her towards medical research, so that when she entered Hunter College she planned to major in science and especially chemistry. (Again, as with Yalow, the New York public educational system of the time showed its strength, demonstrating that it could be an effective substitute for the educational and cultural privilege of class.)

After college Elion had a bleak time searching for support to do graduate work or even merely to get a laboratory job. She describes a world of systematic and taken-for-granted discrimination in which progress was painfully slow. "Jobs were scarce and the few positions that existed in laboratories were not available to women." She describes one teaching job she had—biochemistry for nurses—which ran for three months out of the year. She then describes how a chemist offered to take her into his laboratory for no pay; she accepted for the experience. After eighteen months he was paying her "the magnificent sum of $20 a week." In 1933, some six years after entering undergraduate study, she was able, with the help of her parents, to enter graduate school at New York University. Having completed course work she trained as a teacher, then worked as a substitute teacher by day, researching at nights and weekends, completing her master's by 1941.

It was the outbreak of World War II with its demand for chemists by industrial laboratories which gave Elion, along with many of her generation and gender, the chance of a research job. From the inauspicious foothold of a job in a food industry laboratory she secured an assistantship with George Kitchens at the Burroughs Wellcome research laboratories. This was the first time, now almost ten years after entering Hunter, that Elion had a job where she could develop herself as a scientist. At the same time she also began a PhD at Brooklyn Polytechnic Institute, but the crunch came when she had to choose between going full time to complete the PhD and abandoning it for her industrial research.

Elion tells us that after she had received three honorary doctorates from the Universities of George Washington, Brown and Michigan she felt that she had made the right choice. By carefully reciting this arduous story of getting into research, and of the fact that she has achieved so much without a PhD, she reminds us of just how exceptional her story is. For a man to get so far without a PhD would be surprising; for a woman it is little short of astonishing. It was the context of one of the most powerful US industrial laboratories, not academia with its passion for credentialism, which made this possible.

At Burroughs Wellcome she began that relationship to her work which enabled her to look back and characterize it as both "my vocation and my avocation." Although she began as an organic chemist she was never restricted to the single discipline. She became interested in microbiology and in the biological activities of the compounds she was synthesizing. Thus over the years she worked in biochemistry, pharmacology, immunology and eventually virology. In her Nobel address she sets about reporting forty years of work with no hint of complaint or criticism; instead she describes the research in which she and her colleagues have been engaged as a coherent set of scientific developments achieved over a period of time, which have consistently resulted in producing major therapeutic agents. One of these, Acyclovir, was a pioneering anti-viral for herpes, and also paved the way for other antivirals, not least Retrovir or AZT, also produced by the Wellcome laboratories. It is a matter of some note that she describes this highly innovatory work, often developing drugs for patients with then fatal diseases, without using the language beloved of today's clinical researchers, in which they "aggressively treat." Elion's prose concerning her research, and one feels Elion's laboratory, goes capably on, not minimizing painful matters or glossing over clinical testing, but not glamourizing it with violent metaphor either. Her official portrait echoes this capable good sense.

Such a voice, describing the complex task of basic science directed very closely towards clinical objectives and in active collaboration with the clinical treatment of patients, is rare in the Nobel proceedings. More typically, research, even when it has a considerable pay-off for medicine, is described within the science and not hand-in-hand with its applications. It is often only when the joke is made that had the research not been carried out on such a socially significant compound, then the honour and recognition currently being enjoyed by the researcher would not have been forthcoming, that we can see the boundary line between social and scientific esteem being gently moved around. Elion is refreshing in that she ignores these delicate boundary games between social and scientific prestige systems, and talks about the nature of contemporary medical research when it is done by good scientists committed to patient care in the best industrial laboratories.

WHERE ARE THE FUTURE WOMEN PRIZE-WINNERS?

Perhaps it is not by chance that many women biomedical researchers say that it is easier to work and to be promoted in an industrial laboratory than in an academic setting. Maybe it is there that the Nobel committees should look for more potential women prize-winners, or among the observational sciences such as ethology or astronomy where numbers of women are currently eminent.[29] Can committees and procedures predominantly composed of men scientists under immense pressure to recognize other men scientists acknowledge the contribution of women unless they open their committee structures themselves to women, who are in an age of gender consciousness less likely to be gender blind? Otherwise it seems that the Nobel Prize system is unlikely to escape an even more age-linked construction of women of gold than operated in Plato's Republic. For today's Nobel committees it seems that women have to be at least 70, the age of the wise woman, the symbolic grandmother, to achieve recognition. Is there an unstated anxiety that, by recognizing women at the height of their creativity and with the social and political commitments of their generation, the committee might begin to disturb the networks of power? Perhaps women scientists in a period of feminist consciousness cannot be trusted to sustain the politics of prudential acquiescence which have become increasingly the hallmark of the scientific elite?

And in fifty years' time, when the descendants of today's feminists have access to the records of this past decade, how will the correspondence and debates of the 1980s compare with the unambiguous evidence of the double standard deployed (too late and unsuccessfully) against Marie Curie at the beginning of the century, or the manipulation of women's access to the Royal Society during the 1940s? Will they have been equally "man"aged?

Notes

1 The award of the 1992 medicine and physiology Nobel Prize to two men biochemists in their seventies, E. Krebs and M. Fischer, caused considerable surprise and concern for just this reason.
2 Harriet Zuckerman's pioneering study covered five women laureates. She notes that the interval between the work and its recognition is longer for women than men, yet those women Nobelists whose awards followed her study were older and the gap greater. Her study draws attention to the achievements of Jewish scientists who are strongly represented among the US Nobel laureates. Both Zuckerman and her mentor Robert K. Merton, the pioneering sociologist of science, are Jewish and sensitive to the history of institutionalized anti-Semitism in US academic life. Like

many of his generation, Merton had found it necessary to change his name to one sounding more acceptably Anglo. Zuckerman, *Scientific Elite.*

3 Levi-Montalcini, *Le Prix Nobel*, 1986, p. 276 (hereafter as *LPN* plus date).
4 Zuckerman, *Scientific Elite*, p. 82.
5 For a sustained examination, see H. Rose and S. Rose, "The Incorporation of Science", in H. Rose and S. Rose (eds.), *The Political Economy of Science.*
6 However, the papers of individuals not infrequently have fifty-year restrictions placed on them.
7 The biographies range from popular accounts, such as Opfell, *The Lady Laureates*; Phillips, *The Scientific Lady*, to scientific and biographical accounts written by feminist scientists and historians as part of a project to write women back into the "his"tory of science, to bibliographic guides: Kohlstedt, "In from the Periphery," *Signs*, 4 (1 1978); Schiebinger, "The History and Philosophy of Women in Science," *Signs*, 12 (2 1987); Kass-Simon and Fames (eds.), *Women of Science*; Alic, *Hypatia's Heritage*; Amir-Am and Outram (eds.), *Uneasy Careers and Intimate Lives*; Schiebinger, *The Mind Has No Sex*; Searing (ed.), *The History of Women in Science, Technology and Medicine*; Ogilvie, *Women in Science*; Herzenberg, *Women Scientists from Antiquity to the Present*; Siegel and Finley, *Women in the Scientific Search.*
8 His stuffed horse and bloody clothes are displayed in the historical museum in Stockholm.
9 Senta Troemel-Ploetz draws attention to a little-known biography by Desamka Trbuhovic Gjuric, herself a mathematician acquainted with the Swiss milieu where the Einsteins lived and worked. This has been republished, but rather heavily edited, in German. "Mileva Einstein Maríc: The Woman Who Did Einstein's Mathematics," *W's Stud. Int. Forum*, 13 (5 1990).
10 Ibid., p. 418.
11 Walker, "Did Einstein Espouse His Spouse's Ideas?" *Physics Today*, February (1989). However, more disturbingly, John Hackel, editor of *The Collected Papers of Albert Einstein*, Vols. I and II, ignores this evidence. Despite my feeling that historians of science have recently been more willing to accept the contribution of women scientists, it seems that in the case of Einstein the myth of the unaided male genius must be preserved.
12 Aaron Klug, a crystallographer, a Nobel laureate, and also Jewish, consistently wrote Franklin's contribution back into the scientific record.
13 Franklin was not, however, the only gifted woman in the Randall laboratory, a fact used by Maurice Wilkins to suggest that the laboratory was not hostile to women.
14 Perutz was complex. Though this action was no help to Franklin, he also lobbied the influential Swede Gunner Haag for Dorothy Hodgkin as a candidate for the Nobel Prize. Personal communication, C. Haag.
15 Biologist Richard Lewontin entertainingly deconstructs "Honest Jim Watson's Big Thriller about DNA," in Stent (ed.), *James D. Watson.*
16 Keller, *A Feeling for the Organism.*
17 From Florence Nightingale to Virginia Woolf the plea for privacy is constant. Even today few women have a space within a family home which is exclusively theirs.
18 Balbo, "Crazy Quilts," in Sassoon (ed.), *Women and the State.*
19 Barbara McClintock, *LPHN*, 1983, p. 192.
20 Her invisible college celebrated her achievements in the last year of her life. Fedora and Botstein (eds.), *The Dynamic Genome.*
21 Levi-Montalcini in Knudsin (ed.), *Successful Women in the Sciences.*
22 Levi-Montalcini, *LPN*, 1986, p. 277.
23 Levi-Montalcini, *In Praise of Imperfection.*
24 The culture of biomedical research in Italy has long been relatively unregulated by ethical debate. Even today "smart" drugs are tested on Alzheimer patients by pharmacologists with industrial connections without reference to ethical committees or controls. Personal communication, Steven Rose.
25 Because the Italian highly educated elite is relatively small, women's participation in science has been acceptable as a cultural activity for upper-class women. Thus campaigning for a woman scientist created fewer problems than in other countries, although it unquestionably required much patience.
26 Levi-Montalcini, *LPN*, 1986, p. 283.
27 This is the only Nobel laureate, caught perhaps by constructions of femininity, who does not give a birthdate.
28 Gertrude Elion is the least well documented of the women Nobel Prize winners. As an industrial scientist she has a less public profile; for example, she has published no books, unlike all the other women laureates. The *LPN* 1988 publication is in this case particularly important.

29 There is a problem about the "capturing" of an institution like the Nobel Prize by a discipline or even by a school, which then acts as a self-recruiting oligarchy. This latter is displayed to absurdity in the case of the Nobel Prize for Economics and Chicago Economics.

References

Alic, Margaret. *Hypatia's Heritage: A History of Women in Science from Antiquity through the Nineteenth Century*. Boston: Beacon Press, 1986.

Amir-Am, Pnina and Dorinda Outram, eds. *Uneasy Careers and Intimate Lives: Women in Science 1789–1989*. New Brunswick: Rutgers University Press, 1987.

Balbo, Laura. "Crazy Quilts: Women's Perspectives on the Welfare State Crisis." In *Women and the State*, ed. Ann Showstack Sassoon. London: Hutchinson, 1987.

Fedora, Nina and David Botstein, eds. *The Dynamic Genome: Barbara McClintock's Achievements in the Century of Genetics*. Cold Spring Harbor: Cold Spring Harbor Press, 1982.

Hackel, John, ed. *The Collected Papers of Albert Einstein*. Vols. I and II. Princeton, NJ: Princeton University Press, 1987, 1989.

Herzenberg, Caroline. *Women Scientists from Antiquity to the Present: An International Reference History and Biographical Directory of Some Notable Women Scientists from Ancient to Modern Times*. West Cornwall: Locust Hill Press, 1986.

Kass-Simon, G. and Patricia Fames, eds, *Women of Science: Righting the Record*. Bloomington: Indiana University Press, 1990.

Keller, Evelyn Fox. *A Feeling for the Organism: The Life and Work of Barbara McClintock*. San Francisco: Freeman, 1983.

Kohlstedt, Sally. "In from the Periphery: American Women in Science 1830–1880." *Signs: Journal of Woman in Culture and Society* 4 (1 1978):81–96.

Levi-Montalcini, Rita. *In Praise of Imperfection: My Life and Work*. New York: Basic Books, 1988.

Levi-Montalcini, Rita. In *Successful Women in the Sciences*, ed. Ruth Kudsin. New York: New York Academy of Sciences, 1973.

Levi-Montalcini, Rita. *Le Prix Nobel*. Stockholm: Nobel Foundation, 1986.

Lewontin, Richard. "Honest Jim Watson's Big Thriller about DNA." In *James D.Watson: The Double Helix*, ed. Gunther Stent. London: Weidenfeld and Nicolson, 1981.

Ogilvie, Marilyn Bailey. *Women in Science: Antiquity through the Nineteenth Century: A Biographical Dictionary with Annotated Bibliography*. Cambridge, MA.: MIT Press, 1986.

Opfell, Olga. *The Lady Laureates: Women Who Have Won the Nobel Prize*. Metuchen, NJ: Scarecrow Press, 1986.

Phillips, Patricia. *The Scientific Lady: A Social History of Women's Scientific Interests 1580–1981*. London: St Martin's Press, 1990.

Rose, Hilary and Steven Rose, eds. *The Political Economy of Science*. London: Macmillan, 1976.

Schiebinger, Londa. *The Mind Has No Sex: Women in the Origins of Modern Science*. Cambridge, MA: Harvard University Press, 1989.

Schiebinger, Londa. "The History and Philosophy of Women in Science: A Review Essay." *Signs: Journal of Women in Culture and Society* 12 (2 1987):305–32.

Searing, Susan, ed. *The History of Women in Science, Technology and Medicine: A Bibliographic Guide to the Disciplines and Professions*. Madison: University of Wisconsin Women's Studies Library, 1987.

Siegel, Patricia and Kay Thomas Finley. *Women in the Scientific Search: An American Bio-bibliography 1724–1979*. Metuchen, NJ: Scarecrow Press, 1985.

Troemel-Ploetz, Senta. "Mileva Einstein Maríc: The Woman Who Did Einstein's Mathematics." *Women's Studies International Forum* 13 (5 1990):415–32.

Walker, Harris. "Did Einstein Espouse His Spouse's Ideas?" *Physics Today* (February 1989):9–10.

Zuckerman, Harriet. *The Scientific Elite: Nobel Laureates in the United States*. New York: Free Press, 1977.

7

Asian Americans and the Gender Gap in Science and Technology

Jaekyung Lee

It is well known that Asian Americans are extraordinary educational achievers with higher levels of educational attainment and achievement than other racial groups (Flynn, 1991; Kao, 1995; Sue & Okazaki, 1990). While this phenomenon is often generalized to all Asian Americans regardless of their gender and attributed to home-related factors (e.g., child-rearing and socialization practices), very little is known about an Asian American gender gap that may arise from social stereotyping and institutional discrimination as well as differential parental expectations and family support for boys versus girls at home. Clearly, there is a gender gap across the board in traditionally male fields such as science and technology, where male and female students tend to show markedly different levels of academic interest, engagement, and achievement while in school, and ultimately different occupational choices after graduation. Asian Americans may not be free from such a pervasive gender gap problem, despite their relatively greater participation and accomplishment in science and technology.

Two major questions are addressed in this chapter: Are Asian male college students different from their female counterparts in their academic paths toward careers in traditionally male-dominated science, math, and engineering (SME) fields? If there is any gender gap in Asian American college students' educational and occupational trajectories, is it different from the patterns shown by other racial and ethnic groups? In light of these questions, I will review previous research and then present the findings of my own original research.

FACTORS INFLUENCING GENDER GAP IN SME FIELDS

The gender gap in math and science and the newer gap in technology (i.e., computer science and engineering) have received much attention over the last decade (Meyers, 2003). In 1991, the American Association of University Women (AAUW) released the results of its national survey of gender-related inequalities in the nation's schools (AAUW, 1991). While the survey emphasizes the importance of students' perceived ability in math and science and self-esteem for both boys and girls, it also revealed that girls' relatively lower self-esteem, and its association with poor representation in math and science, dampened their career aspirations. In 1998, the AAUW conducted a follow-up study and found that although the overall gender gap in math and science has narrowed, a gap remains in the level of courses taken; there is a particularly large gap in physics and computer science (AAUW, 1998).

Previous national surveys of college students reported gender differences across race and ethnicity in characteristics of college students including their undergraduate major, grade-point average (GPA), and educational aspirations (Clune, Nuñez, Choy, & Carroll, 2001). With respect to their undergraduate experiences, women were more likely than men to major in certain fields, most notably education (18% vs. 6%) and health professions (10% vs. 4%). Men, in contrast, were more likely than women to major in business and management (26% vs. 19%) and engineering (12% vs. 2%). Women graduated with higher GPAs than men: 61 percent of women had GPAs of 3.0 or higher, compared with 49 percent of men.

What are the key factors that contribute to these uneven gender differences? Previous research has shown them to be sex-stereotyping of subjects, courses, and careers. Subjects and courses were often classified as masculine or feminine according to whether they have been taken by a majority of one sex over the other (Harvey, 1984; Schweigardt, Worrell, & Hale, 2001; Stables & Stables, 1995; Whitehead, 1996; Williams, 1994). Traditionally masculine subject areas included math, science, and engineering, while traditionally feminine subject areas included writing and literature, languages, and social sciences (Schweigardt, Worrell, & Hale, 2001).

This gender difference in course-taking patterns was found to be related to perception of ability in the subject. In a study of high school students, boys chose advanced math classes more than girls, and females' choices were affected by their perceptions of lack of ability in the subject (Pedro, Wolleat, Fennema, & Becker, 1981). Further, studies also reported gender differences among high school students in their perceived efficacy for sex-stereotypical occupations: Female students reported greater self-efficacy for occupations such as social worker, teacher, nurse, or secretary (Lapan & Jingeleski, 1992). The differences in occupational preferences between males and females are also related to a gender-segregated labor market (Gaskell, 1984). Even when girls enter traditionally male-dominated professional fields such as medicine and law, they tend to enter female-oriented, less prestigious areas such as pediatrics and family law (Meyers, 2003). In technology-oriented fields, females tend to enter office-type jobs where they are directed by the technology while males enter fields such as engineering in which they direct the technology.

Are there any earlier gender differences in the level of academic ability and achievement in K-12 math and science, the subjects that are critical for success in SME fields in higher education and the job market? National Assessment of Educational Progress (NAEP) showed that there were few gender differences in average math and science achievement (Coley, 2001). Data from the National Education Longitudinal Survey (NELS) also showed no gender differences in math achievement on average. However, significant male advantages were found in the high end of math achievement (Fan, Chen, & Matsumoto, 1997; Hedges & Nowell, 1995). The observed gender differences for higher-achieving high school students raised concerns about the gender imbalance in the flow of new students into science and engineering careers because these students are likely to purse careers in SME (Fan, Chen, & Matsumoto, 1997).

According to a study, few females serve on faculty despite Title IX's discrimination ban; science and engineering faculties remain overwhelmingly White and male across the nation, including the top fifty universities in the United States (Nelson, 2003). The study found that the percentages of women among full professors in those disciplines ranges from 3 percent to 15 percent. While there is little overt discrimination, the study said that the cycle of male dominance tends to perpetuate itself. The lack of female faculty undermines efforts to attract girls to science and engineering, and particularly into the classroom on the college level—largely due to a lack of role models.

RACIAL AND ETHNIC DIFFERENCES IN GENDER INEQUALITIES

Is there evidence of significant interaction between gender and race factors in terms of participation and performance in SME fields? In other words, does the gender gap in SME fields vary among different racial and ethnic groups? While prior research clearly suggests a gender gap in SME fields, the status of female performance may vary significantly among different racial and ethnic groups. It is necessary to see how race/ethnicity factors interact with gender factors to affect inequalities in SME fields. The literature review suggests mixed findings, depending on the type of outcomes measured, the nature of samples selected, and the stage of one's education and career examined.

As far as math and science achievement of students as measured by standardized test scores is concerned, one argues that "there are more similarities than variations in gender differences among racial and ethnic groups" (Coley, 2001). Despite the prevailing gender gap in math and science, Asian females still perform better than other minority groups' males who outperform their female counterparts in math and science. At the same time, a gender gap also exists for all racial groups in terms of their academic participation and choice, but the status of females tends to vary among racial groups. Hsia (1988) reported that the proportion of Asian American female college students who major in math, physical science, or engineering is comparable to males of all other ethnic groups.

A national survey of college students reported racial differences in their undergraduate and graduate experiences. Consistent with historical trends, persistence rates for students pursuing a bachelor's degree tended to be higher for Asian/Pacific Islander students than for Hispanic and Black, non-Hispanic students (Horn & Maw, 1995). Asians also tended to pursue professional degrees after college graduation more than any other racial group (Clune et al., 2001). However, there was very little difference in the percentage of Asian male and female college graduates who go to graduate school. At the same time, similar proportions of Asian men and women were employed following graduation. This provides some evidence of overall gender equality among Asian college graduates in their educational opportunities and outcomes in the United States. Nevertheless, studies also show that Asian Pacific American women are underrepresented in higher education in doctoral studies, as faculty, and at higher levels of academic administration (Hune, 1998).

Previous studies demonstrated that Asian Americans are more likely to have professional and technical occupations than their White counterparts. While Asian Americans achieve a better occupational status through higher education, some studies suggest the influence of Asian Americans' status in society on educational attainment and choice of careers. Sue and Okazaki (1990) argue that the educational attainments of Asian Americans are highly influenced by the opportunities present for upward mobility. They suggest that, in certain career activities including leadership, entertainment, sports, politics, and so forth, many Asian Americans perceive limitations in their career choices or upward mobility because of English language skills or social discrimination. According to this view, Asian Americans tend to choose majors in college that require quantitative skills (e.g., math and computer science) as opposed to fields requiring more cultural knowledge and English proficiency (e.g., social sciences and humanities), and they also tend to pursue professional and technical occupations in which they have a greater chance of upward mobility.

Although this view of cultural and language barriers may be truer for recent immigrants, it hardly explains why similar patterns of educational and occupational choices persist among the third generation and afterward. Hsia (1988) suggested that the overrepresentation of Asian Americans in certain occupations, such as science and engineering, has always been found primarily in the foreign born, and that the high occupational profile of Chinese and Japanese Americans did not persist in the American environment among those born and educated in the United States. However, Flynn (1991) refuted the argument by showing that native-born Chinese and Japanese are not different from their foreign-born counterparts in terms of occupational achievement.

There is also a view of possible gender differences among immigrant Asian Americans in their motivation and acculturation (Brandon, 1991). This view suggests that since Asian Americans emigrate from countries in which females' status is relatively more inferior than in the United States, the contrast of opportunities may serve as a potent motivator for Asian American immigrant girls. It also suggests that the socialization of Asian American females may allow them to adopt traditional American male sex roles and that Asian American males might acculturate less easily or quickly to American culture than Asian American women. While it remains unclear how such cultural and language factors might affect Asian female representation and performance in SME fields, the currently available data does not render statistical differentiation of Asian American females by their generation status and primary language for possible investigation of the role of culture and language.

RESULTS

Undergraduate major, courses, and grades in SME fields

More Asian males and females choose an undergraduate major in science, math, or engineering than their Black, Hispanic, and White counterparts (see figure 7.1). Particularly, Asian females significantly exceed females in other racial groups in terms of their representation in those SME fields of major: 1.8 times greater than Blacks and 2.7 times greater than Hispanics or Whites. However, there is still a significant gender gap in these fields across racial groups, and Asians are not an exception to this pattern. Asian males chose SME fields 1.7 times more than their female counterparts. This Asian gender gap is relatively larger than the Black gender gap, but smaller than Hispanic and White gender gaps.

These racial differences in undergraduate major are reflected in course-taking patterns. The average number of total course credits in math, science, and engineering by race and gender are shown in figure 7.2. Asian males take more course credits in math, science, and engineering than Black, Hispanic, or White males (the Asian advantage relative to other racial groups is in the range of 1.3 to 1.5 in ratio). The same is true of Asian females (1.2 to 1.7 in ratio). Therefore, the effect of being Asian on course-taking patterns exists regardless of gender, and it is consistent with the expectation based on previous studies. However, there were significant differences between Asian males and females; Asian males take 1.5 times more course credits in math, science, and engineering than their female counterparts. The other racial groups show significant gender gaps, although the degree of the gap varies from group to group: about 1.3 for Blacks and 1.7 for Whites and Hispanics. It is worth noting that Asian females still lag behind Asian males, although the Asian females' level of academic engagement in the fields of math, science, and engineering is highly comparable to other racial groups' males.

Asian male students have higher academic achievement in SME fields than their Black, Hispanic, and White male counterparts. The same is true of Asian females who outperformed the females of other racial groups in SME courses. Nevertheless, female students tended to perform better than males in most cases; the only exception was Asian Americans in mathematics. Figures 7.3 and 7.4 show the average grade point averages (GPA) by race and gender in science and engineering fields and in math, respectively. Among students who took courses in science and engineering, Asian students also tended to attain significantly better grades, on average, than the other racial groups except for Whites. Asian

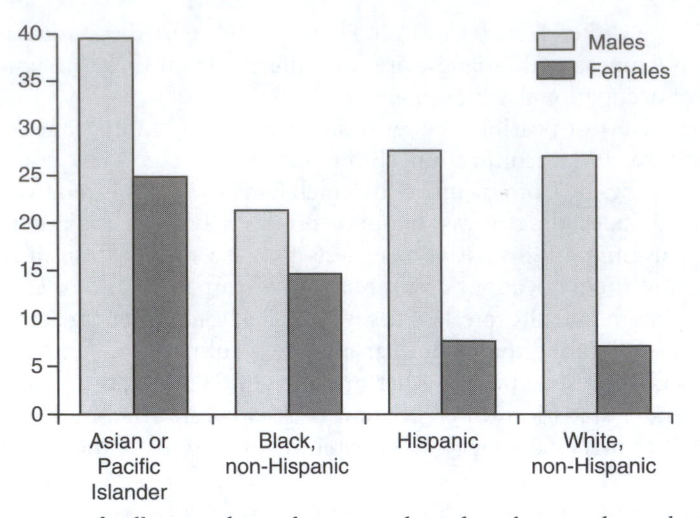

Figure 7.1 Percentages of college graduates by race and gender whose undergraduate major was in science, math, or engineering fields

Data Sources: Baccalaureate and Beyond Longitudinal Study (B&B) 1993/1994. This variable in the survey identifies a respondent's undergraduate major field of study.

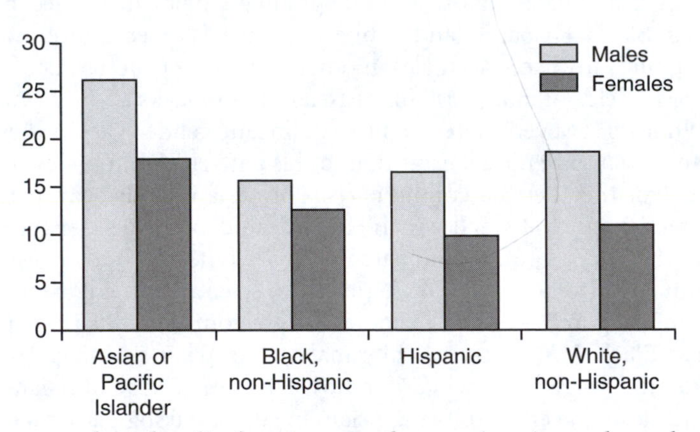

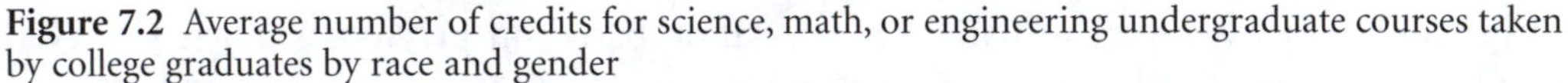

Figure 7.2 Average number of credits for science, math, or engineering undergraduate courses taken by college graduates by race and gender

Data Sources: Baccalaureate and Beyond Longitudinal Study (B&B) 1993/1994. This variable in the undergraduate transcript identifies total number of course credits taken at the sample school.

students also performed significantly better than all other racial groups, including Whites, in mathematics. Both male and female Asian students' GPAs were above 2.5 and their differences were not statistically significant. In contrast, female Blacks and Hispanics had higher GPAs than their male counterparts in math, science, and engineering. Female Whites performed better than males in math but not in science and engineering. The tendency to higher academic achievement of females in SME fields may be attributable to their selectivity; there is underrepresentation of females compared with their male counterparts, and those females who choose the SME fields are likely to be a more selective group, having a higher motivation and ability for learning those subjects than normal expectations.

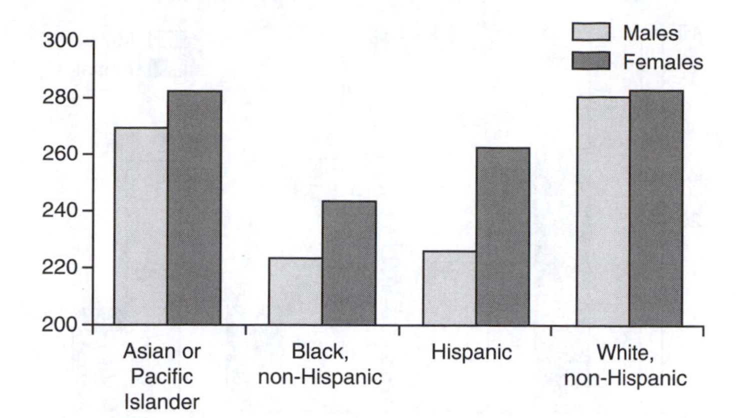

Figure 7.3 Grade point average by race and gender in undergraduate science and engineering courses

Data Sources: Baccalaureate and Beyond Longitudinal Study (B&B) 1993/1994. This variable in the undergraduate transcript identifies grade point averages (GPA) of all science and engineering courses as multiplied by 100.

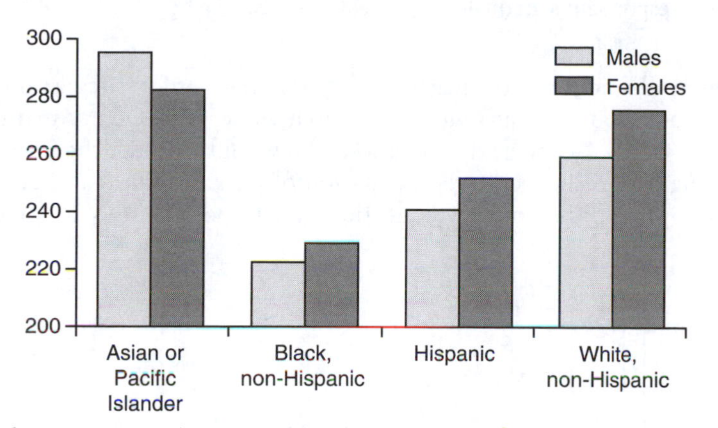

Figure 7.4 Grade point average by race and gender in undergraduate math courses

Data Sources: Baccalaureate and Beyond Longitudinal Study (B&B) 1993/1994. This variable in the undergraduate transcript identifies grade point averages (GPA) of all math courses as multiplied by 100.

Graduate major and faculty career in SME fields

Does the gender gap in undergraduate education persist through graduate education? Very similar patterns of gender and racial gaps are observed in students' choices of math, science, and engineering for graduate majors (see figure 7.5). Both male and female Asians major in those fields more than their Black, Hispanic, or White counterparts. The gender gap for Asians is 1.7, which is about the same as the Black gap (1.5) but smaller than the Hispanic gap (2.3) and the White gap (2.3). Female Asian students' representation is comparable to that of other racial groups' males, and their predominance in these fields continues throughout graduate school.

Students who have a graduate level education may seek more professional and technical careers than those who have only an undergraduate level education. While graduate students' actual choice of occupation depends on their personal aptitude and circumstances,

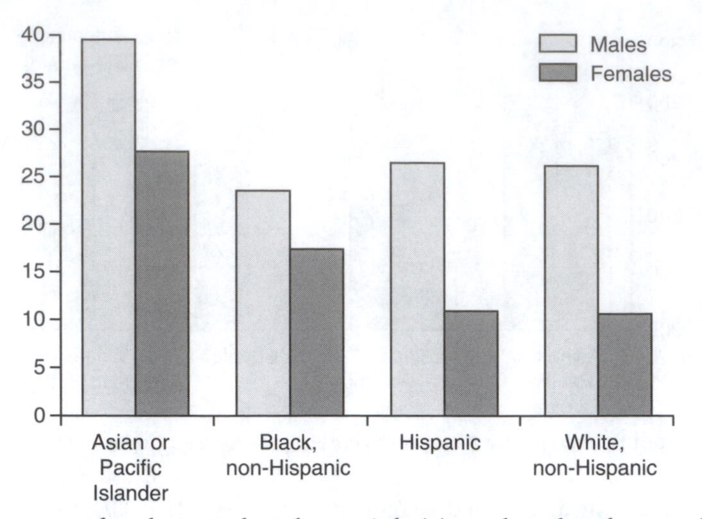

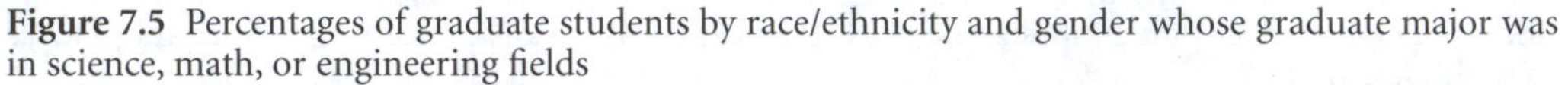
Figure 7.5 Percentages of graduate students by race/ethnicity and gender whose graduate major was in science, math, or engineering fields

Data Sources: Baccalaureate and Beyond Longitudinal Study (B&B) 1993/1997. This variable in the survey identifies a respondent's graduate major field of study.

their choice of academic jobs at colleges and universities is of particular interest, because those females who choose a teaching and research career as a faculty member can serve directly as a role model for the next generation who would consider the SME fields as their career choice. Figure 7.6 shows the representation of different racial and gender groups in SME fields among the postsecondary education faculty population who picked university

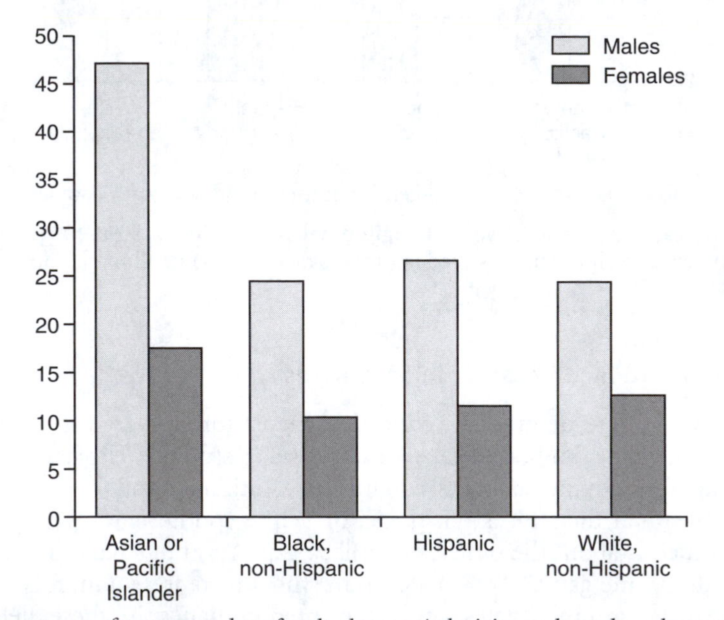

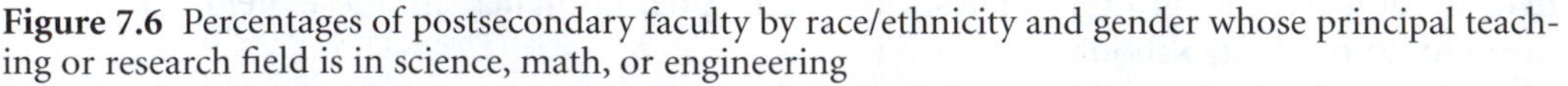
Figure 7.6 Percentages of postsecondary faculty by race/ethnicity and gender whose principal teaching or research field is in science, math, or engineering

Data Sources: 1998–99 National Study of Postsecondary Faculty (NSOPF:99). This variable in the survey identifies the general program area of a respondent's principal field of teaching. If no field of teaching was specified, this variable reflects the field of research.

teaching or research as their career. Gender differences are still evident across racial groups. Both Asian males and females still exceed their counterparts from other racial groups.

GENDER GAP IN ASIAN AMERICAN PATH FROM LEARNING TO TEACHING IN SME FIELDS

The Baccalaureate and Beyond Longitudinal Study (B&B) provides information concerning education and work experiences after completion of bachelor's degrees. It should be noted that the B&B survey did not provide longitudinal information on the employment of graduate students as college or university faculty in SME fields. On the other hand, the National Study of Postsecondary Faculty (NSOPF) provides cross-sectional survey information of the faculty status in SME fields. Despite the lack of direct linkages between the two databases, they both include nationally representative samples of college students and faculty, respectively, and linking the results of two surveys can give us some insight into college students' trajectories from undergraduate education to graduate education, and then to teaching and careers in academia.

Figure 7.7 combines the results of the previous analyses as shown in figure 7.1, figure 7.5, and figure 7.6. As the NSOPF data includes all faculty and instructors across rank, there exists a significant time gap between the B&B cohort group, some of whom may have just gotten a graduate degree to join higher education faculty, and the NSOPF sample, most of

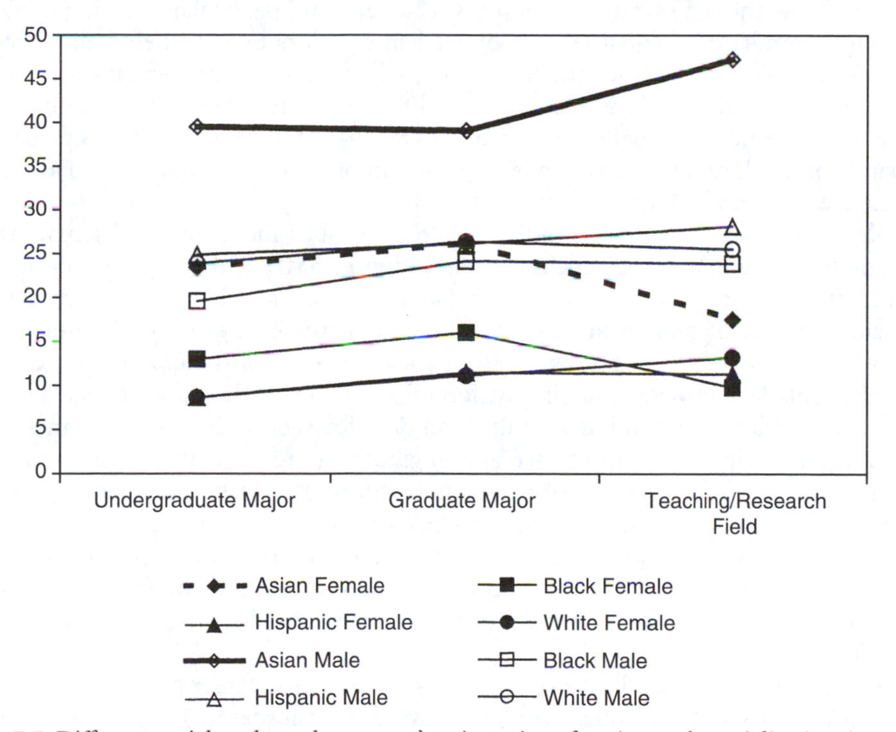

Figure 7.7 Different racial and gender groups' trajectories of major and specialization in science, math, and engineering (SME) fields

Data Sources: Baccalaureate and Beyond Longitudinal Study (B&B) 1993/1997; 1998–99 National Study of Postsecondary Faculty (NSOPF:99). This line graph shows each group's representation at the three stages: undergraduate major (Percent Undergraduate Students' Major in SME Fields), graduate major (Percent Graduate Students' Major in SME Fields), and teaching/research field (Percent Post-secondary Faculty's Teaching/Research in SME Fields).

whom must have become a faculty member much earlier. Nevertheless, when the analysis of the teaching/research field in figure 7.7 was restricted to the category of assistant professors, very similar patterns were found for junior faculty who had entered academia recently.

When the rates of choosing SME fields are compared among those eight groups as classified by race/ethnicity and gender in figure 7.7, it appears that there are three distinct clusters. First, Asian males surpass all other groups, and they stand out with as high as a 40–50 percent range. Second, White, Black, and Hispanic males and Asian females form the second tier in the range of 20s. Finally, White, Black, and Hispanic females cluster around the 10 percent range. Figure 7.7 shows that all race and gender groups maintain their representation in SME fields from undergraduate education to graduate education level, so that their gaps hardly change over the course of postsecondary education. Moreover, most groups also continue their initial trajectory of educational specialization once they get into academia. The notable exception is found among Asians at their critical transition from the stage of learning as an undergraduate or graduate student to the stage of teaching and research as a faculty member at the university. Asian males' representation in postsecondary faculty who specialize in SME fields significantly increases, while Asian females' representation rate drops at the same time.

DISCUSSION

It is evident that a gender gap prevails in traditionally male-dominated SME fields. The gap appears to grow throughout the period of K-12 education, particularly during adolescent years, and it results in unequal choices of academic majors in SME fields during college education. Previous research identified the number of advanced mathematics and science courses taken in high school as the strongest direct influence on choice of a quantitative undergraduate major. Unequal representation of female students in SME fields persists into graduate school education and has an even greater impact on their choice of a teaching and research career in academia.

While prior research generally indicated strong gender inequality and female underrepresentation throughout the pipeline of education in SME fields, a study indicated that the problem of no critical mass of women faculty is not only directly attributable to an insufficient supply of female Ph.D.s but is also exacerbated by the underutilization of women Ph.D.s for faculty positions (Nelson, 2003). In some instances, the percentage of female students far outweighs the proportion of professors of the same gender. The study also showed, with the same kind of utilization data for racial/ethnic groups, that underrepresented minority Ph.D.s in science and engineering are generally placed into faculty positions less often. However, the study did not address the questions of how gender and race factors interact to affect inequalities, and how the inequalities evolve from undergraduate to graduate school level toward an academic career. Moreover, the literature did not pay attention to Asians viewed as a model minority group or explain why female Asians, who outperform even males of other racial groups, still lag behind their male counterparts in SME fields.

This research attempts to fill the void by tracking diverging career paths among different racial and gender groups of college students toward their academic majors and specializations in SME fields. Given the study's focus on Asian Americans, its finding on the widening gap between Asian males and females in their representation in SME fields for postsecondary teaching/research career raises many questions that need further investigation and explanation. What are the barriers in one's transition from the role of learning as a graduate student to the role of teaching as faculty? Is this widening gender gap unique to SME fields? As figure 7.7 shows, it is only Black and Asian females who dropped their representation from graduate major to faculty teaching/research. Is this change in gender gap due simply to females' different career interests and opportunities (e.g., jobs in the government or industry) or more closely related to their different qualifications for university teaching/research jobs (e.g., fewer Ph.D.s or weaker graduate GPAs/research records)? Or is it attributable to any systematic gender discrimination in academic job placement and promotion in the colleges and universities?

There is no evidence that Asian female students do not prefer academic jobs in higher education, particularly teaching and doing research as faculty members. There is also no evidence that Asian females do not perform as well as their male counterparts in graduate school. According to the literature, female students often do not obtain the highest level of achievement because of their low confidence and not their ability level, and confidence plays a large role in the persistence students possess when faced with a challenge. The challenge may come from the stereotype that girls do not perform well in mathematics-related fields (Manning, 1998; Steele, 1999). Encountering these negative stereotype threats may generate anxiety and fear for some students, particularly those who are highly motivated academically and who identify strongly with their chosen school and field of study. However, those female students who chose SME fields and successfully entered graduate school may already have overcome such stereotype threats. Finally, gender discrimination in academic job placement and promotion might be another plausible explanation. Indeed, many universities have been slow in hiring and promoting more women since thirty years ago, when the federal government banned gender discrimination in all academic programs that receive federal aid. However, this simple explanation is challenged by different patterns of gender gap among racial and ethnic groups as shown in figure 7.7.

Further research is needed to examine the interaction of psychological, cultural, and institutional factors that result in different career paths in SME fields for male versus female college students of different racial and ethnic backgrounds. I would suggest that subsequent research pay attention to the simultaneous influences of multiple forces on Asian females in SME fields, particularly "double-edged stereotype threats" and "lack of role models." Asian females in SME fields are likely to face dual, conflicting stereotype threats: a negative stereotype of females as low achievers in SME fields, on one hand, but a positive stereotype of Asians as high achievers in SME fields, on the other hand. Asian female students are likely to struggle hard to cope with the negative stereotyped image of females and to live up to the positive stereotype of Asians while in school. However, they are less likely to find professors of the same gender and race in their department who may serve as role models. There is no critical mass of Asian female faculty in SME fields; for example, only 1.1 percent of chemical engineering faculty at the top U.S. engineering schools are Asian females, whereas 11.3 percent are Asian males (Nelson, 2003). The combination of those factors, the double-edged stereotype threats, and the lack of female Asian faculty as mentors and role models in SME fields could set "glass ceiling" barriers for Asian female students who otherwise may stay and survive in academia.

Longitudinal research may be designed to empirically test the above or other related hypotheses, explaining why Asian female students pick SME fields for undergraduate and graduate education, but their choice of major does not translate into careers in academia. This line of research may help us become more aware of the unique problem with the Asian female group in SME fields that has been obscured because of the aggregate racial statistics reinforcing a widely held group image of high performance and over-representation in SME fields. This research may have implications for other minority female groups as well, to improve their opportunities and reduce barriers throughout the pipeline of postsecondary education and academic career in SME fields.

Note

Research assistance for this publication was provided by Maggie Stoutenburg and Reva M. Fish in the Department of Counseling, School and Educational Psychology of the University at Buffalo.

References

American Association of University Women. (1991). *Shortchanging girls, shortchanging America.* Washington, DC: Author.

American Association of University Women. (1998). *Gender gaps: Where schools still fail our children.* Washington, DC: Author.

Baccalaureate and Beyond Longitudinal Study (B&B:1993/1997). Retrieved November 1, 2004, from www.nces.ed.gov/surveys.

Brandon, P. R. (1991). Gender differences in young Asian Americans' educational attainments. *Sex Roles, 25,* 45–61.

Clune, M. S., Nuñez, A., Choy, S. P., & Carroll, C. D. (2001). *Competing choices: Men's and women's paths after earning a bachelor's degree* (NCES 2001–154). Washington, DC: U.S. Department of Education, National Center for Education Statistics.

Coley, R. (2001). *Differences in the gender gap: Comparisons across racial/ethnic groups in education and work.* Princeton, NJ: Educational Testing Service, Policy Information Center.

Fan, X., Chen, M., & Matsumoto, A. R. (1997). Gender differences in mathematics achievement: Findings from the National Education Longitudinal Study of 1988. *The Journal of Experimental Education, 65,* 229–242.

Flynn, J. R. (1991). *Asian Americans: Achievement beyond IQ.* Hillsdale, NJ: Erlbaum.

Gaskell, J. (1984). Gender and course choice: The orientation of male and female students. *Journal of Education, 166*(1), 89–102.

Harvey, T. J. (1984). Gender differences in subject preference and perception of subject importance among third-year secondary school pupils in single-sex and mixed comprehensive schools. *Educational Studies, 10,* 243–253.

Hedges, L. V., & Nowell, A. (1995). Sex differences in mental test scores, variability, and numbers of high scoring individuals. *Science, 269*(5220), 41–45.

Horn, L., & Maw, C. (1995). *Minority undergraduate participation in postsecondary education* (NCES 95–166). Washington, DC: U.S. Department of Education, National Center for Education Statistics.

Hsia, J. (1988). *Asian Americans in higher education and work.* Hillsdale, NJ: Erlbaum.

Hune, S. (1998). *Asian Pacific American women in higher education: Claiming visibility and voice.* Washington, DC: Association of American Colleges and Universities.

Kao, G. (1995). Asian Americans as model minorities? A look at their academic performance. *American Journal of Education, 103,* 121–159.

Lapan, R. T., & Jingeleski, J. (1992). Circumscribing vocational aspirations in junior high school. *Journal of Counseling Psychology, 39,* 81–90.

Manning, M. L. (1998). Gender differences in young adolescents' mathematics and science achievement. *Childhood Education, 74*(3), 168–171.

Meyers, M. (2003). *The high school experiences of female engineering majors.* Unpublished research paper.

National Study of Postsecondary Faculty (NSOPF:99). Retrieved November 1, 2004, from www.nces.ed.gov/surveys.

Nelson, D. J. (2003). The standing of women in academia. *Chemical Engineering Progress, 99*(9), 38S–41S. Retrieved October 27, 2004, from http//cheminfo.chem.ou.edu/faculty/djn/diversity/Pubs/CEPo3Aug/p38S–41S.html

Pedro, J. D., Wolleat, P., Fennema, E., & Becker, A. D. (1981). Election of high school mathematics by females and males: Attributions and attitudes. *American Educational Research Journal, 18,* 207–218.

Schweigardt, W. J., Worrell, F. C., & Hale, R. J. (2001). Gender differences in the motivation for and selection of courses in a summer program for academically talented students. *Gifted Child Quarterly, 45*(4), 283–292.

Stables, A., & Stables, S. (1995). Gender differences in students' approaches to A-level subjects: A study of first-year A-level students in a tertiary college. *Educational Research, 37,* 39–51.

Steele, C. (1999). Thin ice: "Stereotype threat" and black college students. *Atlantic Monthly, 284*(2), 44–54.
Sue, S., & Okazaki, S. (1990). Asian-American educational achievement: A phenomenon in search of an explanation. *American Psychologist, 45*(8), 913–920.
Whitehead, J. M. (1996). Sex stereotypes, gender identity, and subject choice at A-level. *Educational Research, 38*, 147–160.
Williams, J. E. (1994). Gender differences in high school students' efficacy-expectation/performance discrepancies across four subject matter domains. *Psychology in the Schools, 31*, 232–237.

8

Walking a Tightrope

The Feminist Life of a *Drosophila* Biologist

Marta Wayne

Despite widely reported success in increases of the number of women enrolling in graduate school, the androcentric focus of science remains present in biology at every level: from what questions are asked, to what answers may be considered, to who may ask/answer the questions. This is an increasing problem for me both personally, as a woman who is a scientist and a feminist, and politically, because of the ever-increasing presence of science (particularly my field, evolutionary genetics) in people's lives. There are continuities among the ways that assumptions of the male as norm influence my field, from the training of women scientists to the interpretation of data. The growing scholarship in feminist science studies offers the hope of a better science and a better climate for feminist scientists, but communication between women's studies and life sciences professionals is as yet at an early stage.

The last decade of feminist scholarship on the sciences has produced an impressive array of new research on the reciprocal relationship between scientific knowledge and gender inequalities.[1] Though much of this work has been undertaken by researchers in the humanities and social sciences, some of this work has been informed by feminists in the sciences who are writing for publications outside of the sciences, about topics that scientists usually do not address. Their work has added scientific specificity to feminist critiques of the methods and theories of a masculinist science. Yet the work that would extend feminist theoretical insights into revising and/or rebuilding contemporary scientific theory, which some would call "feminist science," is just beginning.[2] As a scientist who is also a feminist, such an effort promises an intellectual home, but the task is daunting nonetheless.

It is often the case that matters related to the "climate" for women in science are set apart from a focus on the substantive influences of gender constructs on knowledge. I propose that these are more entangled than such a clean division suggests. It is not only that the methods and theories of a masculinist science are at play in shaping content; it is also that those theories and methods contribute to the working environment of women in science. In my field, evolutionary genetics in *Drosophila* (fruit flies), male-centered gender norms are naturalized through interpretations of *Drosophila* behaviors and then reasserted as paradigmatic dichotomized sex differences that legitimate the major and minor insults of a chilly climate.

I do research on structures in the ovary in *Drosophila*, and on sex differences in gene expression. Though my research is grounded in a challenge to the assumption that the developmental processes of male *Drosophila* represent the developmental processes of all

Drosophila, I reject the belief that these sex differences are relevant to understanding human social behavior, though I do see my research as potentially contributing to understandings of women's bodies. I take this position in a context in which it is taken for granted that it is scientifically reasonable and useful to extrapolate to human behavior from a *Drosophila* model, and that the male is the norm.[3]

In my field, models for evolutionary adaptation are developed from *Drosophila* experiments. For example, I conducted a study of viability in *Drosophila* (Wayne *et al.* 2000). Viability in this context means whether or not the insect can survive from an egg to an adult, metamorphosed fly. Viability is a classic trait of central importance to evolution, and it has been studied for decades. However, no one had ever asked whether or not females and males have the same genes for viability. When I designed my experiment, I collected the data separately for the two sexes, and interestingly enough, the genetic architecture of viability is different between females and males.

Scientists have known for many years now that there are genes which cause death in one sex and not the other: genes with names like *daughterless*, *sex lethal*, *sisterless*, etc. (and the negative association is surely not coincidental). In fact, the discovery and analysis of these mutants is how one of the crown jewels of developmental genetics, the *Drosophila* sex determination pathway, was elucidated. But the connection between the existence of these mutations and the possible differences in the genetic architecture of viability between the sexes had not been made because the conventions of the field presumed that males were "relevant" and females were not. Those conventions limited the experiments. They limited what could and could not be embraced as valid data and they limited the interpretive options available to scientists. The conventions so limited the field that they drove interpretations of contrary evidence.

For example, in 1948 Bateman wrote a paper about how often female flies will mate in the laboratory (Bateman 1948). The paper is used in textbooks (Krebs and Davis 1993; Drickamer *et al.* 1996; Futuyma 1998; Freeman and Herron 2004, p. 376) as the classic study demonstrating a central tenet of animal behavior, that females will be the choosy or "coy" sex, while males will mate with anything that crosses their path. The idea is that since eggs are larger and fewer than sperm, females make a greater investment in their offspring than males do, so they have to make sure that their fewer offspring get the best possible father. Bateman's data are said to show that a female fly will mate only as often as necessary to ensure fertilized eggs for the rest of her life—once or twice. However, a careful review of the paper reveals that Bateman's data do not support this at all—female flies mate far more often than is necessary to ensure a lifetime supply of sperm. Furthermore, Bateman's interpretation of his data makes a much more limited claim. Yet his paper is widely miscited by scientists and is described as demonstrating "Bateman's Principle" (e.g., Arnold 1994) or "Bateman's Rule" (e.g., O'Connell and Johnston 1998).

In her landmark essay on the myth of the coy female, Sarah Blaffer Hrdy discusses Bateman in the broader context of sexual selection and primatology (Hrdy 1987). She emphasizes that it was the increased presence of women as researchers that prompted a substantive revision of the coy female paradigm, because these women asked different questions. Hrdy argues that "women scientists were less likely than male scientists to identify with authority and with the scientific status quo. . . . They may have been more willing to entertain unorthodox ideas about sex roles" (Hrdy 1999, p. xix). But since Hrdy did not admit to or endorse a specifically feminist agenda, her critique of Bateman addressed only the scientific inadequacy and inaccuracy of the "coy female" paradigm, but not the enduring commitment to it. She raised questions about the paradigm but did not posit alternatives to it.

Since Hrdy first critiqued Bateman's rule in 1987, there has been a plethora of work that shows that female flies are anything but coy. In the wild, they'll mate more than ten times (Imhof *et al.* 1998). In fact, female flies will mate so readily that male flies have evolved special qualities to ensure that their sperm are the ones that fertilize the eggs, not the male before or the male after. Despite these additional observations, many researchers fail to note the (continued) contradiction between *Drosophila* data and Bateman's Principle or fail to grasp the full import of it, in that they continue to interpret mating in terms of male-centered concepts.

For example, one metaphor refers to "sperm competition" to capture a biological outcome of females' mating practices. A football analogy is widely used to describe two phases: offense, to get rid of the sperm of the preceding male, and defense, to resist the offense of the following male. Some researchers prefer to call it "sperm precedence" instead (Baker and Bellis 1988; Bellis *et al.* 1990), but either way the female's central contribution disappears; she is reduced to a passive "playing field."

Conflict theory is another popular metaphor in evolutionary biology. According to this theory, the interests of females and males often, necessarily, conflict because of the differential investment in gametes (eggs and sperm) and in offspring. The crux of conflict theory is that males should have as many offspring as possible, since sperm are small and "cheap" to produce from a biological energetics point of view. However, females cannot have as many offspring as males, since eggs are large and biologically "expensive" to produce. According to this theory, because females are limited in this way, they must be more choosy (coy) than males, making sure that their fewer gametes get the best possible sperm. The theory expands to include differential parental care between the sexes, gestation time, and "costs" of mating. Conflict theory itself is influenced by Bateman (1948), because persistent distortions of his interpretation sustain the belief that while males will re-mate as often as possible, females will not.

An emerging theory about "female resistance" is perhaps a first step away from male-centered interpretations. Female resistance has to do with how the female deals with sperm from more than one male inside her reproductive tract. In this model, some females use the sperm of males that mate first, and some use the sperm of males that mate second, and so on; in other words, paternity is strongly influenced by some capacity of the female. Indeed there is a large body of recent literature on the subject, with titles like *Female Control: Sexual Selection by Cryptic Female Choice* (Eberhard 1996). Yet even here there are persisting and pernicious assumptions about the centrality of males to reproduction. Again females are defined in relation to males; they are *responding to* a male strategy rather than *acting* independently ("females are normally evolving *in response* to male variation . . ."). Female response is graphed as an interaction, rather than as a main effect, with the focus remaining on "the relative success of sperm" (Clark *et al.* 1999).

Rather than developing models of female response to sperm precedence, researchers could focus directly on female behavioral physiology; but the field of scientifically credible ideas does not include this as an option. It may take a feminist political stance to stretch to the "unimaginable." If science were the product of a woman-centered society rather than a patriarchal one, we might be modeling strategies of the female fly to use the sperm she wanted while participating in as many matings as she pleased, and males would respond to *her* manipulation of *them*.

Unfortunately, the interpretive framework for sperm precedence is not limited to flies but is extrapolated to *people*, in part because flies are a model organism for humans.[4] The concept of model organisms is that since genes in flies and humans ultimately come from a common ancestor, by studying genes in a simpler, more malleable system such as the fly, we can learn about the function of related genes in a more complex system: *homo sapiens*. Developing model systems in insects and animals is a hallmark of contemporary biology, and this approach has proven medically useful for some narrowly biological topics. It becomes rather more problematic when scientists extrapolate behavior and its genetic underpinnings from the simple brain and genome of a fly to the complex brain and genome of people. Yet *Drosophila* research has given rise to empirical studies on sperm competition in humans (Baker and Bellis 1989; Baker and Bellis 1993a, 1993b). Thus, conceptual limitations in interpreting data from fruit flies translate directly into myopic research and interpretation of data in humans.

Bateman's rule has been used to explain rape (because men, like male flies, are "selected" to mate with as many females as possible, and to interpret no as yes, and women, like female flies, are "programmed" to say no when they mean yes) (e.g., Thornhill and Thornhill 1992); sexual harassment (same general ideas); gendered theories of jealousy (males will be jealous of promiscuity on the part of females, but females will not care if males are promiscuous as long as they continue to provide resources) (e.g., Buss 1991); and a lot of other ugly cultural norms thanks to the "new" science of evolutionary psychology. These publications have found an outlet in a variety of arenas, from apparently feminist compilations (Wilson *et al.* 1997) to law publications (Kennan 1998; Wertheimer 2003) to popular science books (Pinker 1998, 2003). In these contexts, conflict theory becomes an apology for inequality, a statement that biologically speaking, the interests of females and males are inherently in conflict even at the level of the cell, because to phrase things in terms of conflict is to imply winners and losers, or at best, détente.

It would be simplistic to argue that conflict theory in evolutionary biology underwrites the educational context for scientists, but there is a resonance. Competition for resources is a characteristic of contemporary scientific research. Undergraduate, graduate, and postdoctoral training all involve learning how to negotiate interpersonal and intellectual conflicts. Such conflicts are understood as necessary and unavoidable components of the process of identifying and cultivating the "winners." As a woman, I notice, I have noticed, I will continue to notice the symbolic resonance between male-centered biological theory, where "females" are understood as secondary or irrelevant, and male-centered education, where male and female students are taught the biological theory.

As a graduate student, I survived my training by learning to interpret my day-to-day experiences with colleagues in the lab in feminist political terms, but I interpreted my research in a "less political" way. As my exposure to feminist critiques of science has grown, I have come to understand this interpretation as an accommodation to the belief that science is value-neutral. I have learned that feminist perspectives are crucial to transforming scientific knowledge—that my feminist perspective has refreshed and invigorated my science.[5] I have also concluded that the androcentric bias that presumes the male as norm, the only sex worthy of study, is the same one that creates a climate hostile to women scientists.

In retrospect, this shift occurred in four distinct phases, beginning with acceptance of the status quo, then a questioning of the gender politics in my social environment, and then an increasing awareness of how deeply beliefs about gender differences infect scientific knowledge. I am now presented with phase four, where unsettling questions are emerging about how my research could be used by others to fuel human sex-difference research.

PHASE 1. I LOVE SCIENCE. WHAT PROBLEM?

When I first began as an undergraduate science major, I had no idea of the politics of the lab environment, never mind gender politics. In my first research position, I was supervised by a wonderful man, Dr. S., who shared my most important passions in life: science, science, and surfing. I was the only other person in his laboratory. He taught me all kinds of complicated and demanding biochemical techniques. Dr. S. was full of positive reinforcement. He answered all of my questions seriously and made me feel smart for asking them. Together we solved technical problems as they arose and improved the experimental protocol based on the results of the early experiments.

We worked hard, but we also had fun together. Every day at lunch we surveyed the surf and wind conditions, and if the waves were especially good we would take off and surf during the afternoon, returning to finish the day's work in the evening. Dr. S. took advantage of these times to talk about the various paths one might take as a scientist, and the pros and cons of each. The work itself went splendidly, yielding great results. In groupwide lab meetings, and when scientists visited from other institutions, Dr. S. tirelessly promoted me and my work, rather than appropriating praise and credit as my supervisor.

Unfortunately, we got along so well, and enjoyed working together so much, that everyone else in the research group assumed we were sleeping together. Because I was quite naïve, and because the other members of the group rarely set foot in our lab, I did not learn about this gossip until the end of the summer. I was very angry and very hurt. Our shared excitement was about science, not sex; a platonic relationship had been sullied by this accusation. The reason that this was so upsetting was because my colleagues apparently thought that a scientist like Dr. S. couldn't possibly find a mere female undergraduate intellectually exciting, and that therefore our relationship must have been based on something else, namely, sex. The fact that Dr. S., equally angry, commented that this would probably happen throughout my career did not exactly alleviate my hurt and anger. Surely it would be different in another laboratory.

During the next school year and the following summer, I worked in another lab, with the person who became my undergraduate advisor, Dr. D. He was everything I thought a scientist should be: brilliant, famous, kind, witty—and grandfatherly. Surely no one would think I was sleeping with HIM. On my first day of work, Dr. D. introduced me to three men (Dr. X., Dr. Y., and Mr. Z.) with whom I would be sharing a very small lab. He explained to them that "since I was a young lady, he hoped that they would behave themselves and watch their language." This had little effect. Dr. X. spoke to me only in a snarl and took every opportunity to make me feel stupid and inadequate. At the time I consoled myself with the hope that he did this because he feared that I was smarter than he was. Dr. Y. swore at me repeatedly and made remarks about my legs as well as making more obscene comments.

My only sometime ally was Mr. Z., who was addressed by the other two as "camel driver" and "rug trader" because he was Middle Eastern. The three of them routinely discussed their plans to get together in the evenings and on the weekends to drink beer or go to baseball games but never asked me to join them. I thought that my excluded status was all part of the game of learning to be a scientist and that I should be tougher. I concluded that I should focus more on my fruit fly experiments. The point was to uncover the secrets of nature, right?

PHASE 2. I MAY LOVE SCIENCE, BUT THERE'S SOMETHING WRONG WITH ME

I was admitted to a prestigious Ivy League graduate school. I thought that by being admitted to Ivy U., I had proven myself. Everyone would treat me like a scientist. I began to work with Dr. A., who was young, enthusiastic, and liked to talk science with me. He showered me with positive feedback for even the simplest laboratory task and made me feel respected and even admired. But Dr. A. only worked in his lab late at night. During the day, I was the only woman in the lab besides the technician, who always worked with headphones on, in a laboratory of 13 men (2 visiting scientists, 4 postdocs, 2 graduate students, and 4 undergraduates as well as Dr. A. himself). The laboratory environment was quite different during the hours I regularly worked, and my colleagues did not welcome me.

Initially, I was eager to talk about science as much as possible, but I was continually snubbed, ignored, and/or treated patronizingly by lab members for my combined enthusiasm and lack of expertise. By the end of my second year, I avoided the lab as much as possible. Thus, I was simultaneously nerdy and not dedicated enough. I did not flirt, not being interested in sleeping with anyone in the lab nor wishing to be accused of sleeping with anyone in the lab. Thus, I was snobbish and cold. The only person in the lab who would talk to me claimed he was doing me a favor by telling me what people were saying. There was little I could do to remedy these perceptions of me. The plainer I dressed, the more frigid I was perceived to be; the less time I spent talking about science, the more obvious it was to my detractors that I was ignorant. If I tried to flirt I was indeed a slut, and if I worked hard, it was obvious that I was trying to make up for my deficiencies.

Since I continued to get along well with Dr. A., despite my unhappy relations with members of his lab, the gossip circulated we were sleeping together. I began to take pains to not be seen interacting with him outside the lab, and our relationship was damaged as a result. When I explained my behavior to him, he told me he thought that my concerns were "asinine" but sadly also withdrew. Thus I was beset at every turn, yet I worried that I was being hypersensitive.

I became involved in the graduate women's alliance, an openly feminist organization. In that context, I began to realize that being the only woman in the lab had something to do with my troubles. Without these women I never would have finished graduate school, because they helped me put my experiences into a coherent framework other than the only one that had occurred to me, which was that I was not cut out to be a scientist after all.

Four years later, however, I was no happier in the lab group. The technician had quit after her accusations of sexual harassment from someone in the lab went unresolved. Dr. A. had received a great job offer at a major midwestern university which included funding for all of us, so most of us moved. Two women graduate students joined us, later to drop out of the lab and graduate school entirely. The controversy around the sexual harassment accusation, the unhappiness of my women colleagues, and the move to another university all took their toll on me; but I stayed, and stayed, and stayed. I looked in the mirror one day and did not like what I saw: a calculating, hardened person. I began to think of abandoning my plans for a science career. Was the research I loved worth my sense of self?

PHASE 3. THE PERSONAL IS POLITICAL IS PROFESSIONAL

At this point, I sought out a woman mentor and feminist colleagues. Both provided critical support for my scientific career. Even though I was again outnumbered by men among my colleagues, there were many women with whom I could be completely open, and my mentor, Dr. L., proved to be organized, kind, socially adept, brilliant, famous, and witty. She even had a Real Life outside the lab. I also had some supportive male colleagues inside and outside the lab, and a few even tolerated my openly feminist stance. Still, the pattern of harassment had not completely evaporated. There was a very unpleasant instance of inappropriate physical behavior by a senior colleague which resulted in a severe rift and subsequent period of isolation for me. There was yet another colleague whom I trusted who did his best to convince me that I was not dedicated enough.

The feminist community in the area included someone in women's studies who was working with a group of feminist scientists. I looked her up, joined the group's seminars, and began to get a serious education in feminist theory and feminist science studies. Since that time, I've worked with a team who developed and taught a course in women in science and technology, presented at feminist conferences, and provided one another with support, encouragement, and friendship. Intellectually, I am more aware of the biases that my colleagues and myself bring to our science. I can analyze the results of my colleagues better for it, and when my gut contorts over some obfuscated gender-based assumption, I can articulate the problem clearly. I am now comfortable with the awareness that at first was so painful to me: science is after all a social enterprise, not an exercise practiced by disembodied brains.[6] But given that most of my colleagues are not comfortable with this idea, how do I behave as a scientist while remaining true to my beliefs?

PHASE 4. CLAIMING A POLITICAL STANCE

I am in phase four as I write this essay, since putting my ideas to paper for a feminist audience is itself a political act. However exhilarating it is to share my views with you, there may be costs, since my colleagues are unlikely to endorse my account. In short, taking a step toward feminist community requires taking a step away from my scientific one. Because my work on the structure of the ovary in *Drosophila* could be appropriated for use in sex difference research, I find myself walking a tightrope between illuminating basics in female biology and contributing to essentialist, male-centered, sociobiological agendas. The persistent misreading of Bateman's work is a case in point. If the scientific community has focused on describing the male and the not-male, research on females could be seen as a corrective, a realignment of prevailing assumptions. Yet an open discussion within science about the inadequacy of the male-centered model of evolution seems unlikely, at least in my field. A discussion about female biology that does not include a focus on sex differences, given the context, is unimaginable.

Some are arguing that the rigidly dichotomized heterosexual model of the natural world is inadequate and inaccurate. I agree, but at the same time, there are continuities among animals with ovaries that have a much overlooked contribution to make to medical knowledge. Can we foster the growth of knowledge about women's biologies without reinforcing sex difference arguments, especially given the masculinist culture of contemporary science? And if women are excluded/marginalized from science as researchers, and females are excluded/marginalized from science as subject matter, is any change in that culture possible? Given that this is going on in my field, is it also going on in other fields, particularly those that use animal models? I do not have any answers to the questions I have raised, but I hope that this article may serve as a starting point to a conversation in feminist science studies about the long shadow of androcentrism on both climate and construction of knowledge.

Acknowledgments

I gratefully acknowledge M. B. Wyer for encouragement, critical thinking, and invaluable editorial input; M. Barbercheck, D. Cookmeyer, H. Ö. Öztürk, and B. Subramaniam for helpful comments on the manuscript; and S. V. Rosser for helpful discussion.

Notes

1 For an overview of this work, see the introductory essays to this volume.

2 For a discussion of the varying visions of feminist science, see Sandra Harding's *Whose Science? Whose Knowledge?* Ithaca, NY: Cornell University Press, 1991.

3 Extrapolation from flies to people is presented explicitly in most genetics textbooks: Griffiths *et al.* 2008, pp. 20–21; Hartwell *et al.* 2008, pp. 6–7; Klug and Cummings 2000, with respect to behavior genetics, pp. 612–622. However, rather than mentioning sex, it is generally assumed that data are universal for males and females.

4 For model organisms in general, see http://www.nih.gov/science/models/index.html; for *Drosophila* in particular, see http://www.nih.gov/science/models/nmm/appb3.html.
5 For different perspectives on the value of feminism for science, see Fedigan (1997) and Hrdy (1999).
6 Most feminist theorists would take this for granted, but scientists are less willing to grasp its importance. For a debate about this topic, see Sokal (1996).

References

Arnold, Steven J. 1994. "Bateman Principles and the Measurement of Sexual Selection in Plants and Animals." *American Naturalist* 144: S126–S149 Suppl. S.

Baker, Robin R. and M. A. Bellis. 1988. "Kamikaze Sperm in Mammals?" *Animal Behavior* 36: 867–869.

Baker, Robin R. and M. A. Bellis. 1989. "Number of Sperm in Human Ejaculates Varies in Accordance With Sperm Competition Theory." *Animal Behavior* 37: 867–869.

Baker, Robin R. and M. A. Bellis. 1993a. "Human Sperm Competition: Ejaculate Adjustment by Males and the Function of Masturbation." *Animal Behavior* 46: 861–885.

Baker, Robin R. and M. A. Bellis. 1993b. "Human Sperm Competition: Ejaculate Manipulation by Females and a Function for the Female Orgasm." *Animal Behavior* 46: 887–909.

Bateman, A. J. 1948. "Intra-sexual Selection in *Drosophila.*" *Heredity* 2: 349–368.

Bellis, Mark A., Robin R. Baker, and Matthew J. G. Gage. 1990. "Variation in Rat Ejaculates Consistent with the Kamikaze-Sperm Hypothesis." *Journal of Mammalogy* 713: 936–939.

Bleier, Ruth. 1984. *Science and Gender: A Critique of Biology and its Theories on Women.* New York: Pergamon Press.

Buss, David M. 1991. "Evolutionary Personality Psychology." *Annual Review of Psychology* 42: 459–491.

Clark, Andrew G., David J. Begun, and Timothy Prout. 1999. "Female × male interactions in Drosophila sperm competition." *Science* 283: 217–220.

Davis, Cinda-Sue and Sue V. Rosser. 1996. "Program and Curricular Interventions." In *The Equity Equation: Fostering the Advancement of Women in the Sciences, Mathematics, and Engineering,* edited by Cinda-Sue Davis, Angela B. Ginorio, Carol S. Hollenshead, Paula M. Rayman, and Barbara B. Lazarus. San Francisco: Jossey-Bass Publishers, pages 232–264.

Drickamer, Lee C., Stephen H. Vessey, and Doug Meikle. 1996. *Animal Behavior: Mechanisms, Ecology, Evolution.* Dubuque: Wm. C. Brown Publishers.

Eberhard, William G. 1996. *Female Control: Sexual Selection by Cryptic Female Choice.* Princeton, NJ: Princeton University Press.

Fedigan, Linda H. 1997. "Is Primatology a Feminist Science?" In *Women in Human Evolution,* edited by L. Hager. New York: Routledge, pages 56–76.

Freeman, Scott and Jon C. Herron, 2004. *Evolutionary Analysis.* Upper Saddle River, NJ: Pearson Prentice Hall.

Futuyma, Douglas J. 1998. *Evolutionary Biology.* Sunderland, MA: Sinauer Associates Inc.

Griffiths, Anthony F. J., Susan R. Wessler, Richard C. Lewontin, and Sean B. Carroll. 2008. *Introduction to Genetic Analysis.* New York: W. H. Freeman and Company.

Hager, Lori D., ed. 1997. *Women in Human Evolution.* New York: Routledge.

Hartwell, Leland H., Leroy Hood, Michael L. Goldberg, Ann E. Reynolds, Lee M. Silver, and Ruth C. Veres. 2008. *Genetics from Genes to Genomes.* New York: McGraw-Hill.

Hrdy, Sarah B., ed. 1987. "Empathy, Polyandry, and the Myth of the Coy Female." In *Feminist Approaches to Science,* edited by Ruth Bleier. New York: Teachers College Press.

Hrdy, Sarah B. 1999. "Preface: On Raising Darwin's Consciousness." In *The Woman That Never Evolved,* xiii–xxxii. Cambridge, MA: Harvard University Press.

Imhof, Marianne, Bettina Harr, Gottfried Brem, and Christian Schlötterer. 1998. "Multiple mating in wild *Drosophila melanogaster* revisited by microsatellite analysis." *Molecular Ecology* 7: 915–917.

Keller, Evelyn Fox. 1985. *Selections on Gender and Science.* New Haven, CN: Yale University Press.

Keller, Evelyn Fox and Helen E. Longino, eds. 1996. *Feminism and Science.* Oxford Readings in Feminism. Oxford: Oxford University Press.

Kennan, Brian. 1998. "Evolutionary Biology and Strict Liability for Rape." *Law and Psychology Review* 22: 131–177.

Klug, William S. and Michael R. Cummings. 2000. *Concepts of Genetics.* Upper Saddle River, NJ: Prentice Hall.

Krebs, John R. and Nicholas B. Davis. 1993. *An Introduction to Behavioral Ecology*. Oxford: Blackwell Scientific Publications.
Longino, Helen. 1990. *Science as Social Knowledge*. Princeton, NJ: Princeton University Press.
O'Connell, Lisa M. and Mark O. Johnston. 1998. "Male and Female Pollination Success in a Deceptive Orchid, a Selection Study." *Ecology* 79: 1246–1260.
Pinker, Steven. 1998. *How the Mind Works*. New York: W. W. Norton.
Pinker, Steven. 2003. *The Blank Slate: The Modern Denial of Human Nature*. New York: Penguin.
Sokal, Alan. 1996. "A Physicist Experiments with Cultural Studies + the Conventions of Academic Discourse." *Lingua Franca* 6: 62–64.
Spanier, Bonnie. 1995. *Im/Partial Science: Gender Ideology and Molecular Biology*. Bloomington: Indiana University Press.
Thornhill, Randy and Nancy Wilmsen Thornhill. 1992. "The Evolutionary Psychology of Men's Coercive Sexuality." *Behavioral and Brain Sciences* 152: 363–375.
Wayne, Marta L., J. Brant Hackett, Sergey V. Nuzhdin, Elena G. Pasyukova, and Trudy F. C. Mackay. 2000. "Quantitative Trait Locus Mapping of Fitness-related Traits in *Drosophila melanogaster*." *Genetical Research* 77: 107–116.
Wertheimer, Alan, 2003. *Consent to Sexual Relations*, Cambridge Studies in Philosophy and Law. Cambridge, UK: Cambridge University Press.
Wilson, Margo, Martin Daly, and Joanna E. Scheib. 1997. "Femicide: An Evolutionary Psychological Perspective." In *Feminism & Evolutionary Biology*, edited by Patricia Gowaty. New York: Chapman and Hall Publisher, pages 431–458.

Section II

Stereotypes, Rationality, and Masculinity in Science and Engineering

As we revised material for the second edition of *Women, Science, and Technology*, we faced a dearth of information about the lives of women engineers. Though women are now relatively well-represented among degree earners at all levels in the biological sciences, and so have become more visible in the last few decades, engineering fields have been slower to change. In the entire United States in the decade 1950–1959, for instance, only 13 women received doctorates in engineering. This represented just one-fifth of 1% of doctorates in engineering, and it was, astonishingly, an all-time high. By the 1990–1999 decade, women received just 11% of the doctorates in engineering.[1] Slow change, indeed. However, degrees in engineering are not the only measure of ability to be technologically expert and innovative, as the historical record reveals. The human capacity for invention has left us a legacy of sometimes sublime, often curious, incredibly useful, and terribly destructive tools and weapons.

As a field of study, engineering is rooted in the demands of warfare. In the United States, engineering education began at the U.S. Military Academy at West Point in 1802, to meet needs associated with moving troops, weapons, and supplies.[2] Engineering, in other words, emerged within social and historical contexts in order to address specific goals, values, and priorities. Thus, engineering is indelibly and inescapably marked by beliefs and practices related to women, men, and gender differences. New technologies produced through engineering innovation are of interest not only because of their impact on social life, but also because of the social goals, values, and priorities embedded in their invention and production. Even the definition of "technology" betrays social influences, since contemporary engineering research emphasizes the development of patentable devices, products, and systems that can be exploited for profit. We equate "technology" with "machines," "electronics," "computers," "home entertainment systems," and "information systems."

In the United States, we have a long history of equating technological competence with masculinity and technological incompetence with femininity.[3] The inventions are only a part of the larger picture of human activity and agency, one that includes individual, social, and symbolic level dynamics.[4] At the individual level, there are the people who train in (or are excluded from) the skills to invent. At the social level, there are the educational organizations and training practices that shape who will be included (or excluded), as well as the economic and business practices that shape who will benefit from (or be exploited by) the inventions. At the symbolic level, there are stereotypes, media images, metaphors, and motifs that shape the meanings and interpretations of daily events and interactions. It is thus no coincidence that the most lucrative research, training, and education fields are the ones in which women are least represented, since women have been systematically marginalized, dismissed, and disadvantaged at all three levels.

Yet, technologies are pervasive elements of our daily lives, so much so that they are taken for granted even while necessary to meet our basic needs. When technological invention is redefined as a human activity, rather than one for which only men are suited, the ingenuity of our foremothers becomes apparent. If we assume that women were agents of their own destinies, then women would have been among those who developed digging sticks for harvesting and planting, the hoe, the plow, hand pollination, spindle whorls for making thread, looms for weaving cloth, potters' wheels, and grain mills, to name just a few examples. Some have argued that inventions by women historically have been widely and systematically appropriated by men when these inventions garnered social, economic, and political clout.[5]

Despite the systematic exclusion of women from formal training for careers in engineering professions, despite cultural norms that presume women's technological incompetence, and despite the lost history of women's talents and achievements, women have designed, developed, and utilized new technologies. They have operated machines in factories and mills, used typewriters and telephones in offices, built and flown airplanes, maintained and driven automobiles, and developed, designed, operated, and operated as, computers. They often did so surrounded initially by controversy, and usually for lower wages than those of men with the same skills.[6]

Against this historical backdrop, the biographical and autobiographical accounts of women, like those included in Section I essays, reveal the psychological costs of transgressing cultural norms for women. As the stories of even the most successful women show, male colleagues ignored, discounted, and dismissed women's commitments and achievements. In both the distant and recent past, dispassionate and rational decision-making about women's abilities is not possible when women are presumed to be inferior to men by virtue of the natural order. The increasing representation of women in science and engineering since the 1970s suggests that women are interested, able, and competitive to men in intellectual ability in scientific, mathematical, and engineering fields. So what about today's images of scientists and engineers as professionals? Have the years since the second wave of the women's movement provoked changes in the cultural assumption that scientific and engineering expertise is somehow especially masculine?

Children's drawings of scientists have provided some evidence that stereotypes about scientists are learned relatively early in childhood. In these stereotypes, the scientist is a white man of reason, driven by mind over body, someone who strains his eyes in intense observation, who works such long and unusual hours that he cannot find time to shave. Science is a one-gender and one-race world, where tools of destruction, violence, and pain are created and morality is irrelevant.[7] In advertising and the popular press, there are also occasional and somewhat more positive versions of the familiar stereotype. For instance, there is the Mathematical Marlboro Man who is a rugged individualist, self-sufficient and immersed in mathematics, "toying with it day and night, devoting every scrap of available energy to understanding it."[8] There is also the image of the Scientist Athlete who is driven to compete in order to win, as Mary Barbercheck describes in her article in this section. There is also the image of the Engineering Professor, who rushes through a 12-hour day packed with top-notch teaching and cutting-edge research, works late into the night writing a grant proposal, and then falls asleep at his desk in exhaustion, a photograph of his wife and child in the background.[9]

What is common to these images is that the scientist or engineer is represented as a man doing the work that is culturally constructed as men's—that is, work described as at once objective, obsessive, rational, competitive, and asocial. Girls and women are not socialized into the habits of mind and behavior that are necessary in this kind of work, it is often said, nor do the exigencies of most women's lives present them with experiences that might cause them to acquire such a mindset. Women are associated in American culture with the ongoing details and moral ambiguities of daily life—work that involves taking care of children, teaching them "right and wrong," nurturing family relationships, and providing emotional support. This set of ideas about women and men, about their qualities and characteristics, is called symbolic gender, "a central organizing discourse of culture, one that not only shapes how we experience and understand ourselves as men and women but also interweaves with other discourses and shapes *them.*"[10] Though symbolic gender reveals empirical information about the everyday lives of women and men only indirectly and partially, its historical durability speaks to its power to represent an enduring cultural fit between conceptions of men and those of science and engineering—a fit that discourages if not precludes the full participation of women as researchers, inventors, and educators. Popular culture images of women as either whores or virgins, as either villains or victims, prevail, most recently on the internet as a cyber-Barbie or cyber-warrior. If women are represented as scientists or engineers at all, they are portrayed as routinely compromised by the conflict between their obligations to femininity and their commitments as professionals.[11] Though competent women are present today among television characters and in motion pictures in general and as professionals, there are too few who are represented as both emotionally and professionally competent.[12]

However, recent alternative representations serve only to place the persisting images in bold relief. The repetition of the images and ideas equating women with emotion and intuition, and men with reason and rationality, constitute stereotypes that capture, shape, and reproduce cultural attitudes. Stereotypes function to provide organizing interpretive frameworks that are necessary elements of human cognition, but stereotypes can also be oppressive when they reiterate negative images, attitudes, and beliefs about social groups.[13] This is especially critical as an educational issue since educators have an obligation to ensure that they are not reproducing stereotypes that advantage one social group at the expense of others.

Researchers have built a compelling case that social beliefs about women's abilities and interests are related to women's underrepresentation in STEM. Since occupations are culturally stereotyped by gender, and since these stereotypes influence occupational choices, there is substantial empirical evidence that undergraduate students perceive science, technology, engineering and mathematics (STEM) professions in light of stereotypes about women and men. Students make gender-appropriate choices of majors associated with those professions accordingly and regardless of the accuracy of the stereotypes.[14]

Undergraduate students' academic performance and persistence are key determinants of their educational attainment, especially in professions that require advanced training. However, students' academic performance and persistence take place within an educational environment filled with racial, ethnic, and gender stereotypes. Negative stereotypes have well-established negative effects on undergraduate students' performance on standardized tests.[15] Indeed, negative stereotypes have their most pronounced effects on those who are high achievers. In environments where negative stereotypes are widespread, removing allusions to these stereotypes within a classroom boosts performance of those who identify with the target group in the stereotype.[16] There are few studies on positive stereotypes, but they have been linked to "boosts" in performance and persistence. New research demonstrates that teaching students about the social and economic origins and impacts of negative stereotypes significantly improves their academic performance when they are a member of a targeted group.[17]

Stereotypes about scientists and engineers embed social messages about who can, or cannot, be a scientist or engineer. Much of the empirical evidence about the function of stereotypes points to how negative stereotypes suppress the performance of high-achieving undergraduate students. Though stereotypes are often understood as pervasively negative in content and effect, they also can carry positive and affirming images.

There is ample evidence of gender differences in the delivery of education and in educational achievement in primary and secondary schools in the United States. In 1992 the American Association of University Women issued a summary report on literally hundreds of studies on the ways students' gender influences the education they receive. The report, *How Schools Shortchange Girls*, concluded that "girls do not receive equitable amounts of teacher attention, that they are less apt than boys to see themselves reflected in the materials they study, and that they often are not expected or encouraged to pursue higher level mathematics and science courses."[18] When they do, interestingly, they do well. High school females and males with the same SAT scores do not do equally well in college, since the females get *better* grades in their college courses.[19] In short, based on well-documented gender differences in the delivery of basic science and math education at primary and secondary levels, and based on persisting stereotypes that women do not "fit" into science, mathematics, and engineering, it seems a foregone conclusion that women would not seek careers in science and engineering. If women do not seek to become scientists and engineers, then it seems as if there may be some truth in the stereotypes. Thus, every individual woman who becomes a scientist or engineer is engaging in a political and social act of resistance to the status quo, bringing her experiences, values, and priorities into her profession. As Sandra Harding has argued, this means that efforts to recruit and retain women in science and engineering professions go beyond mere equity and instead promote the advancement of frontiers of knowledge.[20]

The most comprehensive review of research about women's status in science, mathematics, and engineering to date is a report from the National Academy of Sciences, titled *Beyond Bias and Barriers*.[21] The report was issued by a committee that included university presidents, chancellors, provosts, leading policy analysts, and distinguished scientists and engineers. After exhaustive review of the data, the committee determined that barriers to women's full participation rest solely on social biases and ignorance. There are no significant differences between women and men in terms of biology, ability, interests, or motivation to achieve in science, engineering, and mathematics careers that can account for the degree and persistence of gender differences in educational and professional outcomes. The report was focused in particular on academic careers and the committee's unambiguous conclusion was that explicit discrimination and implicit biases are slowing progress toward the full participation of women in academic science and engineering education.

The data on women's interest in scientific and engineering careers shows steady but uneven growth across all of the fields. Among undergraduates, the percentage of degree earners who were women steadily increased between 1983 and 2004 in almost every field tracked by the National Science Foundation, including chemistry (from 33.8% to 51%), oceanography (from 11.7% to 53.4%), and chemical engineering (from 20.9% to 35.4%). An important exception was in computer sciences (declined from 36.4% to 25%).[22] Among graduate students, there was a similar increase between 1983 and 2004 in the percentage of doctorates awarded to women, with increases in every field tracked by the National Science Foundation, including computer sciences (from 12.6% to 20.5%). Indeed, a slightly higher percentage of doctorates were awarded to women in the biological sciences in 2004 (46.3%) than in the social sciences (44%). Overall, women earned 37.4% of the science and engineering doctorates awarded in 2004.[23] Only in engineering have the percentages remained low, but even here professional organizations tout gains in recent years.

WOMEN IN A MAN'S WORLD

Articles about women's experiences in the general interest science press sometimes portray the difficulties women have in negotiating a balance between being seen as serious about science but not appearing to be too serious about being a woman. In 1984, the widely distributed Association of American Colleges publication *Out of the Classroom: A Chilly Campus Climate for Women?* documented women's experiences being discounted and discouraged by their male colleagues because they were women.[24] Though there has been an important "warming" in the climate for women, contemporary research continues to document the reality of persisting obstacles to women's achievement and success in science. Women are excluded from professional activities by male colleagues, have more negative interactions with colleagues than do men, have more difficulty with processes of evaluation in the promotion and tenure process, are assigned less laboratory space than male colleagues with comparable levels of funding, experience more stress than male colleagues in negotiating work/life balance issues, and carry heavier teaching and advising loads than do their male colleagues.[25]

DIVERGING FROM THE NORM

Many of the readings in our book refer to the "culture" or "climate" in science and engineering. These common terms reduce the great diversity of practices across labs, departments, and individuals in science and engineering to a single "type" of culture or climate that adheres to a common set of values, beliefs, and behaviors. It is important to remember that though there are commonalities across fields and across time in the association between stereotypes about scientists' behavior and masculinity, there are also significant differences. Many of these differences relate to the range of ways in which men can be "masculine."[26] Biologists can have stereotypes about chemists, and vice versa, for instance.

Some differences between the cultures of disciplines relate to the historical presence and contributions of women within a discipline, where new subfields emerged because of their work. It was a collaboration between two women, Adrienne Zihlman and Nancy Tanner, that led to a challenge of the prevailing "Man the Hunter" model in human evolution. Empirical data from fossil, nonhuman primate, and ethnographic records led them to argue that the conventional model was an inaccurate portrayal of the organization of prehistoric human society. Though Zihlman denies that they offered their model as a feminist reinterpretation, she makes the point that the feminist movement created a context in which the question "Where are the women?" could be asked. The willingness of Zihlman and Tanner to look for the answer led to a reconsideration of the earlier model and still provides a touchstone in continuing debates about human origins.[27]

Similarly, Margaret Mead's interest in Ruth Benedict's work led her to explore the relationship between variations in sex roles across cultures, work that represents an early challenge to biological explanations for women's subordination to men. Their work broke new ground in anthropology where, according to an interview with Benedict in *Time* (March 16, 1933) jobs were rare but jobs for women were rarer still.[28]

Historian Margaret Rossiter points to aggregate data from the first half of the twentieth century to suggest that a critical mass of women in a field, once it was reached, led to continuing high participation in the field by women. Her data show that 60% of women in science in 1921 were clustered in just three fields, botany, zoology, and psychology. These fields today continue to have uniquely high percentages of women participating at all levels, with 78% of bachelor's and 67% of doctorates in psychology awarded to women in 2004.[29]

GENDER AND POWER IN SCIENCE AND ENGINEERING

Perhaps the answer to the question "Do women opt out of science and engineering or are they pushed out?" is that both processes contribute to the underrepresentation of women in science and engineering. The articles in this section of *Women, Science, and Technology* examine the cultural backdrop in which these processes take place, with an emphasis on the symbolic work of gender in language, images, and assumptions about professional skill and ability. Though research about stereotypes comes from the empirical traditions of psychology, and research about the symbolic and institutional power of language draws from postmodern theory, literary criticism, and interdisciplinary perspectives, this section draws different approaches together because they share a common goal: to disrupt the repetition of oppressive values and attitudes.

Carol Cohn's essay describes how twentieth-century notions of masculinity threaten the future of the human race, where the concept of "rationality" is used to mask the epitome of the irrational—weapons of mass destruction. Her uniquely autobiographical and yet fully social perspective on the language of defense intellectuals exposes a phenomenon that is so pervasive and taken for granted as to be invisible. She argues that the hyper-rational language they use masks and displaces the deeply irrational work of defense strategy and planning. Mary Barbercheck takes a somewhat different tack, exploring the images in the advertising in the American Association for the Advancement of Science (AAAS) masthead journal, *Science*. She uncovers distinct patterns in the activities and characteristics of the people portrayed in the advertisements. These portrayals, she argues, draw on cultural assumptions about science as a new frontier, women and people of color as the beneficiaries (and not creators) of scientific knowledge, and white men as figures who exercise power and authority.

Banu Subramaniam's article, on the language used to describe "invasive species" in research about ecology and the environment, builds on both of these articles to make the point that the language of the research betrays the perspective of the research*er* on the topic of study. When scientists focus on the "alien" or "nonnative" plant that gains a foothold in a new area, their emphasis is on the characteristics of the outsider plant. The language suggests the threat that the outsider poses to what is assumed to be the natural, stable, original environment. This language taps the rhetoric of cultural fears about immigrants to add to the drama of the scientific scenario. There is a cost associated with that, argues Subramaniam, in the reiteration of racist and Euro-centered stereotypes about immigrants. There is also a scientific cost, she points out, in the displacement of the complex and ever-changing environment from the center of analysis. Researchers may learn a great deal about morning glories, for instance, but they will learn little about how natural environments change over time, with changes in stream and river beds, shifts in wildlife populations and migration patterns, and increases or reductions in insect species and activity. Her article represents a novel approach to educating a next generation of scientists to be conscious of the ways in which social biases can infect portrayals of natural phenomenon.

The concluding reading in this section is an article about masculinity in the formulation of what it means to be an engineer. Wendy Faulkner, like the other contributors in this section, focuses on the powerful influence of seemingly intangible symbols on human activity; in this case it is ideas, values, and commitments about masculinity that shape the meanings that engineering professionals give and take from their work. Faulkner's highly theoretical approach is a rare effort to conceptualize how a field famously committed to disembodied and machine-like rational inquiry nonetheless can be understood to motivate through the ever-so-human search for pleasure. It offers us a way to appreciate why so many people for so long, women and men, have dedicated their lives to human invention.

Notes

1 M. F. Fox, "Women, Men, and Engineering," in *Women, Gender, and Technology*, ed. M. F. Fox, D. G. Johnson, and S. V. Rosser, pp. 47–59. Urbana: University of Illinois Press, 2006.

2 Ibid., p. 54.

3 R. Oldenziel, *Making Technology Masculine: Men, Women, and Modern Machines in America, 1870–1945*, Amsterdam: Amsterdam University Press, 1999. In addition to charting the history of the relationship between engineering and masculinity in the United States, Oldenziel explains that at the turn of the century, the autobiography was the literary genre by which white middle-class men in engineering shaped their emerging profession. None of the pioneering women in civil or industrial engineering in this era published their stories.

4 S. Harding, *Whose Science, Whose Knowledge?* Ithaca, NY: Cornell University Press, 1991.

5 A. Stanley, "Women Hold Up Two-Thirds of the Sky: Notes for a Revised History of Technology," in *Sex/Machine: Readings in Culture, Gender and Technology,* ed. Patrick Hopkins, pp. 16–32, Bloomington: Indiana University Press, 1998. Stanley points in particular to the appropriation by the (male) medical profession of women's knowledge about the medicinal qualities of plants. She also points to historical evidence of women's contributions to the development of the reaper, the sewing machine, small engines, and the printing press.

6 J. Wosk, *Women and the Machine: Representations from the Spinning Wheel to the Electronic Age*, Baltimore: Johns Hopkins University Press, 2001; V. Scharff, *Taking the Wheel: Women and the Coming of the Motor Age*, Albuquerque: University of New Mexico Press, 1992; B. A. Toole, "Ada Byron, Lady Lovelace," *IEEE Annals of the History of Computing* 18 (1996): 4–12; J. Light, "Programming," in *Gender and Technology*, ed. N. Lerman, R. Oldenziel, and A. Mohun, Baltimore: Johns Hopkins University Press, 2003; A. Kessler-Harris, *A Woman's Wage: Historical Meanings and Social Consequences*, Lexington: University Press of Kentucky, 1990.

7 D. Chambers, "Stereotypic Images of the Scientist: The Draw the Scientist Test," *Science Education* 67, no. 2 (1983): 255–65, quote on p. 256. There is evidence that components of this image persist; see, e.g., B. L. Sherriff and L. Binkley, "The Irreconcilable Images of Women, Science, and Engineering," *Journal of Women and Minorities in Science and Engineering* 3 (1997): 21–36; C. Barman, "Students Views About Scientists and School Science: Engaging K-8 Teachers in a National Study," *Journal of Science Teacher Education* 10, no. 1 (1999): 43–54; and G. Jones, A. Howe, and M. Rua, "Gender Differences in Students' Experiences, Interests, and Attitudes toward Science and Scientists," *Science Education* 84, no. 2 (2000): 180–92.

8 C. Herrion, *Women in Mathematics: The Addition of Difference*, quote on p. 5. Bloomington: Indiana University Press, 1997.

9 T. Grose, "21st Century Professor," *ASEE Prism*, January 2007, pp. 26–31.

10 C. Cohn, "War, Wimps, and Women," in *Gendering War Talk*, ed. M. Cooke and A. Woollacott, quote on p. 228, emphasis in original. Princeton, NJ: Princeton University Press, 1993.

11 A. S. Owen, S. R. Stein, and L. R. Vande Berg, *Bad Girls: Cultural Politics and Media Representations of Transgressive Women*. New York: Peter Lang, 2007.

12 For an account of these images from the 1980s, see "Fatal and Fetal Visions: The Backlash in the Movies," in Susan Faludi's *Backlash: The Undeclared War Against American Women*, New York: Doubleday, 1991.

13 K. Kumashiro, "Against Repetition: Addressing Resistance to Anti-Oppressive Change in the Practices of Learning, Teaching, Supervising, and Researching," *Harvard Educational Review* 72, no. 1 (2002): 67–93.

14 J. Hughes, "Gender Attributions of Science and Academic Attributes: An Examination of Undergraduate Science, Mathematics, and Technology Majors," *Journal of Women and Minorities in Science and Engineering* 8 (2002): 53–65; S. Beyer, "The Accuracy of Academic Gender Stereotypes," *Sex Roles* 40 (1999): 787–813; P. Lightbody and A. Durndell, "The Masculine Image of Careers in Science and Technology," *British Journal of Educational Psychology* 66 (1996): 231–46.

15 D. M. Quinn and S. J. Spencer, "The Interference of Stereotype Threat with Women's Generation of Mathematical Problem-Solving Strategies," *Journal of Social Issues* 57 (2001): 55–71; C. M. Steele, "A Threat in the Air: How Stereotypes Shape Intellectual Identity and Performance," *American Psychologist* 52 (1997): 613–29.

16 S. C. Wheeler and R. E. Petty, "The Effects of Stereotype Activation on Behavior: A Review of Possible Mechanisms," *Psychological Bulletin* 127 (2001): 797–826; D. L. Oswald and R. D. Harvey, "Hostile Environments, Stereotype Threat, and Math Performance among Undergraduate Women," *Current Psychology* 19 (2001): 338–56; P. G. Davies, S. J. Spencer, and R. Gerhardstein, "Consuming Images: How Television Commercials that Elicit Stereotype Threat Can Restrain Women Academically and Professionally," *Personality and Social Psychology Bulletin* 28, no. 12 (2002): 1615–28.

17 M. Johns, T. Schmader, and A. Martens, "Knowing Is Half the Battle: Teaching Stereotype Threat as a Means of Improving Women's Math Performance," *Psychological Science* 16 (2005): 175–79; M. Shih, T. L. Pittinsky, and N. Ambady, "Stereotype Susceptibility: Identity Salience and Shifts in Quantitative Performance," *Psychological Science* 10 (1999): 80–83.

18 American Association of University Women (AAUW), *How Schools Shortchange Girls*, Washington, DC, AAUW Educational Foundation and National Education Association, 1992; M. Sadker and D. Sadker, *Failing at Fairness: How America's Schools Cheat Girls*, New York: Charles Scribner's, 1994.

19 T. Tindall and B. Hamil, "Gender Disparity in Science Education: The Causes, Consequences, and Solutions," *Education* 125, no. 2 (2004): 282–95.

20 Harding, *Whose Science, Whose Knowledge?*

21 Committee on Science Engineering and Public Policy (COSEPUP), *Beyond Bias and Barriers: Fulfilling the Potential of Women in Academic Science and Engineering*. Washington DC: National Academies Press, 2007.

22 W. Grossman, "Cyber View: Access Denied," *Scientific American*, August 1998, p. 38.

23 National Science Foundation (NSF), *Women, Minorities, and Persons with Disabilities in Science and Engineering*. Washington, DC: U.S. Government Printing Office, 2007.

24 R. Hall and B. Sandler, *Out of the Classroom: A Chilly Campus Climate for Women?* Washington, DC: Project on the Status and Education of Women, Association of American Colleges, 1984, p. 5.
25 C. Hult, R. Calister, and K. Sullivan, "Is There a Global Warming toward Women in Academia?" *Liberal Education* (Summer/Fall 2005): 50–57; A. Lawler, "Princeton Study Strikes Sad but Familiar Chord," *Science* 302 (2003): 33; Committee on Women Faculty in the School of Science at MIT, "A Study of the Status of Women Faculty in Science at MIT," *The MIT Faculty Newsletter* 11, no. 4 (1999): 1–17.
26 See, e.g., R. W. Connell, *Masculinities*. Los Angeles: University of California Press, 1995.
27 L. Hager, "Sex and Gender in Paleoanthropology," in *Women in Human Evolution*, ed. Lori Hager, pp. 1–28. New York: Routledge, 1997.
28 A. Kuper, *The Chosen Primate*, Cambridge, MA: Harvard University Press, 1994, p. 187; M. Rossiter, *Women Scientists in America: Struggles and Strategies to 1940*, Baltimore: Johns Hopkins University Press, 1982, p. 151.
29 Rossiter, *Women Scientists in America*, pp. 150–53; National Science Foundation, *Science and Engineering Indicators, 2006*, Washington, DC: U.S. Government Printing Office, 2006.

9

Sex and Death in the Rational World of Defense Intellectuals

Carol Cohn

. . . My close encounter with nuclear strategic analysis started in the summer of 1984. I was one of forty-eight college teachers (one of ten women) attending a summer workshop on nuclear weapons, nuclear strategic doctrine, and arms control, taught by distinguished "defense intellectuals." Defense intellectuals are men (and indeed, they are virtually all men) "who use the concept of deterrence to explain why it is safe to have weapons of a kind and number it is not safe to use."[1] They are civilians who move in and out of government, working sometimes as administrative officials or consultants, sometimes at universities and think tanks. They formulate what they call "rational" systems for dealing with the problems created by nuclear weapons: how to manage the arms race, how to deter the use of nuclear weapons, how to fight a nuclear war if deterrence fails. It is their calculations that are used to explain the necessity of having nuclear destructive capability at what George Kennan has called "levels of such grotesque dimensions as to defy rational understanding."[2] At the same time, it is their reasoning that is used to explain why it is not safe to live without nuclear weapons. In short, they create the theory that informs and legitimates American nuclear strategic practice.

For two weeks, I listened to men engage in dispassionate discussion of nuclear war. I found myself aghast, but morbidly fascinated—not by nuclear weaponry, or by images of nuclear destruction, but by the extraordinary abstraction and removal from what I knew as reality that characterized the professional discourse. I became obsessed by the question, How can they think this way? At the end of the summer program, when I was offered the opportunity to stay on at the university's center on defense technology and arms control (hereafter known as "the Center"), I jumped at the chance to find out how they could think "this" way.

. . . But as I learned their language, as I became more and more engaged with their information and their arguments, I found that my own thinking was changing. Soon, I could no longer cling to the comfort of studying an external and objectified "them." I had to confront a new question: How can *I* think this way? How can any of us? . . .

STAGE I: LISTENING

Clean bombs and clean language

Entering the world of defense intellectuals was a bizarre experience—bizarre because it is a world where men spend their days calmly and matter-of-factly discussing nuclear weapons, nuclear strategy, and nuclear war. The discussions are carefully and intricately reasoned, occurring seemingly without any sense of horror, urgency, or moral outrage—in fact, there seems to be no graphic reality behind the words, as they speak of "first strikes," "counterforce exchanges," and "limited nuclear war," or as they debate the comparative values of a "minimum deterrent posture" versus a "nuclear war-fighting capability."

Yet what is striking about the men themselves is not, as the content of their conversations might suggest, their cold-bloodedness. Rather, it is that they are a group of men unusually endowed with charm, humor, intelligence, concern, and decency. Reader, I liked them. At least, I liked many of them. The attempt to understand how such men could contribute to an endeavor that I see as so fundamentally destructive became a continuing obsession for me, a lens through which I came to examine all of my experiences in their world.

In this early stage, I was gripped by the extraordinary language used to discuss nuclear war. What hit me first was the elaborate use of abstraction and euphemism, of words so bland that they never forced the speaker or enabled the listener to touch the realities of nuclear holocaust that lay behind the words.

Anyone who has seen pictures of Hiroshima burn victims or tried to imagine the pain of hundreds of glass shards blasted into flesh may find it perverse beyond imagination to hear a class of nuclear devices matter-of-factly referred to as "clean bombs." "Clean bombs" are nuclear devices that are largely fusion rather than fission and that therefore release a higher quantity of energy, not as radiation, but as blast, as destructive explosive power.[3]

"Clean bombs" may provide the perfect metaphor for the language of defense analysts and arms controllers. This language has enormous destructive power, but without emotional fallout, without the emotional fallout that would result if it were clear one was talking about plans for mass murder, mangled bodies, and unspeakable human suffering. Defense analysts talk about "countervalue attacks" rather than about incinerating cities. Human death, in nuclear parlance, is most often referred to as "collateral damage"; for, as one defense analyst said wryly, "The Air Force doesn't target people, it targets shoe factories."

Some phrases carry this cleaning-up to the point of inverting meaning. The MX missile will carry ten warheads, each with the explosure power of 300–475 kilotons of TNT: *one* missile the bearer of destruction approximately 250–400 times that of the Hiroshima bombing.[4] Ronald Reagan has dubbed the MX missile "the Peacekeeper." While this renaming was the object of considerable scorn in the community of defense analysts, these very same analysts refer to the MX as a "damage limitation weapon."

These phrases, only a few of the hundreds that could be discussed, exemplify the astounding chasm between image and reality that characterizes technostrategic language. They also hint at the terrifying way in which the existence of nuclear devices has distorted our perceptions and redefined the world. "Clean bombs" tells us that radiation is the only "dirty" part of killing people.

To take this one step further, such phrases can even seem healthful/curative/corrective. So that we not only have "clean bombs" but also "surgically clean strikes" ("counterforce" attacks that can purportedly "take out"—i.e., accurately destroy—an opponent's weapons or command centers without causing significant injury to anything else). The image of excision of the offending weapon is unspeakably ludicrous when the surgical tool is not a delicately controlled scalpel but a nuclear warhead. And somehow it seems to be forgotten that even scalpels spill blood.[5]

White men in ties discussing missile size

Feminists have often suggested that an important aspect of the arms race is phallic worship, that "missile envy" is a significant motivating force in the nuclear build-up.[6] I have always found this an uncomfortably reductionist explanation and hoped that my research at the Center would yield a more complex analysis. But still, I was curious about the extent to which I might find a sexual subtext in the defense professionals' discourse. I was not prepared for what I found.

I think I had naively imagined myself as a feminist spy in the house of death—that I would need to sneak around and eavesdrop on what men said in unguarded moments, using all my subtlety and cunning to unearth whatever sexual imagery might be underneath how they thought and spoke. I had naively believed that these men, at least in public, would appear to be aware of feminist critiques. If they had not changed their language, I thought that at least at some point in a long talk about "penetration aids," someone would suddenly look up, slightly embarrassed to be caught in such blatant confirmation of feminist analyses of What's Going On Here.[7]

Of course, I was wrong. There was no evidence that any feminist critiques had ever reached the ears, much less the minds, of these men. American military dependence on nuclear weapons was explained as "irresistible, because you get more bang for the buck." Another lecturer solemnly and scientifically announced "to disarm is to get rid of all your stuff." (This may, in turn, explain why they see serious talk of nuclear disarmament as perfectly resistable, not to mention foolish. If disarmament is emasculation, how could any real man even consider it?) A professor's explanation of why the MX missile is to be placed in the silos of the newest Minuteman missiles, instead of replacing the older, less accurate ones, was "because they're in the nicest hole—you're not going to take the nicest missile you have and put it in a crummy hole." Other lectures were filled with discussion of vertical erector launchers, thrust-to-weight ratios, soft lay downs, deep penetration, and the comparative advantages of protracted versus spasm attacks—or what one military adviser to the National Security Council has called "releasing 70 to 80 percent of our megatonnage in one orgasmic whump."[8] There was serious concern about the need to harden our missiles and the need to "face it, the Russians are a little harder than we are." Disbelieving glances would occasionally pass between me and my one ally in the summer program, another woman, but no one else seemed to notice.

If the imagery is transparent, its significance may be less so. The temptation is to draw some conclusions about the defense intellectuals themselves—about what they are *really* talking about, or their motivations; but the temptation is worth resisting. Individual motivations cannot necessarily be read directly from imagery; the imagery itself does not originate in these particular individuals but in a broader cultural context.

Sexual imagery has, of course, been a part of the world of warfare since long before nuclear weapons were even a gleam in a physicist's eye. The history of the atomic bomb project itself is rife with overt images of competitive male sexuality, as is the discourse of the early nuclear physicists, strategists, and SAC commanders.[9] Both the military itself and the arms manufacturers are constantly exploiting the phallic imagery and promise of sexual domination that their weapons so conveniently suggest. A quick glance at the publications that constitute some of the research sources for defense intellectuals makes the depth and pervasiveness of the imagery evident.

Air Force Magazine's advertisements for new weapons, for example, rival *Playboy* as a catalog of men's sexual anxieties and fantasies. Consider the following, from the June 1985 issue: emblazoned in bold letters across the top of a two-page advertisement for the AV-8B Harrier II—"Speak Softly and Carry a Big Stick." The copy below boasts "an exceptional thrust to weight ratio" and "vectored thrust capability that makes the . . . unique rapid response possible." Then, just in case we've failed to get the message, the last line reminds us, "Just the sort of 'Big Stick' Teddy Roosevelt had in mind way back in 1901. . . ."[10]

Given the degree to which it suffuses their world, that defense intellectuals themselves use a lot of sexual imagery does not seem especially surprising. Nor does it, by itself, constitute grounds for imputing motivation. For me, the interesting issue is not so much the imagery's psychodynamic origins, as how it functions. How does it serve to make it possible for strategic planners and other defense intellectuals to do their macabre work? How does it function in their construction of a work world that feels tenable? Several stories illustrate the complexity.

During the summer program, a group of us visited the New London Navy base where nuclear submarines are homeported and the General Dynamics Electric Boat boatyards where a new Trident submarine was being constructed. At one point during the trip we took a tour of a nuclear powered submarine. When we reached the part of the sub where the missiles are housed, the officer accompanying us turned with a grin and asked if we wanted to stick our hands through a hole to "pat the missile." *Pat the missile?*

The image reappeared the next week, when a lecturer scornfully declared that the only real reason for deploying cruise and Pershing II missiles in Western Europe was "so that our allies can pat them." Some months later, another group of us went to be briefed at NORAD (the North American Aerospace Defense Command). On the way back, our plane went to refuel at Offut Air Force Base, the Strategic Air Command headquarters near Omaha, Nebraska. When word leaked out that our landing would be delayed because the new B-1 bomber was in the area, the plane became charged with a tangible excitement that built as we flew in our holding pattern, people craning their necks to try to catch a glimpse of the B-1 in the skies, and climaxed as we touched down on the runway and hurtled past it. Later, when I returned to the Center I encountered a man who, unable to go on the trip, said to me enviously, "I hear you got to pat a B-1."

What is all this "patting?" What are men doing when they "pat" these high-tech phalluses? Patting is an assertion of intimacy, sexual possession, affectionate domination. The thrill and pleasure of "patting the missile" is the proximity of all that phallic power, the possibility of vicariously appropriating it as one's own.

But if the predilection for patting phallic objects indicates something of the homoerotic excitement suggested by the language, it also has another side. For patting is not only an act of sexual intimacy. It is also what one does to babies, small children, the pet dog. One pats that which is small, cute, and harmless—not terrifyingly destructive. Pat it, and its lethality disappears.

Much of the sexual imagery I heard was rife with the sort of ambiguity suggested by "patting the missiles." The imagery can be construed as a deadly serious display of the connections between masculine sexuality and the arms race. At the same time, it can also be heard as a way of minimizing the seriousness of militarist endeavors, of denying their deadly consequences. A former Pentagon target analyst, in telling me why he thought plans for "limited nuclear war" were ridiculous, said, "Look, you gotta understand that it's a pissing contest—you gotta expect them to use everything they've got." What does this image say? Most obviously, that this is all about competition for manhood, and thus there is tremendous danger. But at the same time, the image diminishes the contest and its outcomes, by representing it as an act of boyish mischief. . . .

Domestic bliss

Sanitized abstraction and sexual and patriarchal imagery, even if disturbing, seemed to fit easily into the masculinist world of nuclear war planning. What did not fit, what surprised and puzzled me most when I first heard it, was the set of metaphors that evoked images that can only be called domestic.

Nuclear missiles are based in "silos." On a Trident submarine, which carries twenty-four multiple warhead nuclear missiles, crew members call the part of the submarine where the missiles are lined up in their silos ready for launching "the Christmas tree farm." What could be more bucolic—farms, silos, Christmas trees?

In the ever-friendly, even romantic world of nuclear weaponry, enemies "exchange" warheads; one missile "takes out" another; weapons systems can "marry up"; "coupling" is sometimes used to refer to the wiring between mechanisms of warning and response, or to the psychopolitical links between strategic (intercontinental) and theater (European-based) weapons. The patterns in which a MIRVed missile's nuclear warheads land is known as a "footprint."[11] These nuclear explosives are not dropped; a "bus" "delivers" them. In addition, nuclear bombs are not referred to as bombs or even warheads; they are referred to as "reentry vehicles," a term far more bland and benign, which is then shortened to "RVs," a term not only totally abstract and removed from the reality of a bomb but also resonant with the image of the recreational vehicles of the ideal family vacation.

These domestic images must be more than simply one more form of distancing, one more way to remove oneself from the grisly reality behind the words; ordinary abstraction is adequate to that task. Something else, something very peculiar, is going on here. Calling the pattern in which bombs fall a "footprint" almost seems a willful distorting process, a playful, perverse refusal of accountability—because to be accountable to reality is to be unable to do this work.

These words may also serve to domesticate, to *tame* the wild and uncontrollable forces of nuclear destruction. The metaphors minimize; they are a way to make phenomena that are beyond what the mind can encompass smaller and safer, and thus they are a way of gaining mastery over the unmasterable. The fire-breathing dragon under the bed, the one who threatens to incinerate your family, your town, your planet, becomes a pet you can pat.

Using language evocative of everyday experiences also may simply serve to make the nuclear strategic community more comfortable with what they are doing. "PAL" (permissive action links) is the carefully constructed, friendly acronym for the electronic system designed to prevent the unauthorized firing of nuclear warheads. "BAMBI" was the acronym developed for an early version of an antiballistic missile system (for Ballistic Missile Boost Intercept). The president's Annual Nuclear Weapons Stockpile Memorandum, which outlines both short- and long-range plans for production of new nuclear weapons, is benignly referred to as "the shopping list." The National Command Authorities choose from a "menu of options" when deciding among different targeting plans. The "cookie cutter" is a phrase used to describe a particular model of nuclear attack. Apparently it is also used at the Department of Defense to refer to the neutron bomb.[12]

The imagery that domesticates, that humanizes insentient weapons, may also serve, paradoxically, to make it all right to ignore sentient human bodies, human lives.[13] Perhaps it is possible to spend one's time thinking about scenarios for the use of destructive technology and to have human bodies remain invisible in that technological world precisely because that world itself now *includes* the domestic, the human, the warm, and the playful—the Christmas trees, the RVs, the affectionate pats. It is a world that is in some sense complete unto itself; it even includes death and loss. But it is weapons, not humans, that get "killed." "Fratricide" occurs when one of your warheads "kills" another of your own warheads. There is much discussion of "vulnerability" and "survivability," but it is about the vulnerability and survival of weapons systems, not people.

Male birth and creation

There is one set of domestic images that demands separate attention—images that suggest men's desire to appropriate from women the power of giving life and that conflate creation and destruction. The bomb project is rife with images of male birth.[14] In December 1942, Ernest Lawrence's telegram to the physicists at Chicago read, "Congratulations to the new parents. Can hardly wait to see the new arrival."[15] At Los Alamos, the atom bomb was referred to as "Oppenheimer's baby." One of the physicists working at Los Alamos, Richard Feynman, writes that when he was temporarily on leave after his wife's death, he received a telegram saying, "The baby is expected on such and such a day."[16] At Lawrence Livermore, the hydrogen bomb was referred to as "Teller's baby," although those who wanted to disparage Edward Teller's contribution claimed he was not the bomb's father but its mother. They claimed that Stanislaw Ulam was the real father; he had the all-important idea and inseminated Teller with it. Teller only "carried it" after that.[17]

Forty years later, this idea of male birth and its accompanying belittling of maternity—the denial of women's role in the process of creation and the reduction of "motherhood" to the provision of nurturance (apparently Teller did not need to provide an egg, only a womb)—seems thoroughly incorporated into the nuclear mentality, as I learned on a subsequent visit to U.S. Space Command in Colorado Springs. One of the briefings I attended included discussion of a new satellite system, the not yet "on line" MILSTAR system. The officer doing the briefing gave an excited recitation of its technical capabilities and then an explanation of the new Unified Space Command's role in the system. Self-effacingly he said, "We'll do the motherhood role—telemetry, tracking, and control —the maintenance."

In light of the imagery of male birth, the extraordinary names given to the bombs that reduced Hiroshima and Nagasaki to ash and rubble—"Little Boy" and "Fat Man"—at last become intelligible. These ultimate destroyers were the progeny of the atomic scientists—and emphatically not just any progeny but male progeny. In early tests, before they were certain that the bombs would work, the scientists expressed their concern by saying that they hoped the baby was a boy, not a girl—that is, not a dud.[18] General Grove's triumphant cable to Secretary of War Henry Stimson at the Potsdam conference, informing him that the first atomic bomb test was successful, read, after decoding: "Doctor has just returned most enthusiastic and confident that the little boy is as husky as his big brother. The light in his eyes discernible from here to Highhold and I could have heard his screams from here to my farm."[19] Stimson, in turn, informed Churchill by writing him a note that read, "Babies satisfactorily born."[20] In 1952, Teller's exultant telegram to Los Alamos announcing the successful test of the hydrogen bomb, "Mike," at Eniwetok Atoll in the Marshall Islands, read, "It's a boy."[21] The nuclear scientists gave birth to male progeny with the ultimate power of violent domination over female Nature. The defense intellectuals' project is the creation of abstract formulations to control the forces the scientists created—and to participate thereby in their world-creating/destroying power.

The entire history of the bomb project, in fact, seems permeated with imagery that confounds man's overwhelming technological power to destroy nature with the power to create—imagery that inverts men's destruction and asserts in its place the power to create new life and a new world. It converts men's destruction into their rebirth.

William L. Laurence witnessed the Trinity test of the first atomic bomb and wrote: "The big boom came about a hundred seconds after the great flash—the first cry of a new-born world. . . . They clapped their hands as they leaped from the ground—earthbound man symbolising the birth of a new force."[22] Watching "Fat Man" being assembled the day before it was dropped on Nagasaki, he described seeing the bomb as "being fashioned into a living thing."[23] Decades later, General Bruce K. Holloway, the commander in chief of the Strategic Air Command from 1968 to 1972, described a nuclear war as involving "a big bang, like the start of the universe."[24]

God and the nuclear priesthood

The possibility that the language reveals an attempt to appropriate ultimate creative power is evident in another striking aspect of the language of nuclear weaponry and doctrine—the religious imagery. In a subculture of hard-nosed realism and hyperrationality, in a world that claims as a sign of its superiority its vigilant purging of all nonrational elements, and in which people carefully excise from their discourse every possible trace of soft sentimentality, as though purging dangerous nonsterile elements from a lab, the last thing one might expect to find is religious imagery—imagery of the forces that science has been defined in *opposition to*. For surely, given that science's identity was forged by its separation from, by its struggle for freedom from, the constraints of religion, the only thing as unscientific as the female, the subjective, the emotional, would be the religious. And yet, religious imagery permeates the nuclear past and present. The first atomic bomb test was called Trinity—the unity of the Father, the Son, and the Holy Spirit, the male forces of Creation. The imagery is echoed in the language of the physicists who worked on the bomb and witnessed the test: "It was as though we stood at the first day of creation." Robert Oppenheimer thought of Krishna's words to Arjuna in the *Bhagavad Gita:* "I am become Death, the Shatterer of Worlds."[25]

Perhaps most astonishing of all is the fact that the creators of strategic doctrine actually refer to members of their community as "the nuclear priesthood." It is hard to decide what is most extraordinary about this: the easy arrogance of their claim to the virtues and supernatural power of the priesthood; the tacit admission (*never* spoken directly) that rather than being unflinching, hard-nosed, objective, empirically minded scientific describers of reality, they are really the creators of dogma; or the extraordinary implicit statement about who, or rather what, has become god. If this new priesthood attains its status through an inspired knowledge of nuclear weapons, it gives a whole new meaning to the phrase "a mighty fortress is our God."

STAGE 2: LEARNING TO SPEAK THE LANGUAGE

Although I was startled by the combination of dry abstraction and counterintuitive imagery that characterizes the language of defense intellectuals, my attention and energy were quickly focused on decoding and learning to speak it. The first task was training the tongue in the articulation of acronyms.

Several years of reading the literature of nuclear weaponry and strategy had not prepared me for the degree to which acronyms littered all conversations, nor for the way in which they are used. Formerly, I had thought of them mainly as utilitarian. They allow you to write or speak faster. They act as a form of abstraction, removing you from the reality behind the words. They restrict communication to the initiated, leaving all others both uncomprehending and voiceless in the debate.

But, being at the Center, hearing the defense analysts use the acronyms, and then watching as I and others in the group started to fling acronyms around in our conversation revealed some additional, unexpected dimensions.

First, in speaking and hearing, a lot of these terms can be very sexy. A small supersonic rocket "designed to penetrate any Soviet air defense" is called a SRAM (for short-range attack missile). Submarine-launched cruise missiles are not referred to as SLCMs, but "slick'ems." Ground-launched cruise missiles are "glick'ems." Air-launched cruise missiles are not sexy but magical—"alchems" (ALCMs) replete with the illusion of turning base metals into gold.

TACAMO, the acronym used to refer to the planes designed to provide communications links to submarines, stands for "take charge and move out." The image seems closely related to the nicknames given to the new guidance systems for "smart weapons"—"shoot and scoot" or "fire and forget."

Other acronyms work in other ways. The plane in which the president supposedly will be flying around above a nuclear holocaust, receiving intelligence and issuing commands for the next bombing, is referred to as "kneecap" (for NEACP—National Emergency Airborne Command Post). The edge of derision suggested in referring to it as "kneecap" mirrors the edge of derision implied when it is talked about at all, since few believe that the president really would have the time to get into it, or that the communications systems would be working if he were in it, and some might go so far as to question the usefulness of his being able to direct an extended nuclear war from his kneecap even if it were feasible. (I never heard the morality of this idea addressed.) But it seems to me that speaking about it with that edge of derision is *exactly* what allows it to be spoken about and seriously discussed at all. It is the very ability to make fun of a concept that makes it possible to work with it rather than reject it outright.

In other words, what I learned at the program is that talking about nuclear weapons is fun. I am serious. The words are fun to say; they are racy, sexy, snappy. You can throw them around in rapid-fire succession. They are quick, clean, light; they trip off the tongue. You can reel off dozens of them in seconds, forgetting about how one might just interfere with the next, not to mention with the lives beneath them.

I am not describing a phenomenon experienced only by the perverse, although the phenomenon itself may be perverse indeed. Nearly everyone I observed clearly took pleasure in using the words. It mattered little whether we were lecturers or students, hawks or doves, men or women—we all learned it, and we all spoke it. Some of us may have spoken with a self-consciously ironic edge, but the pleasure was there nonetheless.

Part of the appeal was the thrill of being able to manipulate an arcane language, the power of entering the secret kingdom, being someone in the know. It is a glow that is a significant part of learning about nuclear weaponry. Few know, and those who do are powerful. You can rub elbows with them, perhaps even be one yourself.

That feeling, of course, does not come solely from the language. The whole setup of the summer program itself, for example, communicated the allures of power and the benefits of white male privileges. We were provided with luxurious accommodations, complete with young black women who came in to clean up after us each day; generous funding paid not only our transportation and food but also a large honorarium for attending; we met in lavishly appointed classrooms and lounges. Access to excellent athletic facilities was guaranteed by a "Temporary Privilege Card," which seemed to me to sum up the essence of the experience. Perhaps most important of all were the endless allusions by our lecturers to "what I told John [Kennedy]" and "and then Henry [Kissinger] said," or the lunches where we could sit next to a prominent political figure and listen to Washington gossip.

A more subtle, but perhaps more important, element of learning the language is that when you speak it, you feel in control. The experience of mastering the words infuses your relation to the material. You can get so good at manipulating the words that it almost feels as though the whole thing is under control. Learning the language gives a sense of what I would call cognitive mastery, the feeling of mastery of technology that is finally *not* controllable but is instead powerful beyond human comprehension, powerful in a way that stretches and even thrills the imagination.

The more conversations I participated in using this language, the less frightened I was of nuclear war. How can learning to speak a language have such a powerful effect? One answer, I believe, is that the *process* of learning the language is itself a part of what removes you from the reality of nuclear war.

I entered a world where people spoke what amounted to a foreign language, a language I had to learn if we were to communicate with one another. So I became engaged in the challenge of it—of decoding the acronyms and figuring out which were the proper verbs to use. My focus was on the task of solving the puzzles, developing language competency—not on the weapons and wars behind the words. Although my interest was in thinking about nuclear war and its prevention, my energy was elsewhere.

By the time I was through, I had learned far more than a set of abstract words that refers to grisly subjects, for even when the subjects of a standard English and nuke-speak description seem to be the same, they are, in fact, about utterly different phenomena. Consider the following descriptions, in each of which the subject is the aftermath of a nuclear attack:

> Everything was black, had vanished into the black dust, was destroyed. Only the flames that were beginning to lick their way up had any color. From the dust that was like a fog, figures began to loom up, black, hairless, faceless. They screamed with voices that were no longer human. Their screams drowned out the groans rising everywhere from the rubble, groans that seemed to rise from the very earth itself.[26]
>
> [You have to have ways to maintain communications in a] nuclear environment, a situation bound to include EMP blackout, brute force damage to systems, a heavy jamming environment, and so on.[27]

There are no ways to describe the phenomena represented in the first with the language of the second. Learning to speak the language of defense analysts is not a conscious, cold-blooded decision to ignore the effects of nuclear weapons on real live human beings, to ignore the sensory, the emotional experience, the human impact. It is simply learning a new language, but by the time you are through, the content of what you can talk about is monumentally different, as is the perspective from which you speak.

In the example above, the differences in the two descriptions of a "nuclear environment" stem partly from a difference in the vividness of the words themselves—the words of the first intensely immediate and evocative, the words of the second abstract and distancing. The passages also differ in their content: the first describes the effects of a nuclear blast on human beings; the second describes the impact of a nuclear blast on technical systems designed to assure the "command and control" of nuclear weapons. Both of these differences may stem from the difference of perspective: the speaker in the first is a victim of nuclear weapons; the speaker in the second is a user. The speaker in the first is using words to try to name and contain the horror of human suffering all around her; the speaker in the second is using words to ensure the possibility of launching the next nuclear attack. Technostrategic language can be used only to articulate the perspective of the users of nuclear weapons, not that of the victims.

Thus, speaking the expert language not only offers distance, a feeling of control, and an alternative focus for one's energies; it also offers escape—escape from thinking of oneself as a victim of nuclear war. I do not mean this on the level of individual consciousness; it is not that defense analysts somehow convince themselves that they would not be among the victims of nuclear war, should it occur. But I do mean it in terms of the structural position the speakers of the language occupy and the perspective they get from that position. *Structurally*, speaking technostrategic language removes them from the position of victim and puts them in the position of the planner, the user, the actor. From that position, there is neither need nor way to see oneself as a victim; no matter what one deeply knows or believes about the likelihood of nuclear war, and no matter what sort of terror or despair the knowledge of nuclear war's reality might inspire, the speakers of technostrategic language are positionally allowed, even forced, to escape that awareness, to escape viewing nuclear war from the position of the victim, by virtue of their linguistic stance as users, rather than victims, of nuclear weaponry.

Finally, then, I suspect that much of the reduced anxiety about nuclear war commonly experienced by both new speakers of the language and long-time experts comes from characteristics of the language itself: the distance afforded by its abstraction, the sense of control afforded by mastering it, and the fact that its content and concerns are that of the users rather than the victims of nuclear weapons. In learning the language, one goes from being the passive, powerless victim to the competent, wily, powerful purveyor of nuclear threats and nuclear explosive power. The enormous destructive effects of nuclear weapons systems become extensions of the self, rather than threats to it.

STAGE 3: DIALOGUE

It did not take very long to learn the language of nuclear war and much of the specialized information it contained. My focus quickly changed from mastering technical information and doctrinal arcana to attempting to understand more about how the dogma was rationalized. . . .

Since underlying rationales are rarely discussed in the everyday business of defense planning, I had to start asking more questions. At first, although I was tempted to use my newly acquired proficiency in technostrategic jargon, I vowed to speak English. I had long believed that one of the most important functions of an expert language is exclusion—the denial of a voice to those outside the professional community. I wanted to see whether a well-informed person could speak English and still carry on a knowledgeable conversation.

What I found was that no matter how well-informed or complex my questions were, if I spoke English rather than expert jargon, the men responded to me as though I were ignorant, simpleminded, or both. It did not appear to occur to anyone that I might actually be choosing not to speak their language.

A strong distaste for being patronized and dismissed made my experiment in English short-lived. I adapted my everyday speech to the vocabulary of strategic analysis. I spoke of "escalation dominance," "preemptive strikes," and, one of my favorites, "subholocaust engagements." Using the right phrases opened my way into long, elaborate discussions that taught me a lot about technostrategic reasoning and how to manipulate it.

I found, however, that the better I got at engaging in this discourse, the more impossible it became for me to express my own ideas, my own values. I could adopt the language and gain a wealth of new concepts and reasoning strategies—but at the same time as the language gave me access to things I had been unable to speak about before, it radically excluded others. I could not use the language to express my concerns because it was physically impossible. This language does not allow certain questions to be asked or certain values to be expressed.

To pick a bald example: the word "peace" is not a part of this discourse. As close as one can come is "strategic stability," a term that refers to a balance of numbers and types of weapons systems—not the political, social, economic, and psychological conditions implied by the word "peace." Not only is there no word signifying peace in this discourse, but the word "peace" itself cannot be used. To speak it is immediately to brand oneself as a soft-headed activist instead of an expert, a professional to be taken seriously.

If I was unable to speak my concerns in this language, more disturbing still was that I found it hard even to keep them in my own head. I had begun my research expecting abstract and sanitized discussions of nuclear war and had readied myself to replace my words for theirs, to be ever vigilant against slipping into the never-never land of abstraction. But no matter how prepared I was, no matter how firm my commitment to staying aware of the reality behind the words, over and over I found that I could not stay connected, could not keep human lives as my reference point. I found I could go for days speaking about nuclear weapons without once thinking about the people who would be incinerated by them.

It is tempting to attribute this problem to qualities of the language, the words themselves —the abstractness, the euphemisms, the sanitized, friendly, sexy acronyms. Then all we would need to do is change the words, make them more vivid; get the military planners to say "mass murder" instead of "collateral damage" and their thinking would change.

The problem, however, is not only that defense intellectuals use abstract terminology that removes them from the realities of which they speak. There *is* no reality of which they speak. Or, rather, the "reality" of which they speak is itself a world of abstractions. Deterrence theory, and much of strategic doctrine altogether, was invented largely by mathematicians, economists, and a few political scientists. It was invented to hold together abstractly, its validity judged by its internal logic. Questions of the correspondence to observable reality were not the issue. These abstract systems were developed as a way to make it possible to "think about the unthinkable"—not as a way to describe or codify relations on the ground.[28]

So the greatest problem with the idea of "limited nuclear war," for example, is not that it is grotesque to refer to the death and suffering caused by *any* use of nuclear weapons as "limited" or that "limited nuclear war" is an abstraction that is disconnected from human reality but, rather, that "limited nuclear war" is itself an abstract conceptual system, designed, embodied, achieved by computer modeling. It is an abstract world in which hypothetical, calm, rational actors have sufficient information to know exactly what size nuclear weapon the opponent has used against which targets, and in which they have adequate command and control to make sure that their response is precisely equilibrated to the attack. In this scenario, no field commander would use the tactical "mini-nukes" at his disposal in the height of a losing battle; no EMP-generated electronic failures, or direct attacks on command, and control centers, or human errors would destroy communications networks. Our rational actors would be free of emotional response to being attacked, free of political pressures from the populace, free from madness or despair or any of the myriad other factors that regularly affect human actions and decision making. They would act solely on the basis of a perfectly informed mathematical calculus of megatonnage.

So to refer to "limited nuclear war" is already to enter into a system that is de facto abstract and removed from reality. To use more descriptive language would not, by itself, change that. In fact, I am tempted to say that the abstractness of the entire conceptual system makes descriptive language nearly beside the point. In a discussion of "limited nuclear war," for example, it might make some difference if in place of saying "In a counter-force attack against hard targets collateral damage could be limited," a strategic analyst had to use words that were less abstract—if he had to say, for instance, "If we launch the missiles we have aimed at their missile silos, the explosions would cause the immediate mass murder of 10 million women, men, and children, as well as the extended illness, suffering, and eventual death of many millions more." It is true that the second sentence does not roll off the tongue or slide across one's consciousness quite as easily. But it is also true, I believe, that the ability to speak about "limited nuclear war" stems as much, if not more, from the fact that the term "limited nuclear war" refers to an abstract conceptual system rather than to events that might take place in the real world. As such, there is no need to think about the concrete human realities behind the model; what counts is the internal logic of the system.

This realization that the abstraction was not just in the words but also characterized the entire conceptual system itself helped me make sense of my difficulty in staying connected to human lives. But there was still a piece missing. How is it possible, for example, to make sense of the following paragraph? It is taken from a discussion of a scenario ("regime A") in which the United States and the USSR have revised their offensive weaponry, banned MIRVs, and gone to a regime of single warhead (Midgetman) missiles, with no "defensive shield" (or what is familiarly known as "Star Wars" or SDI):

> The strategic stability of regime A is based on the fact that both sides are deprived of any incentive ever to strike first. Since it takes roughly two warheads to destroy one enemy silo, an attacker must expend two of his missiles to destroy one of the enemy's. A first strike disarms the attacker. The aggressor ends up worse off than the aggressed.[29]

"The aggressor ends up worse off than the aggressed"? The homeland of "the aggressed" has just been devastated by the explosions of, say, a thousand nuclear bombs, each likely to be ten to one hundred times more powerful than the bomb dropped on Hiroshima, and the aggressor, whose homeland is still untouched, "ends up worse off"? How is it possible to think this? Even abstract language and abstract thinking do not seem to be a sufficient explanation.

I was only able to "make sense of it" when I finally asked myself the question that feminists have been asking about theories in every discipline: What is the reference point? Who (or what) is the *subject* here?

In other disciplines, we have frequently found that the reference point for theories about "universal human phenomena" has actually been white men. In technostrategic discourse, the reference point is not white men, it is not human beings at all; it is the weapons themselves. The aggressor thus ends up worse off than the aggressed because he has fewer weapons left; human factors are irrelevant to the calculus of gain and loss. . . .

The fact that the subjects of strategic paradigms are weapons has several important implications. First, and perhaps most critically, there simply is no way to talk about human death or human societies when you are using a language designed to talk about weapons. Human death simply *is* "collateral damage"—collateral to the real subject, which is the weapons themselves.

Second, if human lives are not the reference point, then it is not only impossible to talk about humans in this language; it also becomes in some sense illegitimate to ask the paradigm to reflect human concerns. Hence, questions that break through the numbing language of strategic analysis and raise issues in human terms can be dismissed easily. No one will claim that the questions are unimportant, but they are inexpert, unprofessional, irrelevant to the business at hand to ask. The discourse among the experts remains hermetically sealed.

The problem, then, is not only that the language is narrow but also that it is seen by its speakers as complete or whole unto itself—as representing a body of truths that exist independently of any other truth or knowledge. The isolation of this technical knowledge from social or psychological or moral thought, or feelings, is all seen as legitimate and necessary. The outcome is that defense intellectuals can talk about the weapons that are supposed to protect particular political entities, particular peoples and their way of life, without actually asking if weapons *can* do it, or if they are the best *way* to do it, or whether they may even damage the entities you are supposedly protecting. It is not that the men I spoke with would say that these are invalid questions. They would, however, simply say that they are separate questions, questions that are outside what they do, outside their realm of expertise. So their deliberations go on quite independently, as though with a life of their own, disconnected from the functions and values they are supposedly to serve.

Finally, the third problem is that this discourse has become virtually the only legitimate form of response to the question of how to achieve security. If the language of weaponry was one competing voice in the discussion, or one that was integrated with others, the fact that the referents of strategic paradigms are only weapons would be of little note. But when we realize that the only language and expertise offered to those interested in pursuing peace refers to nothing but weapons, its limits become staggering, and its entrapping qualities—the way in which, once you adopt it, it becomes so hard to stay connected to human concerns—become more comprehensible.

STAGE 4: THE TERROR

As a newcomer to the world of defense analysts, I was continually startled by likeable and admirable men, by their gallows humor, by the bloodcurdling casualness with which they regularly blew up the world while standing and chatting over the coffeepot. I also *heard* the language they spoke—heard the acronyms and euphemisms and abstractions, heard the imagery, heard the pleasure with which they used it.

Within a few weeks, what had once been remarkable became unnoticeable. As I learned to speak, my perspective changed. I no longer stood outside the impermeable wall of technostrategic language and, once inside, I could no longer see it. Speaking the language, I could no longer really hear it. And once inside its protective walls, I began to find it difficult to get out. The impermeability worked both ways.

I had not only learned to speak a language; I had started to think in it. Its questions became my questions, its concepts shaped my responses to new ideas. Its definitions of the parameters of reality became mine. Like the White Queen, I began to believe six impossible things before breakfast. Not because I consciously believed, for instance, that a "surgically clean counterforce strike" was really possible, but instead because some elaborate piece of doctrinal reasoning I used was already predicated on the possibility of those strikes, as well as on a host of other impossible things.

My grasp on what *I* knew as reality seemed to slip. I might get very excited, for example, about a new strategic justification for a "no first use" policy and spend time discussing the ways in which its implications for our force structure in Western Europe were superior to the older version. And after a day or two I would suddenly step back, aghast that I was so involved with the military justifications for not using nuclear weapons—as though the moral ones were not enough. What I was actually talking about—the mass incineration caused by a nuclear attack—was no longer in my head.

Or I might hear some proposals that seemed to me infinitely superior to the usual arms control fare. First I would work out how and why these proposals were better and then work out all the ways to counter the arguments against them. But then, it might dawn on me that even though these two proposals sounded so different, they still shared a host of assumptions that I was not willing to make (e.g., about the inevitable, eternal conflict of interests between the United States and the USSR, or the desirability of having some form of nuclear deterrent, or the goal of "managing," rather than ending, the nuclear arms race). After struggling to this point of seeing what united both positions, I would first feel as though I had really accomplished something. And then all of a sudden, I would realize that these new insights were things I actually knew *before I ever entered* this community. Apparently, I had since forgotten them, at least functionally, if not absolutely.

I began to feel that I had fallen down the rabbit hole—and it was a struggle to climb back out.

CONCLUSIONS

Suffice it to say that the issues about language do not disappear after you have mastered technostrategic discourse. The seductions remain great. You can find all sorts of ways to seemingly beat the boys at their own game; you can show how even within their own definitions of rationality, most of what is happening in the development and deployment of nuclear forces is wildly irrational. You can also impress your friends and colleagues with sickly humorous stories about the way things really happen on the inside. There is tremendous pleasure in it, especially for those of us who have been closed out, who have been told that it is really all beyond us and we should just leave it to the benevolently paternal men in charge.

But as the pleasures deepen, so do the dangers. The activity of trying to out-reason defense intellectuals in their own games gets you thinking inside their rules, tacitly accepting all the unspoken assumptions of their paradigms. You become subject to the tyranny of concepts. The language shapes your categories of thought (e.g., here it becomes "good nukes" or "bad nukes," not nukes or no nukes) and defines the boundaries of imagination (as you try to imagine a "minimally destabilizing basing mode" rather than a way to prevent the weapon from being deployed at all). . . .

Other recent entrants into this world have commented to me that, while it is the cold-blooded, abstract discussions that are most striking at first, within a short time "you get past it—you stop hearing it, it stops bothering you, it becomes normal—and you come to see that the language, itself, is not the problem."

However, I think it would be a mistake to dismiss these early impressions. They can help us learn something about the militarization of the mind, and they have, I believe, important implications for feminist scholars and activists who seek to create a more just and peaceful world.

Mechanisms of the mind's militarization are revealed through both listening to the language and learning to speak it. *Listening*, it becomes clear that participation in the world of nuclear strategic analysis does not necessarily require confrontation with the central fact about military activity—that the purpose of all weaponry and all strategy is to injure human bodies.[30] In fact, as Elaine Scarry points out, participation in military thinking does not require confrontation with, and actually demands the elision of, this reality.[31]

Listening to the discourse of nuclear experts reveals a series of culturally grounded and culturally acceptable mechanisms that serve this purpose and that make it possible to "think about the unthinkable," to work in institutions that foster the proliferation of nuclear weapons, to plan mass incinerations of millions of human beings for a living. Language that is abstract, sanitized, full of euphemisms; language that is sexy and fun to use; paradigms whose referent is weapons; imagery that domesticates and deflates the forces of mass destruction; imagery that reverses sentient and nonsentient matter, that conflates birth and death, destruction and creation—all of these are part of what makes it possible to be radically removed from the reality of what one is talking about and from the realities one is creating through the discourse.

Learning to speak the language reveals something about how thinking can become more abstract, more focused on parts disembedded from their context, more attentive to the survival of weapons than the survival of human beings. That is, it reveals something about the process of militarization—and the way in which that process may be undergone by man or woman, hawk or dove.

Most often, the act of learning technostrategic language is conceived of as an additive process: you add a new set of vocabulary words; you add the reflex ability to decode and use endless numbers of acronyms; you add some new information that the specialized language contains; you add the conceptual tools that will allow you to "think strategically." This additive view appears to be held by defense intellectuals themselves; as one said to me, "Much of the debate is in technical terms—learn it, and decide whether it's relevant later." This view also appears to be held by many who think of themselves as antinuclear, be they scholars and professionals attempting to change the field from within, or public interest lobbyists and educational organizations, or some feminist antimilitarists. Some believe that our nuclear policies are so riddled with irrationality that there is a lot of room for well-reasoned, well-informed arguments to make a difference; others, even if they do not believe that the technical information is very important, see it as necessary to master the language simply because it is too difficult to attain public legitimacy without it. In either case, the idea is that you add the expert language and information and proceed from there.

However, I have been arguing throughout this article that learning the language is a transformative, rather than an additive, process. When you choose to learn it you enter a new mode of thinking—a mode of thinking not only about nuclear weapons but also, de facto, about military and political power and about the relationship between human ends and technological means.

Thus, those of us who find U.S. nuclear policy desperately misguided appear to face a serious quandary. If we refuse to learn the language, we are virtually guaranteed that our voices will remain outside the "politically relevant" spectrum of opinion. Yet, if we do learn and speak it, we not only severely limit what we can say, but we also invite the transformation, the militarization, of our own thinking.

I have no solutions to this dilemma, but I would like to offer a few thoughts in an effort to reformulate its terms. First, it is important to recognize an assumption implicit in adopting the strategy of learning the language. When we assume that learning and speaking the language will give us a voice recognized as legitimate and will give us greater political influence, *we are assuming that the language itself actually articulates the criteria and reasoning strategies upon which nuclear weapons development and deployment decisions are made.* I believe that this is largely an illusion. Instead, I want to suggest that technostrategic discourse functions more as a gloss, as an ideological curtain behind which the actual reasons for these decisions hide. That rather than informing and shaping decisions, it far more often functions as a legitimation for political outcomes that have occurred for utterly different reasons. If this is true, it raises some serious questions about the extent of the political returns we might get from using technostrategic discourse, and whether they can ever balance out the potential problems and inherent costs.

I do not, however, want to suggest that none of us should learn the language. I do not believe that this language is well suited to achieving the goals desired by antimilitarists, yet at the same time, I, for one, have found the experience of learning the language useful and worthwhile (even if at times traumatic). The question for those of us who do choose to learn it, I think, is what use are we going to make of that knowledge?

One of the most intriguing options opened by learning the language is that it suggests a basis upon which to challenge the legitimacy of the defense intellectuals' dominance of the discourse on nuclear issues. When defense intellectuals are criticized for the cold-blooded inhumanity of the scenarios they plan, their response is to claim the high ground of rationality; they are the only ones whose response to the existence of nuclear weapons is objective and realistic. They portray those who are radically opposed to the nuclear status quo as irrational, unrealistic, too emotional. "Idealistic activists" is the pejorative they set against their own hard-nosed professionalism.

Much of their claim to legitimacy, then, is a claim to objectivity born of technical expertise and to the disciplined purging of the emotional valences that might threaten their objectivity. But if the surface of their discourse—its abstraction and technical jargon—appears at first to support these claims, a look just below the surface does not. There we find currents of homoerotic excitement, heterosexual domination, the drive toward competency and mastery, the pleasures of membership in an elite and privileged group, the ultimate importance and meaning of membership in the priesthood, and the thrilling power of becoming Death, shatterer of worlds. How is it possible to hold this up as a paragon of cool-headed objectivity?

I do not wish here to discuss or judge the holding of "objectivity" as an epistemological goal. I would simply point out that, as defense intellectuals rest their claims to legitimacy on the untainted rationality of their discourse, their project fails according to its own criteria. Deconstructing strategic discourse's claims to rationality is, then, in and of itself, an important way to challenge its hegemony as the sole legitimate language for public debate about nuclear policy.

I believe that feminists, and others who seek a more just and peaceful world, have a dual task before us—a deconstructive project and a reconstructive project that are intimately linked.[32] Our deconstructive task requires close attention to, and the dismantling of, technostrategic discourse. The dominant voice of militarized masculinity and decontextualized rationality speaks so loudly in our culture, it will remain difficult for any other voices to be heard until that voice loses some of its power to define what we hear and how we name the world—until that voice is delegitimated.

Our reconstructive task is a task of creating compelling alternative visions of possible futures, a task of recognizing and developing alternative conceptions of rationality, a task of creating rich and imaginative alternative voices—diverse voices whose conversations with each other will invent those futures.

Notes

1 Thomas Powers, "How Nuclear War Could Start," *New York Review of Books* (January 17, 1985), 33.
2 George Kennan, "A Modest Proposal," *New York Review of Books* (July 16, 1981), 14.
3 Fusion weapons' proportionally smaller yield of radioactive fallout led Atomic Energy Commission Chairman Lewis Strauss to announce in 1956 that hydrogen bomb tests were important "not only from a military point of view but from a humanitarian aspect." Although the bombs being tested were 1,000 times more powerful than those that devastated Hiroshima and Nagasaki, the proportional reduction of fallout apparently qualified them as not only clean but also humanitarian. Lewis Strauss is quoted in Ralph Lapp, "The 'Humanitarian' H-Bomb," *Bulletin of Atomic Scientists* 12, no. 7 (September 1956):263.
4 "Kiloton" (or kt) is a measure of explosive power, measured by the number of thousands of tons of TNT required to release an equivalent amount of energy. The atomic bomb dropped on Hiroshima is estimated to have been approximately 12 kt. An MX missile is designed to carry up to ten Mk 21 reentry vehicles, each with a W-87 warhead. The yield of W-87 warheads is 300 kt, but they are "upgradable" to 475 kt.
5 Conservative government assessments of the number of deaths resulting from a "surgically clean" counterforce attack vary widely. The Office of Technology Assessment projects 2 million to 20 million immediate deaths. (See James Fallows, *National Defense* [New York: Random House, 1981], 159.) A 1975 Defense Department study estimated 18.3 million fatalities, while the U.S. Arms Control and Disarmament Agency, using different assumptions, arrived at a figure of 50 million (cited by Desmond Ball, "Can Nuclear War Be Controlled?" Adelphi Paper no. 169 [London: International Institute for Strategic Studies, 1981]).
6 The phrase is Helen Caldicott's in *Missile Envy: The Arms Race and Nuclear War* (Toronto: Bantam Books, 1986).
7 For the uninitiated, "penetration aids" refers to devices that help bombers or missiles get past the "enemy's" defensive systems, e.g., stealth technology, chaff, or decoys. Within the defense intellectual community, they are also familiarly known as "penaids."
8 General William Odom, "C³I and Telecommunications at the Policy Level," Incidental Paper, Seminar on C³I: Command, Control, Communications and Intelligence (Cambridge, Mass.: Harvard University, Center for Information Policy Research, Spring 1980), 5.
9 This point has been amply documented by Brian Easlea, *Fathering the Unthinkable: Masculinity, Scientists and the Nuclear Arms Race* (London: Pluto Press, 1983).
10 *Air Force Magazine* 68, no. 6 (June 1985):77–78

11 MIRV stands for "multiple independently targetable re-entry vehicles." A MIRVed missile not only carries more than one warhead; its warheads can be aimed at different targets.

12 Henry T. Nash, "The Bureaucratization of Homicide," *Bulletin of Atomic Scientists* (April 1980), reprinted in E. P. Thompson and Dan Smith, eds., *Protest and Survive* (New York: Monthly Review Press, 1981), 159.

13 For a discussion of the functions of imagery that reverses sentient and insentient matter, that "exchange[s] . . . idioms between weapons and bodies," see Elaine Scarry, *The Body in Pain: The Making and Unmaking of the World* (New York: Oxford University Press, 1985), 60–157, esp. 67.

14 For further discussion of men's desire to appropriate from women the power of giving life and death, and its implications for men's war-making activities, see Dorothy Dinnerstein, *The Mermaid and the Minotaur* (New York: Harper & Row, 1977). For further analysis of male birth imagery in the atomic bomb project, see Evelyn Fox Keller, "From Secrets of Life to Secrets of Death" (paper delivered at the Kansas Seminar, Yale University, New Haven, Conn., November 1986), and Easlea (n. 9 above), 81–116.

15 Lawrence is quoted by Herbert Childs in *An American Genius: The Life of Ernest Orlando Lawrence* (New York: E. P. Dutton, 1968), 340.

16 Feynman writes about the telegram in Richard P. Feynman, "Los Alamos from Below," in *Reminiscences of Los Alamos, 1943–1945*, ed. Lawrence Badash, Joseph O. Hirshfelder, and Herbert P. Broida (Dordrecht: D. Reidel Publishing Co., 1980), 130.

17 Hans Bethe is quoted as saying that "Ulam was the father of the hydrogen bomb and Edward was the mother, because he carried the baby for quite a while" (J. Bernstein, *Hans Bethe: Prophet of Energy* [New York: Basic Books, 1980], 95).

18 The concern about having a boy, not a girl, is written about by Robert Jungk, *Brighter Than a Thousand Suns*, trans. James Cleugh (New York: Harcourt, Brace & Co., 1956), 197.

19 Richard E. Hewlett and Oscar E. Anderson, *The New World, 1939/46: A History of the United States Atomic Energy Commission*, 2 vols. (University Park: Pennsylvania State University Press, 1962), 1:386.

20 Winston Churchill, *The Second World War*, vol. 6., *Triumph and Tragedy* (London: Cassell, 1954), 551.

21 Quoted by Easlea, 130.

22 William L. Laurence, *Dawn over Zero: The Study of the Atomic Bomb* (London: Museum Press, 1974), 10.

23 Ibid., 188.

24 From a 1985 interview in which Holloway was explaining the logic of a "decapitating" strike against the Soviet leadership and command and control systems—and thus how nuclear war would be different from World War II, which was a "war of attrition," in which transportation, supply depots, and other targets were hit, rather than being a "big bang" (Daniel Ford, "The Button," *New Yorker Magazine* 61, no. 7 [April 8, 1985], 49).

25 Jungk, 201.

26 Hisako Matsubara, *Cranes at Dusk* (Garden City, N.Y.: Dial Press, 1985). The author was a child in Kyoto at the time the atomic bomb was dropped. Her description is based on the memories of survivors.

27 General Robert Rosenberg (formerly on the National Security Council staff during the Carter Administration), "The Influence of Policymaking on C^3I," Incidental Paper, Seminar on C^3I (Cambridge, Mass.: Harvard University, Center for Information Policy Research, Spring 1980), 59.

28 For fascinating, detailed accounts of the development of strategic doctrine, see Fred Kaplan, *The Wizards of Armageddon* (New York: Simon & Schuster, 1983), and Gregg E. Herken, *The Counsels of War* (New York: Alfred A. Knopf, 1985).

29 Charles Krauthammer, "Will Star Wars Kill Arms Control?" *New Republic*, no. 3,653 (January 21, 1985), 12–16.

30 For an eloquent and graphic exploration of this point, see Scarry (n. 13 above), 73.

31 Scarry catalogs a variety of mechanisms that serve this purpose (ibid., 60–157). The point is further developed by Sara Ruddick, "The Rationality of Care," in *Thinking about Women, War, and the Military*, ed. Jean Bethke Elshtain and Sheila Tobias (Totowa, N.J.: Rowman & Allanheld, in press).

32 Sandra Harding and Merrill Hintikka, eds., *Discovering Reality: Feminist Perspectives on Epistemology, Metaphysics, Methodology and the Philosophy of Science* (Dordrecht: D. Reidel Publishing Co., 1983), ix–xix, esp. x.

10

Science, Sex, and Stereotypical Images in Scientific Advertising

Mary Barbercheck

Our culture is filled with messages that tell us not only how we should behave, but also who we should be. We receive these messages throughout our lives, and so their meaning seems natural. One of the most pervasive messages is that of the profound differences between men and women (Valian 1998). Most work in our culture is segregated by sex, and most professional fields in the sciences and engineering are traditionally male-dominated; it is only recently that women have entered the scientific workforce in appreciable numbers (Rossiter 1982, 1995). Although discrimination on the basis of sex has been outlawed in the United States since the 1970s, impediments for women scientists continue to exist, but in increasingly subtle patterns (Sonnert and Holton 1995; Valian 1998). Only a small fraction of the young men and women who receive their first academic degree in science, mathematics, or engineering will become research scientists. Research that tracks the flow of girls and women through the educational pipeline on their way to scientific careers reveals that women are awarded more than half of all undergraduate degrees. However, as aspiring scientists move through the training system, more women than men are diverted from the science career pipeline. Women currently account for about 38 percent of the graduate science and engineering degrees awarded in the United States, but only about 22 percent of employed scientists and engineers are women (National Science Foundation 1999).

There are two dominant models that explain why women are less likely to stay in science fields and, if awarded advanced degrees, are less likely than men to be successful in scientific careers (Sonnert and Holton 1995). The deficit model attributes the lower number of women scientists to their differential treatment. This model emphasizes structural barriers—legal, political, and social—that exist or existed in the social system of science. An assumption of the deficit model is that the goals of women and men are similar, but barriers to advancement keep women from accomplishing these goals at rates similar to those of men. Formal barriers (laws) have been removed, but subtle informal barriers remain: less access to resources such as space, equipment, and money; isolation from mentors, power brokers, key researchers and administrators, and "old-boy" networks (A Study on the Status of Women Faculty in Science at MIT, http://web.mit.edu/faculty/reports/, 1999; Wenneras and Wold 1997). In the deficit model it is assumed that women do not learn the "hidden" skills essential for a successful science career. Formal and informal structural barriers can directly affect the careers of women scientists but may also discourage women from choosing a science career. If women perceive that structural barriers in the social system of science will impede the potential rewards of a scientific career, they may choose another career in which the potential costs do not appear to outweigh the potential benefits. Interventions based on the deficit model include legal and policy changes, outreach programs to introduce young people to science and scientists, and mentoring programs.

The second model, the difference model, suggests that there are fewer women than men in science because women act differently (Sonnert and Holton 1995). The difference model assumes that there are fundamental differences in the outlook and goals of men and women. The obstacles of career achievement lie within women: they are innate or the result of gender-role socialization and cultural values. Particular attitudes about science may define it as a male field and therefore encourage male participation and discourage female participation. Even as rigid gender-role socialization declines, social practices still reinforce the image of the aggressive, successful man and the nurturing, supportive woman (Valian 1998). Whereas girls are socialized to interact in a style that deemphasizes aggressiveness and competitiveness, those characteristics are encouraged in boys and become embedded in a male interaction style. Culturally sanctioned typical female traits, in the current social system of science, are likely to put women at a disadvantage. These socialization patterns tend to distance women from the very characteristics that the social system of science rewards and reinforces: ambition, self-confidence, resilience, aggressiveness, and competitiveness. Interventions based on the difference model would seek to change the social norms of science to include qualities perceived as feminine. Instead of training women scientists to act more like men scientists, perhaps a social and epistemological reform of science is needed that will accommodate a wider range of behaviors and communication styles (Malcom 1999).

These two models take an either/or approach in that either obstacles are imposed on individuals by a system that constrains their choices or individuals place constraints on themselves that aggregate to social patterns. In this sense, then, neither model adequately takes into account the force of stereotypes, gender schemas, and images that surround us but do not seem to have the direct or concrete power to shape an individual's choices.

It is almost impossible not to know the difference between norms for masculinity and femininity in our society. Cultural beliefs about the masculine nature of scientists and science abound and may distance women from the field. The majority of Americans describe a scientist as a man who wears a white coat, works in a laboratory, and has facial hair and an "extravagant" hairdo (Mirsky 1997). In films, scientists are often portrayed as mad or overreaching, occasionally as heroic, although Ribalow (1999) has identified a new type of Hollywood scientist, the brainy babe. Scientific textbooks have reinforced the notion that science is a masculine endeavor by mentioning and picturing men almost exclusively and by showing the few women who do appear in gender-stereotypical roles. It is difficult to imagine that the constant pairing of "scientist" and "man" has no effect on women and men.

The mass media, especially advertising, sends strong cultural messages about desirable roles for women and men. The images of women and men in advertising are varied yet present a familiar range of stereotypes. The images of women are not self-reflective but describe women as they are ideally seen by men, and therefore the locus of power is with men (MacCurdy 1994). In advertisements and the mass media in general, male characters tend to be independent, assertive, athletic, important, technical, and responsible; they show ingenuity, leadership, bravery, or aggression. Female characters are more likely than men to be warm, emotional, romantic, affectionate, sensitive, frail, domestic, or helpless; they tend to need advice or protection, or serve others. It has been suggested that advertisements act as "achievement scripts" for women, providing them with models of appropriate behavior (Geis et al. 1984).

The task of the advertiser is to favorably dispose viewers to a product. Advertisements are intentionally designed to be unambiguous about a subject. The powerful ability of the advertiser to use a few models and props to evoke an understandable scene with just a quick glance is due not to art or technology, but to institutionalized arrangements in social life that allow us to understand the meanings of the images (Goffman 1976). The textual material outside of the picture also provides information about the product, but this is commonly redundant; the picture itself is often designed to tell a story without much textual assistance.

When people are shown in advertisements, we see the advertisers' views of how people of different sexes and ethnicities can be profitably pictured. Although the images shown in advertisements do not need to be representative of the social behavior of any particular person in real life, the pictures are usually not seen as peculiar or unnatural. For example, simply imagining a switch in the sex of the person in an advertisement often reveals the cultural stereotypes portrayed in them. Advertisers do not create the stereotypes they employ; they draw upon the cultural norms that are known to all who participate in social situations (Goffman 1976). Like movie images of scientists (Ribalow 1999), print advertising images are a mirror that reflects the aspirations, hopes, and beliefs of the audience.

> As advertising agencies compete among one another to influence us, we become the final referees in this contest and, in turn end up ourselves influencing in a significant way the content of the advertising page. . . . The images and appeals that persuade us are by no means alien invasions into our psyche imposed from without, but rather welcomed reinforcements of our own original values. (Manca and Manca 1994)

The relationship between images of science and images of masculinity could reveal more than one model of masculinity, since the images of masculinity are varied (Lie and Sorensen 1996). Even if real individual men and women are different, there are fairly consistent ideas concerning masculinity and femininity. These stereotypes do not need to correspond to the personalities of real scientists, because the stereotypes are as real as living persons. Therefore, these stereotypes are not necessarily literally true, but they should not be devalued or ignored. Individuals are susceptible to stereotypes, and identification with a stereotype can influence performance (Steele 1997; Shih, Pittinsky, and Ambady 1999). Stereotypes may significantly impact decisions of great import in the lives of aspiring scientists, for example, in choosing to enter or remain in a field of study or on a career path, or in influencing those who make hiring, tenure, and promotion decisions (Valian 1998). In this essay I provide examples of messages that are sent to scientists—men and women, aspiring and established—during what one would assume to be a positive professional activity, reading a scientific journal.

Immersed in the broader culture and also in the culture of science, scientists are often not aware of biases that may arise from the stereotypes of men and women. A prevailing assumption is that gender is irrelevant in doing science, and therefore is irrelevant in the culture of science. But if gender is truly irrelevant, then one would expect that the images of people in professional publications would be quite different from those in the broader culture. I was curious to examine some current images of both men and women in science—more specifically, the images of people in the journal *Science*—to determine if the images there connect, either overtly or subtly, to gender stereotypes from the broader culture. *Science*, a weekly publication of the American Association for the Advancement of Science (AAAS), is one of the foremost scientific journals in the United States and is widely read across scientific disciplines. One of the missions of the AAAS is to promote diversity in the scientific workplace. In a news article entitled "Association Denounces Discrimination in the Workplace," the AAAS Board and Council issued a joint statement calling on the scientific community to be active in ensuring that discrimination "finds no comfort in our education and research institutions" (Pabst 1995). The joint statement was released to sensitize the scientific community to the fact that although discriminatory practices are illegal, they still exist in the sciences. AAAS's position is that discrimination is contrary to the very core of what it is to do science or be a scientist:

> The AAAS is committed to a work environment free from unlawful and unjust discrimination and condemns such discrimination in any form. . . . All scientific inquiry—from creative generation of ideas, through investigation, and to interpretation of findings—benefits from different points of view. . . . Discrimination creates an atmosphere that is not conducive to the advancement of science. It diverts attention from the rigors of science and undermines an individual's work or academic performance. . . . It also contributes to the loss of talent in science and engineering and to the alienation of capable individuals from the scientific and engineering professions. (Pabst 1995)

METHODS

To explore the extent of gender stereotypical content in *Science*, I examined the advertisements in all issues published from January 1995 through December 1997. I coded the frequency of each advertisement greater than a third of a page in size that carried a pictorial image in the following seven categories: (1) issue type: normal or with special focus/advertising supplement aimed at minorities or women; (2) people: with or without people in picture; (3) profession: scientist or not scientist; (4) sex: male, female, or both in the same picture; (5) race: white, African American, Asian, other, or more than one race in the same picture; (6) image: authority/expert, athlete, nerd/nonconformist, or other; and (7) special qualities: the words *easy/simple* or *efficient/fast/reliable* emphasized in font size larger than bulk text. I scored the advertisements only in categories where I could unambiguously and immediately recognize the category; therefore, not all advertisements were coded in every category. I scored all advertisements regardless of the number of times a given advertisement appeared. Because my goal was to examine the use of stereotypical images rather than specific advertisements, each advertisement was considered a separate use.

A total of 141 issues of *Science* were examined. Nineteen (13.5 percent) of the issues were special issues that focused on women or people of color in science or contained advertising supplements aimed at women and people of color. In the 141 issues there were 4,070 pictorial advertisements greater than a third of a page in size. In 1,094 (26.8 percent) of the advertisements, bodies or body parts recognizable as belonging to either a man or a woman were depicted.

ADVERTISING RACE, GENDER, AND SCIENCE

The numbers of advertisements depicting women in regular issues was approximately half that of advertisements depicting men, whereas in special issues the numbers of advertisements depicting men and women were nearly equal. The proportion of advertisements depicting men, women, and both sexes in the same picture was different in the two types of issues (table 10.1). In regular issues men, women, and both sexes together were portrayed in 63.5, 29.7, and 6.8 percent of the advertisements depicting people, respectively. In special issues the percentage of advertisements with men and women decreased to 30.5 percent and 28.2 percent, whereas advertisements depicting both sexes together increased to

Table 10.1 People in Advertisements in Regular and Special Issues of *Science*, 1995–1997

	Regular Issues ($n = 122$)			*Special Issues* ($n = 19$)		
	Males ($n = 528$, 63.5%)	Females ($n = 247$, 29.7%)	Both ($n = 57$, 6.8%)	Males ($n = 80$, 30.5%)	Females ($n = 74$, 28.2%)	Both ($n = 108$, 41.2%)
White	87.1%	80.9%	24.6%	82.5%	54.1%	39.8%
African American	5.1%	9.3%	1.7%	10.0%	16.2%	0.9%
Asian	6.8%	8.9%	0.0%	6.3%	4.1%	0.0%
Multiple Races	0.9%	0.8%	73.6%	0.8%	25.7%	59.3%
Portrayed as Scientist	72.5%	63.5%	78.9%	66.3%	70.2%	59.3%

Figures are for percentage of advertisements portraying whites, African Americans, Asians, multiple races, and scientists calculated as percentages of total number of advertisements depicting men, women, or both sexes together.

41.2 percent. In 1995 women comprised approximately 50 percent of the U.S. population, and so were underrepresented in the advertisements (U.S. Bureau of the Census 1996).

The proportion of advertisements depicting people of color varied with issue type. In 1995 African Americans comprised about 14 percent of the U.S. population. In regular issues African Americans were underrepresented, comprising about 6.5 percent of the people depicted in the advertisements containing only one person. They were proportionately represented in special issues, comprising about 14 percent of the people shown in advertisements. Asians, who in 1995 comprised about 4 percent of the U.S. population, were overrepresented in both special (5.9 percent) and regular (7.5 percent) issues of *Science*.

The proportion of advertisements depicting whites, African Americans, Asians, or multiple races in advertisements with men, women, or both sexes was different in regular and special issues of *Science*. The percentage of advertisements depicting white people was lower in special issues than in regular issues, where there was a slight decrease in the percentage of white men, from 87.1 percent in regular issues to 82.5 percent in special issues. However, most of the change in race was due to the decrease of the proportion of advertisements with white women, from 80.9 percent in regular issues to 54.1 percent in special issues. White men and women were more frequently shown together (24.6 and 39.8 percent of advertisements depicting white people in regular and special issues, respectively) than were Asian (0 percent) or African American (<2 percent) men and women. African Americans occurred in a higher proportion of advertisements in special issues than in regular issues. This was due mainly to the increase of the percentage of African American women, from 9.3 percent of advertisements depicting women in regular issues to 16.2 percent in special issues. There was a similar increase (from 5.5 to 11 percent of advertisements depicting men) in percentage of advertisements with African American men between the regular and special issues.

Issue type and sex were related to the percentage of advertisements depicting multiple races together. When multiple races were depicted, it was most common for both women and men to be shown in the same advertisement (regular issues, 73.6 percent; special issues, 59.3 percent). Depiction of women in advertisements with multiple races increased from 0.8 percent in regular issues to 25.7 percent in special issues. Men and women of color were usually paired with white people. The depiction of only men in multiple-race groups was low in both types of issues (regular, 0.9 percent; special, 0.8 percent). The most acceptable way to portray diversity under the current social system of science may be through an increase in images of women of color rather than through an increase in images of men of color. The combination of the increased percentage of African American females and the general absence of racially diverse male-only images suggests that diversity itself is a gendered image.

As *Science* is a scientific journal, it is interesting to look at the proportions of men and women portrayed as the scientific workforce. In the world of *Science* advertisements in regular issues, women make up about 18 percent of the total scientific workforce, which is slightly lower than the real percentage of 22 percent of women in the U.S. scientific workforce in 1995 (National Science Foundation 1999). In advertisements in special issues, women increased as members of the scientific workforce to about 20 percent, still lower, but more realistically reflecting the actual proportion in the United States.

THE STEREOTYPES

The hero

Science is often mythologized in terms of the masculine hero ethic, and the advertisements in this study are no exception. The most common image was of the scientist as a hero or authority at the frontiers of progress. The word *discovery* often appears in a large font in these advertisements. The image is usually of a serious, lab-coat-clad scientist alone, often in quarter face, gazing at the object of his study, some type of apparatus, or into the unknown (fig. 10.1). He is unaware of and detached from the viewer. Exactly what the expert is gazing at is often unseen and remains a mystery to the viewer. Mastery of science bestows power in relation to others who lack this knowledge (Wacjman 1991). Sometimes the science and the scientist are one, with data or formulae projected onto the profile of a male head, an image of unity between the knowledge seeker and the knowledge—and perhaps a representation of the perfect flow of knowledge from nature through the scientist, without distortions of interpretation. The percentage of advertisements depicting the hero/authority image was related to sex and type of issue. When a woman was shown in the same advertisement with a man, she was more likely to be depicted as an authority than if she was shown alone. In the regular issues men, women, or both men and women in the same advertisements were depicted as heroes/authorities in 63.2, 44. 5, and 56.2 percent of the ads, respectively. In the special issues the depiction of males as heroes/authorities decreased to 52.8 percent, whereas the depiction of females and both sexes as heroes/authorities increased to 55.6 and 68.7 percent, respectively. Thus one effect of the special issues was to equalize the representation of women and men as heroes.

The nerd

A common portrayal in the advertisements is of the science nerd or its more benign variant, the social nonconformist. Nerds can be recognized by their unkempt hair, heavy-framed glasses, and pocket protectors. Unlike the experts, whose serious gaze is directed elsewhere, the nerd or nonconformist is smiling and looks directly out of the page at the viewer. The nerd displays just a little too much enthusiasm for his work, for example, jumping for joy. This is analogous to Seely's (1994) nonscientific "wild and crazy guy," which emphasizes a somewhat bizarre person free from the shackles of strict societal norms. Stemming from an antiscience theme in popular culture, the common stereotype of the nerd or egghead is depicted as an undersocialized person whose scientific pursuits leave him detached from the "real" world (Sonnert and Holton 1995). The nonconformist is a scientist but often is not associated with the stereotypical signifiers—a lab coat or heavy glasses. The nonconformist

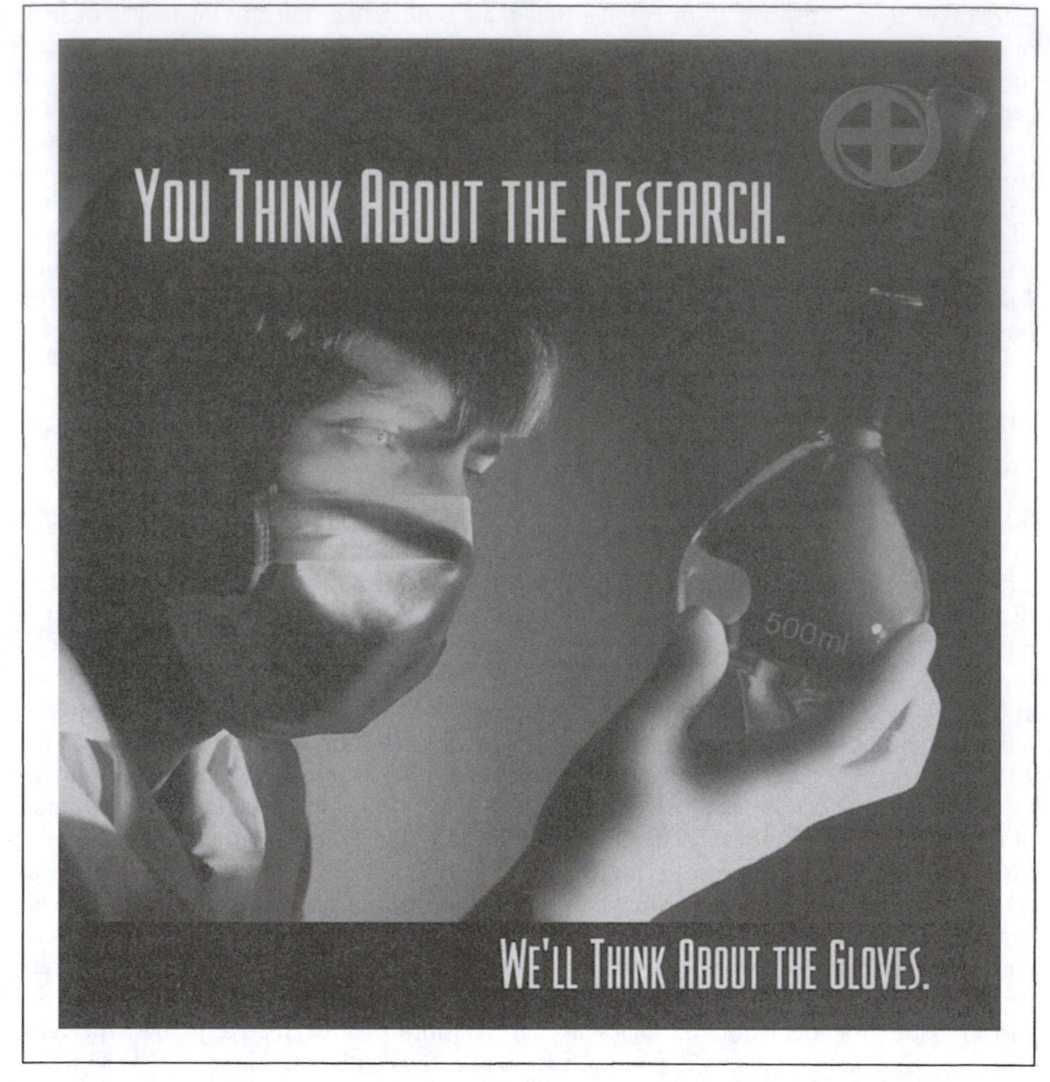

Figure 10.1 The hero/expert. This expert wears the signifiers of the scientist—gloves, lab coat, and face mask. His gaze and attention are focused away from the viewer, and he is totally absorbed by his research. © Microflex Corp. Used by permission

is often young, hip, and a rebel. One advertisement depicted an award-winning young male scientist wearing a flannel shirt and jeans, barefooted, and squatting (rather than sitting) on a stool. Males (18.6 percent) were depicted as nerds/nonconformists in a higher percentage of advertisements than were women (6.6 percent) or both sexes in the same advertisement (2.2 percent).

Women with scientific aspirations may face a double marginalization: entering the stigmatized subculture of nerds and then being an oddity among her colleagues because of her sex. Fortunately, the nerd image appears to be reserved largely for men. However, one type of nonconformist image in which only females are depicted is the witch. The nerd or nonconformist is a social misfit because of his devotion to and enthusiasm for science and technology, which in this interpretation can be seen as a positive characterization. Technology is a medium of power, and technological competence is considered a valuable asset. In contrast, the witch is engaged in magic—an activity that is defined in opposition to science and technology. This association of gender with technological competence in the advertisements has a corollary in the sex-segregated workforce. For instance, women are more likely than men to work in jobs or careers that require little technical competence, and men tend to be found in a wide range of occupations that require technological training (Cockburn 1985).

Men at work and play

Several nonscientific images are employed for men in scientific advertisements, including construction workers, astronauts, businessmen, and technical representatives. However, the most common nonscientist image of men in advertisements is an athlete or explorer. This image in scientific advertising is analogous to Seely's (1994) characterization of the "man's man" in nonscientific advertising. In this image the masculine ideals of competition, rugged individualism, independence, strength, and quest for high quality are emphasized, and correspond well to characteristics valued in the current social culture of science. The man's man needs other men to validate his masculinity—he does manly stuff with other equally manly men. The man's man exists in an exclusive sphere, which affirms the belief that real men cannot enjoy serious challenges in the company of women. Indeed, in this stereotype the participation of women would devalue the achievement, so no true camaraderie can exist between men and women. In the *Science* advertisements, when men are not depicted as a hero or authority, they are most often (14.2 percent) depicted as being engaged in competitive sports or in physically challenging or risky recreational activities, alone or with other men. This image is used less frequently for women (2.8 percent) or in ads with both sexes (2.2 percent). Women are not portrayed with this image even when sports where females excel are depicted, such as gymnastics and track. Athletic activities of women or of both sexes together consisted of less strenuous, noncompetitive activities—for example, walking—and were depicted in an advertisement for life insurance rather than in advertisements for products used to do science.

SCIENCE MADE SIMPLE

As I proceeded with the content analysis of the advertisements, a striking pattern emerged regarding the large-font text that often accompanied the image and smaller bulk text. Words commonly appearing in large text included *accurate, fast, precise, economical, easy*, and *simple*. In regular issues women (13.3 percent) were more frequently associated with the words *easy* or *simple* than were men (6.7 percent) or both sexes (2. 7 percent). Men (19.7 percent) were more frequently associated with the words *fast, reliable*, or *accurate* than were women (10.9 percent) or both sexes (12.8 percent). In special issues men were more frequently associated with *easy* (10.8 percent) than were women (4.8 percent) or both sexes together (1.7 percent), but women (4.8 percent) and men and women together (1.7 percent) were still less associated with *reliable, fast*, or *accurate* than men alone (22.9 percent).

In several advertisements, real or cartoon women are shown using equipment whose chief selling point appears to be ease of use (fig. 10.2). In one advertisement, a smiling blond woman is portrayed similarly to a "car model," merely pointing to, rather than using, a piece of equipment that is described as easy to use. In one advertisement that invites multiple interpretations, a soft-focus woman engaged in lab work provides the backdrop for a piece of equipment embedded in a puzzle shape, accompanied by an ambiguous phrase in large text: "The missing piece to simplify your day!" The double entendre implied by the sexual slang *piece* is clearly intended. We are left to guess at which piece is simplifying whose

Unique New Features!
The missing piece to simplify your day!
The new BioRobot™ 9600 is the missing tool to automate your molecular biology tasks, increase your productivity, and ensure consistent high-quality results.
The new High-Speed Pipetting System means increased throughput and shorter run-times, and the new Tip-Change

Figure 10.2 Science made simple. Women portrayed as scientists are often paired with the message of ease-of-use of technical or analytical equipment. © Quiagen Corp. Reprinted with permission

day. Whether the woman pictured is an instrument of simplification and objectification for the unseen (male) head of the lab or the advertised apparatus is simplifying a difficult technical task for the woman, neither interpretation has a positive meaning.

In contrast, in similar advertisements where a man is shown using a piece of scientific equipment, the word *easy* is generally not used. *Easy* is replaced by the words *fast, accurate,* or *reliable* in large text. Men are portrayed as having the ability to master technology and tackle the challenging problems in science, whereas women are portrayed as doing science with the aid of technology that makes the job easy. Perhaps this distinction allows the depiction of women's participation in science without devaluing the achievements of men.

WOMEN AT WORK AND NOT AT PLAY

Women were portrayed as nonscientists in 36.5 percent of the advertisements in which they occurred in the regular issues, and in 29.8 percent of the advertisements in the special issues of *Science.* Nonscientist images of men were dominated by sports images—men at play. The masculine sports images reflect activities largely engaged in for pleasure or thrill seeking. In contrast, the most common nonscientist depiction of women is as a beneficiary or consumer of science, often as a mother with a child or engaged in a domestic activity such as grocery shopping. This image is well described by MacCurdy's (1994) image category of the Virgin Mary, which depicts a nurturing, sexless, and selfless mother who exists primarily to care for others and who needs protection from the world outside the home. The women in these images are presented in ways that suggest vulnerability in an intimate setting, for example, wrapped in towels, wearing a bathrobe, or nude. She is often shown cheek to cheek with her child, smiling blissfully and gazing directly at the viewer or at her child. This is quite a contrast to the serious scientist, who, his attention and gaze focused elsewhere, is detached from the viewer.

Consumption has been defined as women's work (Bartel 1988), and since the 1920s the gender norm of division between production (male) and consumption (female) has been used by advertisers. The image of woman as a caregiver or as a consumer rather than producer of scientific knowledge has also been commonly used in pharmaceutical advertising (Craig 1992; Mosher 1976). Whereas the role of caregiver is a real and worthwhile one for women (and men), are there any compelling reasons why these images should appear in a scientific journal? Unlike the sports images used for men, the characteristics culturally associated with caregiving and housekeeping are not the qualities usually deemed necessary for a successful career in science (teaching, perhaps, but not conducting research). These domestic images reinforce the idea that the appropriate activity for a woman is to stay at home and act as caregiver to her family, and affirm that a woman is dependent, which conflicts with the image of the independent scientist.

Even when depicted as scientists, women retain their stereotypical roles as nurturers. A recruitment advertisement for a prestigious midwestern university shows a female student

Figure 10.3 The emotional woman. This portrayal conflicts with the notion that science requires the masculine attributes of intellectual objectivity and emotional detachment. © 2000 Pan Vera Corporation. Reprinted with permission

with someone who is presumably her female adviser—both are wearing lab coats and standing in front of apparatus used in molecular biology research. In a large font over the picture are the words "It's much easier to learn from someone who cares that you are learning." Not only must the depicted female faculty member attend to all of her regular duties; it is assumed that she will mother her students and at the same time make science easy to learn.

In a series of advertisements that occurred only in special issues, a number of cleansers and cosmetics produced by the company are shown, accompanied by large text suggesting that of course women will want to work on these products, deemed to be of special interest to women. One may wonder what this means for women scientists who are not particularly nurturing or interested in working on or using the types of products that all women are supposed to be interested in—those associated with the "appropriate" roles for women. The message relayed is that if you can't be at home cleaning the house, you should want to work on products that other women can use to clean the house. Do these images of women as caregivers and housekeepers imply that male scientists don't care about students, or that men shouldn't be interested in domestic activities or do housework? The intellectual and detached aspects of the ideal scientist/hero image appear to be irreconcilable with the feminine caregiver images presented in the advertisements.

MOTHER NATURE

In the history and philosophy of science, nature has long been depicted as a woman, and science as the method by which to know and control nature. The notion that women are closer to nature and are more emotional, less analytical, and weaker than men is prevalent in Western culture. In the industrialized world, where scientific and technical rationality are highly valued, these associations play a powerful role in the ideological construction of women (nonscientists) as inferior (Plumwood 1993). The logical structure of dualism forms a major basis for the connection between nature and women, for example, man/woman, culture/nature, master/servant, reason/emotion, scientific/ignorant, civilized/primitive, public/private, independent/dependent, objective/subjective, competitive/nurturing, intellectual/intuitive. To read the first side of each pair is to read a list of stereotypical qualities traditionally attributed to men and to the human, while the second side presents qualities traditionally excluded from male ideals and associated with women. The stereotype is that scientists are white men who think objectively, rationally, and with suppressed emotions with the object of transcending and controlling nature. Women in real life and in advertisements can now be scientists (especially if the science is associated with caregiving or housekeeping), but the image of science itself and its orientation to masculine ideals and the domination of nature remains unchanged.

Although occurring infrequently, the image of women as a symbol of nature or as object of study is used to sell scientific products. In an image of the most overarching "mother" of all, we see a female hand with long, red-polished nails holding the earth. In another advertisement, we see only fragments of a woman—a made-up eye, a lipsticked and pursed mouth—interspersed with fragments of nature—flowers and seashells —accompanied by the word *discovery* in a large font (fig. 10.4). This image seems to suggest that, like nature, women can be fragmented, reduced, understood, and controlled by science. In what is perhaps the most explicit conflation of nature and woman in an advertisement for a biotechnology company, a nude woman, in profile, is overlaid by a representation of the double helix of DNA and an antique map of Africa (fig. 10.5). These are all images representing the unknown, ripe for exploitation—Freud's dark continent.

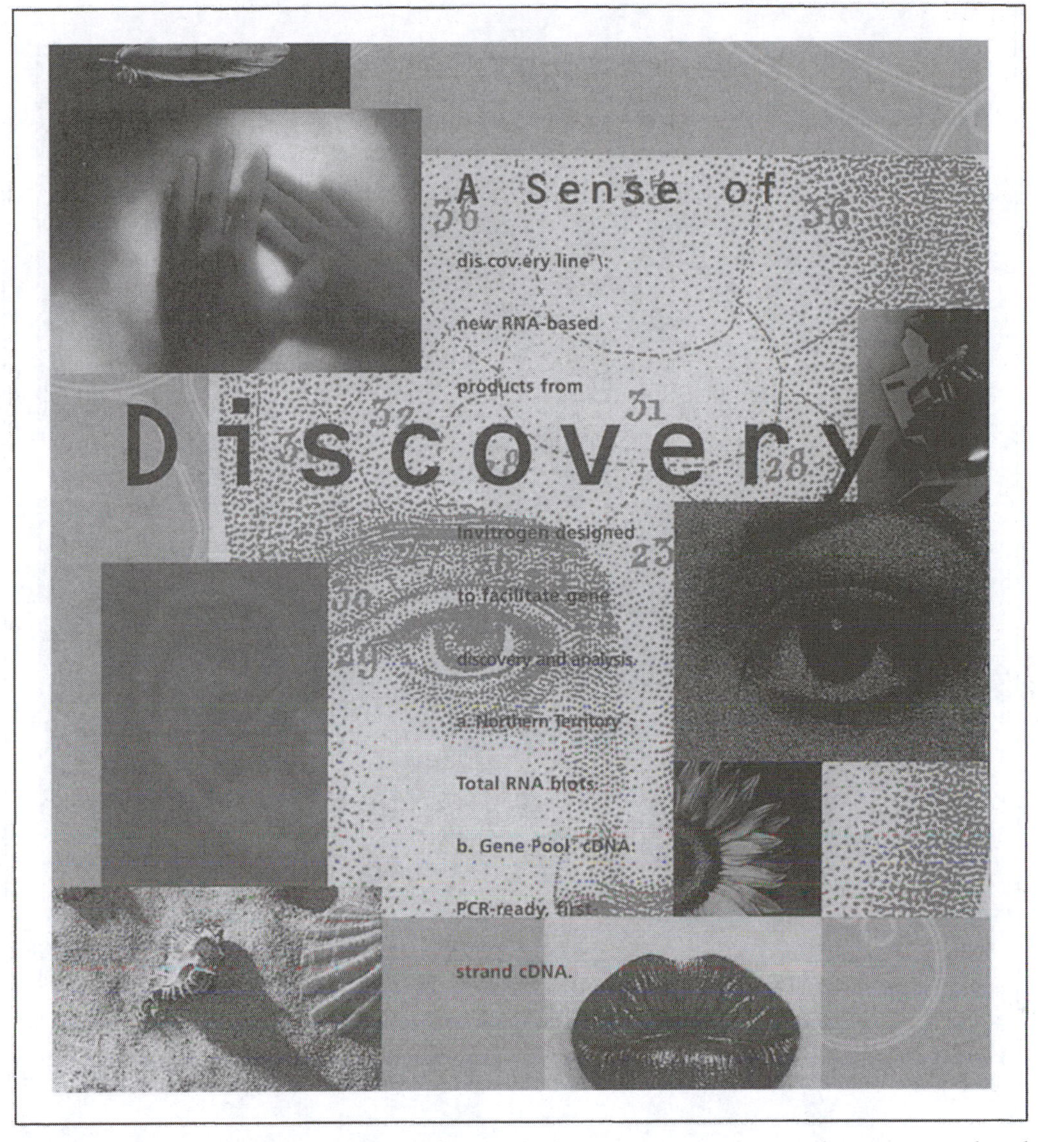

Figure 10.4 Woman as nature as object of scientific discovery. Here women and nature are reduced to fragments and identified with each other. Copyright 2000, Invitrogen Corporation. All rights reserved. Used by permission

WHY WE SHOULD CARE

The point of examining the advertisements in *Science* is not to show that they are a problem per se. I am not necessarily advocating a sort of affirmative-action program for advertisers. The advertisements are most useful as a tool for revealing prevailing general attitudes and stereotypes about science and who can do science. We may hope, perhaps idealistically, that when our societal attitudes about people and their abilities become more equitable, the images described here will no longer appeal and sell products to us, and advertisers and advertising editors will provide images that reflect more equitable beliefs.

The stereotypes portrayed in the advertisements are not gender-neutral. Historically the

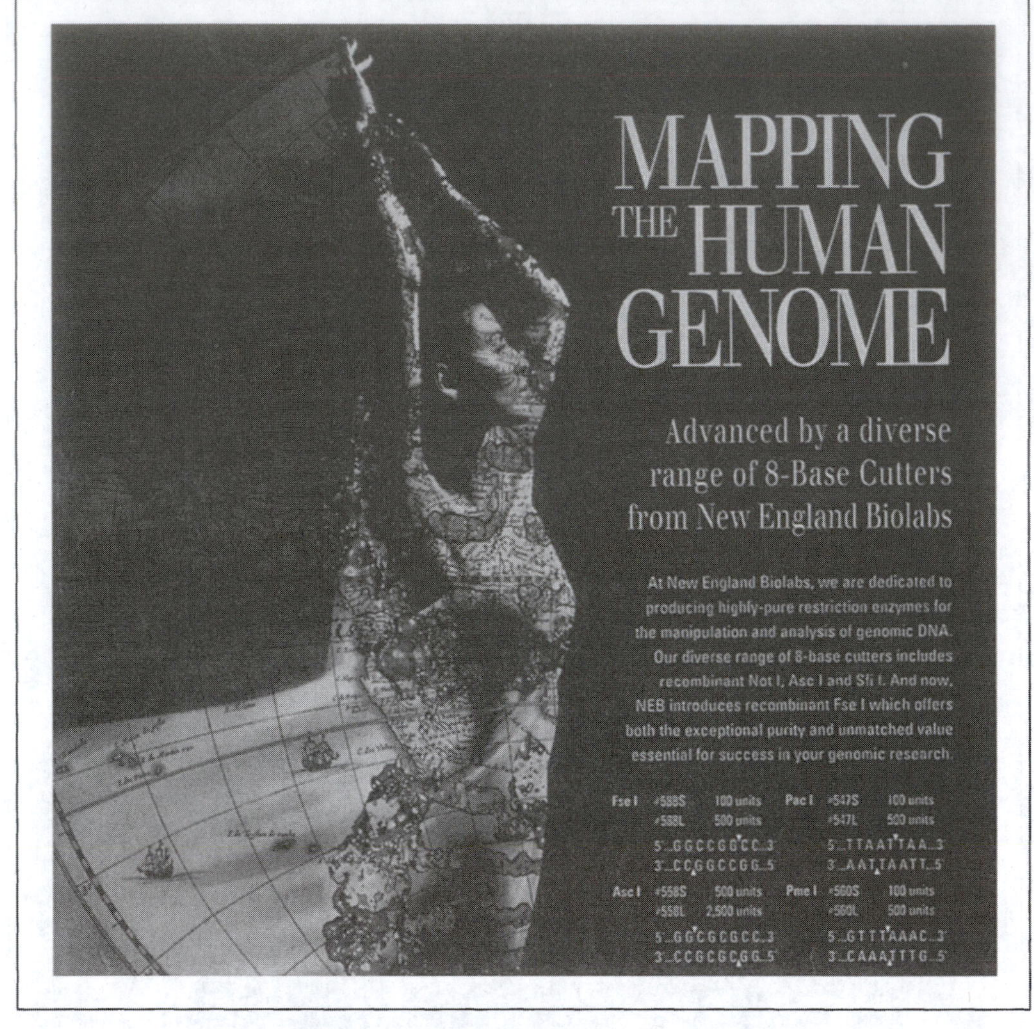

Figure 10.5 Women represented as nature and the unknown. Women, DNA, and Africa as territory to be explored and exploited. © New England Biolabs. Used by permission

mass media have portrayed a strongly negative image of women scientists as atypical women and atypical scientists, and increased numbers of women in the scientific workforce have not been matched by comparable improvements in the presentation of women (LaFollette 1988; Valian 1998). The advertisements in *Science* share biases similar to those in the popular media, where science is portrayed most often as an inappropriate activity for women. What is problematic is that such conventional stereotypes occur in the pages of *Science*, a prestigious and highly respected journal published by a scientific society whose goals include the active promotion of diversity and elimination of discrimination in science (Pabst 1995). What one would hope and expect to find in the pages of a journal with these stated goals would be images that reflect the goals that the members of the scientific society deem important. That advertisers and advertising editors are aware of gender and scientific stereotypes is evidenced by the greater proportion and more positive portrayals of women and people of color in the special issues compared with the regular issues of *Science*. Do editors and advertisers think that women and people of color read only the special issues of *Science?* Do they think that only women and people of color read the special issues?

The goal of the advertiser is to appeal to the conscious and subconscious desires and value systems of the consumer to create a need for a specific product, and that need must be translated into a badge of group membership. We live in a culture in which both men and women "know" that science is masculine. The emphasis on white men and stereotypes of masculinity in scientific advertising borrows from cultural assumptions about white male superiority to reaffirm the importance and superiority of science, and vice versa. At the same time this emphasis reinscribes the inferiority of the "other"—people of color and women—and underscores their outsider status and the inappropriateness of science as a career for them. In the current social system of science, to present an image of women and people of color as full participants in science would be to devalue science. The pervasiveness of links among cultural beliefs about men and women and science as portrayed by the advertisements suggests that the attainment of diversity in science still presents a considerable challenge.

My purpose here has been to alert hopeful and practicing scientists to some of the neglected social aspects of a science career. Once outsiders are aware of the potential problems, these issues can be counteracted, and individuals can make informed decisions that can increase chances for success and equity. Although I do not know how the advertisements in *Science* affect those who read them, we can be reasonably sure that these familiar images do little to challenge our stereotypes and biases. The images are devoid of any competing set of cultural messages and values that may be needed to change our stereotypes about science and scientists. The most significant message of this study may be that there is more cultural content between the covers of *Science* than we would care to acknowledge. The change of images in scientific advertising to more positive ones for women and people of color may help us slowly change our stereotypes about men, women, race, and science, and therefore make a contribution to the attainment of greater diversity in science.

References

Bartel, D. 1988. *Putting on Appearances: Gender and Advertising*. Temple Univ. Press, Philadelphia, 219 pp.

Cockburn, C. 1985. *Machinery of Dominance: Women, Men, and Technical Know-how*. Northeastern Univ. Press, Boston, 282 pp.

Craig, R.S. 1992. Women as home caregivers: Gender portrayal in OTC drug commercials. *J. Drug Education* 22: 303–12.

Day, J. Cheeseman. 1995. US Bureau of Census. USGPO, P25–1104.

Geis, E., Brown, V., Jennings (Walstedt), J., and Porter, N. 1984. TV commercials as achievement scripts for women. *Sex Roles* 10:513–25.

Goffman, E. 1976. Gender Advertisements. *Studies in the Anthropology of Visual Communication* 3(2):65–154.

LaFollette, M. 1988. Eyes on the stars: Images of women scientists in popular magazines. *Science, Technol And Human Values* 13:262–75.

Lie, M., and Sorensen, K. 1996. *Making Technology Our Own? Domesticating Technology into Everyday Life*. Scandinavian Univ. Press, Oslo, 244 pp.

MacCurdy, M. 1994. The four women of the apocalypse: Polarized feminine images in magazine advertisements. Pp. 31–48 in *Gender and Utopia in Advertising*, L. Manca and A. Manca, eds. Procopian Press, Lisle, Ill., 168 pp.

Malcom, S.M. 1999. Fault lines. *Science* 284:127

Manca, L., and Manca, A. 1994. Adam through the looking glass: Images of men in magazine advertisements of the 1980s. Pp. 111–31 in *Gender and Utopia in Advertising*, L. Manca and A. Manca, eds. Procopian Press, Lisle, Ill., 168 pp.

Mirsky, S. 1997. The big picture. *Scientific American*, November, p. 28.

Mosher, E.H. 1976. Portrayal of women in drug advertising: A medical betrayal. *J. Drug Issues* 6: 72–78.

National Science Foundation. 1999. *Women, Minorities, and Persons with Disabilities in Science and Engineering, 1998.* NSF 99–338. Washington, D.C.
Pabst, D. 1995. Association denounces discrimination in the workplace. *Science* 268:590.
Plumwood, V. 1993. *Feminism and the Mastery of Nature.* Routledge, New York.
Ribalow, M.Z. 1999. Swashbucklers and brainy babes? *Science* 284:2089–90.
Rossiter, M.W. 1982. *Women Scientists in America: Struggles and Strategies to 1940.* Johns Hopkins Univ. Press, Baltimore, 439 pp.
——. 1995. *The Women Scientist in America: Before Affirmative Action, 1940–1972.* Johns Hopkins Univ. Press, Baltimore, 584 pp.
Seely, P. 1994. The mirror and the window on the man of the nineties: Portrayals of males in television advertising. Pp. 95–110 in *Gender and Utopia in Advertising,* L. Manca and A. Manca, eds. Procopian Press, Lisle, Ill., 168 pp.
Shih, M., Pittinsky, T.L., and Ambady, N. 1999. Stereotype susceptibility: Identity salience and shifts in quantitative performance. *Psychological Science* 10:80–83.
Sonnert, G., and Holton, G. 1995. *Who Succeeds in Science? The Gender Dimension.* Rutgers University Press, New Brunswick, N.J., 215 pp.
Steele, C.M. 1997. A threat in the air: How stereotypes shape intellectual identity and performance. *American Psychologist* 52:613–29.
Thompson, E.L. 1979. Sexual bias in drug advertisements. *Soc. Sci. Med.* 13A:187–91.
U.S. Bureau of the Census. 1996. *Population Projections of the United States by Age, Race, Sex, and Hispanic Origin: 1995–2050.* Document no. P25–1130. Government Printing Office, Washington, D.C.
Valian, V. 1998. *Why So Slow? The Advancement of Women.* MIT Press, Cambridge, Mass.
Wajcman, J. 1991. *Feminism Confronts Technology.* Pennsylvania State Univ. Press, University Park, 184 pp.
——. 1991. The built environment: Women's place, gendered space. Pp. 110–36 in *Feminism Confronts Technology,* Pennsylvania State University Press, University Park.
Wennerás, C., and Wold, A. 1997. Nepotism and sexism in peer review. *Nature* 387:341–43.

11

The Aliens Have Landed!

Reflections on the Rhetoric of Biological Invasions

Banu Subramaniam

Two years ago, in a special issue entitled *Biological Invaders* by the prestigious journal *Science*, an article begins as follows:

> One spring morning in 1995, ecologist Jayne Belnap walked into a dry grassland in Canyonlands National Park, Utah, an area that she has been studying for more than 15 years. "I literally stopped and went, 'Oh my God!' " she recalls. The natural grassland —with needle grass, Indian rice grass, saltbush, and the occasional pinyon-juniper tree—that Belnap had seen the year before no longer existed; it had become overgrown with 2-foot-high Eurasian cheatgrass. "I was stunned," says Belnap, "It was like the aliens had landed."[1]

One of the ironies in the world today is that in an era of globalization, there is a renewed call for the importance of the "local" and the protection of the indigenous. With the increased permeability of nations and their borders, and the increased consumption and celebration of our common natures and cultures, we begin to obsess about our different natures and cultures with a fervent nationalism, stressing the need to close our borders to those "outsiders."[2] The anxieties around the free movement of capital, commodities, and entertainment and the copious consumption of natural and cultural products have reached fever pitch. In the realm of culture and the economy, nationalisms,[3] fundamentalisms,[4] World Trade Organization protests, and censorship of "foreign" influences, calls for the preservation of national cultures abound.[5] In the realm of nature, there is increasing attention to the destruction of forests, conservation, preservation of native forests and lands, the commodification of organisms, and concern over the invasion and destruction of native habitats through alien plant and animal invasions.[6] "Development" is one area[7] in which both the natural and cultural worlds implode.[8] At the heart of the critiques is the fundamental question of what we mean by nature and culture. Who gets to define it? Are nature and culture static and unchanging entities? If nature and natural processes shift and change over time, as most biologists believe, how do we characterize and accommodate these evolutions?

Over the last two decades, feminist and postcolonial critics of science have elaborated the relationship between our conceptions of nature and their changing political, economic, and cultural contexts. Nature and culture, they argue, are co-constituted, simultaneously semiotic and material. Through Donna Haraway's "material-semiotic worlds"[9] can emerge a history of "naturecultures,"[10] tracing and elaborating the inextricable interconnections between natures and cultures.

Nowhere is this more apparent than in the growing panic about alien and exotic plants and animals. Newspaper articles, magazines, journals, and websites have all sprung up demanding urgent action to stem the rise of exotic flora and fauna. For anyone who is an immigrant or is familiar with the immigration process, the rhetoric is unmistakable. First consider the terminology: A species that enters the country for the first time is called an "alien" or an "exotic" species; after an unspecified passage of time, it is considered a resident; after a greater unspecified passage of time, it is considered a naturalized species.[11]

As Nancy Tomes argues, our anxieties about social incorporation (associated with expanding markets, increasingly permeable borders and boundaries, growing affordability of travel and mass immigration) have historically spilled into our conceptions of nature. For example, she documents how our panic about germs has historically coincided with periods of heavy immigration to the United States, of groups being perceived as "alien" and difficult to assimilate. She documents these germ panics in the early twentieth century in response to the new immigration from eastern and southern Europe and those in the late twentieth century to the new immigration from Asia, Africa, and Latin America. "Fear of racial impurities and suspicions of immigrant hygiene practices are common elements of both periods," she writes. "These fears heightened the germ panic by the greater ease and frequency with which immigrants travel back and forth between their old, presumably disease ridden countries and their new, germ obsessed American homeland."[12]

I argue in this chapter that the recent hyperbole about alien species is similar to the germ panics and is in response to changing racial, economic, and gender norms in the country. The globalization of markets and the real and perceived lack of local control feed nationalist discourse.[13] Despite the supposed low unemployment rates and a great economy, the search of companies for cheap labor abroad and the easing of immigration into the country have increasingly been perceived as threats to local employment. These shifts continue to be interpreted by some elements of both the Right and the Left as a problem of immigration. Immigrants and foreigners, the product of the "global," are perceived to be one of the reasons for the problems in the "local." These shifts and trends are evident in the national rhetoric surrounding alien and exotic plants and animals.

THEY CAME, THEY BRED, THEY CONQUERED[14]

Newspapers and magazines introduce the topic of biological invasions with the sound of alarm. Consider some of the titles:

Alien Invasion: They're Green, They're Mean, and They May Be Taking Over a Park or Preserve Near You[15]
Aliens Reeking Havoc; the Invasion of the Woodland Soil Snatchers[16]
Native Species Invaded[17]
Bio-invasions Spark Concerns[18]
It's a Cancer[19]
Creepy Strangler Climbs Oregon's Least-Wanted List[20]
Biological Invaders Threaten U.S. Ecology[21]
U.S. Can't Handle Today's Tide of Immigrants[22]
Alien Threat[23]
Biological Invaders Sweep In[24]
Stemming the Tide of Invading Species[25]

Congress Threatens Wild Immigrants[26]
Invasive Species: Pathogens of Globalization[27]

The majority of these titles do not specify that the article is about plants and animals but rather present a more generalized classic fear of the outsider, the alien that is here to take over the country. An opening line of an article reads, "The survey is not even halfway done, yet it has already revealed a disturbing trend: immigrants are forcing old-timers out of their homes."[28] Invaders are reported to be "racing out of control" causing "an explosion in slow motion."[29] Aliens, they claim, are redrawing the global landscape in ways no one imagined. Exotic plants, they argue, are irreversibly altering waterways and farmlands. The "irreversibility" is highlighted as a way to stress the sharp departure from the past—a vision of how we are moving from a peaceful, coevolved nature in perfect harmony and balance to an uncertain future with alien and exotic plants and animals. They argue that we cannot recapture the glorious past, our nostalgia for a pure and uncontaminated nature in harmony and balance, if we do not act *now* to stem the tide of outsiders.

The parallels in the rhetoric surrounding foreign plants and that of foreign peoples are striking. Like the earlier germ panic surrounding immigration and immigrants, questions of hygiene and disease haunt exotic plants and animals. Similar to the unhygienic immigrants, alien plants are accused of "crowd(ing) out native plants and animals, spread(ing) disease, damag(ing) crops, and threaten(ing) drinking water supplies."[30] The xenophobic rhetoric that surrounds immigrants is extended to plants and animals.

The first parallel is that aliens are "other." One *Wall Street Journal* article quotes a biologist's first encounter with an Asian eel: "The minute I saw it, I knew it wasn't from here," he said.[31] Second is the idea that aliens and exotic plants are everywhere, taking over everything: "They're in national parks and monuments. In wildlife refuges and coastal marine sanctuaries. In wilderness areas that were intended to remain living dioramas of our American paradise lost."[32] "Today, invasive aliens afflict almost every habitat in the country, from farms and pastures to forests and wetlands—and as every homeowner knows, gardens, flower beds and lawns."[33]

The third parallel is the suggestion that aliens and exotic plants are silently growing in strength and number. So even if you haven't noticed it, be warned about the alien invasion. And if you haven't heard about biological invasions, the reason is that the "invasion of alien plants into natural areas has been stealthy and silent, and thus largely ignored." E. O. Wilson states, "Alien species are the stealth destroyers of the American environment."[34] Articles remind us that alien plants are "evil beauties"—that while they may appear to look harmless and even beautiful, they are evil because they destroy native plants and habitats.[35] The fourth parallel is that aliens are difficult to destroy and will persist because they can withstand extreme situations. In an article on the invasion of the Asian eel in Florida,

> The eel's most alarming trait, though, is its uncanny ability to survive extreme conditions. In one study by a Harvard zoologist, an Asian swamp eel lived seven months in a damp towel without food or water. The olive-brown creature prefers tropical waters, yet it can flourish in subzero temperatures. It prefers fresh water but can tolerate high salinity. It breathes under water like a fish, but can slither across dry land, sometimes in packs of 50 or more, sucking air through a two-holed snout. . . . Even more of a riddle is how to kill the eel: It thus far appears almost immune to poisons and dynamite.[36]

The fifth parallel is that aliens are "aggressive predators and pests and are prolific in nature, reproducing rapidly."[37] This rhetoric of uncontrollable fertility and reproduction is another hallmark of human immigrants. Repeatedly, alien plants are characterized as being aggressive, uncontrollable, prolific, invasive, and expanding. One article summarizes it as "They Came, They Bred, They Conquered."[38] Alien species are characterized as destroyers of everything around. A park warden is quoted as saying, "To me, the nutria [a swamp rat] are no different than somebody taking a bulldozer to the marsh."[39]

Sixth, once these plants gain a foothold, they never look back.[40] Singularly motivated to take over native land, the plants, the articles imply, have become disconnected to their homelands, will never return, and are therefore "here to stay." Finally, like that of human immigrants, the greatest focus is on the economic costs because it is believed that exotic plants consume resources and return nothing. "Exotic species are a parasite on the U.S. economy, sapping an estimated $138 billion annually, nearly twice the annual state budget of New York or a third more than Bill Gates' personal fortune."[41] Not only are aliens invading rural and natural habitats, but they are also endangering the cities. "Cities invaded," articles cry. From historical sites to urban hardwoods, alien bugs are reported to be causing millions of dollars' worth of damage.[42]

> Just as human immigrants may find more opportunities in an already over-crowded city than in a small town, invasive plants take advantage of the constant turnover and jockeying for position that characterizes species-rich ecological communities. The classical dictum that "diversity begets stability," Stohlgren says, is simply not true in some ecosystems. Communities with high diversity tend to be in constant flux, creating openings for invasives. . . . From a conservation perspective, the results of these multi-site, multi-scale studies are disturbing. The invasions may threaten some of the last strongholds of certain biologically rich habitats, such as tall-grass prairie, aspen woodlands, and moist riparian zones.[43]

THE OVERSEXED FEMALE

One of the classic metaphors surrounding immigrants is the oversexualized female. Foreign women are always associated with superfertility—reproduction gone amuck. Such a view suggests that the consumption of economic resources by invaders today will only multiply in future generations through rampant overbreeding and overpopulation. Consider this:

> Canada thistle is a classic invasive. One flowering stem can produce as many as 40,000 seeds, which can lie in the ground for as long as 20 years and still germinate. And once the plant starts to grow, it doesn't stop. Through an extensive system of horizontal roots, a thistle plant can expand as much as 20 feet in one season. Plowing up the weed is no help; indeed, it exacerbates the problem; even root fragments less than an inch long can produce new stems. . . . The challenge posed by thistle is heightened because, like other troublesome aliens, it has few enemies.[44]

Along with the superfertility of exotic and alien plants is the fear of miscegenation. There is much concern about the ability of exotic plants to cross-fertilize and cross-contaminate native plants and produce hybrids. Native females are, of course, in this story, passive helpless victims of the sexual proclivity of foreign and exotic males.

RESPONDING TO ALIEN SPECIES

Journalists and scientists borrow the images of illegal immigrants arriving in the country by means of difficult, sometimes stealthy journeys when they describe the entry of exotic plants and animals. Alien plant and animal movements are described with the same metaphors of illegal, unwelcome, and unlawful entry—for example, "Exotic species—from non-native fish to various plants, bugs and shellfish—have found their way into the country in numerous ways, such as *clinging* to ships, *burrowing* into *wooden shipping crates*, in food, *aboard aircraft* or *in water discharged* from foreign freighters" (italics mine).[45]

So how do we respond to these unlawful and stealthy entrants? Paralleling images of armed guards patrolling borders are images of the nation responding in kind to plants and animals.

"The alien species invasion—Interior Secretary Bruce Babbitt of the Clinton Administration calls it an 'explosion in slow motion'—is turning even staunch conservationists into stone killers."[46] Like our so-called solution to immigration and the drug problem, our answer to the problem of exotic and alien plants and animals is to "fight" and to wage wars against them. In 1999, President Clinton signed an executive order creating the national Invasive Species Management Plan, directing federal agencies to "mobilize the federal government to defend against these aggressive predators and pests."[47] Thus, the "Feds" were called on to "fight the invaders" and defend the nation against the "growing threat from non-native species."[48] It is implied that the situation is so dire and the number of invaders so great that even the most humane individuals (conservationists) cannot help but turn into killers in order to respond to the violations of alien species that are just not "welcome" into the country.[49]

One magazine published an article titled "When Ecologists Become Killers," allegedly transforming life lovers into "killer conservationists."[50] Like those of the human immigration problem, the resources are scant and the strategies often futile for the exotic species problem. "Two dozen federal agencies have stitched together a crazy quilt of detection and eradication efforts with state and local authorities. But much of the effort is aimed at ports, borders and threats to crops. There is little left over to combat emergencies."[51]

A recent review sponsored by the Ecological Society of America published in the *Issues of Ecology* concludes that the current strategy of denying entry only to species already proven noxious or detrimental should change.[52] Instead of "innocent until proven guilty," we should instead adopt "guilty until proven innocent." This strategy is further racialized when a biologist rephrases it by suggesting that we ought to replace our current system of "blacklisting" imported species (where a species must be proved to be harmful before it is banned) with a "whitelist" (where a species has to proved to be safe before entry).[53] Thus, exotic and alien plants are marked as "guilty," foreigners, and black and therefore kept out purely by some notions of the virtue of their identity.

NATIVES

What is tragic in all this is, of course, the impact on the poor natives. "Native Species Invaded," "Paradise Lost," and "Keeping Paradise Safe for the Natives" are the repeated cries.[54] Native species are presented as hapless victims who are outcompeted and outmaneuvered by exotic plants. Very often, exotic plants are credited with (and by implication, native species are denied) basic physiological functions, such as reproduction and the capacity to adapt. For example, "When an exotic species establishes a beachhead, it can proliferate over time and spread to new areas. It can also adapt—it tends to get better and better at exploiting an area's resources, and at suppressing native species."[55] Invaders, interior secretary Bruce Babbitt says, "are racing out of control as the nonnative species in many cases overpower native species and alter regional ecosystems."[56] Experts warn of the growing invasion of foreigners into the nation's aquatic systems, threatening native species, waterways, and ecology. Not only do they crowd out native plants and animals, but they also endanger food production through the spread of disease and damage to crops, and they affect humans through threatening drinking water supplies. Consider this:

> English ivy joins 99 plants on a state list of botanical miscreants that includes Himalayan blackberry, Scotch thistle and poison hemlock. With dark green leaves and an aristocratic heritage, however, it looks like anything but a menace.
>
> Don't be fooled.
>
> The creeper loves Oregon, where it has no natural enemies.
>
> It needs little sunlight. It loves mild, wet climates.
>
> Robust and inspired, English ivy jumps garden borders, spreading across forest floors, smothering and killing ferns, shrubs and other plants that support elaborate ecosystems and provide feeding opportunities for wildlife. Insatiable, English ivy then climbs and wraps trees, choking off light and air.[57]

Articles invariably end with a nostalgic lament to the destruction of native forests and the loss of nature when it was pure, untainted, and untouched by the onslaught of foreign invasions. At the end of one article, a resident deplores the dire situation: "I grew up on the blackwater," he declares, "and I'm watching it disappear, it's really sad." And the article concludes, "Spoken like a true native."[58]

THE RHETORIC OF BIOLOGICAL INVASIONS

In this chapter, I have traced the striking similarities in the qualities ascribed to foreign plants and animals and foreign people. The xenophobic rhetoric is unmistakable. The point of my analysis is not to suggest that we are not losing native species or that we should allow plants and animals to flow freely across habitats in the name of modernity or globalization. Instead, it is to suggest that we are living in a cultural moment where the anxieties of globalization are feeding nationalisms through xenophobia. The battle against exotic and alien plants is a symptom of a campaign that misplaces and displaces anxieties about economic, social, political, and cultural changes onto outsiders and foreigners.

In his article "Nativism and Nature," Jonah Paretti persuasively argues that the language of exotic and alien plant and animal "invasions" reflects a pervasive nativism in conservation biology that makes environmentalists biased against alien species.[59] Nativism strongly grounds most of the literature against biological invasions. For example, the final chapter of one of the many recent books on the topic is entitled "Going Local: Personal Actions for a Native Planet." Such rhetoric conjures up a vision where everything is in its "rightful" place in the world and where everyone is a "native."[60]

The "natives," however, are, of course, the white settlers who reached the Americas to displace the original natives, to become its new, true natives. In this chapter it is the white settlers that come to be the "local" and the "native." The chapter includes many suggestions for how ordinary citizens can help toward a quest for a native planet by eliminating the exotics—from drawing public attention to the issue of exotics by writing op-ed pieces on biological invasions to the local newspaper, pressuring local conservation groups to take up this important yet unpublicized issue, to planting native plants in one's garden.

I want to be clear that I am not without sympathy or concern about the destruction of habitats, which is alarming. Indeed, we need to publicize and spread awareness about the destruction of species and habitats. However, in their zeal to draw attention to the loss of habitats, such journalists and scientists feed on the xenophobia rampant in a changing world. They focus less on the degradation of habitats and more on alien and exotic plants and animals as the main and even sole problem. With humans, the politics of class and race are essential, thriving on the fear of not all immigrants but immigrants from particular places and of particular places and classes; however, the language of biological invasions renders *all* outsiders,[61] even the familiar albeit nonlocal outsiders, into the undesirable alien.[62] Conservation of habitats and our flora and fauna need not come at the expense of immigrants.

Instead, let us consider exotic and alien species in their diversities. Mark Sagoff points out that the broad generalizations of exotic and alien plants obscure the heterogeneity of the life histories, ecologies, and contributions of native and exotic plants.[63] For example, he points out that nearly all the U.S. crops are exotic plants while most of the insects that cause crop damage are native species. It seems to me the height of irony that alongside a national campaign in the United States to keep out all exotic and alien plants in order to preserve the purity and sanctity of native habitats, there is simultaneously another campaign that promotes the widespread use of technologically bred, genetically modified organisms for agricultural purposes.[64] In these cases, the ecological dangers of growing genetically modified crops in large fields are presented as being minimal. Concerns of cross-fertilization with native and wild plants for which there is little empirical evidence are dismissed as antiscience and antitechnology. Ultimately, it would seem that it is a matter of control, discipline, and capital. As long as exotic and alien plants know their rightful place as workers, laborers and providers, and controlled commodities, their positions manipulated and controlled by the natives, their presence is tolerated. Once they are accused of unruly practices that prevent them from staying in their subservient place, they threaten the natural order of things.[65]

What is most disturbing about displacing anxieties attending contemporary politics onto alien and exotic plants is that other potential loci of problems are obscured. For example, some scholars point to the fact that exotic and alien plants are most often found on disturbed sites.[66] Perhaps the increase in exotic and alien plants is less about their arrival and more about the shifts in the quality of natural habitats through the process of development that allow their establishment. When habitats are degraded by humans, the change causes a shift in the selection pressures on plants at those sites. A displacement of the problem on the intrinsic "qualities" of exotic and alien plants and not on their degraded habitats may produce misguided management policies. Rather than preserving land and checking development, we instead put resources into policing boundaries and borders and blaming foreign and alien plants for an ever-increasing problem. Unchecked development, weak environmental controls, and the free flow of plants and animals across nations all serve certain economic interests in contemporary globalization. Displacement of blame onto foreigners does not solve the problem of the extinction of species and the degradation of habitats.

More central to issues of native and exotic plants are questions of what gets to be called a "native" species. Given that the majority of U.S. Americans are immigrants themselves, the reinvention of the "native" as the white settlers and not the "Native Americans" is striking. The systematic marginalization and disenfranchisement of "Native Americans" makes the irony all the more poignant. In southern California, where my project is based, questions of what are deemed native and exotic are deeply fraught. How do we develop dynamic models of "nature," which do not need to be artificially managed to remain the same year after year? How do we understand the human species as part of nature, in all its shifts and evolutions? These important questions can guide biologists in the development of experimental research. Is it possible to characterize exotic and native plants? Do they all share common life-history parameters and ecological traits? How heterogeneous and diverse are the species within those categories? How static and coevolved are native communities? What is the relationship of plants and their soil communities, and what impact do exotic plants have on them? Do they destroy and degrade these communities? As ecologists, we can test these theories, intervene and participate in the national conversation not only on exotic plants but also on immigration and race relations.

As feminists, we must intervene in the global circulations of science. Feminist and postcolonial critics of science have shown us repeatedly how larger political, economic, and cultural factors inform and shape scientific questions, answers, practices, and rhetoric. While I have largely focused on science that has reached and been popularized in mainstream culture, the scientific community is much more heterogeneous.[67] There is a long tradition of dissent and alternative views in most scientific fields. Many ecologists and conservation biologists have developed alternate models and disagree sharply with the dominant framework of conservation biology.[68] Feminists in the humanities and social sciences can and must build alliances with progressive scientists in the natural and physical sciences. Further, women's studies programs must make it a goal to produce a scientifically and technologically proficient group of students and faculty who are not relegated only to the role of "critics" (important though this is) but are also members of the scientific enterprise, producing knowledge about the natural world, a world that is deeply embedded in its social and cultural histories. Studying "naturecultures" means being cognizant of how science is embedded in these cultural contexts. Just as science does not mirror nature, we must not reduce science to mirroring politics either—Right or Left. Living in naturecultures means developing a self-reflexivity, continually wrestling with the interconnections of natures and cultures, politics and science, the humanities and the sciences, and feminisms and science.

Notes

An earlier version of this chapter was published in the journal *Meridians: feminism, race, transnationalism* 2, no. 1 (Bloomington: Indiana University Press, 2001): 26–40. My thanks to Indiana University Press for giving permission to reprint it here.

This work would not have been possible without my wonderful collaborators James Bever and Peggy Schultz. Most of the arguments in this piece have been developed in discussions with the two of them. I am deeply indebted to Kum-Kum Bhavnani and Geeta Patel for their encouragement to develop this piece for publication. I would like to thank Natalie Joseph for her painstaking research in unearthing many of the articles cited. Comments and advice of James Bever, S. Hariharan, Geeta Patel, and three anonymous reviewers have considerably strengthened the arguments and the text of this chapter.

1 Martin Enserink, "Biological Invaders Sweep In," *Science* 285, no. 5435 (September 17, 1999): 1834–36.
2 Connections between the body as fortress and nation as fortress, the body and nation in late capitalism, can be seen in Emily Martin, "The End of the Body?" in *The Gender Sexuality Reader*, ed. Roger N. Lancaster and Micaela di Leonardo (New York: Routledge, 1997), 543–59. For thinking about the persistence of national states in an age of globalization, see Jean Comaroff and John L. Comaroff, "Millennial Capitalism: First Thoughts on a Second Coming," *Public Culture* (Comaroff and Comaroff, guest eds.) 12, no. 2 (2000): 291–343.
3 For provocative thoughts about the production of ethnicity and civil society/nation, botanical taxonomies, and immigration policies in the United States, see Minoo Moallem and Iain A. Boal, "Multicultural Nationalism and the Poetics of Inauguration," in *Between Women and Nation*, ed. Caren Kaplan, Norma Alarcón, and Minoo Moallem (Durham, N.C.: Duke University Press, 1999), 243–63.
4 For some recent work on ethnocentrism and nationalism produced as a certain politics, see Paola Bachetta's article on xenophobia and the Hindu right in India: Paola Bachetta, "When the (Hindu) Nation Exiles Its Queers," *Social Text* 17, no. 4 (Autumn 1999): 141–66.
5 For a study of cultures, including national cultures and cultural nationalism, see Marshall Sahlins, " 'Sentimental Pessimism' and Ethnographic Experience: Or, Why Culture Is Not a Disappearing

Object," in *Biographies of Scientific Objects*, ed. Lorraine Daston (Chicago: University of Chicago Press, 2000), 158–202. For cultural nationalism and new modes of citizenship, see Aihwa Ong, *Flexible Citizenship: The Cultural Logis of Transnationality* (Durham, N.C.: Duke University Press, 1999).

6 For an excellent discussion of how environmentalists in governmental and nongovernmental organizations, corporations, and financial institutions have sought to fashion a new environmentalism based on markets, see *People, Plants, and Justice: The Politics of Nature Conservation*, ed. Charles Zerner (New York: Columbia University Press, 2000).

7 For discourses of development under current conditions of International Monetary Fund regulations and so forth, see the work of Arturo Escobar. In particular see Escobar, "Cultural Politics and Biological Diversity: State, Capital, and Social Movements in the Pacific Coast of Columbia," in *The Politics of Culture in the Shadow of Capital*, ed. Lisa Lowe and David Lloyd (Durham, N.C.: Duke University Press, 1997).

8 I use "implode" in Donna Haraway's sense of "heterogeneous and continual construction through historically located practice, where the actors are not all human." See Haraway, *Modest_Witness@Second_Millennium.FemaleMan_Meets_OncoMouse: Feminism and Technoscience* (New York: Routledge, 1997), 68.

9 See Haraway, *Modest_Witness*, 68.

10 Thyrza Nichols Goodeve, *How Like a Leaf: An Interview with Donna Haraway* (New York: Routledge, 2000).

11 Earthwatch, "The Aliens Have Landed," *Earthwatch: The Journal of Earthwatch Institute* 15, no. 6 (November/December 1996): 8.

12 Nancy Tomes, "The Making of a Germ Panic, Then and Now," *American Journal of Public Health* 90, no. 2 (February 2000): 191–99.

13 For standard discussions on globalization, see works by David Harvey and Saskia Sassen. See Sassen, "Spatialities and Temporalities of the Global: Elements for a Theorization," *Public Culture* 12, no. 1 (2000): 215–32. An interesting reflection on nationalism and movement can be seen in Arjun Appadurai, "Sovereignty without Territoriality: Notes for a Postnational Geography," in *The Geography of Identity*, ed. Patricia Yeager (Ann Arbor: University of Michigan Press, 1996), 40–58.

14 Section title in Christopher Bright, "Invasive Species: Pathogens of Globalization," *Foreign Policy* 116, no. 51 (1999): 14.

15 Mark Cheater, "Alien Invasion: They're Green, They're Mean, and They May Be Taking Over a Park or Preserve Near You," *Nature Conservancy* 42, no. 5 (September/October 1992): 24–29.

16 Barbara Stewart, "The Invasion of the Woodland Soil Snatchers," *New York Times*, April 24, 2001, 1B.

17 United Press International, "Native Species Invaded," *ABC News*, March 16, 1998.

18 "Bio-invasions Spark Concerns," *CQ Researcher 9*, no. 37 (October 1, 2000): 856.

19 Quote by refuge biologist Keith Weaver in Joseph B. Verrengia, "When Ecologists Become Killers," *MSNBC News*, October 4, 1999, www.msnbc.com/news (accessed December 20, 2000).

20 Jonathan Brinckman, "Creepy Strangler Climbs Oregon's Least-Wanted List," *Oregonian*, February 28, 2001, 1A.

21 Kim A. McDonald, "Biological Invaders Threaten U.S. Ecology," *Chronicle of Higher Education* 45, no. 23 (February 12, 1999): A15.

22 Ling-Ling Yeh, "U.S. Can't Handle Today's Tide of Immigrants," *Christian Science Monitor* 87, no. 81 (March 23, 1995): 19.

23 Christopher Bright, "Alien Threat," *World Watch* 11, no. 6 (November/December 1998): 8.

24 Jocelyn Kaiser, "Stemming the Tide of Invading Species," *Science* 285, no. 5435 (September 17, 1999): 1836–40.

25 Kaiser, "Stemming the Tide."

26 H. Weiner, "Congress Threatens Wild Immigrants," *Earth Island Journal* 11, no. 4 (1996).

27 Bright, "Invasive Species."

28 Stewart, "Woodland Soil Snatchers."

29 Josef Hebert, "Feds to Fight Invaders," *ABC News*, February 3, 1998.

30 Verrengia, "When Ecologists Become Killers."

31 Mark Robichaux, "Alien Invasion: Plague of Asian Eels Highlights Damage from Foreign Species," *Wall Street Journal*, September 27, 2000, 12A.

32 Verrengia, "When Ecologists Become Killers."

33 Cheater, "Alien Invasion."

34 McDonald, "Biological Invaders," A15.

35 Cheater, "Alien Invasion."
36 Robichaux, "Alien Invasion," 12A.
37 Joseph B. Verrengia, "Some Species Aren't Welcome," *ABC News*, September 27, 1999, www.abcnews.go.com (accessed December 20, 2000).
38 Bright, "Invasive Species."
39 Verrengia, "Some Species Aren't Welcome."
40 Cheater, "Alien Invasion."
41 Verrengia, "When Ecologists Become Killers."
42 Verrengia, "When Ecologists Become Killers."
43 U.S. Geological Survey Midcontinent Ecological Science Center, "USGS Research Upsets Conventional Wisdom on Invasive Species Invasions" (news release, Fort Collins, Colo., May 13, 1999).
44 Cheater, "Alien Invasion."
45 Hebert, "Feds to Fight Invaders."
46 Verrengia, "When Ecologists Become Killers."
47 Hebert, "Feds to Fight Invaders."
48 Hebert, "Feds to Fight Invaders."
49 Verrengia, "Some Species Aren't Welcome."
50 Verrengia, "When Ecologists Become Killers."
51 Verrengia, "When Ecologists Become Killers."
52 Richard N. Mack et al., "Biotic Invasions: Causes, Epidemiology, Global Consequences and Control," *Issues in Ecology* 15, no. 5 (Spring 2000): 12.
53 Daniel Simberloff, quoted in Kim Todd, *Tinkering with Eden: A Natural History of Exotics in America* (New York: W. W. Norton, 2001), 253.
54 Respectively, United Press International, "Native Species Invaded"; Verrengia, "Some Species Aren't Welcome"; and Richard Stone, "Keeping Paradise Safe for the Natives," *Science* 285, no. 5435 (September 17, 1999): 1837.
55 Stone, "Keeping Paradise Safe," 1837.
56 Hebert, "Feds to Fight Invaders."
57 Brinckman, "Creepy Strangler."
58 Verrengia, "Some Species Aren't Welcome."
59 Jonah H. Paretti, "Nativism and Nature: Rethinking Biological Invasions," *Environmental Values* 7, no. 2: 183–92.
60 Jason Van Driesche and Roy Van Driesche, *Nature out of Place: Biological Invasions in the Global Age* (Washington, D.C.: Island Press, 2000).
61 Robert Devine, *Alien Invasion: America's Battle with Non-native Animals and Species* (Washington, D.C: National Geographic Society, 1999).
62 I am indebted to an anonymous reviewer for this insight.
63 Mark Sagoff, "Why Exotic Species Are Not as Bad as We Fear," *Chronicle of Higher Education* 46, no. 42 (June 23, 2000): B7.
64 The campaigns are, of course, not by the same groups. Many ecologists have expressed reservations about genetically modified food. My point, however, is about the rhetoric that circulates in the mainstream United States.
65 Anna Lowenhaupt Tsing makes a similar point in her analysis of native and exotic bees. Tsing, "Empowering Nature, or Some Gleanings in Bee Culture," in *Naturalizing Power: Essays in Feminist Cultural Analysis*, ed. Sylvia Yanagisako and Carol Delaney (New York: Routledge, 1994), 113–43.
66 Mack et al., "Biotic Invasions."
67 In this piece, I have largely focused on the popular press and those scholarly articles and scientists who have been publicized by the mainstream press. Scholarly articles, most often, do not share the sensationalism of the popular press. However, the same biologists employ different rhetoric in scientific and popular writings. The relationship of the popular and scholarly press is a complex one and beyond the scope of this chapter.
68 For example, see David R. Keller and Frank B. Golley, eds., *The Philosophy of Ecology: From Science to Synthesis* (Athens: University of Georgia Press, 2000).

12

The Power *and* the Pleasure?

A Research Agenda for "Making Gender Stick" to Engineers

Wendy Faulkner

This article is structured around four interrelated themes. The first briefly explores some of the equations between masculinity and technology that arguably contribute to the continued exclusion of women from engineering. It asks why this equation is so durable when there are frequent and visible mismatches between masculine images of technology and actual practice. The second theme examines more closely the mismatch between image and practice with findings that point to contradictory constructions of masculinity in the detail of engineering knowledge and practice. It also highlights the tendency for engineering to be conceived in dichotomous terms and considers how this relates to gender dualisms, if at all. The third theme explores the contested suggestion, from both "diversity" politics and standpoint epistemologies, that a stronger presence of women in engineering design could change the shape of artifacts. This debate takes us to the heart of wider debates about gender essentialism, but the evidence is limited. Finally, the fourth theme picks up the challenge in the title of this article—to explore subjective experiences of engineers, particularly their pleasures in technology and how this may be linked with their ambivalent relationships with power. I conclude by encouraging further research to "write gender in" to our understanding of engineers and engineering practice—as necessary steps in better understanding, and destabilising, the technology-masculinity nexus.

MASCULINITY AND TECHNOLOGY: MISMATCHES BETWEEN IMAGE AND PRACTICE

It is notable that whereas women are pushing open the doors of other powerful institutions (including science) and despite nearly two decades of government- and industry-backed "women into engineering" campaigns, the numbers entering engineering are still derisory in most countries. Whether or not they are subject to discrimination and/or discouragement, most girls and women are voting with their feet: it does not even occur to them to get into design roles in technology; they just are not interested. This crucial point is frequently missed by initiatives to get more women into engineering, which typically start from the assumption that women (their socialization, etc.) have to be modified to fit into engineering, not the other way around (Henwood 1996). The virtual failure of these initiatives suggests that the symbolic association of masculinity and technology must be operating strongly.

Within this association, one can discern a series of highly gendered dichotomies, of which three stand out for me. First and foremost is the assumed-to-be mutually exclusive distinction between being people focused and machine focused—one version of the sociological distinction between feminine expressiveness and masculine instrumentalism. Sherry Turkle (1988) shows that women starting out in computing are often reticent about computing because they see hobbyist hackers as the only model for intimacy with computers, and so many hackers appear to eschew or be incapable of human intimacy. Similarly, Tove Håpnes and Bente Rasmussen (1991) demonstrate that a central reason for the declining intake of young women into computer science in Norway is girls' rejection of the "nerd" image of computer hackers. It seems that for a women to opt to work so closely with technology is potentially to reject any meaningful engagement in the social world and so face "gender inauthenticity" (Keller 1985; Cockburn 1985).

Of course, most women routinely interact with people and technologies; some even develop strong emotional attachments to particular artifacts.[1] As feminist scholars of technology have long argued, however, women's everyday encounters with technological artifacts are rarely recognized as such (Berg and Lie 1995). Our most common cultural images of technology—industrial plants, space rockets, weapon systems, and so on—are large technological systems associated with powerful institutions. Here we meet a second interesting dualism—in this case, implicitly rather than explicitly gendered. "Hard" technology is inert and powerful like the examples above; this is real technology. "Soft" technology is likely to be smaller scale, like kitchen appliances, or more organic, like drugs; most people do not readily identify such products as "technology." So the world of technology is made to feel remote and overwhelmingly powerful because the hard-soft dualism factors out those other technologies that we all meet on a daily basis and can in some sense "relate to."[2]

Third, the hard-soft dualism also finds expression in relation to styles of thought (Edwards 1996, 167–72) since the association of engineering with science brings with it long-standing gender dualisms. On the masculine side of those dualisms, we have an objectivist rationality associated with emotional detachment and with abstract theoretical (especially mathematical) and reductionist approaches to problem solving. On the feminine side, we have a more subjective rationality associated with emotional connectedness and with concrete, empirical, and holistic approaches to problem solving. As we will see under the next two themes, abstract styles of thinking and working are often associated more with men and concrete ones with women. Yet both sides of the concrete-abstract dualism are required within engineering. I return to these familiar dualisms for the same reason Evelyn Fox Keller (1990) does in relation to science: they are widely held as truths by technical and nontechnical people, women and men alike.[3] The fact that popular images of both science and technology are strongly associated with the masculine side of these dualisms must be one of the reasons why, in a deeply gender divided world, most girls and women do not even consider a career in engineering.

Evidence from studies of technology education in schools tends to confirm the operation of all three sets of dualisms addressed here (namely, people vs. technology focused, soft vs. hard technology, and concrete vs. abstract approaches). Early differences in interests and role-playing developed outside school shape how girls and boys respond to, and are interpreted as responding to, technology inside. For example, girls are more likely than boys to feel confident about, and to succeed in, working with tables of data concerning health, reproduction, or domestic situations but anticipate failure—"I don't know anything about that"—when faced with tables of data on machinery, building sites, or cars (Murphy 1990). The reverse holds for most boys; the task is the same, but the content is gendered. Girls are usually less confident than boys in handling "real technology"—and this extends to the use of all sorts of equipment in school, which boys tend to monopolize. The greater people-centeredness of most girls also is reflected in how they approach technical tasks. Recent surveys undertaken in U.K. schools reveal that teenage girls in design and technology classes are more likely than boys to "identify the issues that underlie tasks in empathizing with users and evaluating products and systems in terms of how well they might perform for the user," whereas boys are more likely to approach technical tasks in isolation and judge the context to be irrelevant (Kimbell, Stables, and Green 1996, 94; also Murphy 1990).[4] This is an astounding result: it seems that girls demonstrate greater potential in precisely those holistic and heterogeneous approaches so necessary to success in technological design, as well as a thirst for what educationalists call "deep understanding," yet their different "learning styles" are read by teachers as indicating a lack of confidence or ability.

Of course, many of the ways of thinking and doing that we stereotypically deem feminine are useful if not essential in technical work: linguistic abilities in computer programming, for instance. And plenty of women now do jobs that are extremely technical, just as plenty of men are technically incompetent. In short, there are huge mismatches between the image and practice of technology with respect to gender. This crucial point is often missed. Yet I believe it obliges us to look more closely at why the gendered dualisms, so palpable in cultural images of technology, are so effective in keeping it a male domain. Specifically, we need to examine the relationship between the continued male dominance of engineering and masculine images of technology and how these images are sustained. For example, how much do engineers contribute to promoting these images, and how are they influenced by them? What influence do these images have on young people more generally? How much do "we" nontechnical people "need" to believe in the objectivity of our experts and thus contribute to sustaining these images? What other interests do they serve?

CONTRADICTORY GENDERING IN ENGINEERING KNOWLEDGE AND PRACTICE

The mismatch between image and practice highlighted above invites us to bring a gender gaze to the "black box" of engineering knowledge and practice. Though there is a crying need for detailed empirical work, we do have a small number of studies that together appear to suggest that the gendering taking place at this level is both more complex than conventionally assumed and highly contradictory.

A useful way to view this complexity is to focus on some of the dichotomous ways in which engineering work is often categorized. The most obvious and perhaps pivotal of these is the distinction between the manual labor of the craft or technician engineer,[5] who works directly on the artifact in a greasy workshop, and the mental labor of the professional (graduate) engineer, who frequently works remotely from the artifact (via a computer) in an almost clinically clean office. As Judy Wajcman (1991, chap. 6) argues, these two versions of masculinity are essentially class based and embody the often gendered dualism of mind-body. But the distinction is also reproduced with professional engineering practice since it nevertheless involves hands-on "tinkering" work as well as mathematical analysis. So, the dichotomy from science, which labels concrete, empirical approaches as feminine, is clearly at odds with the importance of hands-on work in technician and professional engineering. Men are deemed to be more "natural" technologists because they possess both the appropriate rationality and good mechanical skills.

Paradoxically, while men's relationships with artifacts can be cast as instrumental in the sense of being task rather than people centered, it is also expressive. Like others, McIlwee and Robinson (1992) found that most men engineers display a deep fascination with tools, machinery, and gadgets. They conclude that

> the culture of engineering involves a preoccupation with tinkering that goes beyond the requirements of the job. Vocation becomes avocation, and, in turn, devotion.

> It is not enough to be competent in the hands-on aspects of engineering: one should be obsessed with them. It is not enough to know the difference between a piston and a rod: one should take obvious joy in this knowledge. The engineers must be ready not only to engage in technical exchanges during work periods, but interested in participating in them during breaks as well. To be seen as a competent engineer means throwing one's self into these rituals of tinkering. (McIlwee and Robinson 1992, 139)

The women engineers studied did not share this obsession; they had other topics of conversations and sources of joy. Hobbyist hackers, like tinkering engineers, have a very expressive relationship with their artifacts (which, as we saw earlier, often alienates aspiring women computer specialists). To a lesser extent, so do "ordinary" male users of home computers, where their wives are quite instrumental—they use them as a tool (Aune 1996). I believe the pleasures that men engineers take in technology are a very important element of "what makes them tick," and I will return to this point in the last of my four themes.

Juxtapositions of apparently dualistic concrete and abstract approaches are also found in computing. Software developers often draw a distinction between top-down planning approaches to programming and more "bottom-up" approaches involving trial and error. The term *hacking* is often used to describe the latter approach, yet hobbyist hackers are generally boys and young men. So once again we see one version of masculinity associated with concrete approaches coexisting with another version of masculinity based on more abstract approaches. Håpnes and Rasmussen (1991) found that while computer science teachers favor the "dedicated" computer science students who adopt the more formal approaches they teach, they nonetheless respect the hackers. Moreover, Håpnes (1996) provides some rich evidence to show that many familiar and often gendered dichotomies are ambiguously but intimately and necessarily combined in the course of programming work. So hobbyists consciously use judicious mixtures of concrete and abstract approaches because while logical planning and command approaches are effective for many tasks, interaction with the system—"muddling through," as one called it—is often essential, for example, in identifying bugs or in finding one's way around unfamiliar software.

Within engineering education (at least until recently in the United States), higher status and credit attach to the more mathematical and abstract analytical work and less to hands-on concrete work—even though it is widely recognized in the profession that those who become the best engineers are often not those who perform the best academically (Hacker 1989, chap. 3). Sally Hacker's (1989) account of life as an engineering student is reminiscent of the "hazing experience" that new recruits to the military forces are put through.[6] Engineering education is characterized by seemingly endless and repetitive drills of mathematical problem solving (Hacker 1989, chap. 3). As well as cementing a mathematical approach to engineering (see below), Hacker saw this induction as channeling the passions and erotic energies of would-be engineers, creating a profound mind-body split that she argued is alien to many women. To explore this theme further, she interviewed humanities and engineering faculty members at an elite institute about their early life (Hacker 1990, chap. 4). She found that while both groups were strongly cerebral as opposed to corporeal in their interests at high school—most had not excelled at sport and were shy of romantic and sexual relationships—in adult life, the engineers continued to experience far more anxiety about their bodies and about emotionally demanding personal relationships than did the humanities academics. Hacker's academic elite are probably an extreme case in terms of the mind-body split, given the evidence cited above for the prominence of concrete approaches and of displays of technical competence in engineering work in industry. Moreover, the evidence that sport is a primary outside interest for many engineers (Mellström 1995; McIlwee and Robinson 1992) does not tally with the picture of engineers "having a problem with their bodies." But while the mind-body dichotomy should not be overstated, that between rationality and emotionality is very evident.

It is common to find at least some engineers who seek refuge from human relationships in mathematics or technology (e.g., Mellström 1995; Håpnes 1996). More subtle signs of a "problem with emotions" can be found in relation to the use of mathematics in engineering design. Typical of much engineering discourse, literary engineer Samuel Florman (1976, 142) talks of the "pristine realm of pure mathematics" and "the pure beauty of mathematics" and sees human complexity as the other messy but necessary side of this coin in engineering. In a very interesting passage, Louis Bucciarelli (1994, 108) deconstructs a typical university engineering problem to demonstrate just how much of this complexity has to be pared away: "The student must learn to perceive the world of mechanisms and machinery as embodying mathematical and physical principle alone, must in effect learn to *not* see what is there but irrelevant. . . . Reductionism is the lesson." Two things stand out here. First, the emotional detachment of the dry, formal processes of analytical problem solving learned at university stands in stark contrast to the emotionally laden dramas that ethnologists (Bucciarelli 1994; Mellström 1995) and journalists (Kidder 1981; Lovell and Kluger 1994) have observed unfolding in the course of engineering design and problem-solving work in industry. Second, exercises of reductionism exclude much "social" information that is vital to successful design.

Sociologists of technology have long argued that engineers must take a holistic view and integrate heterogeneous technical and nontechnical elements if artifacts are to "work" and meet a "real" need (Sørensen and Levold 1992; Law 1987; Vincenti 1990; Faulkner 1994). This provides an interesting contrast to science, where reductionist approaches and specialisms are generally esteemed far more highly than holistic ones and where holism tends to be gendered feminine.[7] As Knut Sørensen (1997) reminds us, however, the rather heroic model of the heterogeneous engineer—the "captains of industry," such as Thomas Hughes's (1983) portrayal of Thomas Edison—is increasingly at odds with the reality facing most engineers today since engineers are increasingly defined by their technical specialism and only participate in a holistic project to the extent that they are organized into multiskilled collective endeavors. The fragmented labor process in engineering (e.g., Constant 1984) therefore houses a dichotomy between narrowly defined specialist tasks, where reductionism is acceptable, and heterogeneous tasks, where holism is necessary and "the social" cannot be deemed irrelevant. Fieldwork observations in Scotland reveal interesting differences in how the heterogeneous role of user interaction in information systems gets gendered.[8] In universities, where the users are individuals and the role is seen in terms of supporting, we see a higher proportion of women in these roles than in the more technical jobs; in companies, where the users are other businesses and the role has marketing connotations, we see it occupied by men, leaving comparatively more women in the "backroom" technical jobs. By contrast, McIlwee and Robinson (1992) found that in U.S. companies, design roles remain higher status among engineers, and women are generally making greater incursions into engineering marketing and management.

I have shown that many dualistic epistemologies found in engineering practice are gendered in contradictory ways and that many fractured masculinities within engineering are sustained simultaneously—among engineers as a group and, to varying degrees, by individuals: they coexist in tension. Like Håpnes (1996), I am struck by the clear evidence that both sides of these dichotomies—concrete and abstract, specialist and holist, technical and social (one could go on)—are necessary to engineering work. Bucciarelli (1994, 48) comments that "when we look at the contemporary world and see technology, we often oversimplify and split the world in two." Engineering knowledge and practice are conceptualized in dichotomous terms even when there is no necessary hierarchy or obvious gendering implied—for example, formal versus experiential knowledge (Vincenti 1990), visual versus analytical knowledge (Ferguson 1992). Mellström (1995, 76) also notes the primacy of what he calls "binary thinking": "Technical problems are given the character of either-or, plus-minus, negative-positive, and in its most basic technical form: zero or one." He sees this binary thinking as grounded in "a complete faith in cause and effect," epitomized for him by the outburst of a frustrated designer one day: "Either things work or they don't!"

There is nothing inherently gendered about the distinctions addressed here, nor (I suspect) are they intrinsic to engineering. For this reason, it might be worth exploring further why dichotomous or dualistic thinking appears so endemic to engineering (and engineers) and whether this relates at all to gender. Gender is generally conceived of in dichotomous terms—because of the obvious link with sex (as in femaleness and maleness) but also because heterosexuality is usually posited ideologically on an attraction of gendered opposites. I agree with Henwood (1994) that heterosexism is an underresearched theme in the gendering of technology and believe that it may provide at least partial answers. But I would also stress that dualistic ideologies are still ideologies; real women and men do not fit dichotomous assumptions any more readily than do real engineers.

SEX AND GENDER IN ENGINEERING: THE QUESTION OF "STYLES"

This third theme explores another potential but contested aspect of gender in engineering—namely, whether there are gender differences in how engineering is done or whether women and men (*might*) bring different styles, perspectives, and priorities to engineering. Let me establish some "facts" from the outset: there are no innate differences in technical ability between women and men, girls and boys; there are no universal differences in how females and males engage with technical tasks.[9] But there is suggestive evidence of some differences in some settings—for instance, the evidence that girls tend to bring a more heterogeneous approach to design tasks in school.

In their study on the acquisition of programming skills by school and college students, Sherry Turkle and Seymour Papert (1990) found that girls and women tend to adopt an interactive or relational, "bricoleur" approach, while boys and men tend to adopt a formal, linear, and hierarchical planning approach.[10] Both approaches "work," yet the bricoleurs found themselves actively discouraged by their teachers, forced to pursue this approach surreptitiously or unlearn it or give up on computing. Turkle and Papert see a clear link between the two styles and wider patterns of gendering, invoking object-relations psychoanalytical theory to argue that boys will tend to favor abstract approaches, while girls will tend to favor concrete, relational approaches. Personally I am not convinced that appeal to psychology is necessary since early socialization is extremely influential. And I am not convinced that the gender split found by Turkle and Papert obtains among working programmers; many women programmers favor abstract approaches because they lack teenage experience at hacking. However, I am struck by their conclusion that "the computer supports epistemological pluralism, but the computer culture does not" (Turkle and Papert 1990, 132). The point is, then, that in both the design and programming classes, dominant gendered assumptions—about males having an aptitude and about the value of nominally "masculine" styles—are sustained even in the face of counterevidence.

This phenomenon indicates aspects of the exclusion of would-be female technologists rarely grasped by equal opportunity campaigns. It also begs some intriguing questions about the consequences of the gendering and male dominance of engineering on the design of technological artifacts. Might the greater participation of women in engineering in any way change technological design and thus the shape of new technologies in the future? Or, more broadly, could engineering support different epistemological styles of work? These questions are being raised (albeit in different languages) in both liberal discourses within equal opportunities campaigns and radical discourses within technology studies and feminism.

The liberal discourse is essentially a democratic case for inclusion: should such a powerful occupation as engineering be predominantly shaped by a singular set of values and styles? Certainly this is the thrust of much recent equal opportunities campaigning throughout the industrialized world, where the current fashion is to stress the possible benefits to male-dominated areas (quite apart from equity) of the diversity that it is argued would accompany the higher representation of both women and ethnic minorities: diversity in inputs to innovation or marketing, for instance.[11] The "diversity" position builds implicitly on an assumption that women *by being women* bring different approaches and priorities—an assumption that many see as dangerously essentialist. At the very least, it fails to challenge stereotypical constructions of femininity and masculinity or to acknowledge that these constructions are not just different but unequal (in the sense that femininity is associated with subordinate and masculinity with controlling roles in society). For example, a common strategy used to encourage more women into computing and engineering is to emphasize the social elements of technical work—such as the growing importance of organization or communication in information technology—on the assumption that women will both be more attracted to and have more to offer to engineering if it is defined in nontechnical terms. As Hapnes and Rasmussen (1991) argue, this strategy leaves intact the equation of technology and masculinity—and, I might add, the dichotomy between feminine expressiveness and masculine instrumentalism on which this equation partially rests.

The radical discourse within technology studies is an extension of social constructivism, which carries the possibility that technological artifacts, once deconstructed, can be reconstructed (e.g., Bijker and Law 1992) or, to use the language of social shaping, that there are "roads not taken" that could in principle be followed (MacKenzie and Wajcman 1985). In addition, the notion of "styles" is familiar in the sociology of scientific knowledge. Yet, very few studies have given detailed consideration to how the male dominance of engineering affects the actual shape of technological artifacts,[12] and none has really "got inside" the design process to explore the possibility of there being differently gendered "styles" or outcomes.

The radical discourse within feminism is reflected in the insistence of many activists that men's interests, priorities, perceptions, and experiences are bound to be reflected in the design of artifacts. For some, it has long been an article of faith that the better representation of women in technology would, by itself, begin to transform both the products of technology and its modus operandi (e.g., Arnold and Faulkner 1985). Yet feminist standpoint epistemology has not been formally applied to technology—theoretically or empirically. Few would argue with the notion that women designers should be more likely to "see" the needs of particular female users (e.g., for wider gangways on buses, to allow for women with young children, or airbags that are not lethal to short women). More contested is Hilary Rose's (1983, 1994, chap. 2) argument that women are more likely to bring a "caring rationality" to technical work because their position in the sexual division of labor means that they generally do (or are more socialized to do) more caring work than men. This argument is appealing insomuch as it would be "a good thing" if there were a greater "professional ethic of care" in engineering (Andersen and Sørensen 1994).

The question of gendered styles tends to divide women engineers as it does women scientists (Barinaga 1993). The politics of diversity have made it more acceptable for women engineers to talk about the particular contribution they can make, but the pressure to become accepted as "one of the guys" (Cockburn 1985, chap. 6; Hacker 1989, chap. 3; Carter and Kirkup 1990) means that many vehemently deny any differences. Keller (1992) has argued that we need to "learn to count past two"—past the two popular choices that either see women as the same as men, and so deny gender, or see women as different from men, and so treat gender as fixed. In seeking, tentatively, to count past two, I am inclined to adopt the language of "styles"—but without embracing essentialism. I do so with the expectation that epistemological pluralism is in principle possible but that normative pressures of various types will generally result in the suppression of all but a few styles. It would be very useful to explore further which "styles" get suppressed and whether this is gendered at all.

SUBJECTIVITIES IN ENGINEERING: THE POWER AND THE PLEASURE

It will be apparent that for me the durability of the equation between masculinity and technology remains problematic and insufficiently explored—hence my concern with the mismatch between masculine images of technology and technical practice, and with the complex and contradictory gendering that takes place within engineering. Like Fergus Murray (1993), I believe we need to take a closer look at engineers' subjective experiences—my fourth theme—to explore the equation further. In particular, I propose that three subjectivities warrant further attention: the pleasure engineers so palpably take in technology, their ambivalent relationships with power, and the identity work they do.[13] I will argue that engineers' pleasure in technology and their close identification with technology are crucial elements in the individual identities and shared culture of engineers. They provide some solace and reward to engineers whose everyday work and lives often offer only limited excitement or power. And they cement a fraternity that effectively excludes women engineers from important informal networks.

The "glint in the eyes" of obsessive tinkerers and hobbyists is evident to all who have cared to look. This theme became a central focus of Sally Hacker's (1990) later work. She both witnessed and experienced the sensual, even erotic, pleasures to be had in making things work. Similarly, in *The Existential Pleasures of Engineering*, Samuel Florman (1976) extols at length the sensual absorption, spiritual connection, emotional comfort, and aesthetic pleasures to be found in engineers' intimacy with technical artifacts. In all of the studies reported here, engineers manifest sensual and emotional pleasures (if not spiritual or aesthetic ones) in working with technology. And in virtually all cases, such pleasures are first experienced in early childhood, and they strengthen during adolescence—in the classic case of taking cars apart, technology provides a rare focus for bonding between fathers and sons—with the result that engineering is a "self-evident" career choice for most male engineers (Mellström 1995, chap. 7).

Underlying engineers' central mission, Florman (1976) argues, is a more basic pleasure that, in effusively gendered language, he sees as grounded in an instinctive desire to change if not conquer the natural world; "born of human need, [this desire] has taken on a life of its own. Man the creator is by his very nature not satisfied to accept the world as it is" (Florman 1976, 120). Conscious that nature is "not to be tampered with unthinkingly," Florman suggests that the awesome and intimidating scale of nature nevertheless inspires in the civil engineer a "yearning for immensity," an existential impulse for the "vanity of pyramids or dams"—constructions that "inevitably invoke thoughts of the divine" (pp. 122, 126, 124). In another delightful passage, he describes engineers' pride in the machine:

> The machine still stands as one of mankind's most notable achievements. Man is

weak, and yet the machine is incredibly strong and productive. The primordial joy of the successful hunt of the abundant harvest has its modern counterpart in the exhilaration of the man who has invented or produced a successful machine. (p. 130)

Hacker (1989, chap. 3; 1990, chap. 9) also perceived that part of the pleasure of engineering is a pleasure in domination and control—over workers as well as the natural world—but sees this pleasure as echoing prominent themes in present-day eroticism rather than male instincts. The connection with eroticism is frequently hinted at by feminists, not least because it is overwhelming in the language of potency and birth that surrounds military technology (Easlea 1983). In her very thoughtful and illuminating ethnography of defense intellectuals, Carol Cohn ([1987] 1996) suggests that this language does not necessarily reflect individual motivations; rather, it may function to tame or make tenable "thinking the unthinkable"—of nuclear annihilation. In a similar vein perhaps, Hacker suggested that the fun engineers have with technology is a compensation for contributing to larger systems of dominance and control—an especially important "reward" when so many engineers occupy fragmented roles in the labor process and other sources of job satisfaction may be limited (Hacker 1989, chap. 3).

Hacker's (1989, 1990) crucial contribution was to show us how the power and the pleasure of engineering are linked—hence the title of this article. I see a resonance with Florman's (1976) rhetoric about dams and machines: the power of the technology symbolically extends engineers' limited sense of strength or potency. Often the men who appear to take the most pleasure in technology are relatively unpowerful—hackers and other technical hobbyists are obvious examples. Maureen McNeil (1987, 194) asks, "Couldn't the obsessional knowledge of some working class lads who are car buffs, or some of the avid readers of mechanics or computer magazines, be interpreted as evidence of impotence?" Flis Henwood (1993, 41) responds that in such cases, technology offers a symbolic promise of power, as well as the potential to compensate materially for their relative lack of class power by "a strengthening of their gender power" through the acquisition of technical expertise. Perhaps another kind of power promised by engineering is power over wayward bodies and emotions, given Hacker's (1990, chap. 4) evidence, cited under my second theme, about adolescent anxieties. Drawing on Turkle (1984), Paul Edwards (1996, 171) suggests that the "holding power" of computer programming lies in the simulated character of the "microworlds" so created. These microworlds are appealing because, on one hand, they offer "vast powers to transform, refine, and produce information" and, on the other, they are free of unwanted emotional complexity. Within them, "things make sense in a way human intersubjectivity cannot," and "the programmer is omnipotent" (if not omniscient). So, Edwards suggests. "For men, to whom power is an icon of identity and an index of success, a microworld can become a challenging arena for an adult quest for power and control" (p. 172).

It seems that mathematical and technical prowess offers some engineers at least compensation for a lack of prowess or control in various realms. As Downey and Lucena (1995) note, "Engineers routinely feel powerless themselves but are viewed as highly empowered by outsiders"—a rather clear case of many men's contradictory experience of power (cf. Kaufman 1994). Even accomplished professional engineers are prone to feeling powerless in relation to politics—witness the number who, privately at least, harbor technocratic dreams in the belief that the world would be a better place if it were run by people like them. Engineers' ambivalent relationship with power has roots in their structural location within capitalist industry. Situated ambiguously between capital, labor, and the state, they have collectively identified themselves with technology because this self-ascription provided them with a cloak of neutrality (Berner 1992). So throughout their history, engineers have faced tensions between a desire for professional autonomy and the demands of corporate or state employers (e.g., Layton 1971). Mellström's (1995, 54) engineers all complained about clashes between engineering perspectives and business perspectives—over the time required to develop a design, for instance—and saw the latter "as threatening the technical core of their professional identity." The ambivalence is profound: the limited opportunity to do "real engineering" is probably the single largest source of dissatisfaction among engineers (Downey and Lucena 1995; Mellström 1995), yet several studies report that most engineers are interested in acquiring organizational power (Zussman 1985; Whalley 1986; McIlwee and Robinson 1992, 20)—which usually implies moving away from narrowly technical work.

The collective identification with technology attributed to engineers' structural location has a subjective counterpart. Florman's (1976) rhetoric about pleasure captures just how profoundly engineers identify themselves with their technology. Engineers in the car industry are proud to be associated with such a visible artifact and see themselves as "hardware men"; the design jobs that hold the highest status and challenge are those associated with the "hard mechanical core" of the engine, drive line, and chassis (Mellström 1995, chap. 3). Similarly, engineers in microprocessors have an "unmistakable identification with future technology" and experience great professional pride when they can demonstrate their technical virtuosity to their peers, in producing "a beautiful layout," for instance (Mellström 1995, chap. 4).

I suspect that engineers' close identification with technology is precisely what makes it harder to establish a positive identity for themselves among nonengineers; it gives them a "separate reality," to use Murray's (1993) phrase. Indeed, we may view engineering as a fraternity built around this common identity with, and pleasure and pride in, technology. Higher education is typically where engineers meet others like themselves in large numbers for the first time. They survive the "hazing experience" through solidarity: they are "all in it together." In Sweden, the shared social life of engineering students revolves around displays of drinking prowess and a common interest in sports (Mellström 1995, chap. 7). And in the United States, where fewer engineers seem to be athletic, this culture spills over into the "technical locker room" of the workplace populated as it is by "jocks" engaged in ritualistic displays of hands-on technical competence; "the image of technical virtuosity is revered—even where actions of technical virtuosity in the lab are rare" (McIlwee and Robinson 1992, 139). The ethnographic work of both Mellström (1995, chap. 5) and Hacker (1990, chap. 4) shows that this is a recurring theme in the jokes and stories that engineers share with one another. Engineers' humor typically celebrates their technical prowess and ridicules the lack of it in others.[14] By this stroke, technical prowess is what defines them as engineers and what gives them power (since the "outgroup" is defined as less powerful because they lack such prowess). . . .

Murray (1993) points out that the collective identity of engineers draws heavily on the image of the war hero—withdrawing from normal life and sacrificing all for the common good. Dramatic examples are provided in Tom Wolfe's (1980) account of the hardship and privation that early male astronauts went through, aptly titled *The Right Stuff*; Tracy Kidder's (1981) portrayal of computer hardware design in *The Soul of a New Machine*; and Brian Easlea's (1983) study of the Manhattan Project. Murray suggests that the same phenomenon occurs more routinely in the development of business applications software. Here, engineers in project teams experience strong bonding and loyalty, working together on a common objective for which personal commitments, and sometimes their health and sanity, have to be sacrificed to get the new software to the customer on time. It is likely that this experience is repeated in numerous industries as companies compete under the market pressure to "innovate or die" (Freeman 1982, chap. 1). The thrill of the shared mission and its successful completion in a new product launch is short-lived as the new product becomes superseded and the process must begin all over again (often with teams being disbanded and regrouped in the process). Bucciarelli (1994, 195) and Mellström (1995, 99) both observe how lifeless and lacking in human imprint the final artifact seems once all the drama of its design is over. The pain and the pleasure experienced by engineers on this roller coaster thus manifest and reflect the demands of capitalist competition for market share.

In sum, engineers chose their work for more than money or status.[15] The pleasure in technology is a strong motivator and a significant reward. Together with a shared pride in technology and in technical competence, it is a central element in the individual identities and shared culture of engineers—and so acts to demarcate the engineers from outsiders (and often men engineers from women engineers). The symbolic associations of both technology and technical prowess with power may act as compensation in a situation where most engineers perceive themselves to have only limited power.

Finally, this article is offered as a challenge to further unravel the technology-masculinity nexus by bringing a gender gaze to the study of engineers and engineering practice. It has shown that engineering may be gendered in four ways: (1) in symbolic representations and images of engineering available to those outside of engineering, (2) in symbolic gendering of engineering knowledge and practice, (3) in gender differences in how engineering is done, and (4) in engineers' subjective experiences and identities as engineers. I believe further research on these themes is justified because engineers represent such a powerful instantiation of the equation between masculinity and technology, and I believe that one important consequence of such work will be to destabilize that equation (cf. Haraway 1986).

Notes

AUTHOR'S NOTE: I would like to thank the following for taking the time to read and comment on earlier written versions of this article: Frank Bechhofer, Liz Bondi, Annette Burfoot, Cynthia Cockburn, Flis Henwood, Lisa Jacobsen, Anne Kerr, Gill Kirkup, John MacInnes, Donald MacKenzie, Elaine Seymour, Knut Sørensen, Jan Webb, and the anonymous reviewers. Verbal presentations were made to the Gender Studies and Science and Technology Studies networks in Edinburgh in early 1997; to the Network of European Centres in Science and Technology Studies (NECSTS) Workshop on Gender, Science, and Technology, held at NTNU, Trondheim, Norway, in May 1997; and to a colloquium of the Women's Studies Center at the University of Colorado at Boulder in April 1998. I am grateful to all those who shared their thoughts and experiences with me on those occasions, especially Sandi Bish, Ginger Melton, and Jill Tietjen, who between them have spent nearly fifty years in engineering.

1. This is delightfully illustrated by Inga's outpouring of love for her washing machine in Berg (1997, chap. 4).
2. Dominant cultural images of technology may be changing as information and communication technologies become increasingly accessible, but even here the hard-soft dualism is evident.
3. An illustration of this was provided by the 1993 special issue of *Science* on women in science, which gave considerable credence to the notion that women bring different "styles" to science—on the basis of interviews of both female and male scientists (Barinaga 1993) and popular accounts of how early women entrants brought "empathy" and "patience" to the field of primatology (Morell 1993).
4. An example here might be the task of specifying a fabric for use in mountain conditions: girls would be more likely to ask what fabric is being used for. For example, will its wearer be active or stationary, which, of course, has a bearing on whether the fabric used needs to be breathable (Patricia Murphy, private communication, October 1996).
5. The label "craft engineer" may not have meaning outside the United Kingdom: it connotes someone who is not university trained and who usually works in the maintenance, installation, or manufacture of artifacts.

6 The term *hazing* is uniquely American; the parallel was suggested to me during 1998 by Ginger Melton (of the University of Colorado at Boulder) but is widely recognized among U.S. engineers.
7 Within the life sciences, for example, molecular biology is seen as very exciting while whole organism and observational sciences, such as animal behavior and primatology, are seen as decidedly more "girlie."
8 This material was shared with me during 1996 by Jan Webb of the University of Edinburgh's Sociology Department.
9 For example, longitudinal studies show no significant difference in how girls and boys do mathematics—either in how well they perform or in how they approach the task (Walkerdine and The Girls and Mathematics Unit 1989).
10 In her earlier study, Turkle (1984) used the terms *hard* and *soft* mastery. I believe this was unfortunate since, like the terms *hard* and *soft* technology, they infer a gendered hierarchy even where none may exist or be warranted.
11 I take it as significant that a senior policy person in the United Kingdom was willing to endorse my proposal for a pilot study on gender in engineering on the basis that it sought to explore whether there were differences in style and whether companies suppress styles (including "feminine" styles), which might be useful to them.
12 I believe the approach of simultaneously investigating the design and use of technology—to reveal the interaction between them—is extremely worthwhile and warrants more widespread adoption. It has been put to good use in terms of further gender analyses of technology in some recent historical work collected in a special issue of *Technology and Culture* ("Gender Analysis" 1997) and in the European collaborative Vienna project (Cockburn and Dilic 1994: Cockburn and Ormrod 1993), both of which investigate a series of everyday technologies, as well as a recent study on how cockpit design has been geared to "fitting" the average bodily proportions of men, not women (Weber 1997).
13 By "identity work," I understand the work that individuals and groups do to create or sustain an image or identity with which they can feel comfortable.
14 One relatively well-known example is the list often pinned to a door or wall that starts, "Real programmers. . . ." Mellström (1995, 77) found one such list that included the entry, "Real programmers don't write in BASIC. Actually, no programmers write in BASIC, after the age of 12."
15 Engineers in the United Kingdom, uniquely I suspect, enjoy little of either money or status, which makes the resistance to women entrants in that country particularly interesting.

References

Andersen, H. W., and K. H. Sørensen. 1994. *Frankenstein's Dilema: En Bokom Teknologi, Miliø of Verdier*. Oslo, Norway: Ad Notam Glydendal.
Arnold, E., and W. Faulkner. 1985. Smothered by invention: The masculinity of technology. In *Smothered by invention: Technology in women's lives*, edited by W. Faulkner and E. Arnold, 18–50. London: Pluto.
Aune, M. 1996. The computer in everyday life: Patterns of domestication of a new technology. In *Making technology our own? Domesticating technology into everyday life*, edited by M. Lie and K. H. Sørensen, 91–120. Oslo: Scandinavian University Press.
Barinaga, M. 1993. Is there a "female style" in science? *Science* 260: 384–91.
Berg, A.-J. 1997. *Digital Feminism*. Dragvoll, Norway. Senter for Teknologi og Samfunn, Norwegian University of Science and Technology.
Berg, A.-J., and M. Lie. 1995. Feminism and constructivism: Do artifacts have gender? *Science, Technology, & Human Values* 20 (3): 332–51.
Berner, B. 1992. Professional or wage worker? Engineer and economic transformation in Sweden. *Polhem* 2: 131–61.
Bijker, W., and J. Law, eds. 1992. *Shaping technology/building society*. Cambridge: MIT Press.
Bucciarelli, L. L. 1994. *Designing engineers*. Cambridge: MIT Press.
Carter, R., and G. Kirkup. 1990. *Women in engineering: A good place to be?* London: Macmillan.
Cockburn, C. 1985. Caught in the wheels: The high cost of being a female cog in the male machinery of engineering. In *The social shaping of technology*, edited by D. MacKenzie and J. Wajcman, 55–66. Milton Keynes: Open University Press.

Cohn, C. [1987] 1996. Sex and death in the rational world of defense intellectuals. In *Gender and scientific authority*, edited by B. Loslell, S. G. Kohlstedt, H. Longino, and E. Hammonds, 183–214. Reprint, Chicago: University of Chicago Press.

Constant, E. 1984. Communities and hierarchies: Structures in the practice of science and technology. In *The nature of technological knowledge: Are models of scientific change Relevant?* edited by R. Laudan, 27–43. Dordrecht, the Netherlands: Reidel.

Downey, G. L., and J. A. Lucena. 1995. Engineering studies. In *Handbook of science and technology studies*, edited by S. Jasanoff, E. G. Markle, J. C. Petersen, and T. Pinch, 167–88. Thousand Oaks, CA: Sage.

Easlea, B. 1983. *Fathering the unthinkable: Masculinity, scientists and the nuclear arms race.* London: Pluto.

Edwards, P. N. 1996. *The closed world: Computers and the politics of discourse in cold war America.* Cambridge: MIT Press.

Faulkner, W. 1994. Conceptualizing knowledge used in innovation: A second look at the science-technology distinction and industrial innovation. *Science, Technology, & Human Values* 19 (4): 425–58.

Ferguson, E. J. 1992. *Engineering and the mind's eye.* Cambridge: MIT Press.

Florman, S. 1976. *The existential pleasures of engineering.* New York: St. Martin's.

Freeman, C. 1982. *The economics of industrial innovation.* 2d ed. London: Frances Pinter.

Hacker, S. 1989. *Pleasure, power and technology: Some tales of gender, engineering, and the cooperative workplace.* Boston: Unwin Hyman.

——. 1990. *Doing it the hard way: Investigations of gender and technology.* Boston: Unwin Hyman.

Håpnes, T. 1996. Not in their machines: How hackers transform computers into subcultural artefacts. In *Making technology our own? Domesticating technology into everyday life*, edited by M. Lie and K. H. Sørensen, 121–50. Oslo: Scandinavian University Press.

Håpnes, T., and B. Rasmussen. 1991. Excluding women from the technologies of the future? *Futures*, December, 1107–19.

Haraway, D. 1986. Primatology is politics by other means. In *Feminist approaches to science*, edited by R. Blejer, 77–118. Elmsford, NY: Pergamon.

Henwood, F. 1993. Establishing gender perspectives on information technology: Problems, issues and opportunities. In *Gendered by design? Information technology and office systems*, edited by E. Green, J. Owen, and D. Pain, 31–52. London: Taylor & Francis.

——. 1994. *Engineering difference: Discourses on gender, sexuality and work.* Working Paper Series No. 3. London: Innovation Studies, University of East London.

——. 1996. WISE choices? Understanding occupational decision-making in a climate of equal opportunities for women in science and technology. *Gender and Education* 8 (2): 199–214.

Hughes, T. 1983. *Networks of power: Electrification in Western society 1880–1930.* Baltimore, MD: Johns Hopkins University Press.

Kaufman, M. 1994. Men, feminism, and men's contradictory experiences of power. In *Theorizing masculinities*, edited by H. Brod and M. Kaufman, 142–64. Thousand Oaks, CA: Sage.

Keller, E. F. 1985. A world of difference. *Reflections on gender and science*, 158–76. New Haven, CT: Yale University Press.

——. 1990. From secrets of life to secrets of death. In *Body/politics: Women and the discourses of science*, edited by M. Jacobus, E. F. Keller, and S. Shuttleworth, 177–91. New York: Routledge Kegan Paul.

——. 1992. How gender matters, or, why it's so hard for us to count past two. In *Inventing women: Science, technology and gender*, edited by G. Kirkup and L. Smith Keller, 42–49. Cambridge, UK: Polity.

Kidder, T. 1981. *The soul of a new machine.* Boston: Little, Brown.

Kimbell, R., R. Stables, and R. Green. 1996. *Understanding practice in design and technology.* Buckingham: Open University Press.

Law, J. 1987. Technology, closure and heterogeneous engineering: The case of the Portuguese expansion. In *The social construction of technological systems*, edited by W. Bijker, T. Hughes, and T. Pinch, 111–34. Cambridge: MIT Press.

Layton, E. T. 1971. *The revolt of the engineers: Social responsibility and the engineering profession.* Cleveland, OH: Press of Case Western University.

Lovell, J., and J. Kluger, 1994. *Lostmoon: The perilous voyage of Apollo 13.* Boston: Houghton Mifflin.

MacKenzie, D., and J. Wajcman. 1985. Introduction. In *The social shaping of technology*, edited by D. MacKenzie and J. Wajcman, 2–25. Milton Keynes: Open University Press.

McIlwee, J. S., and J. G. Robinson. 1992. *Women in engineering: Gender, power, and workplace culture.* Albany: SUNY.
McNeil, M. 1987. It's a man's world. In *Gender and expertise,* edited by M. McNeil, 187–97. London: Free Association Books.
Mellström, U. 1995. *Engineering lives: Technology, time and space in a male-centred world.* Linköping, Sweden: Department of Technology and Social Change.
Morell, V. 1993. Called "trimates," three bold women shaped their field. *Science* 260: 420–25.
Murphy, P. 1990. Gender differences in pupils' reasons to practical work. In *Practical science,* edited by B. Woolnough, 112–22. Milton Keynes: Open University Press.
Murray, F. 1993. A separate reality: Science, technology and masculinity. In *Gendered by design? Information technology and office systems,* edited by E. Green, J. Owen, and D. Pain, 64–80. London: Taylor & Francis.
Rose, H. 1983. Hand, brain, and heart. *Signs: Journal of Women in Culture and Society* 9 (1): 73–96.
——. 1994. *Love, power and knowledge: Towards a feminist transformation of the sciences.* Cambridge, UK: Polity.
Sørensen K. H. 1997. The masculine mystique: Engineering tales. Paper presented to Network of European Centres in Science and Technology Studies workshop on Gender, Science and Technology, May, NTNU, Trondheim, Norway.
Sørensen, K., and N. Levold. 1992. Tacit networks, heterogeneous engineers, and embodied technology. *Science, Technology, & Human Values* 17 (1): 13–35.
Turkle, S. 1984. *The second self.* New York: Simon & Schuster.
——. 1988. Computational reticence: Why women fear the intimate machine. In *Technology and women's voices,* edited by C. Kramarae, 41–61. London: Routledge Kegan Paul.
Turkle, S., and S. Papert. 1990. Epistemological pluralism: Styles and voices within the computer culture. *Signs: Journal of Women in Culture and Society* 16 (1): 128–58.
Vincenti, W. 1990. *What engineers know and how they know it.* Baltimore, MD: Johns Hopkins University Press.
Wajcman, J. 1991. *Feminism confronts technology.* Cambridge, UK: Polity.
Walkerdine, V., and The Girls and Mathematics Unit. 1989. *Counting girls out.* London: Virago.
Weber, R. 1997. Manufacturing gender in commercial and military cockpit design. *Science, Technology, & Human Values* 22 (2): 235–53.
Whalley, P. 1986. *The social production of technical work: The case of British engineers.* London: Macmillan.
Wolfe, T. 1980. *The right stuff.* London: Picador.
Zussman, R. 1985. *Mechanics of the middle class: Work and politics among American engineers.* Berkeley: University of California Press.

Section III

Technologies Born of Difference

How Ideas about Women and Men Shape Science and Technology

In Section II, we examined how ideas about differences between men and women influence who can and cannot become a scientist or engineer. In this section, we will explore how these ideas inform scientific and technological knowledge and their broader impacts on the society. As we discussed in the introduction, "how" research is done is as important a question as "what" kind of research is done. As illustrated in readings by Ruth Bleier, Andrea Tone, Suzanne Kessler, and Anne Fausto-Sterling, which topics are deemed worthy of study, how research questions are formulated, and how the resulting data is gathered, interpreted, and used to produce new knowledge often depend on when, in what social, economic, and political environment, and by whom research is done.

In these readings, the practice of both scientific and technological innovation is a social activity that is constructed and constrained by the shared practices of a community of highly trained specialists. The directions of scientific and technological innovations are further constrained by the growing call for the commercialization of knowledge ("the knowledge economy") and the influence of industry on academic research programs at colleges and universities (due, in no small part, to the decline in funding from the federal government for basic research). This leads to research questions and technological innovations being influenced by commercial opportunities.[1]

In our feminist analyses of science and technology related to, informed by, and informing distinctions between women and men, we have identified four general and interrelated areas where beliefs about women and men and the differences between them have been shown to influence definitions and interpretations of what is "natural": conceptual premises, meanings attached to "the body" as illustrated in debates about birth control, constructions of research questions as illustrated in sociobiology (and its contemporary cousin evolutionary psychology), organization/classification of natural phenomena as illustrated in medical decisions about assigning "gender" to intersexed infants, and interpretations of data as illustrated in the nature/nurture debate when applied to understanding the biology of human bones.

This is not to say that all scientific research and technological development are distorted by gender biases, since sometimes gender is relevant and sometimes it is not. However, as we will see in the four readings in this section, research describing the influence of gender biases on knowledge documents the limitations of the scientific method in producing the "objective" facts upon which the development, application, and impact of new technology rests. Nor is our approach the only way feminists have examined the influence of gender beliefs and biases on technology.

Sue Rosser has focused on these themes using the lenses of feminist theories in three areas: women in the technology workforce, women as users of technology, and women and technology design.[2] By using examples on a spectrum from liberal feminism to cyber feminism, Rosser points out how different priorities in feminist theory lead to identifying different concerns about technology. However, these perspectives share a critique of the prevailing engineering paradigm of "hard systems," which relies on approaches dominated by concepts of separation and control. Moreover, women are not a monolithic group. Our experiences inside and outside of technological innovation and change vary by race, gender, class. The general question, however one goes about answering it, is: How do we think about the development and implementation of technological innovations when women are at the center?

CONCEPTUAL PREMISES

Scientists identify a research question and develop a hypothesis within the context of a set of assumptions that are often unnamed and unannounced. These assumptions constrain the ability of researchers to respond to new evidence. One example is physical anthropologists' theories about women's and men's roles in human evolution.[3] In their reconstruction of early human life, most scientists have assumed since the 1950s that competition and sexual selection were the driving forces of evolution. Males fought with one another for food and for mates—they were the hunters. Females stayed close to home with the infants, foraging for nuts and berries—they were sexually receptive to whoever won the competition among men, and they were monogamous.

In the 1970s, evidence emerged from several fields—primatology, anthropology, archaeology, and paleontology—that contradicted the idea that hunting by males was central to human evolution. Provoked in part by the feminist movement, researchers in these fields began to examine the role that women's activities may have played in early human activities. Nancy Tanner and Adrienne Zihlman, among others, challenged the assumptions of the sexual division of labor and monogamy upon which the hunting hypothesis was based. They argued that it was more likely that individuals shared a wide range of tasks, including both hunting and foraging. They also argued that women chose from among the men, rather than the reverse, and that contemporary Western notions of mating and marriage were too narrow to explain cross-cultural variation.[4]

Despite their efforts to consider a fuller account of human evolution, a male-centered paradigm reemerged. By the 1990s, and despite an increasing amount of genetic, paleontological, anatomical, behavioral, and archaeological evidence contradicting the hunting hypothesis, the theory that women as gatherers have played a role equal to men's in human evolution has yet to gain wide acceptance. Most recently, Robert Sussman, an anthropologist, presented a new theory based on the fossil evidence and living primate species and argued that primates, including early humans, evolved not as hunters but as prey of many predators.[5] The male-as-hunter hypothesis, however, is securely embedded in the public mind through museum displays and television programs about evolution.

Another example of how descriptions of the natural world can be constrained by cultural beliefs appears in introductory molecular biology and genetics textbooks. As Bonnie Spanier points out, the exchange of genetic material in the ubiquitous single-celled bacteria, *Escherichia coli*, is explained with cultural and scientific terms consistent with the reproductive sex of higher organisms (such as humans). However, these terms are inappropriate and misleading. They describe so-called "male" bacterium that replicate and donate a strand of genetic material (DNA) to "female" bacterium through a narrow tube known as a pillus. The only difference between "male" and "female" bacteria is the presence or absence of this transferable piece of DNA, the F-plasmid. An F+ plasmid is denoted as male and F– plasmid as female. At the end of the transfer, both cells are F+; thus, in terms of the metaphor, both cells are male.[6]

Why this exchange of genetic material is described as sex is a mystery. It is not an act of sexual reproduction, since the "male" and "female" bacterial cells do not fuse, as do the sperm and egg, to produce a new cell. The actual event is more like the donation of a kidney than an act of sex. It does not require designation of male and female cells in order to understand the event, and in fact, that designation makes the outcome of the event (two "males") difficult to explain logically within the initial heterosexual framework. Thus, the use of the language of sexual reproduction to describe the replication and donation of the F-plasmid from one bacterium to another is not only inaccurate but also places constraints on how the phenomenon is further analyzed. Scientists may, for example, look only at how the "active male" bacterium initiates the event, when in fact it may be the so-called "passive female" bacterium that makes the transfer possible. With a biased and anthropomorphized view of "bacterial sex," the scientific questions that are pursued are limited by conventional understandings of an unrelated and complex human social behavior.[7] Like the example of the inextinguishable hypothesis of man the hunter and the evolution of "man," once a hypothesis is firmly rooted in the minds of both scientific "experts" and the public, it is difficult to reexamine, question, and modify, even when there is ample reason to do so.

ORGANIZATION/CLASSIFICATION OF THE NATURAL WORLD

Most people in the United States, even the youngest, who have played the game "Twenty Questions" know something basic about the natural world, that everything is classified into one of three categories—animal, vegetable, or mineral. Yet the specifics of these categories did not spring fully developed from nature; rather, they were created, defined, and defended by a series of scientists over centuries of work. Carolus Linnaeus, the eighteenth-century scientist who developed scientific nomenclature still in use today, created the term *Mammalia* to describe one of the groups, humans, within the animal category. Linnaeus, it would seem, was celebrating the role of women in human reproduction by featuring a distinctive characteristic of females.

Historian Londa Schiebinger points out, however, that Linnaeus did not begin to use the term *Mammalia* in a cultural void.[8] The term arose within a historical moment in which social context and conflicts combined to lend credibility to his classification schema. By focusing on the breast as the physical trait that identified *Homo sapiens* as *Mammalia*, Linnaeus drew on his culture's association of nursing with lower animals—women, cows, pigs, and sheep nurse their young. At the same time, his term emphasized and legitimized a sexual division of labor that assigned politics to men and the home to women. He did not do this arbitrarily, argues Schiebinger, since the physical evidence showed other characteristics that could have been emphasized, including hair, as distinguishing the group he called *Mammalia* from other groups. In France, in the late eighteenth century, his new term provided a cultural fit between notions of women's close connection to nature and what should be considered "natural" in the social order. Because of prevailing social attitudes, the apparent emphasis on a trait of women reinscribed rather than challenged assumptions about men's superiority to women.

The drive to classify all human beings in a rank order from the highest (white European males) to the lowest (criminals, lower classes, and children) fueled work in the nineteenth century that focused on brain size. Scientists assumed that brain size was indicative of brain power, that is, the larger the brain, the smarter the individual. (Indeed, this idea that mass equals intelligence circulates even today with slang insults such as "he's a pea brain" or "she's an airhead.") Paul Broca made a particularly well-known contribution to this literature because of his stature as a prominent and respected scientist. Broca, after careful measurement of the size of brains of 292 males and 140 females during autopsy, concluded that a pattern of difference was evident.

> We might ask if the small size of the female brain depends exclusively upon the small size of her body. . . . But we must not forget that women are, on the average, a little less intelligent than men, a difference which we should not exaggerate but which is, nonetheless, real. We are therefore permitted to suppose that the relatively small size of the female brain depends in part upon her physical inferiority and in part upon her intellectual inferiority.[9]

Broca's *a priori* classification of women as inferior to men constrained his ability to interpret his data as supporting any other conclusion. Though it may seem clear in retrospect that Broca presumed the inferiority of women and so found it in his data, at the time his work was widely accepted as factual. Moreover, it was applied directly to arguments that women should be excluded from higher education.[10]

Broca also tried to determine "the relative position of races in the human series," by measuring the ratio of the radius of the lower arm bone to the upper arm bone, reasoning that longer arms were indicated by a higher ratio, signifying a characteristic of apes. The closer a group was to apes, the less it had evolved. His hierarchical classification system fell apart, however, when his measurements found that darker-skinned groups had lower ratios than lighter-skinned groups. When his measurements of the brains of "yellow people" similarly contradicted the presumed hierarchy, he began to question not the value of the hierarchical classification system, but rather the interpretation of the data. Thus he argued that the brain-size classification system worked best to identify inferior people, since some inferiors had large brains but no superiors had small brains.[11]

Though the theories about racial hierarchies that Broca developed have been dismissed by most of today's scientists, the cultural belief that groups can and should be distinguished from one another persists into contemporary U.S. society. The categories are familiar ones: there are only two sexes, male and female, and people fit into only one racial or ethnic category. One is either black or white, Asian or European, Native American or Hispanic. Feminist scholars have pointed out that these categorizations belie the biological reality of diversity among human beings. Anne Fausto-Sterling argues that there are at least three other sexes in addition to male and female that deserve to be included in our sexual classification scheme.[12] Intersexed individuals are neither strictly male nor female. Acknowledgments of their existence can be traced historically throughout human society and as early as the ancient Greeks, but a combination of legal and social practices has rendered them more or less invisible in contemporary society.

Evelynn Hammonds questions the use of contemporary racial categorizations similarly, pointing out that in the United States, characteristics that are associated with race are often conflated with those associated with ethnicity. Even celebrations of "racial diversity" in the popular press implicitly reinscribe racial boundaries and reinforce notions of male dominance by representing women of color as objects in white males' search for the perfect mate.[13]

CONSTRUCTION OF RESEARCH QUESTIONS

The effort to identify fundamental biological differences between women and men is not in and of itself necessarily biased, because medical knowledge about women's biology is critical to women's health care. However, when researchers begin with the assumption that there are important biological differences between women and men that can explain social arrangements, they are asking an already distorted research question, i.e., "How do the biological differences explain social arrangements?" Today's sophisticated technologies and scientific methods cannot prevent biased interpretations of data that flow from biased questions.

In popular culture, too, the persistent common wisdom is that there are important natural differences that structure social life. Compare arguments made in 1879 to those made in 2005. In 1879, Gustav LeBon, an early social psychologist and supporter of Broca's work, argued that "the day when, misunderstanding the inferior occupations which nature has given her, women leave the home and take part in our battles; on this day a social revolution will begin, and everything that maintains the sacred ties of the family will disappear."[14] A hundred and twenty-six years later, Lawrence Summers, as then president of Harvard University, made the following remarks at the NBER Conference on Diversifying the Science and Engineering Workforce:

There are three broad hypotheses about the sources of the very substantial disparities that

> this conference's papers document and have been documented before with respect to the presence of women in high-end scientific professions. One is what I would call . . . the high-powered job hypothesis. The second is what I would call different availability of aptitude at the high end, and the third is what I would call different socialization and patterns of discrimination in a search. And in my own view, their importance probably ranks in exactly the order that I just described.[15]

In Anne Fausto-Sterling's *Myths of Gender* ([1985] 1992), in her chapter entitled "A Question of Genius," she describes the history of efforts in the twentieth century to establish the biological bases of cognitive differences between women and men, with a focus on studying IQ and math ability. It is no accident that her chapter begins with the words "JOBS AND EDUCATION—that's what it's really all about." Most of the sex difference research, says Fausto-Sterling, "has been and continues to be used to avoid facing up to very real problems in our educational system and has provided a rationale for discrimination against women in the workplace."[16] What is remarkable about the sex difference research is how *little difference* has actually been found. Yet, there is a clear effort in the publication and interpretation of research data to exaggerate and focus on the differences.

Paula Caplan and her colleagues write, in a critique of the work reporting sex-related differences in spatial (math) abilities, that "journal policies encourage the publication of studies in which sex differences *are* found and discourage the converse. Therefore, researchers who find no sex difference often do not submit their work for publications, and any real sex difference that might exist . . . are likely to be rather exaggerated."[17] Neurobiologist Ruth Bleier described her experiences submitting a paper to *Science* showing that there were no significant differences in spatial abilities between men and women. This "negative" data was rejected for publication, according to Bleier, because *Science* had been a major outlet for the very sex difference research she critiqued as methodologically flawed.[18]

Scientific work that appears to prove that women are less intelligent than men because we have less innate mathematical ability than men reinforces prevailing cultural stereotypes about women and men, so it persists relatively unchallenged in the mainstream. This remains so despite counter-evidence that the differences arise from, and can be eliminated by, educational and/or social experiences.[19] Because scientific knowledge about sex differences has been used historically as evidence to deny educational and employment options to women, and because scientific researchers seldom address their background assumptions about social life, feminist scholars often are suspicious of research in this tradition. Indeed, since our culture is fundamentally organized around defining women as different from and inferior to men, it is hard to imagine "objective" or women-centered research in this area. Technologies born from presumed differences result in serious side effects from underdiagnosis to unmet expectations.

Feminists are not critical of all work on sex differences, since there are compelling and meritorious reasons to study the differences between the biology of women and men. An important example is in medicine, where it is crucial that we understand how men and women differ in their physiological responses to medications, in how diseases affect them, and in how they age. Though this would be a logical context in which to study biological differences, research on how women and people of color diverge from the norms of white male biology has been noticeably absent in the medical community. As recently as the 1980s, women were routinely excluded from clinical trials to determine drug efficacy and safety, and ethnic differences were (and still are) too seldom explored. Sometimes, then, dominant cultural assumptions underwrite research questions that then produce inadequate data, and sometimes dominant cultural assumptions mean that important research questions are not asked at all. Elizabeth Siegel Watkins, in her essay "Hormone Replacement", gives a brief history of hormone replacement therapy for postmenopausal women and states that "in 1960s and early 1970s HRT offered promises of continued femininity to aging women in a culture that celebrated youth. In the mid-1970s faced with new medical evidence about the potential danger of long term estrogen use and new feminist critiques of sexist assumptions about women in America, medical attitudes toward HRT became more cautious." In the 1990s, with an additional hormone and promise of protection from osteoporosis and heart disease, HRT claimed the best-selling-drug status in America. All of this changed when in 2002 the federally funded Women's Health Initiative completed a study with 16,000 participants and reported that HRT users were at increased risk for heart disease, strokes, blood clots, and breast cancer.[20]

Because medical research about women's bodies is a relatively new area of research, we have included two readings in this section that explore how ideas about men and women influence the medical profession, the commercialization of medical knowledge, and the directions of new technologies related to women's bodies. Andrea Tone's work explores the struggle to establish that women have a right and entitlement to control our reproductive health decisions, a battle waged over the forms and availability (and legality) of birth control in the early twentieth century (not so long ago!). This was no small task in an era when the use of birth control was considered immoral and when many people had neither access nor the economic means to acquire contraception. Sanger believed that public discussions about birth control should be moved from the realm of morality to that of medicine, and that reproductive decisions belonged in the hands of women. When her efforts initially resulted in an increase in the use of condoms, she began promoting diaphragms manufactured and distributed under the auspices of medical professionals, even though that meant that access would still be limited. After decades of characterizing contraceptives as immoral, dangerous, and even "unnatural," the American Medical Association finally endorsed birth control in 1937.

Tone's focus is on one of the most intimate of issues for women, control of our reproductive health, and few would question that research on birth control should be sex-specific. Anne Fausto-Sterling takes a different, but companion, approach in asking readers to consider an area of medical research where the fundamental science assumes sex differences where they may *not* exist. In her study "The Bare Bones of Sex: Sex and Gender," she points out that culture *and not* nature shapes bones. She supports this observation with compelling evidence, contesting the scientific adequacy of received wisdom. She asks, "Why, if bone fragility is so often considered to be a sex-related trait, do so few studies examine the relationships among childbirth, lactation and bone development?" The osteoporosis literature, for instance, commonly relies on an operational definition provided by the World Health Organization (WHO) that measures the bone mineral densities of young white women only. With many cultural factors missing from the definition, including standardization between instruments and sites, drug companies have designed drugs that will increase bone density even though bone density is not the only factor that results in a fracture. These density-boosting drugs, based on evidence from a narrow, rather than diverse, human population, are then marketed to older women who may or may not benefit from the new drugs in terms of reduced fractures. Fausto-Sterling notes that the absence of men in the osteoporosis research is one limitation on the development of a fuller understanding of human bone health. But she points out that another profoundly disturbing limitation stems from research strategies which fail to recognize the wider cultural and institutional "systems" in which people live as influential in their long-term physical health and well-being. "This article," she challenges, "is a call to arms."

INTERPRETATION OF DATA

Almost every step in scientific inquiry requires interpretation of some kind. A zoologist can look at the way an iguana slowly shuts her eyes when she is being petted and interpret that behavior as a sign that the iguana is enjoying herself. Another zoologist could see the same behavior and interpret it as a sign that the iguana is frightened.[21] One of these two scientists has an interpretation that is more consistent with the prevailing consensus about iguana responses to human touching.

In the case of gender there is consensus within the feminist community, derived from theory and research in a variety of fields, that the social, economic, and intellectual frameworks of Western culture have been influenced by beliefs about women and men and the differences between them. This includes "objective" scientific knowledge. When researchers interpret data through language filled with images and metaphors that evoke gender constructs, they are explaining the data based on social preconceptions that they bring to their work. They are tapping into what Carol Cohn has called "symbolic gender" in order to explain their data in terms that are shared by the wider community of scientists.[22]

Anthropologist Emily Martin has analyzed textbooks in molecular biology to examine how symbolic gender has influenced biologists' understanding of eggs and sperm. One of the primary tenets of theory about human evolution asserts that males (and thus sperm) are more important than females to understanding the process of human evolution. Their evidence for this is derived from an interpretation of the relative value of eggs and sperm to reproduction. Martin argues that this evaluation is based on a male-centered interpretation rather than on the biological evidence.

> The texts have an almost dogged insistence on casting female processes in a negative light. The texts celebrate sperm production because it is continuous from puberty to senescence, while they portray egg production as inferior because it is finished at birth. This makes the female seem unproductive, but some texts will also insist that it is she who is wasteful. In a section heading for *Molecular Biology of the Cell*, a best-selling text, we are told that "Oogenesis is wasteful." The text goes on to emphasize that of the seven million oogonia, or egg germ cells, in the female embryo, most degenerate in the ovary. Of those that do go on to become oocytes, or eggs, many also degenerate, so that at birth only two million eggs remain in the ovaries. Degeneration continues throughout a woman's life: by puberty 300,000 eggs remain, and only a few are present by menopause. "During the 40 or so years of a women's reproductive life, only 400 to 500 eggs will have been released," the authors write. "All the rest will have degenerated. It is still a mystery why so many eggs are formed only to die in the ovaries."
>
> The real mystery is why the male's vast production of sperm is not seen as wasteful. Assuming that a man "produces" 100 million (10^8) sperm per day (a conservative estimate), during an average reproductive life of sixty years, he would produce well over two trillion sperm in his lifetime. Assuming that a women "ripens" one egg per lunar month, or thirteen per year, over the course of her forty-year reproductive life, she would total five hundred eggs in her lifetime. But the word "waste" implies an excess, too much produced. Assuming two or three offspring, for every baby a woman produces, she wastes only around two hundred eggs. For every baby a man produces, he wastes more than one trillion (10^{12}) sperm.[23]

Martin focuses on a specific example from a popular textbook. Feminist analyses of interpretive bias also extend beyond specific topics or interpretations within a subject area.

Ruth Bleier's essay in this section, "Sociobiology, Biological Determinism, and Human Behavior," deftly challenges the primary premises of the field of human sociobiology. E. O. Wilson, a highly regarded entomologist at Harvard University, has formed a school of sociobiological thought in which all human behaviors, social relationships, and organization are assumed to be genetically evolved adaptations. Wilsonian sociobiology takes as a primary premise that males and females have an unequal investment in reproduction, from subtle differences in the biological "cost" of producing eggs and sperm, to the obvious differences in parental involvement in the birth of offspring. This unequal investment is hypothesized to lead inexorably to differences in behavior between males and females that maximize the "return" on investment for each. Bleier is unconvinced, however, since these behaviors are remarkably similar to Western, privileged, nineteenth-century white male beliefs about what is appropriate and "natural" in relations between women and men. Unfortunately, sociobiology cannot simply be dismissed as an old-fashioned idea. The arguments of Wilsonian sociobiology have reappeared recently as the cornerstone of evolutionary psychology, a newly defined field of research.[24]

We have reviewed thus far feminist accounts of the intellectual pitfalls of scientific work that does not recognize how gender biases result in research and analyses that fall short of the promise of scientific, objective, and rational inquiry. The resulting male-centered biases can take a literal as well as analytic form. In "The Medical Construction of Gender: Case Management of Intersexed Infants," Suzanne Kessler examines physicians' and parents' surgical and socialization decisions in cases in which infants are born with ambiguous genitalia. Although the medical decisions are theoretically based on the results of diagnostic tests, Kessler's interviews with physicians reveal the overwhelming influence of cultural understandings of gender. Their primary assumption is that there are only two possible and biologically distinct sexes, and that the most important feature in each case is that they be properly equipped to engage in heterosexual intercourse according to their gender. The greatest focus is on the penis and male-centered understanding of sexuality. Infants without an adequate penis are defined as female by default. The "true sex" of the infant, insofar as that can be determined by genetic or hormonal profiles, is disregarded in the face of the parents' discomfort with the ambiguity. Surgical interventions attempt literally to construct the sex of the infant, and then parents consciously cultivate the child's gender identity accordingly. As the physicians and parents try to prepare the child for a life in heterosexual society, they disregard any forms of sexuality that might not involve vaginal intercourse, because of deeply held cultural beliefs about what is "natural" in human sexual relationships.

Notes

1 J. E. Bekelman, Y. Li, and C. P. Gross, "Scope and Impact of Financial Conflicts of Interest in Biomedical Research: A Systematic Review," *Journal of the American Medical Association* 289 (2003): 454–65; C. D. DeAngelis, "The Influence of Money on Medical Science," *Journal of the American Medical Association* 296 (2006): 996–98; J. P. Kassirer, *On the Take: How Medicine's Complicity with Big Business Can Endanger Your Health*, New York: Oxford University Press, 2004.
2 S. Rosser, "Using the Lenses of Feminist Theories to Focus on Women and Technology," in *Women, Gender and Technology*, ed. M. F. Fox, D. Johnson, and S. Rosser, pp. 13–46. Urbana: University of Illinois Press, 2006.
3 For a more detailed discussion, see A. Zihlman, "The Paleolithic Glass Ceiling: Women in Human Evolution," in *Women in Human Evolution*, ed. L. D. Hager, pp. 91–113, New York: Routledge, 1997.
4 N. Tanner and A. Zihlman, "Women in Evolution. Part I. Innovation and Selection in Human Origins," *Signs: Journal of Women in Culture and Society* 1, no. 3, pt. 1 (1976): 585–608; A. Zihlman, "Women in Evolution. Part II. Subsistence and Social Organization among Early Hominids," *Signs: Journal of Women in Culture and Society* 4, no. 1 (1978): 4–20.
5 D. Hart and R. Sussman, *Man the Hunted: Primates, Predators and Human Evolution*. New York: Westview, 2005.
6 B. Spanier, *Im/Partial Science: Gender Ideology in Molecular Biology*. Bloomington: Indiana University Press, 1995, pp. 56–58.
7 Ibid.
8 L. Schiebinger, "Why Mammals Are Called Mammals: Gender Politics in Eighteenth-Century Natural History," in *Feminism and Science*, ed. E. F. Keller and H. Longino, pp. 137–53. New York: Oxford University Press, 1996.
9 As quoted in S. J. Gould, "Women's Brains," in his *The Panda's Thumb*, p. 154, New York: Norton, 1980.
10 Ibid., p. 155.
11 S. J. Gould, "Measuring Heads," in his *The Mismeasure of Man*, pp. 86–87, New York: Norton, 1981. See also N. Stephan, "Race and Gender: The Role of Analogy in Science," in *The "Racial" Economy of Science*, ed. S. Harding, pp. 359–76, Bloomington: Indiana University Press, 1993.
12 A. Fausto-Sterling, "The Five Sexes: Why Male and Female Are Not Enough," *The Sciences*, March/April 1993, pp. 20–25.
13 E. Hammonds, "New Technologies of Race," in *Processed Lives: Gender and Technology in Everyday Life*, ed. J. Terry and M. Calvert, pp. 107–22. New York: Routledge, 1997.

14 As quoted in Gould, "Women's Brains," p. 154.
15 http://www.president.harvard.edu/speeches/2005/0516_womensci.html.
16 A. Fausto-Sterling, *Myths of Gender*. New York, Basic Books, 1992 [1985].
17 P. J. Caplan, G. M. MacPherson, and P. Tobin, "Do Sex-Related Differences in Spatial Abilities Exist? A Multilevel Critique with New Data," *American Physiologist*, 40, no. 7 (1985): 786–99, quote on p. 786. See also P. Caplan and J. Caplan, *Thinking Critically about Research on Sex and Gender*, New York: Harper Collins, 1994, esp. chapter 5, "Are Boys Better than Girls at Math?" pp. 37–47.
18 Ruth Bleier, "A Decade of Feminist Critiques in the Natural Sciences," ed. J. W. Leavitt and L. Gordon, *Signs: Journal of Women in Culture and Society* 14, no. 1 (1988): 182–95. Recent work points toward the influence of experience on neurophysiology. See Ingrid Wickelgran, "Nurture Helps Mold Able Minds," *Science* 283 (1999): 1832–34.
19 R. Devon, R. Engel, and G. Turner, "The Effects of Spatial Visualization Skill Training on Gender and Retention in Engineering," *Journal of Women in Science and Engineering* 4 (1998): 371–80.
20 E. S. Watkins, "Hormone Replacement 'Educate Yourself': Consumer Information about Menopause and Hormone Replacement Therapy," in *Medicating Modern America*, ed. A. Tone and E. S. Watkins, pp. 63–96. New York: New York University Press, 2007.
21 There is better evidence for the latter. See, e.g., Philippe de Vosjoli, *The Green Iguana Manual,* Lakeside, CA: Advanced Vivarium Systems, 1992, p. 40.
22 C. Cohn, "Wars, Wimps, and Women: Talking Gender and Thinking War," in *Gendering War Talk*, ed. M. Cooke and A. Woollacott. Princeton, NJ: Princeton University Press, 1993.
23 Emily Martin, "The Egg and the Sperm: How Science Has Constructed a Romance Based on Stereotypical Male–Female Roles," *Journal of Women in Culture and Society* 16, no. 3 (1991): 485–501, esp. 488–89.
24 For examples of work in this tradition, see E. O. Wilson, *Consilience; The Unity of Knowledge*, New York: Alfred A. Knopf, 1998; E. O. Wilson, *Sociobiology*, Cambridge, MA: Harvard University Press, 1980, esp. chapter 26; and R. Wright, *The Moral Animal; Evolutionary Psychology and Everyday Life*, New York: Peter Smith Publisher, 1995.

13
A Medical Fit for Contraceptives

Andrea Tone

In a 1952 letter to the physician and fellow birth control proponent Clarence Gamble, Margaret Sanger reflected on her career. Her greatest achievement, she wrote, had been "to keep the movement strictly and sanely under medical auspices."[1]

Beginning in the 1920s, Sanger, Gamble, and a network of dedicated researchers, physicians, and activists made a once-radical movement middle-class and respectable. They established doctor-supervised clinics, promoted laboratory testing of contraceptives, encouraged the physician-fitted diaphragm-and-jelly method, and lobbied the American Medical Association (AMA) to reverse its long-standing ban on birth control. In public, Sanger refused to endorse specific brands or devices, fearing that the inevitable charges of impropriety would discredit the movement as a whole. Behind the scenes, however, her support of medicalized birth control shaped the course of contraceptive commercialization. By the 1930s, thanks largely to Sanger, the diaphragm and jelly had become the most frequently prescribed form of birth control in America, and Holland-Rantos its best-known manufacturer. Consciously distancing the birth control business from manufacturers who made contraceptives for the laity, Sanger helped inaugurate a regime of doctors, diaphragms, and corporate science.

When Sanger opened her first birth control clinic in 1916, she and her sister, Ethel, instructed eight women at a time on how to use over-the-counter contraceptives, including condoms, suppositories, and rubber pessaries. Although she later claimed that she referred women "to a druggist to purchase the necessary equipment," boxes of Mizpah pessaries she and her sister had dispensed were found by police when they raided the two-room clinic. The Mizpah (sometimes spelled Mispah) was a flexible, thimble-sized cervical cap made of corrugated rubber. Described in one 1900 advertisement as a "soft, light, and comfortable" uterine supporter, the Mizpah was inserted vaginally and positioned snugly to create an airtight seal around the cervical opening. Sold by druggists and mail-order vendors to treat the still commonly diagnosed condition of a prolapsed or distended uterus, the cervical cap was an effective over-the-counter contraceptive. Of all American-made contraceptives, the Mizpah, or "temporary French pessary," as it was sometimes called, was Sanger's favorite.[2]

Sanger's pessary education had occurred in Europe. At the urging of her then-husband, William, and her friend the labor leader Big Bill Haywood, she and her family sailed to France in 1913. In Paris, discussions with French radicals about contraceptive techniques nurtured her dreams of sexual emancipation and political revolution achieved through universal access to birth control. In March 1914, three months after her return to the United States, she began publishing *The Woman Rebel*, a feminist journal that demanded legal contraception and full woman's rights. Its June issue first used the term "birth control," a moniker coined by Sanger's friend Otto Bobsein as an alternative to the more awkward-sounding "voluntary motherhood" and "family limitation" then in vogue. After seven issues, the strident *Rebel* was deemed unmailable by the U.S. Post Office. Annoyed but undaunted, Sanger flouted the law again by writing *Family Limitation*, a home guide to contraception. This extraordinary pamphlet discussed douches, condoms, and cervical caps and recommended caps, whose use women could most easily and discreetly control. Sanger distributed 100,000 copies of her pamphlet, which implored women to learn how to insert caps into their own bodies and then to "teach each other" how to use them.[3] Sanger envisioned a world of grassroots birth control where women from all walks of life could use contraceptives without reliance on doctors, a populist approach she would soon abandon.

Family Limitation got Sanger into more trouble. In 1915, she found herself back in Europe dodging American law while continuing her contraceptive education. In London, she did research at the British Museum before embarking for the Netherlands, where she planned to tour the country's contraceptive clinics, learning "from personal observation."[4] The trip across the Atlantic was risky. War had broken out the previous year, and, as Sanger later recalled, crossing the Channel entailed "possible unwelcome encounters with . . . floating bombs [and] submarines."[5] She arrived safely, and immediately sought out the renowned physician and birth control advocate Dr. Aletta Jacobs.

Jacobs had an impressive résumé. One of eleven children, she was the first female physician in Holland. In 1882, four years after graduating from medical studies at Amsterdam University, she opened the first medical birth control clinic in the world, giving free contraceptive information and supplies to working-class women. Her initiative (and possibly her well-known imperiousness) gave rise to a national system of contraceptive clinics, which helped reduce the country's maternal and infant mortality rates.

Jacobs clung fast to two principles. The first was the superiority of the physician-fitted Mensinga diaphragm, also referred to in medical literature as the Dutch cup or Mensinga veil. Manufactured in Holland, the device had been invented in 1842 by the gynecologist. W. P. J. Mensinga, Jacobs's friend and mentor and onetime professor of anatomy at the University of Breslau. Jacobs loyally and actively promoted Mensinga's diaphragm, helping to make it the most popular birth control method in clinics in Holland, Germany, and Russia. (The cervical cap, better suited to over-the-counter use, continued to be favored in England and France, where Sanger had first learned about it.)

Jacobs's second principle was the need for physician control over the distribution of contraceptive information and technology. Jacobs believed, as Sanger would later, that birth control was strictly a medical matter. So adamant was she in her conviction that when Sanger requested a meeting with her, Jacobs refused. The American activist's political credentials could not erase the fact that she was, in Jacobs's eyes, "only" a nurse and thus had no business involving herself in the dissemination of birth control.[6]

Disappointed, Sanger made do. She learned about the Dutch system under the tutelage of Dr. Johannes Rutgers, Jacobs's second in command. Rutgers convinced Sanger of the need for medical clinics that favored the physician-fitted diaphragm rather than over-the-counter methods. The diaphragm's prescription and successful use entailed four distinct steps: a pelvic examination; measurement of the diameter of the vagina and assessment of its contours; selection of a corresponding diaphragm size; and, finally, instruction of the patient. Women were pronounced ready for "home" use only after they had successfully inserted and removed the device under a doctor's watchful eye. According to one physician, the procedure required up to forty-five minutes of consultation and "depends on painstaking work with each patient." Even then, patient compliance was not guaranteed. Some doctors insisted on follow-up appointments to double-check patients' technique.[7]

Sanger's acceptance of Rutgers's position on medical contraception and diaphragms shaped the future of the birth control movement. It encouraged her to give up her earlier vision of grassroots contraception, a laypersons' network in which women learned about birth control from each other and by reading educational pamphlets such as *Family Limitation*. Her acceptance also promoted a contraceptive technology that made access to physicians, a far from universal phenomenon in the United States, a precondition of effective use.

Returning to New York in the fall of 1915, Sanger was eager to start a clinic modeled on those she had visited in Holland, but she faced several formidable obstacles. First, U.S. law still regarded birth control as an obscenity. Indeed, while Sanger was away, Anthony Comstock had arrested her husband, William Sanger, for distributing *Family Limitation*. It would be one of his last arrests. During William Sanger's trial, Comstock caught a cold in the courtroom and died soon after of pneumonia.[8] Second, Sanger knew of no physicians like Jacobs and Rutgers ready to risk prosecution or AMA censure by operating contraceptive clinics on this side of the Atlantic. If she wanted a freestanding clinic, she would have to run it herself. Available supplies were a third problem. The Dutch Mensinga pessary was not available in the United States. When Sanger opened her clinic in the fall of 1916, she dispensed the Mizpah, the most frequently used American pessary, instead.[9]

Sanger's clinic did not remain open for long, but Judge Crane's 1918 ruling gave Sanger license to open a second clinic, this time under medical supervision. In 1921, Sanger established the American Birth Control League (ABCL), an advocacy group, which rented rooms for the new clinic. Recruiting a doctor to run it proved difficult. At the last minute, Dr. Lydia DeVilbiss backed out. Sanger persevered, and by late 1922 she had found her doctor: Dorothy Bocker, a public health advocate employed by the Division of Child Hygiene in Milledgeville, Georgia. Sanger warned Bocker that directing a freestanding birth control clinic was risky and that she might go to jail. But she also promised Bocker that the resulting notoriety would "get you such a good boost of publicity, that we can put you on the platform lecturing throughout the country for the next two years." Bocker knew little about birth control techniques. But she promised Sanger that she was "willing to learn."[10]

At about the same time, Sanger found another ally in James Noah Henry Slee. At first glance, Slee was an unlikely convert to the birth control movement. A widower with three grown children and twenty years Sanger's senior, Slee was a well-heeled member of Manhattan's business elite and its conservative Union Club. As president of New York's Three-in-One Oil Company, he was part of the same business establishment Sanger had vilified in her younger, more radical years. After meeting the incandescent woman rebel at a friend's house, Slee pursued Sanger with the ardor of a man deeply in love. Sanger had been separated from William for seven years, and they had recently divorced. Tracking her movements, placing full-page advertisements for his Three-in-One oils—suitable for typewriters, bicycles, guns, sewing machines, furniture, and razors—in the ABCL's *Birth Control Review*, and offering assistance at every turn, Slee coaxed Sanger to say "I do" once more. The two were married September 18, 1922, in Bloomsbury, London. The marriage fulfilled a wish Sanger had confided to her sister from prison: that she "find a widower with money." Slee fitted the bill, and if it is crude to suggest that Sanger married him only for his money and connections, evidence indicates that, at the very least, she prized these assets greatly. At a time when prenuptial agreements were rare, Sanger insisted on most unusual terms. The two were to keep separate apartments, keys, and bank accounts. Slee was to pledge his unqualified support to her professional activities and personal autonomy. In professional circles, she would keep the name Sanger. Slee accepted her terms. Still, Sanger was ambivalent about the marriage. She kept it a secret for two years, during which time she stayed on close terms with several male suitors.[11]

With Slee's financial backing, Sanger embarked on a new chapter of her career, one that distanced the birth control movement from its radical origins and placed it on a more conservative path. Until his death in 1943, Slee honored their marriage contract with boundless generosity. More important, he contributed heavily to Sanger's cause. Between 1921 and 1926, his donations to the ABCL exceeded fifty thousand dollars, making him the league's single largest benefactor.[12] Spouse, fan, and friend, Slee was also Sanger's financier.

With Slee and Bocker on board, Sanger proceeded with her plans. On January 2, 1923, she opened her second clinic, the Birth Control Clinical Research Bureau (BCCRB), at 104 Fifth Avenue, across the hall from ABCL headquarters. Much had changed in the almost seven years since the Brooklyn clinic had opened, including Sanger's priorities. Rather than challenging obscenity laws, a course that might have made contraceptives universally safe and accessible, she set her sights on the passage of "doctors-only bills" to exempt physicians from criminal prosecution. In 1924, the ABCL made its first attempt to introduce a doctors-only bill into Congress (a strategy that sharply conflicted with that of the Voluntary Parenthood League, a rival organization established in 1919 that sought to repeal the Comstock Act altogether). The ABCL's mission failed when it could not get sponsors, but the attempt illustrated Sanger's new approach. Hoping to quell physician fears, she portrayed birth control not as a woman's right but as a medical prerogative.

Sanger's political transformation was partly a strategic accommodation to the culture of her times. In the 1920s, she remained committed to quality birth control for women, a goal few Americans and organizations openly endorsed. Not even the National Woman's Party, formed in 1921 by the militant Alice Paul, would support it. Sanger had learned through her failures the limits of sexual reform in early-twentieth-century America. She recognized too that medical science enjoyed increasing prestige and political clout. Narrowing her agenda, she sought birth control allies through an ideology that trumpeted women's health over their civil liberties and cast doctors, not patients, as agents of contraceptive choice.

By the end of 1923, the BCCRB, operating without a license (which the New York State Board of Charities had refused to grant), had supplied free contraceptive services to 1,558 patients. With a subsidy from Slee, Bocker published a study of patients' experiences in February 1924. It was the first clinical evaluation of contraceptives published in the United States and was distributed solely to doctors. Here was the BCCRB's chance to discredit quackery while shoring up the clinic's scientific reputation. Bocker made the most of the opportunity, railing against over-the-counter contraceptives, whose use bypassed physician expertise. She criticized condoms not only because they forced women to depend on men for fertility control but also because "devices break" and the "technique is rather difficult." Bocker was harsher with the Mizpah pessary, the diaphragm's chief rival. Sanger had praised the over-the-counter cervical cap in *Family Limitation*, but Bocker dismissed it. "Sold like a patent cure-all," Bocker complained, the Mizpah "is as likely to fail as any nostrum designed to remedy an undiagnosed disease." She cited a failure rate of "100 percent," although later clinical studies declared the Mizpah effective and safe.[13]

Bocker's report made much of the distinction between legitimate and fraudulent contraceptives—a distinction determined by the devices' retail status, not their efficacy. Over-the-counter birth control was inherently suspect. Contraceptives acquired from doctors were not. Hence, although it was the BCCRB's job to evaluate the safety and efficacy of *all* contraceptives, Bocker refused to test those she branded illegitimate. "Every effort was made [by clinic staff] to welcome new devices that were being developed by commercial interests," she reported. But "any device, method, or procedure that appeared irrational or fraudulent was dropped without trial." Hers was a curiously dismissive stance for a researcher. But Bocker's position reflected her eagerness to divide commercial contraception into two distinct realms: the ethical medical market and the fraudulent patent medicine market.[14]

Bocker gave high marks to the Mensinga diaphragm—used almost exclusively at the clinic—which Slee was smuggling into the country for Sanger.[15] Through international contacts, Slee had Mensinga diaphragms shipped from Europe to Montreal, where they were sneaked across the U.S. border in Three-in-One oil drums. Slee did not take lawbreaking lightly and resorted to it only after a customs collector pointedly told him that "there was no possible way of bringing them in lawfully."[16] The collector could not have been aware of how long this prohibition would last. Until 1971, the importation of contraceptives into the United States by laypersons remained illegal.

Slee's contraband operation was time-consuming, impractical, and risky, and Sanger knew she could not count on it indefinitely. To keep her clinic adequately stocked with Mensinga diaphragms, Sanger needed a domestic manufacturer. But who? On the one hand, there were plenty of manufacturers of Mizpah pessaries and rubber condoms around. A few blocks from the BCCRB, Youngs Rubber was manufacturing Trojans. Close, too, was Julius Schmid's company. Schmid was known mainly for its skins and rubbers, but by 1923 it had branched out into the diaphragm trade and was making a Ramses diaphragm in "a great many sizes" whose design was similar to the Mensinga's. Sanger was aware of Schmid's commercial activities and wrote about them later. Schmid might have welcomed a partnership with Sanger, producing diaphragms tailored to her specifications in the same way that the company would later manufacture condoms for the Navy. But Sanger was determined to dissociate her clinics from extant contraceptives and brand names associated with the over-the-counter trade. She wanted a company that would confine its sales to doctors. Rather than forging an alliance with Youngs Rubber or Schmid, Sanger turned to Slee, the man whose bounty and support she had come to know well.[17]

In 1925, Slee funded the establishment of the Holland-Rantos Company, the first all-birth-control firm to sell contraceptives exclusively to members of the medical profession. Established in May and incorporated in October under New York State law, it was headed by Herbert Simonds. Company letterhead described the firm's specialty as "Physicians' and Surgeons' Supplies."[18] Its objective was to make "available to American doctors the best possible contraceptive materials."[19] Slee's investment was bighearted, all the more so because Simonds had once been Sanger's suitor. An engineer, Simonds had courted Sanger during the war, when she was separated from William, taking the indomitable pacifist out dancing "almost every evening" while he waited to be shipped overseas. Still an admirer, Simonds began manufacturing diaphragms "fully expecting to be arrested."[20] But no one at Holland-Rantos was, although company officers used the mail to distribute Mensinga-type diaphragms and tubes of lactic acid jelly to druggists and doctors. Postal inspectors visited company headquarters but "were satisfied," Sanger wrote in May 1930, that "medical people only get supplies." A couple months later, in July 1930, a New York circuit court of appeals exempted from federal prosecution manufacturers who sold contraceptives exclusively to "ethical" vendors: licensed doctors and druggists. The ruling, *Youngs Rubber Corporation, Inc.* v. *C. I. Lee & Co., Inc.*, involved rival condom makers, but it established a shield of legitimacy that protected Holland-Rantos too.[21]

Sanger and Simonds suffered no illusions about how difficult it would be to change doctors' minds. They relied on James Cooper to win physicians' support. Cooper was a gynecologist who had worked as a missionary in China before entering private practice in Boston. His impeccable medical credentials, earnestness, and zeal, not to mention what Sanger thought were his "tall, blond, distinguished" good looks, made him a perfect choice for diaphragm crusader.[22] When Sanger decided in early 1925 that the time was ripe to launch a cross-country campaign to promote the Mensinga, she handpicked Cooper to be the ABCL's publicity agent. Cooper's participation would enable doctors to learn about diaphragms from a less threatening advocate. But a doctor of Cooper's stature needed a hefty salary to lure him from private practice, and Sanger implored Slee to help. She proposed a trade: Slee's financial assistance in exchange for her time and affection. On February 2, 1925, she pleaded:

> 1925 is to be the big year for the break in birth control. If Dr. Cooper's association with us is successful, I feel certain that the medical profession will take up the work. When the medical profession does this in the USA, I shall feel that I have made my contribution to the cause and shall feel that I can withdraw from full-time activity. . . .
>
> It is estimated that Dr. Cooper will cost about $10,000 salary and expenses for 1 year. His work will be to lecture before Medical Societies and Associations—getting their cooperation and influence to give contraceptive information in clinics, private and public. If I am able to accomplish this victory with Dr. Cooper's help, I shall bless my adorable husband, JNH Slee, and retire with him to the garden of love.[23]

With visions of retirement with "darling Margy" dancing in his head, Slee paid Cooper ten thousand dollars a year to spread the good word about medical birth control and the superiority of the Mensinga-diaphragm-and-jelly method. By the end of 1926, Sanger's tireless emissary had given more than seven hundred lectures in big cities and small towns in almost every state; sometimes he just rang doctors' doorbells and delivered an impromptu pitch.[24]

Sanger orchestrated Cooper's campaign, but the respected gynecologist was not simply her mouthpiece (although his salary, high even for a successful specialist, might have suggested otherwise). Cooper was a true convert to diaphragmatic birth control. He had even helped create a lactic-acid-and-glycerine-based jelly to be used with diaphragms.[25] Though she trusted Cooper's loyalty, Sanger kept careful tabs on him, demanding frequent reports. Cooper wrote to his boss regularly, detailing doctors' responses to his talks. Their sole concern, he assured her, was where they could acquire "ethical" supplies. Sanger broached this subject cautiously, fearful of the negative repercussions her open endorsement of Holland-Rantos might bring. In a letter to Cooper, she told him that the newly organized company was "getting supplies now and I believe will be able to take orders very shortly. However, I think your recommendations should be casual and not given unless urged to do so. . . . In this way it will look disinterested, as it must."[26]

Sanger was wise to be wary. Her detractors were plentiful and were constantly on the lookout for missteps that could discredit her and the movement. Sanger also had to cling to the trappings of commercial propriety to reassure doctors, who were forbidden by AMA guidelines and state laws to endorse publicly medical products for profit. Such behavior was condemned as quackery, the antithesis of the respectable image Sanger sought to cultivate. Sanger also had to think about the legal status of birth control organizations. The BCCRB and the ABCL and its successors—the Birth Control Federation of America, established in 1939, and Planned Parenthood of America, established in 1942—were nonprofit organizations whose tax and legal status hinged on commercial neutrality. To offset trouble on all these fronts, Sanger devised a code of business conduct. The BCCRB, operated by the ABCL until 1929, when it became independent, recommended methods but not individual brands, although, when asked, it provided a list of ethical manufacturers who sold only to doctors, which included Holland-Rantos. As the number of ABCL-affiliate clinics grew in the 1920s, Sanger made sure that their staffs understood that they were forbidden to be "affiliated with or subsidized by any commercial manufacturer of contraceptives" or to "derive any profit directly or indirectly from the manufacture, distribution or sale of contraceptives, either chemical or mechanical."[27] Sanger herself depended on fund-raising to bankroll her activism but turned down several lucrative endorsement deals, including one for a quarter of a million dollars to "speak on the radio for a chemical product."[28]

Sanger's dream of physician-controlled contraception thus depended, in part, on two competing principles: commercial disinterest and the success of a product manufactured by Holland-Rantos, a company she had helped create. Until the early 1930s, indigent patients were treated and equipped with diaphragms for free at ABCL clinics, which numbered twenty-eight in 1929. To keep operating expenses down, Sanger counted on gratuitous or at-cost diaphragms from Holland-Rantos, which needed a wide profit margin to sustain its generosity. Holland-Rantos officers understood from the outset the company's mission to supply Sanger's clinics. That, after all, was the primary reason the company had been established. In a letter to Sanger dated February 1926, Herbert Simonds estimated that a sale of 5,000 diaphragms a month at $1.50 each "would show a good profit and still leave 15,000 to *give away*."[29]

Long after its products became popular with private practitioners, Holland-Rantos continued to offer discounted diaphragms to birth control clinics. The arrangement allowed thousands of women to acquire contraceptive supplies they might not have had otherwise, and it set up a model of corporate subsidization of clinics that other "ethical" manufacturers would soon match. But the two-tiered price policy angered surgical supply dealers and druggists, who accused Holland-Rantos of price-cutting at their expense. The company's response, outlined in a 1941 letter, was that it was "simply a matter of company policy to make materials available to birth control clinics serving the indigent . . . as a means of contributing something to the movement . . . so that the poor could be properly and adequately served.[30] The policy helps explain the above-average retail price of Holland-Rantos products. Clinic patients could get the company's Koromex diaphragm for under a dollar, but drugstore patrons and private patients in the late 1930s paid an average of three dollars—the highest price on the diaphragm market.[31] Medical prescriptions were Holland-Rantos's largest source of profit; the company made approximately one dollar in profits for every diaphragm it sold to doctors.[32]

Holland-Rantos enjoyed a lucrative partnership with physicians in private practice. The first manufacturer to sell clinic-approved Mensinga diaphragms exclusively to the medical profession, Holland-Rantos had had first crack at courting doctor loyalty.[33] After 1925, other American firms vied for but never caught up with Holland-Rantos's medical market share. The company also benefited from advance publicity. It introduced its version of the Mensinga diaphragm only after Cooper's advertising blitz and Bocker's (Slee-funded) report celebrated the device's efficacy in clinical trials and promised the medical profession at large that "it will probably be available soon."[34]

In addition, the profitable markup gave doctors a financial incentive to prescribe the company's Koromex brand. Although physicians were forbidden to endorse products, professional ethics did not stop them from benefiting from prescriptions filled in their offices. Because they were prescription-only products, Koromex diaphragms and jelly tubes guaranteed doctors a larger segment of retail-related profits than those sold over the counter. An investigation of the diaphragm industry in the late 1930s found that the average physician markup per device ranged from $.75 to $3.50, depending on design.[35] This could mean significant sales, even in a single practice. One study of contraception in a family practice in a Pennsylvania suburb from 1925 to 1936 found that 94 percent of the 884 white female patients were prescribed the vaginal diaphragm and jelly. Most—95 percent—were upper-middle-class women who could afford the device.[36]

Physicians also benefited from Holland-Rantos's promise of medical management of birth control. The company's prescription-only diaphragms underscored its commitment to the medical profession and boosted physicians' roles as supervisors of women's reproductive health. In a laudatory and self-serving article on the history of Holland-Rantos published in *American Medicine*, a popular health journal read by doctors, Anne Kennedy, the ABCL's former secretary, Sanger's close friend, and the executive secretary and treasurer of Holland-Rantos, emphasized these themes. Kennedy neglected to inform readers that she worked for the company. She praised the Holland-Rantos diaphragm as "the modern professional method" and distinguished it from inferior "lay methods [such] as rubber prophylactics, douching preparations, tablets and liquids, suppositories, sponges, [contraceptive] tampons, and so on." The efficacy of these products was irrelevant to Kennedy. Their over-the-counter status made them inimical to medical interests, for they sought "to do away with the indispensable services of the physicians." Only the Holland-Rantos diaphragm, a scientific, doctor-controlled contraceptive, made the medical grade.[37]

A 1929 pamphlet sent to doctors by Holland-Rantos's Research Department raised similar concerns. It warned doctors "not to be confused with the cervical cap method." And it cautioned against prescribing suppositories (which destroy the vagina's "protective flora") and condoms (which break). "Here and abroad," the pamphlet insisted, "the weight of authority unquestionably favors the diaphragm."[38]

The message that Holland-Rantos was right for physicians was paralleled by studies concluding that Koromex diaphragms were right for patients. Bocker's was the first of many clinical investigations that endorsed the diaphragm-and-jelly method.[39] By the early 1930s, many such reports had been published in mainstream medical journals, increasing doctors' diaphragm consciousness. Most studies were based on medical data kept by clinics, which expanded in number from 357 in 1938 to 794 in 1942 and which were more likely to prescribe diaphragms than other methods. Even rubber shortages during World War II did not alter clinics' preferences. In 1943, Dr. Claude Pierce, the medical director of Planned Parenthood Federation of America, reported that the once-novel Holland pessary was prescribed in over 93.3 percent of all contraceptive clinic cases. The esteemed status of the diaphragm ensured more than a loyal following among public practitioners whose work provided data for study. It also guaranteed that most contraception studies read by private practitioners recommended the "Holland" diaphragm.[40]

Holland-Rantos couldn't have asked for better advertising. Six years earlier, when the AMA had finally endorsed contraceptives, the company inaugurated an aggressive advertising campaign targeting physicians in private practice. Presenting itself as the tried-and-true medical choice, the company reminded doctors of the long history of effective Holland-Rantos use. "Over 50,000 physicians, 234 clinics and 140 hospitals have already used the Koromex method," advertising copy declared. "Get in step with the AMA report."[41]

For physicians, the proven reliability of diaphragms was a real benefit. Many doctors had opposed birth control not on moral or religious grounds but for fear of endorsing a technique that did not work. There were good reasons for caution. Until the late 1930s, when the government started to enforce product standards, doctors were at the mercy of the market. Evidence of diaphragm efficacy thus reassured physicians and broadened their involvement in birth control even as it increased their reliance on a single method.

Like that of any medical technology, the effectiveness of the Holland-Rantos device was not preordained. Although the Mensinga diaphragm had been widely used in Europe, Holland-Rantos had struggled to engineer a version tailored to the idiosyncrasies of the American market. The perfect diaphragm needed to be inexpensive to manufacture, effective after repeated use, aesthetically tolerable, and able to withstand variations of climate: "hot houses and cold winters, Florida dampness and Western dryness."[42] One company officer commented that "making a good diaphragm is like baking a cake. You have to put in the right ingredients in the correct amounts to get the best results."[43] There was also the issue of design, whose perfection necessitated three features:

1. Proper spring tension. The spring must be sufficiently stiff to hold the diaphragm without undue pressure or irritation against the pubic bone but not too stiff to interfere with its proper longitudinal position in the vaginal tract.
2. Dimensional accuracy. This includes diameter of spring, height of dome and size of coil.
3. Correct dome. This must be made of rubber that is soft, pliable and resistant to repeated sterilization.[44]

Doctors also learned about the effectiveness of the diaphragm-and-jelly method in medical school. Course time devoted to contraceptive instruction increased dramatically in the interwar years as medical support of birth control grew. In a 1944 survey of 3,381 doctors—the first large-scale investigation of its kind—Dr. Alan Guttmacher, an obstetrics professor at the Johns Hopkins University, found that only 10 percent of graduates before 1920 had received training on contraception, but fully 73 percent of those who had graduated in 1935 or later had. Medical students benefiting from the more enlightened approach received a focused birth control education, one that relied on clinical data—which favored the diaphragm—to rank individual techniques. Not surprisingly, education influenced private practice. Guttmacher found a strong correlation between prescription practices and a physician's graduation year. Four-fifths (83.7 percent) of post-1935 graduates listed diaphragms as the method they prescribed most, whereas only a little more than half (57.4 percent) of pre-1910 graduates did.[45]

This medical reorientation was important for a number of reasons. It ensured the medical patronage and hence commercial viability of select birth control manufacturers, cultivating a relationship between doctors and business that the contraceptive industry would fortify in subsequent decades. In addition, as it fed the coffers of firms such as Holland-Rantos, this shift discredited companies that bypassed the medical profession, impeding the development of effective and cheap over-the-counter methods accessible to the laity. Moreover, in this new age of diaphragms and (primarily male) doctors, contraceptives meant birth control for women. When Bocker's study was published in 1924, condoms were doctors' most frequently recommended method of birth control, followed by the douche. Womb veils ranked a dismal fifth, even less popular with doctors than withdrawal or suppositories.[46] By the 1940s, however, diaphragms had become the No. 1 doctor-recommended contraceptive in the country. A 1941 survey by the Youngs Rubber Corporation (which acquired Holland-Rantos in 1947) found that 306 out of 453 doctors who recommended birth control prescribed "diaphragm pessaries." Only 26 recommended condoms.[47] Guttmacher's 1944 survey confirmed this reversal. He found that 69.6 percent of doctors ranked the diaphragm-and-jelly method their first choice for birth control. Condoms were a distant rival, accounting for only 9.5 percent of doctors' first recommendations. Significantly, the two surveys were completed *after* FDA regulations instituted in 1937 had radically improved the effectiveness of condoms, the method physicians had once preferred.[48]

It was not that doctors suddenly forgot about condoms. Condom sales boomed in the 1920s and 1930s, and medical reports touted the device's effectiveness at preventing VD. But even as men signaled their willingness to wear condoms by purchasing sheaths in unprecedented numbers, birth control advocates disparaged men's ability to be diligent users. If some medical reports were to be believed, American men were selfish, weak, and irresponsible, as ready to submit to condom use as they were to torture. In her pro-pessary report, Bocker had lambasted the condom as a technique that "places [the] wife at [the] mercy of unkind, careless, indifferent, or alcoholic husbands." Robert Dickinson of the CMH concurred. In a report published in the *American Journal of Obstetrics and Gynecology*, Dickinson warned that the sheath "is very commonly refused by the feebly virile and the selfish."[49] Although the medical malignment of male character was at odds with everyday practice, it marshaled support for female birth control and implicitly disputed the need for better or different male methods.

Also important to the popularity of female methods among doctors were changing attitudes within society and the profession that encouraged medical management of reproductive health. One concern was maternal mortality, death from pregnancy or childbirth. Although the likelihood that a woman would die from pregnancy declined during the interwar years, medical alarm about maternal mortality grew. The Chicago physician Joseph B. DeLee described childbirth in 1920 as a "pathologic process" few women escaped unscathed. "So frequent are these bad effects," he wrote, "that I have often wondered whether Nature did not deliberately intend women should be used up in the process of reproduction, in a manner analogous to that of salmon, which dies after spawning."[50]

Emphasizing the life-threatening risks of pregnancy enabled doctors to make a strong appeal for physician rather than layperson or midwife supervision of a woman's reproductive health. Writing at a time when most American women still gave birth at home, doctors such as DeLee hoped that the sterile conditions, trained staff, drugs, and equipment at physicians' disposal would encourage pregnant women to deliver in the hospital, the doctor's domain. Doctors also promoted medical birth control to prevent the development of pregnancy-related problems. In his 1933 address to the Section on Obstetrics, Gynecology, and Abdominal Surgery at the annual AMA meeting, Barton Cooke Hirst proclaimed pregnancy "incompatible with health or existence in some women." He ranked birth control one of the four most pressing gynecological issues of the day, along with death during childbirth, infertility, and cancer of the reproductive organs.[51]

By dissociating birth control from the morally charged issue of a woman's right to procreative self-determination and framing it as a valid form of disease prevention, the Crane decision of 1918 sanctioned physician involvement. In the new taxonomy of prophylaxis, contraception became therapeutically warranted. After 1918, doctors openly discussed and debated the conditions under which pregnancy was inadvisable and birth control indicated. "It is generally conceded," wrote one Philadelphia doctor in private practice, "that in the presence of tuberculosis, heart disease, chronic nephritis, previous cesarean section, and certain other subacute and chronic ailments, the occurrence of pregnancy is an additional menace to the patient's health."[52] Over time, the list of contraindications to pregnancy grew to include almost all recognized physical and psychological ailments and problems "sufficient to interfere with the discharge of ordinary duties of the earning of a living."[53]

As doctors came increasingly to regard birth control as therapy rather than smut, medical associations acknowledged, even endorsed, its place in preventive medicine. Like Catholic doctors who later prescribed oral contraceptives to treat menstrual disorders, doctors who opposed contraceptives for moral or religious reasons in the interwar years now had reason to accept their scientific benefits. By the mid-1920s, the American Gynecological Society and the Section on Obstetrics, Gynecology, and Abdominal Surgery of the AMA had endorsed therapeutic birth control. In 1937, one year after the Supreme Court upheld the right of a physician to receive by mail contraceptives "which might intelligently be employed by conscientious and competent physicians for the purpose of saving life or promoting the well-being of their patients," and after decades of characterizing contraceptives as immoral and dangerous, the AMA finally endorsed birth control—when prescribed by physicians.[54] Medical thinking had indeed shifted. As one doctor put it, "The large majority of the medical profession of this country has more and more to regard contraceptive practice in its true light; that is, not as a moral issue, but rather as a branch of preventive medicine." A sick woman "should be entitled to medical advice which will protect her from pregnancy just as much as citizens should be told to protect themselves from smallpox, diphtheria, or typhoid fever."[55]

The medical turn increased doctors' power over women's bodies. Only physicians could diagnose disease and determine the circumstances under which birth control was indicated. Then as today, there was a blurred line between what medical professionals considered best for the patient and what they thought benefited all of society. As the public health movement gained momentum in the early twentieth century, the rights and welfare of the individual were often subordinated to the needs of the community. The forced quarantining in 1907 of Mary Mallon, otherwise known as "Typhoid Mary," the first healthy carrier of typhoid in the country; the Supreme Court's ruling in *Jacobson* v. *Massachusetts* in 1905 that citizens could be vaccinated for smallpox against their will; and the Tuskegee experiments on syphilitic African American males in the 1930s: all exemplify how arguments for public welfare and health triumphed over civil liberties in the early twentieth century.[56] They also hint at the nativist and racist character of public health policy. Mary Mallon was an Irish immigrant, the Tuskegee patients poor blacks. In both cases, defending public health meant singling out socially marginalized individuals less able to fight back and more likely to be stigmatized as diseased in the first place.

Similar intellectual currents affected public policies governing reproduction. As doctors in the 1920s claimed greater expertise in managing childbirth, venereal disease, and the entire spectrum of human sexuality—from everyday lust to sexual delinquency—eugenicist policy makers accepted biological explanations of and medical solutions for disease, poverty, crime, and "feeblemindedness." Arguing that these menaces to public welfare were inherited by individuals, they pathologized procreation as the source of social ills and called for the termination of the reproductive capabilities of unfit persons. For birth control advocates, the public welfare argument that anchored sterilization appeals was a double-edged sword. On the one hand, it gave contraceptives added respectability as tools of social engineering. On the other, it categorized them as instruments of social control, weapons in a eugenicist war against criminality and imbecility. Few doubted that these were serious problems. But the public welfare approach yielded a slippery slope toward state control once contraception became a public remedy rather than a private choice. The tension between the two would haunt the birth control movement for decades, victimizing those whom Sanger had initially tried hardest to protect, the underprivileged.

Eugenics was not a new concept in 1920s America. The Englishman Francis Galton first coined the term in 1883 to describe an applied science based on the supposition that intellectual, physical, and behavioral traits are inherited. Its objective was human perfection; its method, selective breeding. Galton, a cousin of Charles Darwin's, defined the term this way: "We greatly want a brief word to express the science of improving stock . . . which, especially in the case of man, takes cognizance of all influences that tend in however remote a degree to give to the more suitable races or strains of blood a chance of prevailing speedily over the less suitable than they otherwise would have had. The word *eugenics* would sufficiently express the idea." It derived from a Greek word meaning "good in birth."[57]

In the United States, the ideology of eugenics framed arguments for and against birth control in the late nineteenth and early twentieth centuries. Although the eugenics movement developed its largest following in the United States after 1910, scores of early enthusiasts linked contraception to recognizably eugenicist beliefs. The suffragist Elizabeth Cady

Stanton and the popular birth control author Edward Bliss Foote, along with many free lovers and social purity advocates, supported female control of reproduction partly on the grounds that it would free women to select mates out of true love. Offspring would be healthier and morally stronger because of the purity of the sexual union from which they arose.[58]

The distinction between "fit" and "weak" babies was central to early-twentieth-century eugenicist philosophy. There were two distinct intellectual camps in the United States. The first, known as positive eugenics, followed Galton's lead and called for the unfettered procreation of the fittest members of society to improve the American gene pool. Its rallying cry was virulently nativist and racist, premised on the assumption that Nordic-Teutonic Americans were genetically superior. Falling birthrates among the white, Protestant, and native born and the widespread emigration of foreigners from southern and eastern Europe (over twenty-three million arrived on America's shores between 1880 and 1920) prompted many Progressives, including Theodore Roosevelt, to condemn the use of birth control by "selfish" middle-class and upper-class women as "race suicide." Roosevelt argued that native-born middle-class women who practiced fertility control were forsaking their natural duties as women and citizens by purging America's stock of its finest elements. Such behavior, Roosevelt warned in a 1911 article published in *The Outlook*, "means racial death."[59]

The second camp was known as negative eugenics, a still more insidious strain that sought to suppress, through coercion if necessary, the procreation of unfit groups. Its appeal mushroomed in the early years of the twentieth century, yielding myriad articles, pamphlets, and monographs, including Charles Davenport's widely read *Heredity in Relation to Eugenics*, published in 1911. A Harvard-trained biologist, Davenport had tracked at his research center in Cold Spring Harbor, New York, the pedigrees of extended families believed to be transmitting "defective" genes. His study sought to prove that almost all undesirable attributes, including insanity, alcoholism, eroticism, pauperism, criminality, retardation, and low intelligence, were inherited. It also associated fixed traits with specific ethnic groups. Americans of Anglo-Saxon or Scandinavian descent were the smartest and tidiest; Serbs, "slovenly"; Italians, predisposed to "crimes of personal violence."[60] Barton Cooke Hirst, chair of the AMA's Section on Obstetrics, Gynecology, and Abdominal Surgery, blamed "our loss of wealth, the venal government of cities and states, the ineptitude of Congress, the prevalence of crime, and the wave of dishonesty that has swept the country" on the millions of immigrants who had "brought to this country some racial strains that were certainly not the best." Likening the causes of the country's woes to mishandled animal husbandry, he proclaimed: "If a breeder of live stock defied the laws of eugenics as we do, he would be ruined."[61] Eugenicists typically positioned African Americans at the bottom of the racial hierarchy. The professor and popular author Paul Popenoe and the petroleum scientist Roswell Johnson argued in their 1926 study, *Applied Eugenics*, that "if the number of original contributions which it has made to the world's civilization is any fair criterion of the relative value of a race, then the Negro race must be placed very near zero on the scale. . . . In comparison with some other races the Negro race is germinally lacking in the higher developments of intelligence."[62]

Cloaking their findings in the mantle of science, proponents of negative eugenics pressed for concrete measures to stem the racial degeneration of America. Like most Americans at the time, eugenicists supported segregation and antimiscegenation statutes. They lobbied successfully for national quota laws to restrict the immigration of southern and eastern Europeans. They criticized proposals to fund programs for retarded children and prenatal and obstetric care for the poor. These, they insisted, would encourage imbecility by increasing the life span and fecundity of defective citizens.[63]

Davenport and other eugenicists believed that the least intelligent members of the human species were, like lower animals, biologically programmed to be the most prolific. Hence it was especially important for society to halt the procreation of the unfit before they bred the human race into degeneracy. For many, the coerced surgical sterilization of unfit men and women was the only permanent, effective prophylaxis against racial decay. As one eugenics pamphlet proclaimed, "Eugenic sterilization, conservatively and sympathetically administered, is a practical, humane and necessary step to prevent race deterioration."[64] By the 1920s, scientists' reports of success with male and female sterilization had helped make the subject a mainstream political issue, with most Americans favoring its use on the institutionalized insane.[65]

Male sterilization referred to a vasectomy, the surgical tying of the seminal ducts. Dr. Harry C. Sharp performed the first male sterilization in Indiana in 1899 on a prisoner who complained of an uncontrollable urge to masturbate. Sharp published the results of the surgery, which he claimed had fixed the problem, in a 1902 article titled "The Severing of the Vasa Deferentia and Its Relation to the Neuropsychopathic Constitution." Sharp's insistence that the operation cured his patient's pathological desire illustrates how social views shaped ostensibly scientific conclusions. Subsequent research has shown that vasectomies do not affect sexual drive and that the desire to masturbate—by a man or woman—is neither a physiological nor a psychological problem. Yet Sharp was captive to the prejudices of his time and insisted that sexual deviance was of biological origin. Claiming victory in his crusade against neuropsychosis one surgery at a time, he went on to perform over 450 vasectomies on inmates and to lobby the Indiana legislature to pass the first eugenic law mandating the coerced sterilization of unfit persons, which it did in 1907.[66]

Sharp's medical involvement and political success encouraged other physicians to endorse compulsory sterilization to stem social degeneracy. Sexual surgery was also recommended for women, who had for decades been subjected to gynecological surgeries such as ovariectomies (the surgical removal of ovaries) to "cure" so-called nymphomania and hysteria.[67] It was not a huge surgical step from ovariectomies to salpingectomies. More complicated than a vasectomy, a salpingectomy (today called tubal ligation) involved incising the abdominal wall and tying the Fallopian tubes.[68] Eugenicists hailed vasectomies and salpingectomies as safe, effective, and simple. They rarely "failed," and no special training was required to perform them. Patient inconvenience and recovery time were purportedly minimal. One booklet circulated for general readership estimated that a vasectomy could be performed under a local anesthetic in fifteen or twenty minutes. The operation "is so simple and easily accomplished," gushed one pamphlet for men, "that the man need not remain away from his employment for a period longer than it takes to have the operation performed.[69] A New York judge promised in 1916 that it was "less serious than the extraction of a tooth."[70]

Sterilizations were also credited with saving taxpayers' money, an argument that increased support once the Great Depression arrived. The Human Betterment Foundation of Pasadena, California, the leading U.S. eugenics society at the time, estimated in 1930 that there were eighteen million Americans "burdened by mental disease or mental defect, and in one way or another a charge and tax upon the rest of the population." The economic burden to healthy Americans was considerable. "A billion dollars a year would be a low estimate of the cost of caring for these unfortunates," the foundation stated.[71]

In 1927, the Supreme Court upheld the constitutionality of state sterilization laws in its ruling in *Buck* v. *Bell.* The case involved a 1924 Virginia eugenics statute that legalized the coerced sterilization of "socially inadequate person[s]." Carrie Buck, the plaintiff, was single, white, pregnant, and only seventeen when she was brought to the Virginia Colony for Epileptics and Feeble Minded in Lynchburg. Although she insisted that her pregnancy was the result of rape, her condition and her status as the "daughter of an imbecile" and the mother of "an illegitimate feeble minded child" supplied the primary evidence substantiating the diagnosis of mental ineptitude. In his momentous ruling, Oliver Wendell Holmes, long admired for his heroic *defense* of civil liberties, declared coercive sterilization a valid exercise of the state's right to protect public health. "The principle that sustains compulsory vaccination," he declared, referring to the Court's 1905 ruling in *Jacobson* v. *Massachusetts*, is broad enough "to cover cutting the Fallopian tubes. . . . Three generations of imbeciles are enough."[72]

The Court's decision licensed other states to pass similar eugenics legislation. By 1932, at least twenty-six states had enacted laws permitting the forced sterilization of individuals considered unfit. Although their provisions varied, most authorized sterilizations on men and women suffering from "feeblemindedness, insanity, epilepsy, idiocy, moral degeneracy, imbecility, habitual criminality, or sexual perversion." By 1937, almost twenty-eight thousand men and women had been forced to undergo eugenic surgery in the United States. Most—more than sixteen thousand—were women.[73] The Virginia act served as the model for Germany's Hereditary Health Law in 1933. During the Nuremberg trials following World War II, accused Nazi war criminals cited *Buck* v. *Bell* to justify the forced sterilization of some two million Germans.

Although eugenic theory began to be discredited as bad science in the 1940s, when the atrocities of the Holocaust were brought to light, it continued to shape policies affecting reproduction, birth control, and social equality. From the outset, support for eugenics and contraception overlapped, but not always as tidily as some have asserted.[74] Many eugenicists opposed birth control for the poor and persons of color. Convinced of the ineptitude of the unfit, they harbored doubts that blacks, immigrants, and the poor had the intellectual wherewithal or self-discipline to use birth control effectively—assuming, of course, that they could be convinced of its need. As one racist researcher callously observed, "The American negro may be supposed not to practice contraception largely. Some devices are expensive and intricate, others are expensive and distasteful to self-indulgent men."[75] Moreover, at a time when the efficacy of various contraceptives was still unclear, coerced sterilization seemed to many a surer bet than birth control. The Human Betterment Foundation had such a low regard for the self-control of the institutionalized insane that it rejected even abstinence "under lock and key" as a viable alternative to surgery. Aside from being expensive for the state, compulsory abstinence "is not 100% successful. Some childbearing occurs in any such institution. Some patients escape."[76]

At the same time, the scientific credibility of the birth control movement was enhanced by the search to limit the procreation of undesirable groups, and its leaders appropriated eugenic language to promote their goals. Like most Americans, Sanger supported sterilization for the incarcerated and considered birth control a necessary component of racial improvement. "Birth control," she stated emphatically in 1920, "is nothing more or less than the facilitation of the process of weeding out the unfit [and] of preventing the birth of defectives."[77] But Sanger also believed that socially structured inequality, especially differential access to contraceptives, caused inferiority. Poverty and criminality were made, not born: "Children who are underfed, undernourished, crowded into badly ventilated and unsanitary houses and chronically hungry cannot be expected to attain the mental development of children upon whom every advantage of intelligent and scientific care is bestowed."[78] This unshakable faith in the environmental rather than hereditary origins of human degradation was one Sanger carried over from her radical years. She earnestly believed that a desire to escape poverty (and its concomitant psychological degradation) would motivate women and men to use contraception. She endorsed the principle of reproductive autonomy and maintained that the establishment of public clinics promoting cross-class access to scientific birth control furthered that goal. But she was not above the paternalism intrinsic to eugenic thought, the conviction that sometimes people needed to be reeducated "for their own good."

This elitism was especially apparent in her dealings with African Americans. In a joint effort between the Birth Control Clinical Research Bureau and the National Urban League, Sanger opened a second New York clinic in Harlem in February 1930. Like its sister clinic, the Harlem site aimed to bring birth control to the poor through the distribution of cheap diaphragm-and-jelly kits. But the all-white staff, and the storefront placard identifying the clinic as a research bureau, immediately raised suspicions within the black community that the clinic's goal was to experiment on and sterilize black people. Blacks' fears were not entirely unfounded. Sanger designated the Harlem clinic the official teaching site for physicians wanting diaphragm training. A fee of twenty-five dollars entitled trainees to "three sessions at the Harlem branch where they will not only be present, but will, under instruction, give examinations."[79] Naked, knees splayed, passive, and probed: being fitted for a diaphragm was not (and is not) what most women consider fun. But to be subjected to the procedure as an African American woman by a novice, likely a white male, in the presence of others who watched and commented on his "work," could only have made a difficult situation worse.

Sanger also initially refused to cede control of the clinic to the Harlem Advisory Council, an autonomous board of black health professionals established to help run the clinic and to raise money. She insisted that the clinic met a need that "the race did not recognize" for itself.[80] When blacks continued to view the clinic suspiciously, she racially integrated its staff and reworded promotional pamphlets to emphasize the temporary and harmless nature of the contraception offered. It was not enough. In 1936, a year after the American Birth Control League took over as managers, the financially ailing clinic closed.

Unfortunately, unsuccessful efforts such as these did not motivate birth control leaders to rethink their racist and nativist assumptions. Before the clinic closed, Harlem Advisory Council member Mabel Staupers wrote Sanger: "If the Birth Control Association wishes the cooperation of Negroes . . . we should be treated with the proper courtesy that is due us and not with the usual childish procedures that are maintained with any work that is being done for Negroes."[81] But it was easier to chalk the clinic's failings up to black disinterest than to white racism.

By the 1930s, charges that public clinics and the diaphragms they dispensed were no match for the idiocy and fecundity of the urban poor had forced Sanger and her colleagues to defend the clinic-diaphragm formula as best for the medical profession *and* the masses. Few scientists and doctors questioned that the diaphragm-and-jelly method worked wonders in the laboratory and in the homes of the white middle class. But everyday use among African Americans, immigrants, and the poor—the very groups whose fecundity American society was most intent on controlling—was an altogether separate matter.

In *Facts and Frauds in Woman's Hygiene*, a 1938 study of how women's health needs were shortchanged in American society, the consumer health advocates Rachel Lynn Palmer and Sarah Koslow Greenberg condemned the profits made by diaphragm-and-jelly manufacturers and demanded the "socialization of the birth control business." Palmer and Greenberg recognized what Sanger would not. A patchwork quilt of birth control clinics was no way of bringing contraceptives to the poor, at least not in a country where profits for manufacturers and medical professionals were more important than health care for the poor and where extramural clinics had to be funded by donations and defended against the argument that it would be cheaper for society to sterilize the indigent. In any society without universal health care, working-class people are systematically denied access to doctors and the services they monopolize. Of all people in the birth control movement, Sanger probably understood this best. To her credit, she never gave up her goal of quality birth control for all. She simply failed to achieve it. Throughout Sanger's life, most Americans got contraceptives where they always had, on the open market.

Notes

1 Sanger to Clarence Gamble, July 9, 1952, folder 3098, box 196, Clarence Gamble Papers, Countway Library of Medicine, Harvard Medical School, Boston, Mass.

2 Advertisement for Mizpah pessary in *National Police Gazette*, March 14, 1900, p. 14; Dorothy Bocker, *Birth Control Methods* (New York Birth Control Clinical Research, 1924), pp. 6, 12–13, 18–20; Margaret Sanger, "Why I Went to Jail," *Together*, February 1960, p. 20; Robert L. Dickinson, "Contraception: A Medical Review of the Situation," *American Journal of Obstetrics and Gynecology* 8 (November 1924): 590–91; *Catalog of Drug Sundries* (Bridgeport, Conn.: Robert Dalton), box 785, Sex Collection, American Medical Association Health Fraud Archives, AMA Headquarters, Chicago; Ellen Chesler, *Woman of Valor: Margaret Sanger and the Birth Control Movement in America* (New York: Simon and Schuster, 1992), p. 151.

3 David M. Kennedy, *Birth Control in America: The Career of Margaret Sanger* (New Haven, Conn.: Yale University Press, 1970), pp. 20–21, 25–27; Chesler, *Woman of Valor*, p. 97; Andrea Tone, *Controlling Reproduction: An American History* (Wilmington, Del.: Scholarly Resources, 1997), pp. 155–56; James Reed, *From Private Vice to Public Virtue: The Birth Control Movement and American Society since 1830* (New York: Basic Books, 1978), pp. 97–99.

4 Margaret Sanger, *The Autobiography of Margaret Sanger* (1938; reprint, Elmsford, N.Y.: Maxwell Reprint Company, 1970), p. 142.

5 Ibid., p. 144.

6 Johannes Rutgers, "Clinics in Holland," *Birth Control Review* 4 (April 1920): 9; Bocker, *Birth Control Methods*, p. 26; Chesler, *Woman of Valor*, p. 145; Reed, *From Private Vice to Public Virtue*, p. 95. Advertisement for Mizpah pessary in *National Police Gazette*, March 14, 1900, p. 14; Sanger, *Autobiography*, pp. 142, 148; "Contraceptive Devices," *Human Fertility* 10 (September 1945): 69–70.

7 Margaret Sanger, "Suggestions for the Establishment of a Birth Control Clinic," p. 7, reel 29, Margaret Sanger Papers, Manuscripts Division, Library of Congress; Lovette Dewees and Gilbert W. Beebe, "Contraception in Private Practice: A Twelve Years Study," *Journal of the American Medical Association* 110 (April 9, 1938): 1169; Reed, *From Private Vice to Public Virtue*, p. 95; Chesler, *Woman of Valor*, p. 145; Dickinson, "Contraception," p. 591.

8 Reed, *From Private Vice to Public Virtue*, p. 97.

9 Bocker, *Birth Control Methods*, p. 19.

10 Quoted in Kennedy, *Birth Control in America*, p. 182; Reed, *From Private Vice to Public Virtue*, p. 113.

11 Reed, *From Private Vice to Public Virtue*, p. 113; Chesler, *Women of Valor*, pp. 247–49; Kennedy, *Birth Control in America*, pp. 98–99.

12 Kennedy, *Birth Control in America*, p. 99; Chesler, *Woman of Valor*, p. 113.

13 R. Christian Johnson, "Feminism, Philanthropy, and Science in the Development of the Oral Contraceptive Pill," *Pharmacy in History* 19 (1977): 65; Bocker, *Birth Control Methods*, p. 6.

14 Bocker, *Birth Control Methods*, pp. 1, 5, 6, 19.

15 Sanger, "Suggestions for the Establishment of a Birth Control Clinic," pp. 6–7.

16 Slee quoted in Reed, *From Private Vice to Public Virtue*, p. 114.

17 Sanger, *Autobiography*, pp. 363–64; Bocker, *Birth Control Methods*, pp. 19–20, 26; James S. Murphy, *The Condom Industry in the United States* (Jefferson, N.C.: McFarland and Company, 1990), p. 10; "List of Reputable Manufacturers," Margaret Sanger Papers, reel 29; U.S. Patent Office, *Official Gazette*, January 13, 1931, p. 290, March 31, 1931, p. 75, May 26, 1931, p. 883; *Youngs Rubber Corporation* v. *C. I. Lee & Co., Inc.*, 45 *Federal Reporter*, 2nd ser., 103 (1930).

18 Anne Kennedy, "History of the Development of Contraceptive Materials in the United States," *American Medicine*, March 1935, p. 160; "Design Key Factor in New H-R Marketing Programs," *Drug and Cosmetic Industry* 94 (April 1964): 520–21; Herbert R. Simonds to Norman E. Himes, April 17, 1929, folder 320, box 29, Norman Himes Papers, Countway Library of Medicine; *New York City Directory 1933/34* (New York: R. L. Polk and Company, 1934), pt. 2., listing for Holland-Rantos Inc.

19 Kennedy, "History of Contraceptive Materials," p. 160.
20 Reed, *From Private Vice to Public Virtue*, p. 115; Lawrence Lader, *The Margaret Sanger Story and the Fight for Birth Control* (Garden City, N.Y.: Doubleday, 1955), p. 225.
21 Sanger marginalia on Percy Clark to Margaret Sanger, May 5, 1930, Sanger Papers, reel 30; *Youngs Rubber Corporation* v. *C. I. Lee & Co., Inc., 45 Federal Reporter*, 2nd ser., 103 (1930).
22 Sanger, *Autobiography*, p. 362.
23 Quoted in Lader, *Margaret Sanger Story*, p. 223.
24 Lader, *Margaret Sanger Story*, pp. 223–24; Chesler, *Woman of Valor*, p. 248.
25 Sanger, *Autobiography*, pp. 362–63; Reed, *From Private Vice to Public Virtue*, p. 161.
26 Margaret Sanger to James Cooper, October 10, 1925, box 6, folder 135, Florence Rose Collection, Sophia Smith Collection, Smith College.
27 Letter of D. Kenneth Rose, April 7, 1941, box 6, folder 134, Florence Rose Collection; Sanger, "Suggestions for the Establishment of a Birth Control Clinic," pp. 8–9.
28 Lader, *Margaret Sanger Story*, p. 225; Johnson, "Feminism, Philanthropy, and Science," p. 65.
29 "In re: Birth Control Movement," unpublished manuscript, circa 1932, box 1, folder 139, Rockefeller Family Archives, RG 2, Series Office, O.M.R., Rockefeller Archives, Tarrytown, New York; Herbert Simonds to Margaret Sanger, February 1926, box 6, folder 134, Florence Rose Collection; McCann, *Birth Control Politics*, p. 215.
30 Harry W. Hicks to Kenneth Rose, October 10, 1941, box 73, folder 696, Planned Parenthood Federation of America Papers, Sophia Smith Collection.
31 Planned Parenthood Federation of America 1943 study of the contraceptive industry by Foote, Cone & Belding, box 65, folder 647 (hereafter cited as PPFA 1943 study), Planned Parenthood Federation of America Papers, Sophia Smith Collection.
32 "The Accident of Birth," *Fortune*, February 1938, p. 108.
33 This ended in 1929 when one of the company's original officers, Dr. Le Mon Clark, a former professor of sociology and economics, defected to establish the Clinic Supply Company, manufacturers of diaphragms. In 1938, Clark's gross sales, including the sales of contraceptive jelly and cream, totaled thirty thousand dollars a year. But Clark's bid for market supremacy came too late. See "Accident of Birth," *Fortune*, p. 112.
34 Bocker, *Birth Control Methods*, p. 26.
35 Contraceptive Industry Report, p. 33, Planned Parenthood Federation of America Papers, Sophia Smith Collection: Holland-Rantos did not make profits a component of its advertisements to doctors, but it did to druggists, enlisting them in the campaign to solidify physician support. A retail drug advertising campaign in 1937 advised druggists "to let the physicians in your neighbourhood know that you can fill all of their Koromex prescriptions. Let us show you how you can become headquarters for the ethical business in your town." In an advertisement in *Drug Store Retailing*, a magazine for druggists, the company specified an ideal profit margin. It suggested that individual tubes of Koromex jelly purchased by the druggist for $.60 be sold to consumers for $1.50—a profit of 150 percent. "Merchandising News," *Drug and Cosmetic Industry* 41 (August 1937): 216; Rachel Lynn Palmer and Sarah Koslow Greenberg, *Facts and Frauds in Woman's Hygiene: A Medical Guide against Misleading Claims and Dangerous Products* (Garden City, N.Y.: Garden City Publishing Company, 1938), p. 256; Consumers Union of United States, "Analysis of Contraceptive Materials," box 65, folder 646, Planned Parenthood Federation of America Papers, Sophia Smith Collection.
36 Dewees and Beebe, "Contraception in Private Practice," p. 1169.
37 Kennedy, "History of Contraceptive Materials," pp. 159–61; *New York City Directory 1933/34*, pt. 2, listing for Holland-Rantos Inc.
38 Holland-Rantos Co., Inc., *Report on Physicians' Replies to Questionnaire concerning Their Experience with the Vaginal Diaphragm and Jelly* (New York, 1929), pp. 16–17, 19; also see "Why the Diaphragm Is Strictly a Prescription Contraceptive," *Holland-Rantos Bulletin 3* (January 1935), folder 319, box 29, Norman Himes Papers, Countway Library of Medicine.
39 Hannah M. Stone, "Maternal Health and Contraception: A Study of 2,000 Cases from Maternal Health Center, Newark, N.J.," *Medical Journal and Record*, April 19, 1933, pp. 7–15, and May 5, 1933, pp. 7–13; Bessie L. Moses, *Contraception as a Therapeutic Measure* (Baltimore, Md.: Williams and Wilkins Company, 1937). Also see Marie E. Kopp, *Birth Control in Practice* (New York: McBride, 1934); Dewees and Beebe, "Contraception in Private Practice," pp. 1169–72; Ruth A. Robishaw, "A Study of 4,000 Patients Admitted for Contraceptive Advice and Treatment," *American Journal of Obstetrics and Gynecology* 31 (March 1936): 426–35; Regine K. Stix, "Birth Control in a Midwesten City: A Study of the Clinics of the Cincinnati Committee on Maternal Health," *Milbank Memorial Fund Quarterly*, January 1939, pp. 69–91, April 1939, pp. 152–71,

October 1939, pp. 392–423; Hannah M. Stone, "Therapeutic Contraception," *Medical Journal and Record*, March 21, 1928.

40 One study published in the *American Journal of Obstetrics and Gynecology* found that 3,514 of the 4,000 patients (88 percent) admitted to the Maternal Health Clinic of Cleveland between March 1928 and January 1934 were given the "diaphragm pessary used in conjunction with lactic acid jelly." Hannah Stone, Bocker's successor, estimated in 1936 that the combination was prescribed for more than 95 percent of patients at medically directed clinics nationwide. Hannah M. Stone, "Occlusive Methods of Contraception," *Journal of Contraception* (May 1937): 162. Her report was first presented at the Conference on Contraceptive Research and Clinical Practice, held in New York on December 29–30, 1936. Claude C. Pierce, "Contraceptive Services in the United States," *Human Fertility* 8 (September 1943): 92. Also see Sanger, "Suggestions for the Establishment of a Birth Control Clinic," p. 7.

41 "Merchandising News," *Drug and Cosmetic Industry* 41 (August 1937): 216; "Contraceptive," *Drug and Cosmetic Industry* 41 (July 1937): 75.

42 Sanger, *Autobiography*, p. 364.

43 Erick Kunnas to Mrs. Merlin Stoutenburgh, April 14, 1961, "Birth Control Products and Research, Holland-Rantos, 1956–1962," folder 8, box 4, Mary Steichen Calderone Papers, Schlesinger Library, Radcliffe Institute, Cambridge, Mass.

44 Kennedy, "History of Contraceptive Materials," p. 160.

45 Alan F. Guttmacher, "Conception Control and the Medical Profession: The Attitude of 3,381 Physicians toward Contraception and the Contraceptives They Prescribe," *Human Fertility 12* (March 1947): 1–10; National Committee on Maternal Health, "Contraception, Sterilization, and Hygiene of Marriage in the Medical Curriculum," *American Journal of Obstetrics and Gynecology* 31 (January 1936): 165–68.

46 Dickinson, "Contraception," pp. 584–87.

47 Youngs Rubber Corporation Survey, pp. 1–2, National Committee on Maternal Health Papers, Countway Library of Medicine.

48 Guttmacher, "Conception Control and the Medical Profession," p. 8. Also see Marie Pichel Warner, "Contraception: A Study of Five Hundred Cases from Private Practice," *Journal of the American Medical Association* 115 (July 27, 1940): 279–85, and Robishaw, "Study of 4,000 Patients," pp. 426–35. Robishaw found that in four thousand birth control patients studied, the condom was recommended only once.

49 Bocker, *Birth Control Methods*, p. 5; Dickinson, "Contraception," p. 589.

50 Quoted in Judith Walzer Leavitt, *Brought to Bed: Childbearing in America, 1750 to 1950* (New York: Oxford University Press, 1986), pp. 179, 184. Also see McCann, *Birth Control Politics*, pp. 66–67.

51 Barton Cooke Hirst, "The Four Major Problems in Gynecology," *Journal of the American Medical Association* 101 (September 16, 1933): 900.

52 Thaddeus Montgomery, "Indications for Contraception from the Point of View of the Obstetrician and Gynecologist," *American Journal of Obstetrics and Gynecology* 32 (1936): 471.

53 The *Journal of Contraception* classified the following disorders as indications for contraception in 1936: systemic diseases (cardiac disease, nephritis, nephritic toxemias, severe anemia, hypertension, thyroid disease, diabetes, gall bladder disease, epilepsy, and so on); nervous and mental diseases (manic depression, anxiety neuroses, schizophrenia, mental defectiveness, constitutional inferiority, and so on); gynecological problems (recent pelvic repair, fistulas, extreme lacerations, repeated abortions [spontaneous or induced], adnexal disease, and so on); obstetrical problems (toxemias other than nephritic, repeated difficult deliveries, eclampsia, repeated cesarean sections, and so on); orthopedic and surgical problems (tuberculosis of spine or hip, osteomyelitis, fractured pelvis, recent operation, one kidney, kidney stones); venereal diseases (gonorrhea, syphilis, central nervous system syphilis); defects and deformities (spina bifida, various paralyses, congenital blindness, double vagina, and so on); multiparity (too many pregnancies or too frequent pregnancies in a short period of time, often combined with general debility, malnutrition, anemia); indications of husbands (tuberculosis, epilepsy, postencephalitic condition, blindness, mental illness, lues, criminal alcoholism, and so on); eugenic problems (repeated defect in children, harelip in three children, three status lymphaticus deaths, family with Friedrich's ataxia, family with numerous institutionalized psychopaths); not strictly medical circumstances (martial disharmony, recent delivery, economic). See "Medical Indications for Contraception," *Journal of Contraception*, November 1936, p. 189.

54 The case was *United States* v. *One Package*, 13 F. Supp. 334 (1936); Birth Control Federation of America, *Questions and Answers about Birth Control* (New York, circa 1939), reel 10 of 26, Records of the National Association of Colored Women's Clubs, National Archives for Black

Women's History, Washington, D.C.; Richard N. Pierson, Robert L. Dickinson, Ira S. Wile, and Woodbridge E. Morris, "Contraceptive Practice," *American Journal of Obstetrics and Gynecology* 41 (January 1941): 174–75; Ralph E. Brown, "The Legal Status of Contraception: A Practical Interpretation for the Doctor's Guidance," *American Medicine*, March 1935, pp. 167–70.

55 Owen Jones Toland, "Contraception—A Neglected Field for Preventive Medicine," *American Journal of Obstetrics and Gynecology* 27 (January 1934): 52. Toland's talk was first presented at a meeting of the Obstetrical Society of Philadelphia on May 4, 1933.

56 See Judith Walzer Leavitt, *Typhoid Mary: Captive to the Public's Health* (Boston: Beacon Press, 1996); James Jones, *Bad Blood: The Tuskegee Syphilis Experiment—A Tragedy of Race and Medicine* (New York: Free Press, 1981); Alan Kraut, *Silent Travelers: Germs, Genes, and the Immigrant Menace* (New York: Basic Books, 1994); Charles McClain, "Of Medicine, Race, and American Law: The Bubonic Outbreak of 1900," *Law and Social Inquiry* 13 (1988): 447–513; Vanessa Northington Gamble, "A Legacy of Distrust: African-Americans and Medical Research," *American Journal of Preventive Medicine* 9 (1993): 35–38. The Supreme Court's case on smallpox is *Jacobson v. Massachusetts*, 25 S.Ct. 358.

57 Galton quoted in F. H. Hankins, "The Interdependence of Eugenics and Birth Control," *Birth Control Review*, June 1931, p. 170; McCann, *Birth Control Politics*, p. 101. For a comprehensive history of the eugenics movement in the United States, see Daniel J. Kevles, *In the Name of Eugenics: Genetics and the Uses of Human Heredity* (New York: Alfred A. Knopf, 1985).

58 Hal Sears, *The Sex Radicals: Free Love in High Victorian America* (Lawrence: University Press of Kansas, 1977), p. 120; D'Emilio and Freedman, *Intimate Matters*, p. 165.

59 Theodore Roosevelt, "Race Decadence," *Outlook*, April 8, 1911; E. S. Goodhue, "Race Suicide," *Medico-Legal Journal* 25 (June 1907): 251–56.

60 McCann, *Birth Control Politics*, pp. 103–5; Kevles, *In the Name of Eugenics*, pp. 44–54; Dorothy Roberts, *Killing the Black Body: Race, Reproduction, and the Meaning of Liberty* (New York: Vintage, 1997), pp. 59–63.

61 Hirst, "Four Major Problems in Gynecology," p. 900.

62 Paul Popenoe and Roswell Johnson, *Applied Eugenics* (New York: Macmillan Company, 1926), pp. 156–59, 284–85, 292, 297. Also see Roberts, *Killing the Black Body*, pp. 60–61.

63 Roberts, *Killing the Black Body*, p. 65.

64 M. A. Horn, ed., *Mother and Daughter* (Wilmington, Ohio: Hygienic Productions, 1947), p. 74.

65 Kevles, *In the Name of Eugenics*, pp. 90–94.

66 Roberts, *Killing the Black Body*, pp. 66–67; "Eugenic Sterilizations in the United States," *Journal of Contraception*, April 1938, pp. 81–83.

67 By 1906, doctors had performed an estimated 150,000 ovariectomies on American women under the guise of protecting their emotional stability and mental health. Tone, *Controlling Reproduction*, p. 65. Also see Barbara Ehrenreich and Deirdre English, *For Her Own Good: 150 Years of the Experts' Advice to Women* (Garden City, N.Y.: Anchor Press/Doubleday, 1978).

68 Norman Haire, "Sterilization of the Unfit," *Birth Control Review*, February 1922, pp. 10–11; "Eugenic Sterilizations in the United States," *Journal of Contraception*, pp. 81–82; E. H. F. Pirkner, "Prophylactic Salpingapotomy," *Medico-Legal Journal* 33 (March 1917): 14.

69 M. A. Horn, ed., *Father and Son* (Wilmington, Ohio: Hygienic Productions, 1947), p. 22.

70 "Contraception," *Medico-Legal Journal* 33 (April 1916): 10.

71 The Human Betterment Foundation, *Human Sterilization* (Pasadena, Calif.: The Human Betterment Foundation, circa 1930), pp. 5–6.

72 *Buck* v. *Bell*, 274 U.S. 200 (1927).

73 Of the 27,869 Americans sterilized between 1907 and 1937, 16,241 were women. "Eugenic Sterilizations in the United States," *Journal of Contraception*, p. 81.

74 For an especially good analysis of the complexities of the relationship between the eugenics and birth control movements, see McCann, *Birth Control Politics*, chap. 4.

75 Caroline Hadley Robinson, *Seventy Birth Control Clinics: A Survey and Analysis Including the General Effects of Control on Size and Quality of Population* (Baltimore, Md.: Williams and Wilkins Company, 1930), p. 246.

76 Human Betterment Foundation, *Human Sterilization*, p. 6.

77 Sanger quoted in McCann, *Birth Control Politics*, p. 107.

78 Sanger quoted in McCann, *Birth Control Politics*, p. 112.

79 "In re: Birth Control Movement," pp. 3–4, Rockefeller Archives; Roberts, *Killing the Black Body*, pp. 87–88; McCann, *Birth Control Politics*, chap. 5.

80 Sanger quoted in Roberts, *Killing the Black Body*, pp. 87–88.

81 Staupers quoted in Roberts, *Killing the Black Body*, p. 43.

14
Sociobiology, Biological Determinism, and Human Behavior

Ruth Bleier

Because Wilsonian Sociobiology is a particularly dramatic contemporary version of biological determinist theories of human behavior, because it is powerful and persuasive, because it is a particularly good example of bad science, because it provides "scientific" support for a dominant political ideology that directly opposes every goal and issue raised by the women's movement, and because it has been aggressively marketed and perceptibly incorporated into our culture, it seems a fitting area with which to begin the examination of science and scientific theories of biological determinism.

While the general field of sociobiology has a long and solid tradition of studying the social behavior of animals, in 1975 E.O. Wilson, whose area of expertise is insect behaviors, sought to establish sociobiology "as the systematic study of the biological basis of all social behavior." He stated his conviction that "it may not be too much to say that sociology and the other social sciences, as well as the humanities, are the last branches of biology waiting to be included in the Modern Synthesis" (Wilson, 1975b, p. 4). Thus, Wilson and those in his school of human sociobiology believe that all human behaviors, social relationships, and organization are genetically evolved adaptations, as I will describe below. Before proceeding, however, to a critique of the work of Wilsonian Sociobiologists, it is important to distinguish it from the general field of sociobiology. There are many other scientists who study the social behaviors and characteristics of animals and are therefore sociobiologists but do not make reckless extrapolations to human social relationships and behaviors. Their observations and interpretations form an important part of the evidence I use to support my arguments concerning the inadequacies and distortions inherent in the "science" that Wilson and his followers popularize.

By reducing human behavior and complex social phenomena to genes and to inherited and programmed mechanisms of neuronal functioning, the message of the new Wilsonian Sociobiology becomes rapidly clear: we had best resign ourselves to the fact that the more unsavory aspects of human behavior, like wars, racism, and class struggle, are inevitable results of evolutionary adaptations based in our genes. And of key importance is the fact that the particular roles performed by women and men in society are also biologically, genetically determined; in fact, civilization as we know it, or perhaps any at all, could not have evolved in any other way. Thus the Sociobiologist and popular writer David Barash says, "There is good reason to believe that we are (genetically) primed to be much less sexually egalitarian than we appear to be" (Barash, 1979, p. 47).

But it is not only that the direct political and social statements and theories of

Sociobiologists are dangerous to the interests and well-being of women and minorities. If Sociobiology were a valid science, by even traditional standards, we should have to find ways to cope with the consequences of incontrovertible "truths." But this Sociobiology is deeply flawed conceptually, methodologically, and logically *as a science*. It is only *because* it concerns itself with the most complex aspects of human behaviors and social relationships, about which we suffer enormous depths of both ignorance and emotion, that Sociobiology achieves acceptance as a science. The same kinds of logical and methodological flaws in the sciences, say, of ant or camel behavior would be immediately obvious and unacceptable.

In this chapter I first review some basic postulates and assumptions of Sociobiological theory and outline the methodologies used for theory building. I then offer a detailed critique of Sociobiologists' theories and methods and indicate some alternative observations and interpretations that contradict their assumptions and conclusions. Finally, since the fundamental scientific issue is the validity of a theory based on the genetic determination of human behavior, I explore the relationship between genes and the fetal environment and between biology and learning.

SOME PREMISES AND APPROACHES OF SOCIOBIOLOGY

Natural selection of behaviors through gene transmission

The basic premise of Sociobiology is that human behaviors and certain aspects of social organization have evolved, like our bodies, through adaptations based on Darwinian natural selection. It is important to understand Darwin's theory of evolution of the *physical forms* of animals by adaptation in order to understand its application by Sociobiologists to *behavior*. In its modern version, the theory assumes that by some genetic recombination or mutation, a particular anatomical characteristic appears anew in a species, let us say gray body color in a family of orange moths. If the gray color in the moths' particular ecological setting permits more gray than orange moths to survive predation and other causes of an early demise and therefore to reach sexual maturity so that more gray moths are reproduced than their relatives of the original orange color, then an increasing proportion of moths will be gray in successive generations. Over time, the genes for gray will be present in increasing numbers of moths and become a predominant feature of moths in *that* ecological setting. The new genetic feature for gray is then considered, in the language of Darwinian evolution, to be adaptive through natural selection, since it contributes to the maximum fitness of the moths, with *maximum fitness* being defined as the ability to leave many healthy descendants that are themselves able to reproduce and thus spread the genes for gray body color.

Sociobiologists suggest and assume that *behaviors* also evolve in similar ways so that "adaptive" and "successful" behaviors become based in our genes, and that certain genetic configurations became selected because they result in behaviors that are adaptive for survival. Our "innate" predispositions to display these behaviors constitute our human *nature*. It is important to note at this point that to be valid the theory requires that human behaviors be represented by a particular genetic configuration, because evolution through natural selection requires genetic variations (that is, mutant forms) from which to select. But Sociobiologists themselves, as well as geneticists, agree that it is not possible to link any specific human behavior with any specific gene or genetic configuration. The only evidence for such a link is that which is provided by Sociobiologists' circular logic. This logic makes a *premise* of the genetic basis of behaviors, then cites a certain animal or human behavior, constructs a speculative story to explain how the behavior (*if* it were genetically based) could have served or could serve to maximize the reproductive success of the individual, and this *conjecture* then becomes evidence for the *premise* that the behavior was genetically determined.

> This is the central principle of sociobiology: insofar as a behavior reflects at least some component of gene action, individuals will tend to behave so as to maximize their fitness. . . . The result is a very strange sort of purposefulness, in which a goal—maximization of fitness—appears to be sought, but without any of the participants necessarily having awareness of what they are doing, or why. (Barash, 1979, pp. 29 and 25)

Notice the *insofar* clause is key and serves to confuse the issue. All behavior of course reflects at least *some* component of gene action. Individuals of any species of animal behave within the limits of the broad range of biological capabilities defined by their genes. Humans walk rather than fly. Birds peck at their food. When we are frightened, our hearts beat faster. But what is really at issue in Sociobiological theory is not the physical capacity for behavior that biology provides but rather the genetic encoding of the entire range of complex human behaviors and characteristics that are expressed in a nearly infinite variety of ways by different individuals and cultures and often not expressed at all, such as altruism, loyalty, dominance, competitiveness, aggressivity. In addition, Sociobiology claims genetic encoding for such arbitrarily chosen and questionably sexually differentiated "traits" as coyness, fickleness, promiscuity, rapaciousness, or maternalism.

Sociobiologists make a passing attempt to acknowledge that learning, culture, or environment plays a role in human behavior, but it is clear that their hearts (and minds) are not engaged by this idea. David Barash clearly states his position on the contribution of learning to behavior:

> Core elements are the essential person, an entity bequeathed by evolution to each of us; they are the *us* upon which experience acts. The great strength of sociobiology is that its conception of the "core" is grounded in evolution. . . . (1979, p. 10)
>
> Biology and culture undoubtedly work together, but it is tempting to speculate that our biology is somehow more real, lying unnoticed within each of us, quietly but forcefully manipulating much of our behavior. Culture, which is overwhelmingly important in shaping the myriad details of our lives, is more likely seen as a thin veneer, compared to the underlying ground substance of our biology. (1979, p. 14)

Richard Dawkins, the Sociobiologist who coined the catchy anthropomorphic phrase *selfish genes*, explains that genes and their expression are unaffected by environment:

> Now they swarm in huge colonies, safe inside gigantic lumbering robots, sealed off from the outside world, communicating with it by tortuous indirect routes, manipulating it by remote control. They are in you and in me; they created us, body and mind; and their preservation is the ultimate rationale for our existence. They have come a long way, those replicators. Now they go by the name of genes, and we are their survival machines. (1976, p. 21)

Mary Midgley, the British philosopher, suggests that "Dawkins' crude, cheap, blurred genetics is not just an expository device. It is the kingpin of his crude, cheap, blurred psychology" (1980a, p. 120). She further notes how the message of such "science" was transmitted to the general public by the cover of *Time* magazine's sociobiology number, which showed two puppets making love "while invisible genes twitch the strings above them . . ." (1980b, p. 26).

Sex differences in reproductive strategies

Since a key concept for Sociobiological theory is that behaviors are programmed to maximize the ability of the body's genes to reproduce themselves, an important area for Sociobiological speculation is that of reproduction itself. The second key postulate, then, is that the two sexes have a different strategy for maximizing their fitness through the reproduction of the largest possible number of offspring, and it is to this difference that Sociobiologists are able to attribute what they consider to be differences in female and male *natures*, behaviors, and social roles. Sociobiologists believe that women and men have different strategies and behaviors for assuring the reproduction and survival of their genes because they have an "unequal" biological investment in each offspring. Their reasoning is that since human males produce millions of sperm a day and can theoretically "sire offspring with different women at hourly or at most daily intervals" (Van Den Berghe and Barash, 1977, p. 814), their investment in the future in terms of the maximum reproduction of their genes in offspring lies in inseminating as many women as possible. Also, their relative investment in any one offspring is small. The human female, however, has a much greater investment in each of her offspring because her egg is 85,000 times larger than a sperm (hence more "expensive" to produce), because she ordinarily produces but one egg at a time and only about 400 in her lifetime, and because she usually produces no more than one offspring a year. Furthermore, since she is the one who gestates the fetus in her body, her expenditure of energy for those months and for the subsequent year or two of lactation and infant care is considerably greater than the father's. Therefore, while the *genetic* contribution from each parent is equivalent (23 chromosomes), the mother contributes a larger proportion of her total reproductive potential and a larger investment of time and energy. These facts, according to Sociobiologists, result in different reproductive strategies in the two sexes: women are selective and choosy—they go for quality; men go for quantity. Thus, E.O. Wilson writes:

> It pays males to be aggressive, hasty, fickle, and undiscriminating. In theory it is more profitable for females to be coy, to hold back until they can identify males with the best genes. . . . Human beings obey this biological principle faithfully. (1978, p. 125)

And Barash explains further:

> The evolutionary mechanism should be clear. Genes that allow females to accept the sorts of mates who make lesser contributions to their reproductive success will leave fewer copies of themselves than will genes that influence the females to be more selective. . . . For males, a very different strategy applies. The maximum advantage goes to individuals with fewer inhibitions. A genetically influenced tendency to "play fast and loose"—"love 'em and leave 'em"—may well reflect more biological reality than most of us care to admit. (1979, p. 48)

The leap to sex differences in human social roles and characteristics

Thus, we can see that Sociobiologists leap from some obvious facts such as the relative sizes and available numbers of eggs and sperm to sweeping and unwarranted generalizations about and explanations for presumed female and male *innate* characteristics: women are coy, choosy, and fussy; males are fickle and promiscuous. These characteristics then are used to ascribe a biological basis to such social phenomena and arrangements as marital fidelity for women and adultery, polygyny (harems), and rape by men. Sociobiologists explain that a woman stands to lose much less by her husband's sexual infidelity and by his fathering of children outside the marriage than a husband stands to lose by his wife's infidelity, since he would, in the latter case, be helping to rear children who do not bear his genes. It is for this reason, they claim, that there is a sexual double standard: a differential valuation of virginity and a differential condemnation of marital infidelity (Van Den Berghe and Barash, 1977).

Sociobiologists derive two other important postulates from the observation that the eggs and sperms that women and men contribute to the process of conception are different. The first is predictable: since a woman has a greater investment in terms of egg size and the time and energy spent in gestation, she also invests the major portion of total parental care in her offspring. She does this in order to protect her biological investment and her genes, since each of her offspring represents a greater proportion of her total reproductive capacity than it does for the father. An added factor is that women know with certainty that their genes have been passed on in their children; men have to take it on faith.

> Throughout their evolutionary history, males have generally been ill advised to devote themselves too strongly to the care of children, since the undertaking might turn out to be a wasted effort. (Barash, 1979, pp. 108–9)

There is a second important Sociobiological postulate derived from the fact that the total number of eggs available for fertilization is far fewer than the number of sperm available to fertilize them: competition among males for females is inevitable, since females, with their limited reproductive potential, are a scarce resource. Because of this competition on the time scale of evolution, the most reproductively successful males came to be those who were larger and more aggressive. It is this inherited male aggressivity that provides the biological basis for male dominance over females, male dominance hierarchies, competitiveness, territoriality, and war.

This, then, is how Sociobiology sees itself as replacing psychology and sociology. It is a social theory in the guise of biology; Sociobiologists provide the biological basis for all social phenomena and, in particular, for the social roles and the cultural representations of women and men. Thus Dawkins blandly declares:

> The female sex is exploited, and the fundamental evolutionary basis for the exploitation is the fact that eggs are larger than sperms. (1976, p. 158)

And Wilson explains:

> In hunter-gatherer societies, men hunt and women stay home. This strong bias persists in most agricultural and industrial societies and, on that ground alone, appears to have a genetic origin. (1975a, p. 47)

This quotation is particularly perplexing in view of Wilson's obvious and known familiarity with the renowned work of his Harvard colleagues Richard Lee and Irven DeVore and their coworkers on hunter-gatherer societies extensively documenting the exact opposite of this claim: that, in fact, women gatherers are away from "home" as much as the men. His knowledge of what women do in agricultural and industrial societies appears similarly based in mythic imagery rather than in modern anthropological scholarship let alone in the real world of agricultural and industrial economies where 50 to 100 percent of women may work outside the home. The most generous interpretation may be that extrapolations to human societies from insects is a hazardous (though not unrewarding) intellectual undertaking even for eminent entomologists. And, finally, to complete the unanimity of the Sociobiological voice, Barash speaks:

> Women have almost universally found themselves relegated to the nursery while men derive their greatest satisfaction from their jobs. . . . Such differences in male-female attachment to family versus vocation could derive in part from hormonal differences between sexes. (1977, p. 301)

I should like to call attention to the last quotation as an example of Sociobiologists' tendency to play loose with both language and logic. Barash speaks of women being *relegated* (assigned, banished) to the nursery, while men *derive satisfaction* from their jobs, hardly equivalent states, conditions, or situations; he then proceeds to base them *both* in biology as though they *were* equivalent. It is like claiming that repeatedly jailed offenders have an innate attachment to their cells.

SOCIOBIOLOGICAL METHODOLOGY IN THEORY BUILDING

Having stated the basic postulates of their theory, Sociobiologists then go on to catalogue the behaviors they consider to be universal and characteristic of humans and thus to be either explainable by or supportive of their theory. These behaviors and characteristics are never defined so that we all can know that we are talking about the same thing, nor are they selected according to any agreed-upon criteria from psychology, anthropology, or sociology. The behaviors and characteristics they choose to discuss and explain as universals of human societies are what upper/middle-class white male North American and English scientists consider to be characteristic: male aggressivity, territoriality, and tribalism; indoctrinability and conformity; male competitiveness and entrepreneurship; altruism and selfishness. The existence of these supposedly genetically determined human characteristics ("traits") then obviously and logically explains such social phenomena as national chauvinism, xenophobia, and war; slavery and capitalism; ethnocentrism and racism; dominance hierarchies and sexism.

In order to establish that these presumed universal human characteristics and social phenomena have evolved genetically, the next step in Sociobiological theory building is to demonstrate their existence throughout the animal world. The methodology consists essentially of flipping through the encyclopedic catalogue of animal behaviors and selecting particular behaviors of fishes, birds, insects or mammals that can be readily made to exemplify the various categories of human "traits" and social arrangements that Sociobiologists claim to be universal and genetically based. It is this step that introduces a number of methodological flaws into a theory already suffering from the conceptual ailments I have described.

But before discussing these flaws, I should like to place this critical next step within the context of the basic postulates and methodology of Sociobiology that I have described thus far. First, a picture is presented of human social organization and relationships. These are said to have universal elements that are based upon the existence of universal human behavioral traits that have evolved through natural selection because they were optimally adaptive; that is, the best alternative for survival from among several genetic variations. This assumes a specific genetic coding for specific behavioral "traits" and characteristics. It is not possible to adduce scientific proof for the presence or absence of specific behavioral traits in evolving hominids since traits leave no fossil record. Therefore, there is no way to identify the possible genetic variations from which current behavioral solutions have been selected. This forces Sociobiologists to demonstrate biological and evolutionary continuity by establishing similarities with other living nonhuman species that are viewed as representing an evolutionary continuum culminating in the human species. This is done by then describing carefully selected behaviors of particular species that represent and demonstrate some presumed human universal, such as female "coyness." But since we also do not know what the environmental, ecological, or reproductive problems were that such behaviors or characteristics were solving over the past several hundred million years, Sociobiologists attempt to reconstruct evolutionary history by inventing plausible stories that attempt to show how a particular behavior or social interaction in humans or other species *could* have or *would* have been adaptive and therefore favored by natural selection and genetically carried through subsequent generations. Basically, the aim is to establish the biological "innateness" and inevitability of present-day human behaviors and forms of social organization.

FLAWS IN SOCIOBIOLOGICAL THEORY AND METHODOLOGY

In the methodology and arguments used by Sociobiologists and other biological determinists, one can detect a number of recurring and interrelated flaws. The problems begin with the categories and definitions of behaviors that they consider characteristic of all people. When they proceed to draw analogies to animal behaviors, the problems are compounded by their selective use of particular animal models and by the language and concepts they apply to their descriptions of animal behaviors. We will find that these problems are intimately interrelated, but I shall try to analyze each, giving examples from important Sociobiological concepts, and then discuss two other kinds of methodological problems: the scientific tests one uses to validate hypotheses, and the classical and recurring issue of gene-environment, biology-culture interactions.

Ethnocentricity of behavioral description

The first problem lies in the Sociobiological descriptions of presumably universal human behaviors and social relationships, which are curiously similar to social organizations in the white Western industrial capitalist world. In this sense, Sociobiology is in fact an anachronism. It incorporates into its methodology the naive ethnocentric, androcentric, and anthropocentric fallacies discarded at least a decade or two ago by most competent and aware anthropologists and primatologists. Throughout Sociobiological writings there is a pervasive sense of the investigator's perception of his own self as a universal reference point, as equivalent to humanity, viewing all others—the other sex, other classes, races, cultures and civilizations, species, and epochs—in the light and language of his own experiences, values, and beliefs. He and his fraternity become the norm against which all *others* are measured and interpreted. (I use the male pronoun since Sociobiologists with few exceptions are male.) Thus, Sociobiologists make unwarranted generalizations about characteristic human behaviors, such as that "men would rather believe than know" (Wilson, 1975b, p. 561) or that women are coy and marry for upward social mobility. This means that much of the argument of Sociobiologists is devised to explain what *they* define as universal behavioral traits, the existence of which is, however, highly problematic to many students of human behavior. As the anthropologist Nancy Howell has said, ". . . they seem to be innocently ignorant of much of the complexity of human social life and cultures that sociobiology sets out to explain" (1979, p. 1295), though one wonders, when they see rape in the reproductive mechanism of flowers and war as a collective expression of individual male's innate aggressivity, just how "innocently ignorant" they can be. At the same time they seem also to be unconscious of any of the methodological problems that pervade attempts to describe human behavior, problems with which social scientists continue to struggle. As Richard Lewontin has pointed out, "Anthropologists have long been acutely conscious of the difficulties of describing human behavior in such a way as not to dictate the analysis by the categories of description" (1976, p. 24). Sociobiologists simply declare what they consider to be categories of behavioral description, for example, entrepreneurship, territoriality, aggression, dominance, without relationship to any cultural or historical context, and then proceed to arbitrarily assign examples of human and animal behavior to that category to demonstrate its universality in the animal world.

The concept of dominance hierarchies is an example of both ethnocentrism of descriptions of human "traits" and the trap of dictating analysis by the use of arbitrary categorization of behavior. Barash asserts that we are "a species organized along distinct lines of dominance" (1979, p. 186). But as Ruth Hubbard points out:

> We in the industrialized countries have grown up in hierarchically structured societies, so that, to us, dominance hierarchies appear natural and inevitable. But it is a mistake to apply the same categories to societies that function quite differently and to pretend that differences between our society and theirs can be expressed merely as matters of degree. . . . To take widely and complexly different social manifestations and scale them along one dimension does violence to the sources and significances of human social behavior. Western technological societies have developed in their ways for their own historical reasons. Other societies have *their* histories that have led to *their* social forms. (1978, p. 134)

. . . Many anthropological studies suggest that dominance hierarchies have not uniformly characterized the organization of human societies either in the past or today. In order to prove both the universality and the evolutionary inevitability of male dominance and dominance hierarchies, Sociobiologists and other biological determinists cite the example of the prototypical primate troop with its chest-pounding leader that has become familiar to us all. I shall discuss the fallacies of this approach in a section to follow on anthropomorphism.

Another example of the ethnocentric and androcentric application of concepts of human behavior to animals can be found in Sociobiological explanations of polygyny (marriage of one man to many wives) and hypergamy (marriage for upward mobility). I have already alluded to the Sociobiological postulate that men, being producers of millions of sperm a day, maximize their fitness by impregnating as many women as possible and, therefore, have traditionally established systems of polygyny, and that women have evolved to be more selective. Van Den Berghe and Barash (1977) describe the fact that in some bird species the females "prefer" polygynous males (here used to mean males that mate with many females) over bachelors. Wondering why, biologists have concluded that it is because the polygynous males command better territory than bachelors, more land providing more food and more protection for the young. This leads Van Den Berghe and Barash (1977) then to another Sociobiological universal of female behavior, hypergamy, marrying males of higher socioeconomic status for upward social mobility:

> Extrapolating to humans, we suggest that men are selected for engaging in male-male competition over resources appropriate to reproductive success, and that women are selected for preferring men who are successful in that endeavor. Any genetically influenced tendencies in these directions will necessarily be favored by natural selection.
>
> It is true, of course, that social advantages of wealth, power, or rank need not, indeed often do not, coincide with physical superiority. Women in all societies have found a way of resolving this dilemma by marrying wealthy and powerful men while

> taking young and attractive ones as lovers: the object of the game is to have the husband assume parental obligations for the lover's children. Understandably, men in most societies do not take kindly to such female strategies on the part of their wives, though they are not averse to philandering with other men's wives. The solution to this moral dilemma is the double standard, independently invented in countless societies. In any case, ethnographic evidence points to different reproductive strategies on the part of men and women, and to a remarkable consistency in the institutionalized means of accommodating these biological predispositions. (pp. 814, 815)

In this way the authors postulate a genetic tendency and a "biological predisposition" for women to marry men of wealth, power, and rank. Yet it is perfectly obvious that this "predisposition" can govern the behavior of only a small percentage of the world's women, since only a tiny minority of men in all countries of the world have any wealth, power, or rank. Thus, the vast majority of women everywhere, who are in lower socioeconomic classes and marry within their class, are excluded from biological universality. Their "universal" hypergamy is what happens only in romantic fiction. Sociobiologists attempt to establish human *species universals* of behavior by using an extraordinarily ethnocentric and class-biased model based on the behavior of a relatively small group of people in their own countries and others where the sexual and marital exploits of the rich and powerful are familiar topics in the international press. Furthermore, they also imply that there exists a related biological predisposition that expresses itself in the sexual double standard "independently invented in countless societies" because of men's unwillingness to assume obligation for the offspring (genes) of their wives' lovers. There is no suggestion that the double standard could have social origins independent of genes, that it may be but one more reflection of the economic and political domination of men over women in "countless" patriarchal societies.

Since even biological determinists recognize that many so-called human characteristics or behaviors are *not* universal, they postulate "predispositions," that is, traits that are genetically determined but not always expressed. It is very difficult, however, to take seriously the existence of a "predisposition" if it is not manifested in the majority of human beings. Just as Sociobiologists claim territoriality to be an evolutionary predisposition even though it is not manifested in a large number, perhaps the majority, of species, one could use their reasoning to argue that the *sharing* of territory is based on a biological predisposition, since the majority of species do just that.

It is a remarkable feature of Sociobiologists' descriptions of human "traits" that there appears to be no recognition of the possibility that there may be something arbitrary, selective, or subjective in their characterizations of females and males, that if some other group, for example, women or black males or American Indian males, were to list what they consider to be characteristics of women and men, the lists would be quite different. There is no acknowledgment, for example, that there are many women who are *not* coy and would use other adjectives to describe women. Also my guess is that it would come as a surprise to Sociobiologists to know that many American women because of *their* experiences would include in their list of male characteristics helplessness, impracticality, and dependence. One is then left to wonder why this kind of list is any less "scientific" than the list of "human" characteristics Sociobiologists have chosen to describe.

Lack of definition of behavioral units

A further difficulty that one encounters in Sociobiological accounts of human behavioral categories is the absence of any precise description or definition of the behaviors

Sociobiologists are seeking to explain. It is a requirement for any science to define the units or the phenomena that are the subjects of its investigations so as to ensure that different scientists, writers, and their readers are using the same terms to mean the same thing. Certainly a theory of social behavior needs to describe the behaviors it explains. But Sociobiologists do not describe or define what they mean, for example, by entrepreneurship or aggressivity. Is aggressivity fighting in bars, getting ahead in business, being creative, being a football star, a Don Juan, a war hero, a professor? Or is it being a mother who pursues City Hall and all of its politicians until a stoplight is installed where her children have to cross the street on their way to school?

Sociobiologists do not provide the answers to these questions. Every person who reads their literature has her/his own impression of what is being discussed, and perhaps that is precisely where Sociobiology's wide appeal and acceptance lies. Its statements can be interpreted in accordance with any person's subjective experiences, expectations, frame of reference, or prejudices rather than needing to be measured or judged against generally accepted standards of meaning or definition. This omission of a definition of the behavioral units that are being "explained" makes for further difficulty when we try to understand how Sociobiologists relate behaviors to genes. For example, if aggressivity is genetic and biological, what is it that is being inherited? Is it a physiological state of high energy; is it overactive adrenal glands with high levels of adrenalin in the blood; is it high intelligence and creativity; is it good body coordination; is it being "too" short and "therefore" insecure; is it "maternalism?" Or, as another example, what exactly do genes "encode" when they encode for hypergamy in females or entrepreneurship in males? Would biological determinists simply have to agree that what is biologically based is the perception of hunger and the drive for survival, and that both hypergamy and the different forms that entrepreneurship takes are simply those among an infinite variety of behavioral strategies that human beings *learn* and *select* as solutions to the problems of hunger and survival in their particular ecological and cultural niche? Or do they really mean that all females inherit a gene or a cluster of genes that drive them to look for and, of course, scheme to marry a rich man? Would they concede the possibility that, rather than genes for "entrepreneurship," the more successful gatherer-hunters may have been distinguished from the rest by their greater inventiveness (of tools), better memory (for plants and fertile sites), quicker intelligence, more energy or speed, or by superior ecological circumstances? Surely to understand the evolution of complex behaviors, a multiplicity of such characteristics can be considered and perhaps profitably analyzed, but invoking a murky concept like entrepreneurship seems useless, in contrast, except perhaps as a means of justifying the inevitability of our economic system.

Anthropomorphizing: The choice of animal models and use of language

Following close on the heels of the first, large problems of Sociobiological methodology that I have just discussed—its subjective and fuzzy conceptualizations and categorizations of human behavioral "traits" and social relationships—is the next great problem: anthropomorphizing, the substitution of human "equivalents" for real or postulated animal behaviors. In efforts to uncover the biological origins of human behavior, some investigators select an animal model that reflects their image of relationships presumed to exist in human society and then impose the language and concepts ordinarily used to describe human behavior upon their observations and interpretations of animal behaviors. The conclusions are inevitable, for the entire structure is a self-fulfilling prophecy. It involves a method, long in disrepute, of reading human motivation and intent into animal behavior. This makes for poor science because it cannot lead to an understanding of an animal species' behaviors or how the behaviors have come to solve the animal's problems of survival in its particular environment; it is also a circular and ineffectual way to approach human behavior even if one could understand human behavior by extrapolating from animals. (For reasons I discuss later, I do not believe one can.) If you initially interpret an animal's behavior in terms of what you believe about human behavior, you cannot then use your interpretation of *that animal's* behavior to explain something about human behavior.

Anthropomorphizing makes for a poor science of animal behavior for several reasons. The one I have discussed is that applying to animals assumptions that one has about human behavior or relationships structures and distorts the actual observations that investigators make as well as those they fail to make and, in so doing, biases the course and outcome of the research. A second related reason is that the technique makes the assumption that simply because an animal and a human behavior *look* alike, they *are* the same. But the two behaviors could have a superficial similarity and at the same time have a totally different significance for the body economy and represent different solutions to two completely different sets of problems of survival in their respective ecological circumstances. To apply human terminology to animals not only totally ignores these distinctions, but in the process circumvents or cancels out all the relevant questions and investigations that could lead one to understand either the animal or the human behavior. . . .

Flowers, ducks, and rape

We can find a particularly extravagant use of human behavioral concepts and language in the descriptions of animals in Barash's second book, *The Whisperings Within*. He claims he does not want to be a "racy modern Aesop," but says he will, nonetheless, be telling many animal stories about "rape in ducks, adultery in bluebirds, prostitution in hummingbirds, divorce and lesbian pairing in gulls, even homosexual rape in parasitic worms" (p. 2). Noteworthy for its relevance to a key contemporary issue for women is Barash's view of the origins of rape. Among Sociobiologists, Barash in particular sees rape rampant in nature. First he cites the work of Daniel Janzen, "one of our most creative ecologists," who has pointed out that even plants "perform courtship displays, rape, promiscuity, and fickleness just as do animals." Barash goes on to describe what he evidently considers to be rape in flowers:

> For example, plants with male flowers will "attempt" to achieve as many fertilizations as possible. How is this done? Among other things, they bombard female flowers with incredible amounts of pollen, and some even seem to have specially evolved capacities to rape female flowers, by growing a pollen tube which forces its way to the ovary within each female. (1979, p. 30)

So by defining the insertion of a pollen tube into a female flower as a rape, Barash begins to set the scene for the naturalness and—yes—the innocence of rape:

> Plants that commit rape . . . are following evolutionary strategies that maximize their fitness. And, clearly, in neither case do the actors know what they are doing, or why. We human beings like to think we are different. We introspect, we are confident that we know what we are doing, and why. But we may have to open our minds and admit the possibility that our need to maximize our fitness may be whispering somewhere deep within us and that, know it or not, most of the time we are heeding these whisperings. (p. 31)

Barash here strongly suggests that rapists are simply unwitting tools of a blind genetic drive, that rape is an unconscious urge for reproductive success and hence, biologically speaking, both advantageous and inevitable. But he seems unaware that there may be a different definition of rape, that most women see it as an act of violence expressing hatred, contempt, and fear of women and also as a weapon of social control that keeps women from asserting autonomy and freedom of movement and forces them to depend on male "protectors." If *that* is the definition of rape, and I would say women have the right and the knowledge to decide that, then it is not relevant to flowers. And *to name what flowers do as "rape" is specifically to deny that rape is a sexual act of physical violence committed by men against women*, an act embodying and enforcing the political power wielded by men over women.

Later in the book, Barash turns to rape among the birds and bees, especially mallard ducks. He explains that mallard ducks pair up for breeding, leaving some males unmated since there are usually more males than females. He then describes how one male or a group of unmated males may copulate with a mated female without the normal preliminary courtship rituals that mated couples engage in and "despite her obvious and vigorous protest. If that's not rape, it is certainly very much like it" (p. 54). But first of all, he gives no indication whether this is a frequent or a rare occurrence, nor does he describe the circumstances of the observation. Secondly, there is again the problem of language, in the use of the word *protest*. Courtship rituals are complex behaviors set in motion as a result of complex interactions between the hormonal and nervous systems of the animal, usually the female, and certain environmental conditions, for example, season of year. The female's state stimulates the male and, in turn, sets in motion the courtship rituals between partners, which further sequentially prime the reproductive systems for biological readiness to mate, ovulate, and fertilize—an intricate, balanced interplay between sight, smell, the brain, hormones, and gonads.

Thus, we could accommodate Barash's description of resisted copulation within the concept of the female's being *biologically not primed* for mating at the time of the bachelor's intrusion, but to impute *rape* and *protest*—intent and motivation—to ducks is again to use words for some purpose other than the clarity and accuracy required of scientific description and analysis. And the next page provides us with a lead to his purpose:

> Rape in humans is by no means as simple, influenced as it is by an extremely complex overlay of cultural attitudes. Nevertheless mallard rape and bluebird adultery may have a degree of relevance to human behavior. Perhaps human rapists, in their own criminally misguided way, are doing the best they can to maximize their fitness. If so, they are not that different from the sexually excluded bachelor mallards. (p. 55)

So Barash completes his portrait of the pitiful rapist: a lonesome fellow, left out of the mainstream of socially acceptable ways to copulate and so spread his genes about, he must force himself upon an unwilling female for the purpose of ensuring their reproduction.

In these examples, then, Barash used the word *rape*, which has a specific connotation in human terms, to describe behavior of a plant and a bird. This serves two purposes for Sociobiology: to establish that rape is biological and hence *natural* and to defuse rape as an urgent political issue, which has at its heart a cultural tradition of misogyny and male violence directed against women.

Harems

Thus far in the discussion of methodology, the basic problem has been the projection of investigators' personal and cultural values and biases about human behavior in their society onto their observations and interpretations of animals' and other cultures' behaviors. Since what is involved in these anthropomorphic and ethnocentric descriptions is language, we see that words become burdened with heavy implications. Language can be used to mold reality to a particular "truth," to impose a particular perception of the world as reality. Sociobiologists use language to mold the truth when they say that courted females are *coy* or that insects have evolved "*rampant machismo*" (Wilson, 1975b, p. 320) or that *aggressivity* is a universal trait of males. When Barash and other Sociobiologists use the word *rape* to describe a male flower's act of pollinating a female flower, they appropriate the word in order to remove rape from its sociopolitical context of male violence against women, to make it an act of sexual desire and of reproductive *need*, and, finally, to claim for rape a biological basis and inevitability because of its universality in the animal world.

The traditional use of the word *harem* in primatology to describe a single-male troop of females is another example of biased language and androcentric fantasy that served to structure observations and conceptualizations concerning the social organization of such troops. In our culture, *harem* has a generally accepted connotation of a group of women who are dependent economically, socially, and presumably sexually on a powerful male whose bodily needs are their central concern and occupation.[1] When that word was then used to signify single-male troops of female primates and their offspring, it automatically carried with it the entire complex of meanings and assumptions stereotypically associated with humans. It was assumed that the male was of central importance, defending the troop, making decisions, having his choice of sex partners, and in return was groomed, fed, and sexed by his harem of dependent females. Language substituted for actual observations, but it served ideology and circular logic by "demonstrating" that human male dominance and polygyny are innate since they are rooted in our primate ancestors. While hierarchical organization around a central male exists for some primate species under some circumstances, for many species, the solitary male is peripheral, functions mainly as a stud, and remains only so long as the females want him (Lancaster, 1975).

The omission of unwelcome animal data

Another problem in Sociobiological writings is the omission of unwelcome data that confound the stereotype. For example, rather than being engaged by redwinged blackbirds that exhibit polygyny and hypergamy, Sociobiologists, in the true scientific spirit of inquiry, could find it challenging to try to understand the South American male rhea bird that incubates and tends the 50 or so eggs that are laid by several females in the nest he builds. Or they could find it fascinating to explore shared parenting by examining the phenomenon of "double clutching," a situation in which female shore birds produce two clutches of eggs in quick succession, one of which becomes her responsibility and the other the male's. Or there is the female South American jacana bird who has a territory where she keeps a "harem" of males. She fills with eggs the nest that each male builds in his own subterritory and leaves him to incubate them and tend the brood (Bonner, 1980). Many bonded seabird pairs take turns sitting on the nest while the partner goes out to sea to bring back fish. Some penguins have an even more elaborate system whereby both partners fish together leaving the young in a huge creche tended by a few adults. The emperor penguin father remains nearly immobile during the two months he incubates his offspring's egg in a fold of skin about his feet, while the mother hunts for food. Bonner notes that monogamy is the main mating system among animals in which both sexes share in parental care (p. 156), and I wonder why Sociobiologists do not use this phenomenon as a "natural" model for human social organization as much as they do examples of male promiscuity and female domesticity.

Other problems with language and logic

There is another way in which writers can manipulate language and logic in order to reach a desired conclusion. This technique is to use words with different meanings as though they were equivalent. As previously described, Sociobiologists attribute mothers' major responsibility for child care to the greater maternal biological investment in conception, gestation, and lactation. Two Sociobiologists explain the inevitability of the situation:

> For a woman, the successful raising of a single infant is essentially close to a fulltime occupation for a couple of years, and continues to claim much attention and energy for several more years. For a man, it often means only a minor additional burden. To a limited extent, sexual roles can be modified in the direction of equalization of parental load, but even the most "liberated" husband cannot share pregnancy with his wife. In any case, most societies make no attempt to equalize parental care; they leave women holding the babies.
>
> Among most vertebrates, female involvement with offspring is obligatory whereas male involvement is more facultative. For example, . . . among orangutans, males on Sumatra typically associate with a female and her young, whereas on Borneo they defend territories and limit their interactions to other adult males. . . . Significantly, predators and interspecific competitors are more abundant on Sumatra. In short, biology dictates that females bear the offspring, although environmental conditions can exert a powerful influence on the extent of male parental investment. Males and females are selected for differing patterns of parental care, and there is no reason to exempt Homo sapiens from this generalization. (Van Den Berghe and Barash, 1977, pp. 813–14)

The authors show in this example that important, presumably genetic characteristics like the nature and quantity of parental care are actually determined by environmental conditions—but only for the male, since they consider the female still biologically committed to parental care. But this is where slippery language and logic intrude because they themselves reduce the mother's necessary or obligatory involvement in parenting to only the *pregnancy* itself, yet they skip from that fact to the conclusion that the mother's involvement in *child care* is biologically obligatory without in any way demonstrating the fact. Clearly the time that animals spend nursing offspring is obligatory but in most species consumes but a fraction of the mother's day. Among some species, for example, the siamang great ape, tamarins and marmosets, many fathers carry and care for the young all day and return it to the mother only for nursing (Snowdon and Suomi, 1982). Among many primate species studied it has been observed that adult male behavior toward infants is highly flexible and influenced by the particular social circumstances within the troop in any period of time (Parke and Suomi, 1981). Thus, whatever biological influences exist, parenting behaviors by both females and males are molded by social and ecological factors and learning as well. For most animal species, the amount of time the females invest in care of the young is also facultative, also related to ecological conditions and tends to be the reciprocal of the father's investment even in the example presented in the paragraph quoted above. Certainly, for humans, where even breast-feeding is not obligatory, the authors have presented no argument for the natural selection of "differing patterns of parental care."

Nonetheless, Sociobiologists have no doubts about what is biologically right, as Barash expresses it:

> Because men maximize their fitness differently from women, it is perfectly good

> biology that business and profession taste sweeter to them, while home and child care taste sweeter to women. (1979, p. 114)

Once again, as in their treatment of rape, Sociobiologists select for their attention an issue of particular vital and current concern to women and try to establish with faulty methodology the genetic origins of the social arrangements our society provides for child care. But in the same discussion quoted above, Barash goes far beyond expressing his biological opinion about the naturalness of the predominant social order that sees woman's proper place to be in the home. He warns that in the recent efforts to find "alternative lifestyles," it is child-care practices that are frequently at issue and "predictably there is a cost in disregarding biology." He cites a study that describes children reared in the counterculture as being neglected, deprived, and emotionally disturbed, and says that women seeking such "liberation" from total responsibility for child care are adopting a male biological strategy and denying their own. Thus, Sociobiology provides its public with an important sociopolitical theory and program: many aspects of modern civilization, however undesirable, are unavoidable, being expressions of our genetic inheritance; if we attempt to eliminate certain obvious social injustices, we tamper with evolution and risk incalculable harm, as Barash warns, "to everyone concerned." What this may mean we can only guess.

The search for evolutionary behavioral continuity: Culture in animals

Throughout this critique of the methodology of biological determinism, one underlying problem has been the particular animals that are chosen as models for human behavior. The reason for Sociobiologists' citing of examples from a variety of animal species is, as I have mentioned, to establish universality and therefore evolutionary continuity. At the outset it can be said that there is no necessary correlation between universality (even if it *could* be demonstrated, and it cannot) and evolutionary continuity, since what are being examined are present-day representatives of species that have evolved independently of each other for the last 15 million to about 500 million years. That is, according to the fossil evidence, the first hominid lines split off from the apes either about 15 or 5 million years ago and continued their own evolutionary course; the apes and monkeys diverged into their independent evolutionary lines about 40 to 50 million years ago; the first primates radiated off from other mammals about 70 or more million years ago; mammals, from the other land vertebrates about 325 million years ago; and over the previous 200 million years the various water and amphibious vertebrate species were evolving in their niches (Pilbeam, 1972). Thus, with independent lines of development for every species over the last millions to hundreds of millions of years, we do not know what, if any, evolutionary relationships similar behaviors of different present-day species have to each other. But certainly no one can seriously maintain that either the behaviors or the brains of present-day species represent a "recapitulation" of the evolutionary pathway that humans have followed. All that we can assume is that each species has evolved in relationship to the series of ecological niches within which it has survived, and today's array of forms and behaviors represents the varied outcomes of those historical relationships. Related to this point is the fact that the kind of faulty use of animal examples being discussed here involves the implicit assumption of a "chain of being" and "ascent" of humans over more primitive animal "precursors." But since contemporary animals are not our precursors, it is no more logical to look at chimpanzees or mice to gain insight into our behavior that it would be to look at our behavior to gain insight into theirs.

A faulty premise underlying some of the studies or observations of animals, particularly primates, either in the laboratory or in the wild, is that such study will reveal biological mechanisms of behavior that have evolved genetically and are "uncontaminated" by culture. There are two questionable assumptions in this premise. The first is that there is such an entity as "basic biological mechanisms" of human behavior that can be *revealed* by stripping off layers of culture, that is, that there is any definition of human behavior that can conceptually or in reality exclude culture. But that is an issue of such importance and complexity that it requires its own chapter, Chapter 3. The other erroneous assumption is that animals themselves have no culture affecting their "basic biological" or genetically influenced mechanisms, that their behaviors express only genes and no learning. In his fine review and analysis of the behaviors of animals, Bonner (1980) describes the various manifestations of the capacity for learning, teaching, and culture among vertebrates. Related to differences in relative size and complexity of the organization of the brain are differences in complexity and flexibility of behavioral responses to environmental challenges and the ability of animals to learn new and adaptive behaviors from one another. It is this transmission of information by behavioral means that Bonner defines as culture and that plays an important role in social behaviors and relationships among animals and in their adaptation to their environments. Thus, we cannot look to most animal behaviors as *instinctual* or *innate* and, therefore, as providing peepholes into the pure genetic core and *nature* of the human species.

Validation by prediction

One of the methodological techniques by which some Sociobiologists attempt to provide scientific validity or substance to their speculation is by making predictions. One criterion of a theory's value is its ability to predict what we will find under particular circumstances if we go and make the observations or conduct the proper experiment. Sociobiologists, Barash in particular, constantly "prove" the validity and predictive values of their theories by "predicting" what they and everyone else already know to be demonstrated fact; for example:

> Sociobiological theory would predict that adults with the most to gain and the least to lose would be the most eager adopters, and certainly this is true in the United States where childless couples are the predominant adopters. (1977, p. 313)

More relevant to this book is Barash's opinion about depression, which he sees as a "cry for help." Since males are genetically selected to be the *providers* of resources and females are those who are provided *for*,

> . . . in all societies, depression is significantly more common in women than in men. Their biology makes it more likely that women should be the sex to attempt care-eliciting behaviors. Males are supposed to be the care providers. Depression is also frequently associated with marital strife, a finding consistent with the suggestion that depression represents an unconscious effort to mobilize concern, attention and resources, in this case from an unresponsive or insufficiently responsive husband. (1979, p. 217)

Then Barash proceeds once again to address directly issues raised by the women's movement in his observation that, while depression is more common among married than among unmarried women, the opposite is true for men:

> The discovery that unmarried men are more likely to be depressed than are married men has been an important weapon for radical feminists, since it suggests that

> marriage itself is a male-designed phenomenon, tending to free men from depression while depressing women, presumably because of the emotionally stressful, sexist demands made upon married women in today's society. There may be much truth in this claim, but the male-female differences in depression associated with marriage also fit well with the sociobiological hypothesis. If men are the resource-providing sex and women are biologically predisposed to be resource receiving, and if depression is in fact a petition for resources (emotional, financial, etc.) it seems reasonable that unmarried men who showed depressive inclinations would be considered unattractive mates, while depressive tendencies in women would not be nearly as undesirable. (1979, pp. 217, 218)

We find in these passages a medley of methodological faults. First, there is the sarcastic dismissal of any suggestion that there may be a sociocultural context for depression among women, particularly among women who are married. Secondly, the explanation for depression is based upon acceptance of a sequence of unsubstantiated premises: "if men are the resource-providing sex," if women are "biologically predisposed to be resource receiving," "if depression is in fact a petition for resources," that depressed men are unattractive to women, and that depression in unmarried men is a cause rather than a result of their unmarried state. No one knows if any of these is true. Furthermore, other key Sociobiological premises posit quite the opposite—that women are the resource *providers* to their families, by *nature*, the nurturers, the givers. Even in an economic sense, so far as we know, women have historically always shared equally in providing material resources for their families through their labors both within and outside the home.

Barash then goes on to secure his argument by making a prediction: "We would also predict that if depression is a care-eliciting behavior, then it should be especially common following the birth of a child . . ." (1979, p. 218). As usual he claims a particular Sociological theory has been confirmed because it is able to "predict" a phenomenon that he, we, everyone already knows to exist, namely, postpartum depression.

If I were for the moment to accept Sociobiological premises, my predictions would be quite different from those proposed by Sociobiologists: Since women have a great biological investment in each pregnancy, which predisposes them to provide most of the parental care in order to protect optimally their genes in their offspring, I would predict:

1. A low incidence or absence of postpartum depression in women, since depression is *not* the optimal mental/physical state for the high energy requirements of postpartum lactation and infant care. In fact, I would further predict that the infants of depressed mothers do not fare as well physically or emotionally as infants of non-depressed mothers.
2. A high incidence of postpartum depression in fathers because they are deprived of a considerable portion of the parental care formerly invested in them by their wives, who, despite their high energy levels and resource-giving capacities, have finite limits and must share their resources equally. The father's depression is, of course, care-eliciting behavior, an unconscious effort to mobilize concern, attention, and resources.
3. A low incidence of depression in women in general, since most of them are fulfilling their biological predispositions to be mothers and nurturers. As Barash has said, life tastes "sweet" to them; in fact, I would predict that most women are manic most of the time. Furthermore, they are sensible enough to realize the futility of engaging in care-eliciting behavior directed toward men whose biological predisposition is toward aggressivity, activity, and competitiveness rather than nurturance.

4. In general, a high incidence of depression among both married and unmarried men. The vast majority of the men in the world in fact have very few resources, are not leaders, and rarely have an opportunity to hunt or go to war. In the face of the fact that they *cannot* provide resources as they are supposed to, are *not* fulfilling their genetic and evolutionary destinies, all that is left for them to do, in despair and frustration, is to cry for help.

Aside from the amusement of this exercise, I have wanted to illustrate two important fundamental flaws that make Sociobiology a very flimsy superstructure. First, premises in science are ordinarily expected to represent a generally accepted statement of current knowledge or at least a statement with some supporting evidence. But Sociobiological premises themselves are arbitrary, subjective, and conjectural even though they are stated as *givens.* Secondly, given any set of premises, whether conjectural or supported by evidence, *any number* of logical predictions or hypotheses may follow, not just the one a Sociobiologist or any particular scientist chooses to propose. The logical next step then is to recognize the importance of challenging the hypotheses by subjecting them to experimentation or to further observations that may tend to support or exclude one or another of the possible alternative hypotheses.

Sociobiologists predict what is already known to be true, then offer that known fact as proof of the validity of the premises from which they claim to be making the prediction. This method precludes the need either to test the prediction or to question the premises on which it was based.

SUMMARY AND CONCLUSION

Sociobiology, the modern version of biological determinist theories of human behavior, attempts to validate the belief that genes determine behaviors and that social relationships and cultures have evolved through the genetic transmission of behavioral traits and characteristics. Of central importance in Sociobiological theory, in keeping with the biological determinist tradition, are its efforts to explain in terms of *biology* the origins of the gender-differentiated roles and positions held by women and by men in modern as well as past civilizations. In so doing, Sociobiologists attempt to assign *natural* causes to phenomena of social origin. It is in part because Sociobiologists specifically address the very social issues that the women's movement has highlighted that Sociobiology functions as a political theory and program. Sociobiologists reinforce ancient stereotypes of women as coy, passive, dependent, maternal, and nurturant and base these temperaments in our genes. At the same time, and despite their liberal protestations, they explain and justify the existence of women's social and physical oppression by asserting the genetic origins, and hence inevitability, of rape, the sexual double standard, the relegation of women to the private world of home and motherhood, and other forms of the exploitation of women. Furthermore, its use of shoddy methodology and incorrect logic to support insupportable claims suggests a motive force other than the dispassionate pursuit of knowledge.

I have demonstrated a number of basic conceptual and methodological flaws in the work of Sociobiologists, which include faulty logic; unsupported assumptions and premises; inappropriate use of language; lack of definitions of the behaviors being explained; and ethnocentric, androcentric, and anthropocentric biases underlying the questions that are asked, the language used, the selection of animal models, and the interpretation of data. The more fundamental scientific problem, however, is the dichotomy that is drawn between genetic and environmental determinants of behavior. From the time of conception genes do *not* act in isolation from their environment, and even fairly stereotypical behaviors in animals, with few exceptions, represent interactions between experience or learning and biological mechanisms. What has evolved in response to environmental challenge is the brain and its capacities for learning and culture, not behaviors themselves. Behaviors are the *products* of the brain's functioning in interaction with the external world, and the innumerable patterns of social behaviors, relationships, and organization that characterize human societies have evolved through cultural transmission within specific historical contexts.

Note

1 For a different and multidimensional view of harems and the Muslim women who inhabit them, see Ahmed (1982).

References

Ahmed, L. Western ethnocentrism and perceptions of the harem. *Feminist Studies*, 1982, 8, 521–534.

Barash, D. *Sociobiology and behavior*. New York: Elsevier, 1977.

Barash, D. *The whisperings within*. New York: Harper & Row, 1979.

Bonner, J.T. *The evolution of culture in animals*. Princeton: Princeton University Press, 1980.

Brown, J.L. *The evolution of behavior*. New York: W. W. Norton, 1975.

Dawkins, R. *The selfish gene*. New York: Oxford University Press, 1976.

Eaton, G.G. Male dominance and aggression in Japanese macaque reproduction. In W. Montagna and W. Sadler (Eds.), *Reproductive behavior*. New York: Plenum, 1974.

Herschberger, R. *Adam's rib*. New York: Harper & Row, 1970.

Hirsch, H.V.B., and Leventhal, A.G. Functional modification of the developing visual system. In M. Jacobson (Ed.), *Handbook of sensory physiology*, 1978, *9. Development of sensory systems*. New York: Springer-Verlag.

Howell, N. Sociobiological hypotheses explored. *Science*, 1979, 206, 1294–1295.

Hrdy, S.B. *The woman that never evolved*. Cambridge: Harvard University Press, 1981.

Hubbard, R. From termite to human behavior. *Psychology Today*, 1978, 12, 124–134.

Hubbard, R. The theory and practice of genetic reductionism—from Mendel's laws to genetic engineering. In S. Rose (Ed.), *Towards a liberatory biology*. London: Allison and Busby, 1982.

Kolata, G.B. Primate behavior: Sex and the dominant male. *Science*, 1976, 191, 55–56.

Lancaster, J.B. *Primate behavior and the emergence of human culture*. New York: Holt, Rinehart & Winston, 1975.

Lappé, M. *Genetic politics*. New York: Simon & Schuster, 1979.

Leavitt, R.R. *Peaceable primates and gentle people: Anthropological approaches to women's studies*. New York: Harper & Row, 1975.

Leibowitz, L. *Perspectives in the anthropology of women*. New York: Monthly Review Press, 1975.

Leibowitz, L. *Females, males, families: A biosocial approach*. North Scituate, MA: Duxbury Press, 1978.

Lewin, R. Seeds of change in embryonic development. *Science*, 1981, 214, 42–44.

Lewontin, R.C. Sociobiology—a caricature of Darwinism. *Journal of Philosophy of Science*, 1976, 2, 21–31.

Midgley, M. Gene-juggling. In A. Montagu (Ed.), *Sociobiology examined*. Oxford: Oxford University Press, 1980. (a)

Midgley, M. Rival fatalism: The hollowness of the Sociobiology debate. In A. Montagu (Ed.), *Sociobiology examined*. Oxford: Oxford University Press, 1980. (b)

Parke, R.D., and Suomi, S.J. Adult male-infant relationships: Human and nonhuman primate evidence. In K. Immelmann, G. Barlow, L. Petrinovich and M. Main (Eds.), *Early development in animals and man*. New York: Cambridge University Press, 1981.

Pilbeam, D. *The ascent of man*. New York: Macmillan, 1972.

Pilbeam, D. An idea we could live without: The naked ape. In A. Montagu (Ed.), *Man and aggression*. Oxford: Oxford University Press, 1973.

Rosenzweig, M.R., Bennett, E.L., and Diamond, M.C. Brain changes in response to experience. *Scientific American*, 1972, 226, 22–30.

Rowell, T. *Social behavior of monkeys*. Baltimore: Penguin, 1972.

Rowell, T. The concept of social dominance. *Behavioral Biology*, 1974, 11, 131–154.

Schaller, G. *The mountain gorilla: Ecology and behavior.* Chicago: University of Chicago Press, 1963.
Snowdon, C.T., and Suomi, S.J. Paternal behavior in primates. In H.E. Fitzgerald, J.S. Mullins and P. Gage (Eds.), *Child nurturance*, New York: Plenum, 1982.
Van Den Berghe, P.L., and Barash, D.P. Inclusive fitness and human family structure. *American Anthropologist*, 1977, 79, 809–823.
Wilson, E.O. Human decency is animal. *New York Times Magazine*, 1975, Oct. 12, 38–50. (a)
Wilson, E.O. *Sociobiology: The new synthesis.* Cambridge, MA: Harvard University Press, 1975. (b)
Wilson, E.O. *On human nature.* Cambridge, MA: Harvard University Press, 1978.
Winick, M. Nutritional disorders during brain development. In D.B. Tower (Ed.), *The clinical neurosciences.* New York: Raven Press, 1975.
Woolf, V. *Three guineas.* New York: Harcourt Brace & World, 1938.
Yerkes, R.M. *Chimpanzees.* New Haven, Conn.: Yale University Press, 1943.

15

The Medical Construction of Gender

Case Management of Intersexed Infants

Suzanne Kessler

The birth of intersexed infants, babies born with genitals that are neither clearly male nor clearly female, has been documented throughout recorded time. In the late twentieth century, medical technology has advanced to allow scientists to determine chromosomal and hormonal gender, which is typically taken to be the real, natural, biological gender, usually referred to as "sex."[1] Nevertheless, physicians who handle the cases of intersexed infants consider several factors beside biological ones in determining, assigning, and announcing the gender of a particular infant. Indeed, biological factors are often preempted in their deliberations by such cultural factors as the "correct" length of the penis and capacity of the vagina.

In the literature of intersexuality, issues such as announcing a baby's gender at the time of delivery, postdelivery discussions with the parents, and consultations with patients in adolescence are considered only peripherally to the central medical issues—etiology, diagnosis, and surgical procedures.[2] Yet members of medical teams have standard practices for managing intersexuality that rely ultimately on cultural understandings of gender. The process and guidelines by which decisions about gender (re)construction are made reveal the model for the social construction of gender generally. Moreover, in the face of apparently incontrovertible evidence—infants born with some combination of "female" and "male" reproductive and sexual features—physicians hold an incorrigible belief in and insistence upon female and male as the only "natural" options. This paradox highlights and calls into question the idea that female and male are biological givens compelling a culture of two genders.

Ideally, to undertake an extensive study of intersexed infant case management, I would like to have had direct access to particular events, for example, the deliveries of intersexed infants and the initial discussions among physicians, between physicians and parents, between parents, and among parents and family and friends of intersexed infants. The rarity with which intersexuality occurs, however, made this unfeasible.[3] Alternatively, physicians who have had considerable experience in dealing with this condition were interviewed. I do not assume that their "talk" about how they manage such cases mirrors their "talk" in the situation, but their words do reveal that they have certain assumptions about gender and that they impose those assumptions via their medical decisions on the patients they treat.

Interviews were conducted with six medical experts (three women and three men) in the field of pediatric intersexuality: one clinical geneticist, three endocrinologists (two of them pediatric specialists), one psychoendocrinologist, and one urologist. All of them have had extensive clinical experience with various intersexed syndromes, and some are internationally known researchers in the field of intersexuality. They were selected on the basis of their prominence in the field and their representation of four different medical centers in New York City. Although they know one another, they do not collaborate on research and are not part of the same management team. All were interviewed in the spring of 1985, in their offices, and interviews lasted between forty-five minutes and one hour. Unless further referenced, all quotations in this article are from these interviews.

THE THEORY OF INTERSEXUALITY MANAGEMENT

The sophistication of today's medical technology has led to an extensive compilation of various intersex categories based on the various causes of malformed genitals. The "true intersexed" condition, where both ovarian and testicular tissue are present in either the same gonad or in opposite gonads, accounts for fewer than 5 percent of all cases of ambiguous genitals.[4] More commonly, the infant has either ovaries or testes, but the genitals are ambiguous. If the infant has two ovaries, the condition is referred to as female pseudohermaphroditism. If the infant has two testes, the condition is referred to as male pseudohermaphroditism. There are numerous causes of both forms of pseudohermaphroditism, and although there are life-threatening aspects to some of these conditions, having ambiguous genitals per se is not harmful to the infant's health. Although most cases of ambiguous genitals do not represent true intersex, in keeping with the contemporary literature, I will refer to all such cases as intersexed.

Current attitudes toward the intersex condition are primarily influenced by three factors. First are the extraordinary advancements in surgical techniques and endocrinology in the last decade. For example, female genitals can now be constructed to be indistinguishable in appearance from normal natural ones. Some abnormally small penises can be enlarged with the exogenous application of hormones, although surgical skills are not sufficiently advanced to construct a normal-looking and functioning penis out of other tissue. Second, in the contemporary United States the influence of the feminist movement has called into question the valuation of women according to strictly reproductive functions, and the presence or absence of functional gonads is no longer the only or the definitive criterion for gender assignment. Third, contemporary psychological theorists have begun to focus on "gender identity" (one's sense of oneself as belonging to the female or male category) as distinct from "gender role" (cultural expectations of one's behavior as "appropriate" for a female or male). The relevance of this new gender identity theory for rethinking cases of ambiguous genitals is that gender must be assigned as early as possible in order for gender identity to develop successfully. As a result of these three factors, intersexuality is now considered a treatable condition of the genitals, one that needs to be resolved expeditiously.

According to all of the specialists interviewed, management of intersexed cases is based upon the theory of gender proposed first by John Money, J.G. Hampson, and J.L. Hampson in 1955 and developed in 1972 by Money and Anke A. Ehrhardt, which argues that gender identity is changeable until approximately eighteen months of age.[5] To use the Pygmalion allegory, one may begin with the same clay and fashion a god or a goddess.[6] The theory rests on satisfying several conditions: the experts must insure that the parents have no doubt about whether their child is male or female; the genitals must be made to match the assigned gender as soon as possible; gender-appropriate hormones must be administered at puberty; and intersexed children must be kept informed about their situation with age-appropriate explanations. If these conditions are met, the theory proposes, the intersexed child will develop a gender identity in accordance with the gender assignment (regardless of the chromosomal gender) and will not question her or his assignment and request reassignment at a later age.

Supportive evidence for Money and Ehrhardt's theory is based on only a handful of repeatedly cited cases, but it has been accepted because of the prestige of the theoreticians and its resonance with contemporary ideas about gender, children, psychology, and medicine. Gender and children are malleable; psychology and medicine are the tools used to transform them. This theory is so strongly endorsed that it has taken on the character of gospel. "I think we [physicians] have been raised in the Money theory," one endocrinologist said. Another claimed, "We always approach the problem in a similar way and it's been dictated, to a large extent, by the work of John Money and Anke Ehrhardt because they are the only people who have published, at least in medical literature, any data, any guidelines." It is provocative that this physician immediately followed this assertion with: "And I don't know how effective it really is." Contradictory data are rarely cited in reviews of the literature, were not mentioned by any of the physicians interviewed, and have not diminished these physicians' belief in the theory's validity.[7]

The doctors interviewed concur with the argument that gender be assigned immediately, decisively, and irreversibly, and that professional opinions be presented in a clear and unambiguous way. The psychoendocrinologist said that when doctors make a statement about the infant, they should "stick to it." The urologist said, "If you make a statement that later has to be disclaimed or discredited, you've weakened your credibility." A gender assignment made decisively, unambiguously, and irrevocably contributes, I believe, to the general impression that the infant's true, natural "sex" has been discovered, and that something that was there all along has been found. It also serves to maintain the credibility of the medical profession, reassure the parents, and reflexively substantiate Money and Ehrhardt's theory.

Also according to the theory, if operative correction is necessary, it should take place as soon as possible. If the infant is assigned the male gender, the initial stage of penis repair is usually undertaken in the first year, and further surgery is completed before the child enters school. If the infant is assigned the female gender, vulva repair (including clitoral reduction) is usually begun by three months of age. Money suggests that if reduction of phallic tissue were delayed beyond the neonatal period, the infant would have traumatic memories of having been castrated.[8] Vaginoplasty, in those females having an adequate internal structure (e.g., the vaginal canal is near its expected location), is done between the ages of one and four years. Girls who require more complicated surgical procedures might not be surgically corrected until preadolescence.[9] The complete vaginal canal is typically constructed only when the body is fully grown, following pubertal feminization with estrogen, although more recently some specialists have claimed surgical success with vaginal construction in the early childhood years.[10] Although physicians speculate about the possible trauma of an early childhood "castration" memory, there is no corresponding concern that vaginal reconstructive surgery delayed beyond the neonatal period is traumatic.

Even though gender identity theory places the critical age limit for gender reassignment between eighteen months and two years, the physicians acknowledge that diagnosis, gender assignment, and genital reconstruction cannot be delayed for as long as two years, since a clear gender assignment and correctly formed genitals will determine the kind of interactions parents will have with the child.[11] The geneticist argued that when parents "change a diaper and see genitalia that don't mean much in terms of gender assignment, I think it prolongs the negative response to the baby. . . . If you have clitoral enlargement that is so extraordinary that the parents can't distinguish between male and female, it is sometimes helpful to reduce that somewhat so that the parent views the child as female." Another physician concurred: parents "need to go home and do their job as child rearers with it very clear whether it's a boy or a girl."

DIAGNOSIS

A premature gender announcement by an obstetrician, prior to a close examination of an infant's genitals, can be problematic. Money and his colleagues claim that the primary complications in case management of intersexed infants can be traced to mishandling by medical personnel untrained in sexology.[12] According to one of the pediatric endocrinologists interviewed, obstetricians improperly educated about intersexed conditions "don't examine the babies closely enough at birth and say things just by looking, before separating legs and looking at everything, and jump to conclusions, because 99 percent of the time it's correct. . . . People get upset, physicians I mean. And they say things that are inappropriate." For example, he said that an inexperienced obstetricians might blurt out, "I think you have a boy, or no, maybe you have a girl." Other inappropriate remarks a doctor might make in postdelivery consultation with the parents include, "You have a little boy, but he'll never function as a little boy, so you better raise him as a little girl." As a result, said the pediatric endocrinologist, "the family comes away with the idea that they have a little boy, and that's what they wanted, and that's what they're going to get." In such cases parents sometimes insist that the child be raised male despite the physician's instructions to the contrary. "People have in mind certain things they've heard, that this is a boy, and they're not likely to forget that, or they're not likely to let it go easily." The urologist agreed that the first gender attribution is critical: "Once it's been announced, you've got a big problem on your hands." "One of the worst things is to allow [the parents] to go ahead and give a name and tell everyone, and it turns out the child has to be raised in the opposite sex."[13]

Physicians feel that the mismanagement of such cases requires careful remedying. The psychoendocrinologist asserted, "When I'm involved, I spend hours with the parents to explain to them what has happened and how a mistake like that could be made, *or not really a mistake but a different decision*" (my emphasis). One pediatric endocrinologist said, "[I] try to dissuade them from previous misconceptions, and say, 'Well, I know what they meant, but the way they said it confused you. This is, I think, a better way to think about it.' " These statements reveal physicians' efforts not only to protect parents from concluding that their child is neither male nor female but also to protect other physicians' decision-making processes. Case management involves perpetuating the notion that good medical decisions are based on interpretations of the infant's real "sex" rather than on cultural understandings of gender. . . .

The diagnosis of intersexed conditions includes assessing the chromosomal sex and the syndrome that produced the genital ambiguity, and may include medical procedures such as cytologic screening; chromosomal analysis; assessing serum electrolytes; hormone, gonadotropin, and steroids evaluation; digital examination; and radiographic genitography.[14] In any intersexed condition, if the infant is determined to be a genetic female (having an XX chromosome makeup), then the treatment—genital surgery to reduce the phallus size—can proceed relatively quickly, satisfying what the doctors believe are psychological and cultural demands. For example, 21-hydroxylase deficiency, a form of female pseudohermaphroditism and one of the most common conditions, can be determined by a blood test within the first few days.

If, on the other hand, the infant is determined to have at least one Y chromosome, then surgery may be considerably delayed. A decision must be made whether to test the ability of the phallic tissue to respond to (HCG) androgen treatment, which is intended to enlarge the microphallus enough to be a penis. The endocrinologist explained, "You do HCG testing and you find out if the male can make testosterone. . . . You can get those results back probably within three weeks. . . . You're sure the male is making testosterone—but can he respond to it? It can take three months of waiting to see whether the phallus responds." If the Y-chromosome infant cannot make testosterone or cannot respond to the testosterone it makes, the phallus will not develop, and the Y-chromosome infant is not considered to be a male after all.

Should the infant's phallus respond to the local application of testosterone or a brief course of intramuscular injections of low-potency androgen, the gender assignment problem is resolved, but possibly at some later cost, since the penis will not grow again at puberty when the rest of the body develops.[15] Money's case management philosophy assumes that while it may be difficult for an adult male to have a much smaller than average penis, it is very detrimental to the morale of the young boy to have a micropenis.[16] In the former case the male's manliness might be at stake, but in the latter case his essential maleness might be. Although the psychological consequences of these experiences have not been empirically documented, Money and his colleagues suggest that it is wise to avoid the problems of both the micropenis in childhood and the still undersized penis postpuberty by reassigning many of these infants to the female gender.[17] This approach suggests that for Money and his colleagues, chromosomes are less relevant in determining gender than penis size, and that, by implication, "male" is defined not by the genetic condition of having one Y and one X chromosome or by the production of sperm but by the aesthetic condition of having an appropriately sized penis.

The tests and procedures required for diagnosis (and, consequently, for gender assignment) can take several months.[18] Although physicians are anxious not to make a premature gender assignment, their language suggests that it is difficult for them to take a completely neutral position and think and speak only of phallic tissue that belongs to an infant whose gender has not yet been determined or decided. Comments such as "seeing whether the male can respond to testosterone" imply at least a tentative male gender assignment of an XY infant. The psychoendocrinologist's explanation to parents of their infant's treatment program also illustrates this implicit male gender assignment. "Clearly this baby has an underdeveloped phallus. But if the phallus responds to this treatment, we are fairly confident that surgical techniques and hormonal techniques will help this child to look like a boy. But we want to make absolutely sure and use some hormone treatments and see whether the tissue reacts." The mere fact that this doctor refers to the genitals as an "underdeveloped" phallus rather than an overdeveloped clitoris suggests that the infant has been judged to be, at least provisionally, a male. In the case of the undersized phallus, what is ambiguous is not whether this is a penis but whether it is "good enough" to remain one. If at the end of the treatment period the phallic tissue has not responded, what had been a potential penis (referred to in the medical literature as a "clitoropenis") is now considered an enlarged clitoris (or "penoclitoris"), and reconstructive surgery is planned as for the genetic female.

The time-consuming nature of intersex diagnosis and the assumption, based on gender identity theory, that gender should be assigned as soon as possible thus present physicians with difficult dilemmas. Medical personnel are committed to discovering the etiology of the condition in order to determine the best course of treatment, which takes time. Yet they feel an urgent need to provide an immediate assignment and genitals that look and function appropriately. An immediate assignment that will need to be retracted is more problematic than a delayed assignment, since reassignment carries with it an additional set of social complications. The endocrinologist interviewed commented: "We've come very far in that we can diagnose eventually, many of the conditions. But we haven't come far enough. . . . We can't do it early enough. . . . Very frequently a decision is made before all this information is available, simply because it takes so long to make the correct diagnosis. And you cannot let a child go indefinitely, not in this society you can't. . . . There's pressure on parents [for a decision] and the parents transmit that pressure onto physicians." A pediatric endocrinologist agreed: "At times you may need to operate before a diagnosis can be made. . . . In one case parents were told to wait on the announcement while the infant was treated to see if the phallus would grow when treated with androgens. After the first month passed and there was some growth, the parents said they gave it a boy's name. They could only wait a month."

Deliberating out loud on the judiciousness of making parents wait for assignment decisions, the endocrinologist asked rhetorically, "Why do we do all these tests if in the end we're going to make the decision simply on the basis of the appearance of the genitalia?" This question suggests that the principles underlying physicians' decisions are cultural rather than biological, based on parental reaction and the medical team's perception of the infant's societal adjustment prospects given the way her/his genitals look or could be made to look. Moreover, as long as the decision rests largely on the criterion of genital appearance, and male is defined as having a "good-sized" penis, more infants will be assigned to the female gender than to the male.

THE WAITING PERIOD: DEALING WITH AMBIGUITY

During the period of ambiguity between birth and assignment, physicians not only must evaluate the infant's prospects to be a good male but also must manage parents' uncertainty about a genderless child. Physicians advise that parents postpone announcing the gender of the infant until a gender has been explicitly assigned. They believe that parents should not feel compelled to tell other people. The clinical geneticist interviewed said that physicians "basically encourage [parents] to treat [the infant] as neuter." One of the pediatric endocrinologists reported that in France parents confronted with this dilemma sometimes give the infant a neuter name, such as Claude or Jean. The psychoendocrinologist concurred: "If you have a truly borderline situation, and you want to make it dependent on the hormone treatment . . . then the parents are . . . told, 'Try not to make a decision. Refer to the baby as "baby." Don't think in terms of boy or girl.' " Yet, when asked whether this is a reasonable request to make of parents in our society, the physician answered: "I don't think so. I think parents can't do it." . . .

The geneticist explained that when directly asked by parents what to tell others about the gender of the infant, she says, "Why don't you just tell them that the baby is having problems and as soon as the problems are resolved we'll get back to you." A pediatric endocrinologist echoes this suggestion in advising parents to say, "Until the problem is solved [we] would really prefer not to discuss any of the details." According to the urologist, "If [the gender] isn't announced people may mutter about it and may grumble about it, but they haven't got anything to get their teeth into and make trouble over for the child, or the parents, or whatever." In short, parents are asked to sidestep the infant's gender rather than admit that the gender is unknown, thereby collaborating in a web of white lies, ellipses, and mystifications.

Even while physicians teach the parents how to deal with others who will not find the infant's condition comprehensible or acceptable, physicians must also make the condition comprehensible and acceptable to the parents, normalizing the intersexed condition for them. In doing so they help the parents consider the infant's condition in the most positive way. There are four key aspects to this "normalizing" process.

First, physicians teach parents normal fetal development and explain that all fetuses have the potential to be male or female. One of the endocrinologists explains, "In the absence of maleness you have femaleness. . . . It's really the basic design. The other [intersex] is really a variation on a theme." This explanation presents the intersex condition as a natural phase of every fetal development. Another endocrinologist "like[s] to show picture[s] to them and explain that at a certain point in development males and females look alike and then diverge for such and such reason." The professional literature suggests that doctors use diagrams that illustrate "nature's principle of using the same anlagen to produce the external genital parts of the male and female."[19]

Second, physicians stress the normalcy of the infant in other aspects. For example, the geneticist tells parents, "The baby is healthy, but there was a problem in the way the baby was developing." The endocrinologist says the infant has "a mild defect, just like anything could be considered a birth defect, a mole or a hemangioma." This language not only eases the blow to the parents but also redirects their attention. Terms like "hermaphrodite" or "abnormal" are not used. The urologist said that he advised parents "about the generalization of sticking to the good things and not confusing people with something that is unnecessary."

Third, physicians (at least initially) imply that it is not the gender of the child that is ambiguous but the genitals. They talk about "undeveloped," "maldeveloped," or "unfinished" organs. From a number of the physicians interviewed came the following explanations: "At a point in time the development proceeded in a different way, and sometimes the development isn't complete and we may have some trouble . . . in determining what the *actual* sex is. And so we have to do a blood test to help us" (my emphasis); "The baby may be a female, which you would know after the buccal smear, but you can't prove it yet. If so, then it's a normal female with a different appearance. This can be surgically corrected"; "The gender of your child isn't apparent to us at the moment"; "While this looks like a small penis, it's actually a large clitoris. And what we're going to do is put it back in its proper position and reduce the size of the tip of it enough so it doesn't look funny, so it looks right." Money and his colleagues report a case in which parents were advised to tell their friends that the reason their infant's gender was reannounced from male to female is that "the baby was . . . 'closed up down there' . . . when the closed skin was divided, the female organs were revealed, and the baby discovered to be, *in fact*, a girl" (emphasis mine). It was mistakenly assumed to be a male at first because "there was an excess of skin on the clitoris."[20]

The message in these examples is that the trouble lies in the doctor's ability to determine the gender, not in the baby's gender per se. The real gender will presumably be determined/proven by testing, and the "bad" genitals (which are confusing the situation for everyone) will be "repaired." The emphasis is not on the doctors creating gender but in their completing the genitals. Physicians say that they "reconstruct" the genitals rather than "construct" them. The surgeons reconstitute from remaining parts what should have been there all along. The fact that gender in an infant is "reannounced" rather than "reassigned" suggests that the first announcement was a mistake because the announcer was confused by the genitals. The gender always was what it is now seen to be.[21]

Finally, physicians tell parents that social factors are more important in gender development than biological ones, even though they are searching for biological causes. In essence, the physicians teach the parents Money and Ehrhardt's theory of gender development. In doing so, they shift the emphasis from the discovery of biological factors that are a sign of the "real" gender to providing the appropriate social conditions to produce the "real" gender. What remains unsaid is the apparent contradiction in the notion that a "real" or "natural" gender can be, or needs to be, produced artificially. The physician/parent discussions make it clear to family members that gender is not a biological given (even though, of course, their own procedures for diagnosis assume that it is), and that gender is fluid. The psychoendocrinologist paraphrased an explanation to parents thus: "It will depend, ultimately, on how everybody treats your child and how your child is looking as a person. . . . I can with confidence tell them that generally gender [identity] clearly agrees with the assignment." Similarly, a pediatric endocrinologist explained: "[I] try to impress upon them that there's an enormous amount of clinical data to support the fact that if you sex-reverse an infant . . . the majority of the time the alternative gender identity is commensurate with the socialization, the way that they're raised, and how people view them, and that seems to be the most critical."

The implication of these comments is that gender identity (of all children, not just those born with ambiguous genitals) is determined primarily by social factors, that the parents and community always construct the child's gender. In the case of intersexed infants, the physicians merely provide the right genitals to go along with the socialization. Of course, at normal births, when the infant's genitals are unambiguous, the parents are not told that the child's gender is ultimately up to socialization. In those cases, doctors do treat gender as a biological given.

SOCIAL FACTORS IN DECISION MAKING

Most of the physicians interviewed claimed that personal convictions of doctors ought to play no role in the decision-making process. The psychoendocrinologist explained: "I think the most critical factors [are] what is the possibility that this child will grow up with genitals which look like that of the assigned gender and which will ultimately function according to gender. . . . That's why it's so important that it's a well-established team, because [personal convictions] can't really enter into it. It has to be what is surgically and endocrinologically possible for that baby to be able to make it. . . . It's really much more within medical criteria. I don't think many social factors enter into it." While this doctor eschews the importance of social factors in gender assignment, she argues forcefully that social factors are extremely important in the development of gender identity. Indeed, she implies that social factors primarily enter the picture once the infant leaves the hospital.

In fact, doctors make decisions about gender on the basis of shared cultural values that are unstated, perhaps even unconscious, and therefore considered objective rather than subjective. Money states the fundamental rule for gender assignment: "Never assign a baby to be reared, and to surgical and hormonal therapy, as a boy, unless the phallic structure, hypospadiac or otherwise, is neonatally of at least the same caliber as that of same-aged males with small-average penises."[22] Elsewhere, he and his colleagues provide specific measurements for what qualifies as a micropenis: "A penis is, by convention, designated as a micropenis when at birth its dimensions are three or more standard deviations below the mean. . . . When it is correspondingly reduced in diameter with corpora that are vestigial . . . it unquestionably qualifies as a micropenis."[23] A pediatric endocrinologist claimed that although "the [size of the] phallus is not the deciding factor . . . if the phallus is less than 2 centimeters long at birth and won't respond to androgen treatments, then it's made into a female."

These guidelines are clear, but they focus on only one physical feature, one that is distinctly imbued with cultural meaning. This becomes especially apparent in the case of an XX infant with normal female reproductive gonads and a perfect penis. Would the size and shape of the penis, in this case, be the deciding factor in assigning the infant "male," or would the perfect penis be surgically destroyed and female genitals created? Money notes that this dilemma would be complicated by the anticipated reaction of the parents to seeing "their apparent son lose his penis."[24] Other researchers concur that parents are likely to want to raise a child with a normal-shaped penis (regardless of size) as "male," particularly if the scrotal area looks normal and if the parents have had no experience with intersexuality.[25] Elsewhere Money argues in favor of not neonatally amputating the penis of XX infants, since fetal masculinization of brain structures would predispose them "almost invariably

[to] develop behaviorally as tomboys, even when reared as girls."[26] This reasoning implies, first, that tomboyish behavior in girls is bad and should be avoided and, second, that it is preferable to remove the internal female organs, implant prosthetic testes, and regulate the "boy's" hormones for his entire life rather than to overlook or disregard the perfection of the penis.[27]

The ultimate proof to these physicians that they intervened appropriately and gave the intersexed infant the correct gender assignment is that the reconstructed genitals look normal and function normally once the patient reaches adulthood. The vulva, labia, and clitoris should appear ordinary to the woman and her partner(s), and the vagina should be able to receive a normal-sized penis. Similarly, the man and his partner(s) should feel that his penis (even if somewhat smaller than the norm) looks and functions in an unremarkable way. Although there is no reported data on how much emphasis the intersexed person, him- or herself, places upon genital appearance and functioning, the physicians are absolutely clear about what they believe is important. The clinical geneticist said, "If you have . . . a seventeen-year-old young lady who has gotten hormone therapy and has breast development and pubic hair and no vaginal opening, I can't even entertain the notion that this young lady wouldn't want to have corrective surgery." The urologist summarized his criteria: "Happiness is the biggest factor. Anatomy is part of happiness." Money states, "The primary deficit [of not having a sufficient penis]—and destroyer of morale—lies in being unable to satisfy the partner."[28] Another team of clinicians reveals their phallocentrism, arguing that the most serious mistake in gender assignment is to create "an individual unable to engage in genital [heterosexual] sex."[29]

The equation of gender with genitals could only have emerged in an age when medical science can create credible-appearing and functioning genitals, and an emphasis on the good phallus above all else could only have emerged in a culture that has rigid aesthetic and performance criteria for what constitutes maleness. The formulation "good penis equals male; absence of good penis equals female" is treated in the literature and by the physicians interviewed as an objective criterion, operative in all cases. There is a striking lack of attention to the size and shape requirements of the female genitals, other than that the vagina be able to receive a penis.[30]

In the late nineteenth century when women's reproductive function was culturally designated as their essential characteristic, the presence or absence of ovaries (whether or not they were fertile) was held to be the ultimate criterion of gender assignment for hermaphrodites. The urologist interviewed recalled a case as late as the 1950s of a male child reassigned to "female" at the age of four or five because ovaries had been discovered. Nevertheless, doctors today, schooled in the etiology and treatment of the various intersex syndromes, view decisions based primarily on gonads as wrong, although, they complain, the conviction that the gonads are the ultimate criterion "still dictates the decisions of the uneducated and uninformed."[31] Presumably, the educated and informed now know that decisions based primarily on phallic size, shape, and sexual capacity are right.

While the prospect of constructing good genitals is the primary consideration in physicians' gender assignments, another extramedical factor was repeatedly cited by the six physicians interviewed—the specialty of the attending physician. Although generally intersexed infants are treated by teams of specialists, only the person who coordinates the team is actually responsible for the case. This person, acknowledged by the other physicians as having chief responsibility, acts as spokesperson to the parents. Although all of the physicians claimed that these medical teams work smoothly with few discrepancies of opinion, several of them mentioned decision-making orientations that are grounded in particular medical specializations. One endocrinologist stated, "The easiest route to take, where there is ever any question . . . is to raise the child as female. . . . In this country that is usual if the infant falls into the hands of a pediatric endocrinologist. . . . If the decision is made by the urologists, who are mostly males, . . . they're always opting, because they do the surgery, they're always feeling they can correct anything." Another endocrinologist concurred: "[Most urologists] don't think in terms of dynamic processes. They're interested in fixing pipes and lengthening pipes, and not dealing with hormonal, and certainly not psychological issues. . . . 'What can I do with what I've got?" Urologists were defended by the clinical geneticist: "Surgeons here, now I can't speak for elsewhere, they don't get into a situation where the child is a year old and they can't make anything." Whether or not urologists "like to make boys," as one endocrinologist claimed, the following example from a urologist who was interviewed explicitly links a cultural interpretation of masculinity to the medical treatment plan. The case involved an adolescent who had been assigned the female gender at birth but was developing some male pubertal signs and wanted to be a boy. "He was ill-equipped," said the urologist, "yet we made a very respectable male out of him. He now owns a huge construction business—those big cranes that put stuff up on the building."

POSTINFANCY CASE MANAGEMENT

After the infant's gender has been assigned, parents generally latch onto the assignment as the solution to the problem—and it is. The physician as detective has collected the evidence, as lawyer has presented the case, and as judge has rendered a verdict. Although most of the interviewees claimed that the parents are equal participants in the whole process, they gave no instances of parental participation prior to the gender assignment. After the physicians assign the infant's gender, the parents are encouraged to establish the credibility of that gender publicly by, for example, giving a detailed medical explanation to a leader in their community, such as a physician or pastor, who will explain the situation to curious casual acquaintances. Money argues that "medical terminology has a special layman's magic in such a context; it is final and authoritative and closes the issue." He also recommends that eventually the mother "settle [the] argument once and for all among her women friends by allowing some of them to see the baby's reconstructed genitalia."[32] Apparently, the powerful influence of normal-looking genitals helps overcome a history of ambiguous gender.

Some of the same issues that arise in assigning gender recur some years later when, at adolescence, the child may be referred to a physician for counseling.[33] The physician then tells the adolescent many of the same things his or her parents had been told years before, with the same language. Terms like "abnormal," "disorder," "disease," and "hermaphroditism" are avoided; the condition is normalized, and the child's gender is treated as unproblematic. One clinician explains to his patients that sex organs are different in appearance for each person, not just those who are intersexed. Furthermore, he tells the girls "that while most women menstruate, not all do . . . that conception is only one of a number of ways to become a parent; [and] that today some individuals are choosing not to become parents."[34] The clinical geneticist tells a typical female patient: "You are female. Female is not determined by your genes. Lots of other things determine being a woman. And you are a woman but you won't be able to have babies." . . .

Technically these physicians are lying when, for example, they explain to an adolescent XY female with an intersexed history that her "ovaries . . . had to be removed because they were unhealthy or were producing 'the wrong balance of hormones.' "[35] We can presume that these lies are told in the service of what the physicians consider a greater good—keeping individual/concrete genders as clear and uncontaminated as the notions of female and male are in the abstract. The clinician suggests that with some female patients it eventually may be possible to talk to them "about their gonads having some structures and features that are testicular-like."[36] This call for honesty might be based at least partly on the possibility of the child's discovering his or her chromosomal sex inadvertently from a buccal smear taken in a high school biology class. Today's litigious climate is possibly another encouragement.

In sum, the adolescent is typically told that certain internal organs did not form because of an endocrinological defect, not because those organs could never have developed in someone with her or his sex chromosomes. The topic of chromosomes is skirted. There are no published studies on how these adolescents experience their condition and their treatment by doctors. An endocrinologist interviewed mentioned that her adolescent patients rarely ask specifically what is wrong with them, suggesting that they are accomplices in this evasion. In spite of the "truth" having been evaded, the clinician's impression is that "their gender identities and general senses of well-being and self-esteem appear not to have suffered."[37]

CONCLUSION

Physicians conduct careful examinations of intersexed infants' genitals and perform intricate laboratory procedures. They are interpreters of the body, trained and committed to uncovering the "actual" gender obscured by ambiguous genitals. Yet they also have considerable leeway in assigning gender, and their decisions are influenced by cultural as well as medical factors. What is the relationship between the physician as discoverer and the physician as determiner of gender? Where is the relative emphasis placed in discussions with parents and adolescents and in the consciousness of physicians? It is misleading to characterize the doctors whose words are provided here as presenting themselves publicly to the parents as discoverers of the infant's real gender but privately acknowledging that the infant has no real gender other than the one being determined or constructed by the medical professionals. They are not hypocritical. It is also misleading to claim that physicians' focus shifts from discovery to determination over the course of treatment: first the doctors regard the infant's gender as an unknown but discoverable reality; then the doctors relinquish their attempts to find the real gender and treat the infant's gender as something they must construct. They are not medically incompetent or deficient. Instead, I am arguing that the peculiar balance of discovery and determination throughout treatment permits physicians to handle very problematic cases of gender in the most unproblematic of ways.

This balance relies fundamentally on a particular conception of the "natural."[38] Although the deformity of intersexed genitals would be immutable were it not for medical interference, physicians do not consider it natural. Instead they think of, and speak of, the surgical/hormonal alteration of such deformities as natural because such intervention returns the body to what it "ought to have been" if events had taken their typical course. The nonnormative is converted into the normative, and the normative state is considered natural.[39] The genital ambiguity is remedied to conform to a "natural," that is, culturally indisputable, gender dichotomy. Sherry Ortner's claim that the culture/nature distinction is itself a construction—a product of culture—is relevant here. Language and imagery help create and maintain a specific view of what is natural about the two genders and, I would argue, about the very idea of gender—that it consists of two exclusive types: female and male.[40] The belief that gender consists of two exclusive types is maintained and perpetuated by the medical community in the face of incontrovertible physical evidence that this is not mandated by biology. . . .

If physicians recognized that implicit in their management of gender is the notion that finally, and always, people construct gender as well as the social systems that are grounded in gender-based concepts, the possibilities for real societal transformations would be unlimited. Unfortunately, neither in their representations to the families of the intersexed nor among themselves do the physicians interviewed for this study draw such far-reaching implications from their work. Their "understanding" that particular genders are medically (re)constructed in these cases does not lead them to see that gender is always constructed. Accepting genital ambiguity as a natural option would require that physicians also acknowledge that genital ambiguity is "corrected" not because it is threatening to the infant's life but because it is threatening to the infant's culture.

Rather than admit to their role in perpetuating gender, physicians "psychologize" the issue by talking about the parents' anxiety and humiliation in being confronted with an anomalous infant. The physicians talk as though they have no choice but to respond to the parents' pressure for a resolution of psychological discomfort, and as though they have no choice but to use medical technology in the service of a two-gender culture. Neither the psychology nor the technology is doubted, since both shield physicians from responsibility. Indeed, for the most part, neither physicians nor parents emerge from the experience of intersex case management with a greater understanding of the social construction of gender. Society's accountability, like their own, is masked by the assumption that gender is a given. Thus, cases of intersexuality, instead of illustrating nature's failure to ordain gender in these isolated "unfortunate" instances, illustrate physicians' and Western society's failure of imagination—the failure to imagine that each of these management decisions is a moment when a specific instance of biological "sex" is transformed into a culturally constructed gender.

Notes

I want to thank my student Jane Welder for skillfully conducting and transcribing the interviews for this article.

1 Suzanne J. Kessler and Wendy McKenna, *Gender: An Ethnomethodological Approach* (1978; reprint, Chicago: University of Chicago Press, 1985).
2 See, e.g., M. Bolkenius, R. Daum, and E. Heinrich, "Pediatric Surgical Principles in the Management of Children with Intersex," *Progressive Pediatric Surgery* 17 (1984):33–38; Kenneth I. Glassberg, "Gender Assignment in Newborn Male Pseudohermaphrodites," *Urologic Clinics of North America* 7 (June 1980):409–21; and Peter A. Lee et al., "Micropenis. I. Criteria, Etiologies and Classification" *Johns Hopkins Medical Journal* 146 (1980):156–63.
3 It is impossible to get accurate statistics on the frequency of intersexuality. Chromosomal abnormalities (like XOXX or XXXY) are registered, but those conditions do not always imply ambiguous genitals, and most cases of ambiguous genitals do not involve chromosomal abnormalities. None of the physicians interviewed for this study would venture a guess on frequency rates, but all agreed that intersexuality is rare. One physician suggested that the average obstetrician may see only two cases in twenty years. Another estimated that a specialist may see only one a year, or possibly as many as five a year.
4 Mariano Castro-Magana, Moris Angulo, and Platon J. Collipp, "Management of the Child with Ambiguous Genitalia," *Medical Aspects of Human Sexuality* 18 (April 1984):172–88.
5 John Money, J.G. Hampson, and J.L. Hampson, "Hermaphroditism: Recommendations concerning Assignment of Sex, Change of Sex, and Psychologic Management," *Bulletin of the Johns Hopkins Hospital* 97 (1955):284–300; John Money, Reynolds Potter, and Clarice S. Stoll, "Sex Reannouncement in Hereditary Sex Deformity: Psychology and Sociology of Habilitation," *Social Science and Medicine* 3 (1969):207–16; John Money and Anke A. Ehrhardt, *Man and Woman, Boy and Girl* (Baltimore: Johns Hopkins University Press, 1972); John Money, "Psychologic Consideration of Sex Assignment in Intersexuality," *Clinics in Plastic Surgery* 1 (April 1974):215–22, "Psychological Counseling: Hermaphroditism," in *Endocrine and Genetic Diseases of Childhood and Adolescence*, ed. L.I. Gardner (Philadelphia: Saunders, 1975):609–18, and "Birth Defect of the Sex Organs: Telling the Parents and the Patient," *British Journal of Sexual Medicine* 10 (March 1983):14; John Money et al., "Micropenis, Family Mental Health, and Neonatal Management: A Report on Fourteen Patients Reared as Girls," *Journal of Preventive Psychiatry* 1, no. 1 (1981):17–27.

6 Money and Ehrhardt, 152.
7 Contradictory data are presented in Milton Diamond, "Sexual Identity, Monozygotic Twins Reared in Discordant Sex Roles and a BBC Follow-up," *Archives of Sexual Behavior* 11, no. 2 (1982):181–86.
8 Money, "Psychologic Consideration of Sex Assignment in Intersexuality."
9 Castro-Magana, Angulo, and Collipp (n. 4 above).
10 Victor Braren et al., "True Hermaphroditism: A Rational Approach to Diagnosis and Treatment," *Urology* 15 (June 1980):569–74.
11 Studies of normal newborns have shown that from the moment of birth the parent responds to the infant based on the infant's gender. Jeffrey Rubin, F.J. Provenzano, and Z. Luria, "The Eye of the Beholder: Parents' Views on Sex of Newborns," *American Journal of Orthopsychiatry* 44, no. 4 (1974):512–19.
12 Money et al. (n. 5 above).
13 There is evidence from other kinds of sources that once a gender attribution is made, all further information buttresses that attribution, and only the most contradictory new information will cause the original gender attribution to be questioned. See, e.g., Kessler and McKenna (n. 1 above).
14 Castro-Magana, Angulo, and Collipp (n. 4 above).
15 Money, "Psychological Consideration of Sex Assignment in Intersexuality" (n. 5 above).
16 Technically, the term "micropenis" should be reserved for an exceptionally small but well-formed structure. A small, malformed "penis" should be referred to as a "microphallus" (Lee et al. [n. 2 above]).
17 Money et al., 26. A different view is argued by another leading gender identity theorist: "When a little boy (with an imperfect penis) knows he is a male, he creates a penis that functions symbolically the same as those of boys with normal penises" (Robert J. Stoller, *Sex and Gender* [New York: Aronson, 1968], 1:49).
18 W. Ch. Hecker, "Operative Correction of Intersexual Genitals in Children," *Pediatric Surgery* 17 (1984):21–31.
19 Tom Mazur, "Ambiguous Genitalia: Detection and Counseling," *Pediatric Nursing* 9 (November/December 1983):417–31; Money, "Psychologic Consideration of Sex Assignment in Intersexuality" (n. 5 above), 218.
20 Money, Potter, and Stoll (n. 5 above), 211.
21 The term "reassignment" is more commonly used to describe the gender changes of those who are cognizant of their earlier gender, e.g., transsexuals—people whose gender itself was a mistake.
22 Money, "Psychological Counseling: Hermaphroditism" (n. 5 above), 610.
23 Money et al. (n. 5 above), 18.
24 John Money, "Hermaphroditism and Pseudohermaphroditism," in *Gynecologic Endocrinology*, ed. Jay J. Gold (New York: Hoeber, 1968), 449–64, esp. 460.
25 Mojtaba Besheshti et al., "Gender Assignment in Male Pseudohermaphrodite Children," *Urology* (December 1983):604–7. Of course, if the penis looked normal and the empty scrotum were overlooked, it might not be discovered until puberty that the male child was XX, with a female internal structure.
26 John Money, "Psychologic Consideration of Sex Assignment in Intersexuality" (n. 5 above), 216.
27 Weighing the probability of achieving a perfect penis against the probable trauma such procedures might involve is another social factor in decision making. According to an endocrinologist interviewed, if it seemed that an XY infant with an inadequate penis would require as many as ten genital operations over a six-year period in order to have an adequate penis, the infant would be assigned the female gender. In this case, the endocrinologist's practical and compassionate concern would override purely genital criteria.
28 Money, "Psychologic Consideration of Sex Assignment in Intersexuality," 217.
29 Castro-Magana, Angulo, and Collipp (n. 4 above), 180.
30 It is unclear how much of this bias is the result of a general, cultural devaluation of the female and how much the result of physicians' greater facility in constructing aesthetically correct and sexually functional female genitals.
31 Money, "Psychologic Consideration of Sex Assignment in Intersexuality," 215. Remnants of this anachronistic view can still be found, however, when doctors justify the removal of contradictory gonads on the grounds that they are typically sterile or at risk for malignancy (J. Dewhurst and D.B. Grant, "Intersex Problems," *Archives of Disease in Childhood* 59 [July-December 1984]:1191–94). Presumably, if the gonads were functional and healthy their removal would provide an ethical dilemma for at least some medical professionals.

32 Money, "Psychological Counseling: Hermaphroditism" (n. 5 above), 613.
33 As with the literature on infancy, most of the published material on adolescents is on surgical and hormonal management rather than on social management. See, e.g., Joel J. Roslyn, Eric W. Fonkalsurd, and Barbara Lippe, "Intersex Disorders in Adolescents and Adults," *American Journal of Surgery* 146 (July 1983):138–44.
34 Mazur (n. 19 above), 421.
35 Dewhurst and Grant, 1193.
36 Mazur, 422.
37 Ibid.
38 For an extended discussion of different ways of conceptualizing "natural," see Richard W. Smith, "What Kind of Sex Is Natural?" in *The Frontiers of Sex Research*, ed. Vern Bullough (Buffalo: Prometheus, 1979):103–11.
39 This supports sociologist Harold Garfinkel's argument that we treat routine events as our due as social members and that we treat gender, like all normal forms, as a moral imperative. It is no wonder, then, that physicians conceptualize what they are doing as natural and unquestionably "right" (Harold Garfinkel, *Studies in Ethnomethodology* [Englewood Cliffs, N.J.: Prentice Hall, 1967]).
40 Sherry B. Ortner, "Is Female to Male as Nature Is to Culture?" in *Woman, Culture, and Society*, ed. Michelle Zimbalist Rosaldo and Louise Lamphere (Stanford, Calif.: Stanford University Press, 1974):67–87.

16

The Bare Bones of Sex

Sex and Gender

Anne Fausto-Sterling

Here are some curious facts about bones. They can tell us about the kinds of physical labor an individual has performed over a lifetime and about sustained physical trauma. They get thinner or thicker (on average in a population) in different historical periods and in response to different colonial regimes (Molleson 1994; Larsen 1998). They can indicate class, race, and sex (or is it gender—wait and see). We can measure their mineral density and whether on average someone is likely to fracture a limb but not whether a particular individual with a particular density will do so. A bone may break more easily even when its mineral density remains constant (Peacock et al. 2002).[1]

Culture shapes bones. For example, urban ultraorthodox Jewish adolescents have lowered physical activity, less exposure to sunlight, and drink less milk than their more secular counterparts. They also have greatly decreased mineral density in the vertebrae of their lower backs, that is, the lumbar vertebrae (Taha et al. 2001). Chinese women who work daily in the fields have increased bone mineral content and density. The degree of increase correlates with the amount of time spent in physical activity (Hu et al. 1994); weightlessness in space flight leads to bone loss (Skerry 2000); gymnastics training in young women ages seventeen to twenty-seven correlates with increased bone density despite bone resorption caused by total lack of menstruation (Robinson et al. 1995). Consider also some recent demographic trends: in Europe during the past thirty years, the number of vertebral fractures has increased three- to fourfold for women and more than fourfold for men (Mosekilde 2000); in some groups the relative proportions of different parts of the skeleton have changed in recent generations.[2]

What are we to make of reports that African Americans have greater peak bone densities than Caucasian Americans (Aloia et al. 1996; Gilsanz et al. 1998),[3] although this difference may not hold when one compares Africans to British Caucasians (Dibba et al. 1999), or that white women and white men break their hips more often than black women and black men (Kellie and Brody 1990)?[4] How do we interpret reports that Caucasian men have a lifetime fracture risk of 13–25 percent compared with Caucasian women's lifetime risk of 50 percent even though once peak bone mass is attained men and women lose bone at the same rate (Seeman 1997, 1998; NIH Consensus Statement Online 2000)?

This article has been abridged. Readers should seek out the original article for the full evidence, detail, and references provided in the original paper.

Such curious facts raise perplexing questions. Why have bones become more breakable in certain populations? What does it mean to say that a lifestyle behavior such as exercise, diet, drinking, or smoking is a risk factor for osteoporosis? Why do we screen large numbers of women for bone density even though this information does not tell us whether an individual woman will break a bone?[5] Why was a major public policy statement on women's health unable to offer a coherent account of sex (or is it gender?) differences in bone health over the life cycle (Wizemann and Pardue 2001)? Why, if bone fragility is so often considered to be a sex-related trait, do so few studies examine the relationships among childbirth, lactation, and bone development (Sowers 1996; Glock, Shanahan, and McGowan 2000)?

Such curious facts and perplexing questions challenge both feminist and biomedical theory. If "facts" about biology and "facts" about culture are all in a muddle, perhaps the nature/nurture dualism, a mainstay of feminist theory, is not working as it should. Perhaps, too, parsing medical problems into biological (or genetic or hormonal) components in opposition to cultural or lifestyle factors has outlived its usefulness for biomedical theory. I propose that already well-developed dynamic systems theories can provide a better understanding of how social categories act on bone production. Such a framework, especially if it borrows from a second analytic trend called "life course analysis of chronic disease epidemiology" (Kuh and Ben-Shlomo 1997; Ben-Shlomo and Kuh 2002; Kuh and Hardy 2002), can improve our approaches to public health policy, prediction of individual health conditions, and the treatment of individuals with unhealthy bones.[6] To see why we should follow new roads, I consider gender, examining where we—feminist theorists and medical scientists—have recently been.

SEX AND GENDER (AGAIN)

For centuries, scholars, physicians, and laypeople in the United States and Western Europe used biological models to explain the different social, legal, and political statuses of men and women and people of different hues.[7] When the feminist second wave burst onto the political arena in the early 1970s, we made the theoretical claim that sex differs from gender and that social institutions produce observed social differences between men and women (Rubin 1975). Feminists assigned biological (especially reproductive) differences to the word *sex* and gave to *gender* all other differences.

"Sex," however, has become the Achilles' heel of 1970s feminism. We relegated it to the domain of biology and medicine, and biologists and medical scientists have spent the past thirty years expanding it into arenas we firmly believed to belong to our ally gender. Hormones, we learn (once more), cause naturally more assertive men to reach the top in the workplace (Dabbs and Dabbs 2001). Rape is a behavior that can be changed only with the greatest difficulty because it is wired somehow into men's brains (Thornhill and Palmer 2001). The relative size of eggs and sperm dictate that men are naturally polygamous and women naturally monogamous. And more. (See Zuk 2002; Travis 2003 for a critique of these claims.) Feminist scholars have two choices in response to this spreading oil spill of sex. Either we can contest each claim, one at a time, doing what Susan Oyama calls "hauling the theoretical body back and forth across the sex/gender border" (2000, 190), or, as I choose to do here, we can reconsider the 1970s theoretical account of sex and gender.

In thinking about both gender and race, feminists must accept the body as simultaneously composed of genes, hormones, cells, and organs—all of which influence health and behavior—and of culture and history (Verbrugge 1997). As a biologist, I focus on what it might mean to claim that our bodies physically imbibe culture. How does experience shape the very bones that support us?[8] Can we find a way to talk about the body without ceding it to those who would fix it as a naturally determined object existing outside of politics, culture, and social change? This is a project already well under way, not only in feminist theoretical circles but in epidemiology, medical sociology, and anthropology as well.

WHY BONES?

Bones are eloquent. Archaeologists read old bone texts to find out how prehistoric peoples lived and worked. A hyperflexed and damaged big toe, a bony growth on the femur, the knee, or the vertebrae, for example, tell bioarchaeologist Theya Molleson that women in a Near Eastern agricultural community routinely ground grain on all fours, grasping a stone grinder with their hands and pushing back and forth on a saddle-shaped stone. The bones of these neolithic people bear evidence of a gendered division of labor, culture, and biology intertwined (Molleson 1994).[9]

Given that modern forensic pathologists also use bones to learn about how people live and die, it seems odd that a report from the National Institute of Medicine, presented as a state-of-the-art account of gender and medicine, deals only superficially with the sexual differentiation of bone disease (Wizeman and Pardue 2001).[10] In a brief three pages on osteoporosis, the monograph cites dramatic statistics on the frequency of osteoporosis in European and Caucasian American women and the dangers of the condition. The report offers a laundry list of factors believed to affect bone health. Jumbled together, with no attempt to understand their interrelationships or their joint, cumulative contributions to bone development and loss, are hormones, diet, exercise, genetic background, vitamin D production, and the bone-destroying effects of drugs such as cortisone, tobacco, and alcohol. In an anemic end-of-chapter recommendation the authors urge researchers to control for all of the above factors as they design their research studies. Indeed, failure to engage the task of formulating new approaches to biology prevented them from making a stronger analysis.

But osteoporosis is a condition that reveals all of the problems of defining sex apart from gender. A close reading of the osteoporosis literature further reveals the difficulties of adding the variable of race to the mix (a point I will develop in a forthcoming paper [Fausto-Sterling in preparation]) while also exemplifying the claim that disease states are socially produced, both by rhetoric and measurement (e.g., Petersen 1998) and by the manner in which cultural practice shapes the very bones in our bodies (Krieger and Zierler 1995).

OF BONES AND (WO)MEN

The accuracy of the claim that osteoporosis occurs four times more frequently in women than in men (Glock, Shanahan, and McGowan 2000) depends on how we define osteoporosis, in which human populations (and historical periods) we gather statistics, and what portions of the life cycle we compare. The NIH (2000) defines osteoporosis as a skeletal disorder in which weakened bones increase the risk of fracture. When osteoporosis first wandered onto the medical radar screen, the only signal that a person suffered from it was a bone fracture. Post hoc, a doctor could examine a person with a fracture either using a biopsy to look at the structural competence of the bone or by assessing bone density.

If one looks at lifetime risks for fracture, contemporary Caucasian men range from 13 to 25 percent (Bilezikian, Kurland, and Rosen 1999) while Caucasian women (who also live longer) have a 50 percent risk. But not all fractures result from osteoporosis. One study looked at fracture incidence in men and women at different ages and found that between the ages of five and forty-five men break more limbs than women.[11] The breaks, however, result from significant work- and sports-related trauma suffered by healthy bones. After the age of fifty, women break their bones more often than men, although after seventy years of age men do their best to catch up (Melton 1988).

The most commonly used medical standard for a diagnosis of osteoporosis no longer depends on broken bones. With the advent of machines called densitometers used to measure bone's mineral density (of which more in a moment), the World Health Organization (WHO) developed a new "operational" definition: a woman has osteoporosis if her bone mineral density measures 2.5 times the standard deviation below a peak reference standard for young (white) women. The densitometer manufacturer usually provides the reference data to a screening facility (Seeman 1998), and thus rarely, if ever, do assessments of osteoporosis reflect what Margaret Lock calls "local biologies" (Lock 1998, 39).[12] With the WHO definition, the prevalence of osteoporosis for white women is 18 percent, although there is not necessarily associated pathology, since now, by definition, one can "get" or "have" osteoporosis without ever having a broken bone. The WHO definition is controversial, since bone mineral density (BMD, or grams/cm^2) accounts for approximately 70 percent of bone strength, while the other 30 percent derives from the internal structure of bone and overall bone size. And while women with lower bone density are 2.5 times more likely to experience a hip fracture than women with high bone densities, high risks of hip fracture emerge even in women with high bone densities when five or more other risk factors are present (Cummings et al. 1995).[13] Furthermore, it is hard to know how to apply the criterion, based on a baseline of young white women, to men, children, and members of other ethnic groups. To make matters worse, there is a lack of standardization between instruments and sites at which measurements are taken.[14] Thus it comes as no surprise that "controversy exists among experts regarding the continued use of this [WHO] diagnostic criterion" (NIH Consensus Statement Online 2000, 3).

There is a complicated mixture at play. First, osteoporosis—whether defined as fractures or bone density—is on the rise, even when the increased age of a population is taken into account (Mosekilde 2000). At the same time, it is hard to assess the danger of osteoporosis, in part due to drug-company-sponsored "public awareness" campaigns. For example, in preparation for the sales campaign for its new drug, Fosamax, Merck Pharmaceuticals gave a large osteoporosis education grant to the National Osteoporosis Foundation to educate older women about the dangers of osteoporosis (Tanouye 1995).[15] Merck also directly addressed consumers with television ads contrasting frail, pain-wracked older women with lively, attractive seniors, implying the urgent need for older women to use Fosamax (Fugh-Berman, Pearson, Allina, Zones, Worcester, and Whatley 2002).

Mass marketing a new drug, however, requires more than a public awareness campaign. There must also be an easy, relatively inexpensive method of diagnosis. Here the slippage between the new technological measure—bone density—and the old definition of actual fractures and direct assessment of bone structure looms large. Merck promoted affordable bone density testing even before it put Fosamax on the market. The company bought an equipment manufacturing company and ramped up its production of bone density machines while at the same time helping consumers find screening locations by giving a grant to the National Osteoporosis Foundation to push a toll-free number that consumers (presumably alarmed by the Merck TV ads) could call to find a bone density screener in a locale near them (Tanouye 1995; Fugh-Berman, Pearson, Allina, Zones, Worcester, and Whatley 2002).

The availability of a simple technological measure for osteoporosis also made scientific research easier and cheaper. The majority of the thousands upon thousands of research papers on osteoporosis published in the ten years from 1995 to 2005 use BMD as a proxy for osteoporosis. This is true despite a critical scientific literature that insists that the more expensive volumetric measure (grams/cm^2) more accurately measures bone strength and that knowledge of internal bone structure (bone histomorphometry) provides essential information for understanding the actual risk of fracture (Meunier 1988).[16] The explosion of knowledge about osteoporosis codifies a new disorder, still called osteoporosis but sporting a newly simplified account of bone health and disease.[17] Ego Seeman (1997) laments the use of the density measure, which, he argues, "affects the way we conceptualize the skeleton (or fail to), and the way we direct (or misdirect) our research," and "blind[s] us to the biology of bone" (510).

Weaving together these threads—increasing lifetime risk, new disease definitions, and easier measurement—produces an epistemological transformation in our scientific accounts of bones and why they break. The transformation is driven by a combination of cultural forces (why are fracture rates increasing?) and new technologies generated by drug companies interested in creating new markets, disseminated with the help of market forces drummed up by the self-same drug companies, and aided by consumer health movements, including feminist health organizations such as the Society for Women's Research, which argue that gender-based differences in disease have been too long neglected.

Analyzing bone development within the framework of sex versus gender (nature vs. nurture) makes it difficult to understand bone health in men as well as women. Those trying to decide on a proper standard to measure fracture risk in men (should they use a separate male baseline or the only one available, which is for young, white women?) struggle with this problem of gender standardization (Melton et al. 1998). There are differences between men and women, although osteoporosis in men is vastly understudied. In a bibliography of 2,449 citations of papers from 1995 to 1999 (Glock, Shanahan, and McGowan 2000), only 47 (2 percent) addressed osteoporosis in men. But making sense of patterns of bone health for either or both sexes requires a dynamic systems approach. A basic starting place is to ask the development question.

For instance, we find no difference in bone mineral density in (Caucasian) boys and girls under age sixteen but a higher bone mineral density in males than in females thereafter (Zanchetta et al. 1995). This difference (combined with others that develop during middle adulthood) becomes important later in life, since men and women appear to lose bone at the same rate once they have reached a peak bone mass; those starting the loss phase of the life cycle with more bone in place will be less likely to develop highly breakable bones. Researchers offer different explanations for this divergence. Some note that boys continue to grow for an average of two years longer than girls (Seeman 1997). The extra growth period strengthens their bones by adding overall size. Others point additionally to hormones, diet, physical activity, and body weight as contributing to the emerging sex (or is it gender?) difference at puberty (Rizzoli and Bonjour 1999).

So differences in bone mineral density between boys and girls emerge during and after puberty, while for both men and women peak bone mass and strength is reached at twenty-five to thirty years of age (Seeman 1999). Vertebral height is the same in men and women, but vertebral width is greater in men. The volume of the inner latticework does not differ in men and women, but the outer layer of bone (periosteum) is thicker in men. Both width and outer thickness strengthen the bone. In general, sex/gender bone differences at peak are in size rather than density (Bilezikian, Kurland, and Rosen 1999).

This life-cycle analysis reveals three major differences in the pattern of bone growth and loss in men compared with that in women. First, at peak, men have 20 to 30 percent more bone mass and strength than women. Second, following peak, men but not women compensate for bone loss with new increases in vertebral width that continue to strengthen the vertebrae. Over time both men and women lose 70 to 80 percent of bone strength

(Mosekilde 2000), but the pattern of loss differs. In men the decline is gradual, barring secondary causes.[18] In women it is gradual until perimenopause, accelerates for several years during and after the menopause, and then resumes a gradual decline.[19] Lis Moskilde (2000) points out that the rush to link menopause to osteoporosis has led to the neglect of two of the three major differences in the pattern of bone growth between men and women. Yet these two factors are specifically linked to physical activity, and thus amenable to change earlier in life.

Indeed, many studies on children and adolescents address the contribution sociocultural components of bone development make to male-female differences that emerge just after puberty. But the overwhelming focus on menopause as the period of the life cycle in which women enter the danger zone steers us away from examining how earlier sociocultural events shape our bones (see Lock 1998). Once menopause enters the picture, the idea that hormones are at the heart of the problem overwhelms other modes of thought.[20] Nor is it clear how hormones affect bone development and loss. In childhood, growth hormone is essential for long bone growth, the gonadal steroids are important for the cessation of bone growth at puberty, and probably both estrogen and testosterone are important for bone health maintenance (Damien, Price, and Lanyon 1998). The details at the cellular level have yet to be understood (Gasperino 1995).

BASIC BONE BIOLOGY

In the fetus, cartilage creates the scaffolding onto which bone cells climb before secreting the calcium-containing bone matrix that becomes the hard bone.[21] The cells that secrete the bone matrix are called osteoblasts. As they grow, bones are shaped by the strains and stresses put on them by the activity of their owner. Osteoblasts deposit matrix at some sites, while another cell type, the osteoclast, can chip away at areas of too much growth. Growing bones change shape through this give and take of osteoblast and osteoclast activity in a process called bone remodeling.[22] Long bones increase in length throughout childhood by adding on new material at their growing ends. These growth sites close as a result of hormonal changes during puberty, but bone reshaping continues over the course of a life (Currey 2002).

Bone contains two important types of tissue, which can be seen if one cuts it across the middle. The outer dense, hard layer is called compact tissue; the inner layer contains cancellous tissue consisting of a latticework of slender fibers.

Osteoblasts can produce new bone in both locations. Osteoblasts can also transform into osteocytes, cells found in large numbers inside the hard bone tissues (Currey 2002). Osteocytes probably play an important role in bone regeneration when they produce chemical signals that tell osteoblasts that the bone is under mechanical strain and needs to grow (Mosley 2000). Osteoblasts cannot form new bone unless the surface on which they sit is under a mechanical strain, which explains why exercise remains such an important component of bone health while weightlessness in space or prolonged bed rest result in the loss of bone thickness.[23]

Bone development, then, is profoundly influenced by what physiologists call functional adaptation. Although a great deal remains to be understood about the biology of use and disuse, some basic principles are already evident. First, both disuse and predictable moderate use result in bone resorption and increased porosity. However, dynamic strain, that is, strain that is unpredictable and of varied impact level, can lead to a linear increase in bone mass (Mosley 2000).[24] Bones may adopt strain thresholds such that only strains above such thresholds induce new bone formation. Strain thresholds may change over the life cycle. Perhaps the decline in estrogen associated with menopause resets the threshold to a higher strain level, thus requiring very high levels of bone stress to stimulate new bone formation. Such dynamic theories allow us to understand how behavior (e.g., changing forms of exercise) and hormonal changes in the body might together produce bone loss or gain (Frost 1986, 1992).

Even such a simplified account of bone development and maintenance shows how hard it can be to understand why people in one group break their bones more often than people in another. Groups may differ in peak bone size even if bone loss later in life is the same. The bone's inside might be thicker in one group than another, or the outside, compact bone layer might be thicker. There could be less bone loss or a reduction in bone turnover (the balance between osteoclast and osteoblast activity). What is most striking about the medical literature on osteoporosis is that "whether these differences in bone size, mass, or structure, or bone turnover among ethnic groups or between men and women even partly account for the corresponding group differences in fracture rates is unknown" (Seeman 1997, 517).

Genes, of course, are involved in all of the events described in the previous few paragraphs. Rather than as causes of bone construction and destruction, however, genes are best understood as mediators, suspended in a network of signals (including their own) that induce them to synthesize new molecules.[25] The molecules they make may help to produce more bone or to break down existing bone. Either action may, in turn, be a direct effect (e.g., making a structural element such as collagen) or an indirect effect (e.g., causing the death or sustaining the life of bone-making cells). Researchers have identified over thirty genes that affect bone development either positively or negatively in mice (Peacock et al. 2002), and scientists continue to identify genetic variants affecting bone density in humans (Boyden et al. 2002; Little et al. 2002; Ishida et al. 2003).

Finally, how do hormones fit into all of this? Part of the initial logic of thinking about osteoporosis as a basic biological (sex) difference between men and women derives from the observation that bone thinning increases dramatically around the time of menopause. Most thus assume that declining estrogen causes bone loss. Since estrogen codes in most people's minds as a quintessentially female molecule, it becomes extraordinarily difficult to conceptualize osteoporosis as a disease with many contributors stretching over the entire life cycle. Here, gender constructs (Fausto-Sterling 2000) combined with the profits derived from selling estrogen replacement have contributed mightily to shaping the course of scientific research in this field. Estrogen, though, is only one of a number of hormones linked to bone physiology.

At least three major hormone systems acting both independently of one another and through mutual influence regulate bone formation and loss. Fascinatingly, at least two of these operate at times through the brain and the sympathetic (involuntary) nervous system.[26] The first system includes three major hormones that maintain proper calcium levels throughout the body, dipping into the bone calcium reservoir as needed.[27] The hormones (the active form of vitamin D; parathyroid hormone [PTH], which is made by a small pair of glands called the parathyroid glands; and calcitonin, which is secreted by the thyroid glands) regulate blood calcium levels and bone metabolism.[28] At low concentrations PTH maintains a stable level of mineral turnover in the bone, but at high levels it stimulates osteoclast activity, thus releasing calcium into the bloodstream.[29] Although calcitonin counteracts the effects of PTH on osteoclasts, its functions and mode of action are still poorly understood, but PTH affects bone, kidney, and intestine using vitamin D as an intermediary—a point that returns us to the contributions of sunlight and diet. Our diets and cellular machinery provide inactive forms of vitamin D, but these require the direct energy from sunlight hitting the skin to change into potentially active forms. Final transformations from inactive to active forms of vitamin D occur in the liver and kidney (Bezkorovainy and Rafelson 1996).

The second hormone system—which includes both estrogens and androgens—is clearly important for bone development and maintenance; how it regulates bone metabolism remains uncertain (Kousteni et al. 2001, 2002). Recently, some fascinating studies done on mice have suggested that both androgens and estrogens operate in a fashion unusual for steroid hormones—by preventing the death of bone-forming cells without stimulating new gene activity. Whether these results will hold for humans remains to be seen.[30] Other information from animal models suggests that bone response to mechanical strain requires the presence of an estrogen receptor on the osteoblast cell surface (Lee et al. 2003), but a clear story of the role of estrogens and androgens in bone formation and maintenance throughout the life cycle remains to be told.

Last but certainly not least a hormone called leptin, announced to the world with great fanfare in 1995 as a possible "magic bullet" for weight control (Roush 1995), also affects bone formation. Like the sex steroids, leptin works via a relay system in the hypothalamus, a part of the brain linked to the pituitary gland. Fat tissue produces leptin, which signals specialized nerve cells in the hypothalamus; these activated neurons produce two effects—lowering the appetite and stimulating basal metabolism (via the sympathetic nervous system). In mice, leptin has a second, apparently independent effect, also mediated through the hypothalamus and the sympathetic nervous system. Increased leptin signals nerves in the bone to depress bone formation. This presents an interpretive paradox: obesity provides some protection against osteoporosis. But the more fat cells, the more leptin is made, which in theory ought to depress bone formation. There are several possible explanations for this paradox. In mice it may be that the very overweight body becomes insensitive to its high leptin levels, just as obesity contributes to insulin insensitivity in type 2 diabetes. Or the stimulation of bone formation from the mechanical stress of increased weight might trump the effects of leptin, and/or leptin physiology in mice and humans might differ in important ways.[31]

In the next decade we will surely learn a lot more about the relationships among bone formation, leptin, and the sympathetic nervous system.[32] But we also must learn how to study the balances and interactions among all of the various factors that impinge on bone formation. How do social systems that influence what we eat, how and when we exercise, whether we drink or smoke, what kinds of diseases we get and how they are treated, and how we age, to name some most relevant to bone formation, produce a particular bone structure in a particular individual with a particular life history? To even begin to set up this problem in a manner that can stimulate future work and ultimately bring us better answers, we need to learn how to handle complex, dynamic systems.

THINKING SYSTEMATICALLY ABOUT BONE

There are better ways to think about gender and the bare bones of sex. One cannot easily separate bone biology from the experiences of individuals growing, living, and dying in particular cultures and historical periods and under different regimens of social gender.[33] But how can we integrate the varied information presented in this essay in a manner that helps us ask better research and public policy questions and that, in posing better questions, allows us to find better answers? By *better*, I mean several things: in terms of the science I want to take more of the "curious facts" about bone into account when responding to public health problems. I favor emphasizing lifelong healthful habits that might prevent or lessen the severity of bone problems in late life, but I would also like us to have a better idea of how to help people whose bones are already thin. What dietary changes, what regimens of exercise and sun exposure, and what body mass index work best with which medications? How do the medications we choose work? What unintended effects do they have? Finally, *better* includes an ability to predict outcomes for individuals, based on their particular life histories and genetic makeups, rather than merely making probability statements about large and diverse categories of people.

How can we get there from here? Below, I outline in fairly general form the possibilities of dynamic systems and developmental systems approaches. Such formulations allow us to work with the idea that we are always 100 percent nature and 100 percent nurture. I further point to important theoretical and empirical work currently under way by social scientists who study chronic diseases using a life-course approach. Before turning to the specifics of bone development, let me offer a general introduction to these complementary modes of thought.

The varied systems approaches to understanding development share certain features in common. All understand that cells, nervous systems, and whole organisms develop through a process of self-organization rather than according to a preformed set of instructions. The varying relationships among system components lead to change, and new patterns are dynamically stable because the characteristics of the system confer stability. But if the system is sufficiently perturbed, instability ensues and significant fluctuations occur until a new pattern, again dynamically stable, emerges. Bone densities, for example, are often dynamically stable in mid-life but destabilize during old age; most medical interventions aim to restabilize the dynamic system that maintains bone density. But we really do not understand how the transition from a stable to an unstable system of bone maintenance occurs.

To address the bare bones of sex, I highlight, in figure 16.1, seven systems that contribute to bone strength throughout the life cycle.[34] I also describe some of the known interrelationships between them.[35] Each of the seven—physical activity, diet, drugs, bone formation in fetal development, hormones, bone cell metabolism, and biomechanical effects on bone formation—can be analyzed as a complex system in its own right. Bone strength emerges from the interrelated actions of each (and all) of these systems as they act throughout the life cycle. As a first step toward envisioning bone from a systems viewpoint we can construct a theoretical diagram of their interactions. The diagram in systems approaches can be thought of as a theoretical model, to be tested in part or whole and modified as needed.[36] As ways to describe each component system using numerical proxies become available, the pictorial model can provide the framework for a mathematical model. Figure 16.1 represents one possible diagram of a life-course systems account of bone development.

This feminist systems account embeds the proposed subsystems within the dimensions of gender, socioeconomic position, and culture.[37] Consider the diet system. Generally, of course, diet is shaped by culture and subculture, including race and ethnicity (Bryant, Cadogan, and Weaver 1999). But gender further influences diet. For example, one study reports that 27 percent of U.S. teenage girls (compared with 10 percent of adolescent boys) who think they weigh the correct amount are nevertheless trying to lose weight (Walsh and Devlin 1998). It may also be true that there are sex/gender differences in basal metabolism rates that influence food intake.

Figure 16.1 also indicates the cumulative effects of diet on bone formation. Key events may be clustered at certain points in the life cycle.[38] For example, adolescent girls in the United States often diet more and exercise less than during earlier childhood. Diseases such as anorexia nervosa, which have devastating effects on bone development, may also emerge during adolescence. As Yoav Ben-Shlomo and Diana Kuh (2002) point out, such clustering of adverse events is common and may be thought of in terms of "chains of risk" (or benefit). In a life-course approach, prior events set the limits on later ones. If girls and women enter into adulthood with weakened bones, therefore, they can rebuild them, but their peak density may be less than if they had built stronger bones in adolescence.[39]

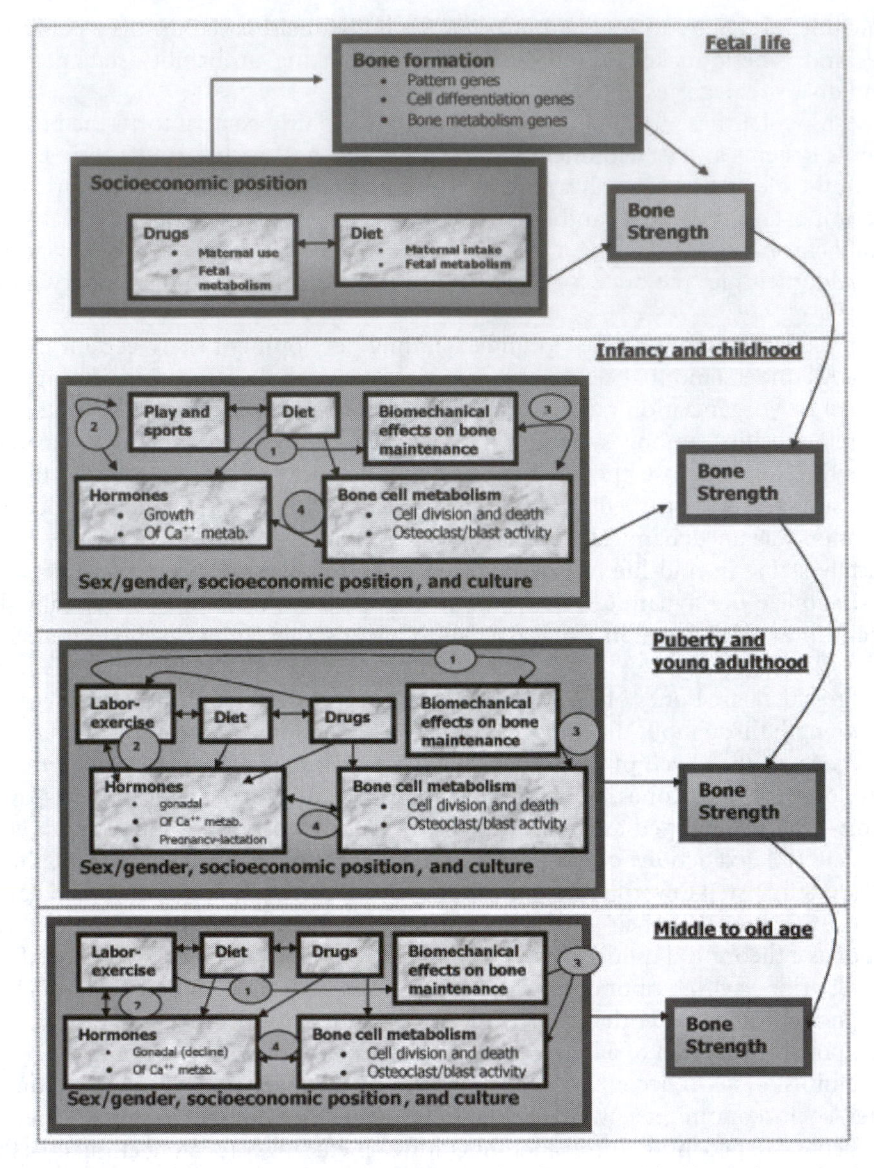

Figure 16.1 A life history-systems overview of bone development. (1) Physical activity has direct effects on bone cell receptors and indirect effects by building stronger muscles, which exert physical strain on bones, thus stimulating bone synthesis. (2) Physical activity that takes place outdoors involves exposure to sunlight, thus stimulating vitamin D synthesis, part of the hormonal system regulating calcium metabolism. (3) Biomechanical strain affects bone cell metabolism by activating genes concerned with bone cell division and bone (re)modeling. (4) Hormones affect bone cell metabolism by activating genes concerned with bone cell division, cell death, bone (re)modeling, and new hormone synthesis.

Alternatively, achieving a safe peak bone density might require more sustained and intense work for a person of one history compared with a person of a different history.

Sex/gender, race, class, and culture also differentiate individuals by forms of play in childhood and beyond (Boot et al. 1997), by choices of formal exercise programs, and, in adulthood, by forms of labor, physical and otherwise. In analyzing the system of physical activity one again applies life-course principles by considering that what happens at any one point builds on what has gone before. Important events with regard to bone development may be clustered and interrelated. For both the diet and physical activity systems, it should be possible to design mathematical models based on some measure of bone strength that would incorporate the effects of each of these social systems on bone development throughout the life cycle; once we have plausible models of each system, we can ask questions about their interactions.

The remaining four systems are often considered within the realm of biology, as if biology were separate from culture, although recent work from some medical epidemiologists challenges this distinction (Ellison 1996; Hertzman 1999; Lamont et al. 2000). The system of biomechanical effects on bone synthesis, for example, requires further investigation of all of its inputs (physical strain, activation of genes that stimulate bone cell development or death, etc. [Harada and Rodan 2003]), but these must then be studied in relationship to the gender-differentiated physical activity system. The different body shapes of adult men and women (related to hormones at puberty among other things) may also affect bone biomechanics, and we need, too, to know more about how growth and development affect the number of bone mechanoreceptors—molecules that translate mechanical stress in biochemical activity (Boman et al. 1998; Pavalko et al. 2003).

The impact of hormones on bone development and maintenance requires research attention of a sort currently lacking in the bone literature. We need to know both about the molecular biology of hormones and bone cell hormone receptors and about life-course effects on hormone systems (Ellison 1996; Worthman 2002). Finally, genes involved in bone cell metabolism, pattern formation, hormone metabolism, drug processing, and many other processes contribute importantly to the development of bone strength (Zelzer and Olsen 2003). Understanding how they function within both the local and global (body and sociocultural) networks contributing to bone development requires a systems-level analysis not yet found in the literature.

CONCLUSION

This article is a call to arms. The sex-gender or nature-nurture accounts of difference fail to appreciate the degree to which culture is a partner in producing body systems commonly referred to as biology—something apart from the social. I introduce an alternative—a life-course systems approach to the analysis of sex/gender. Figure 16.1 is a research proposal for multiple programs of investigation in several disciplines. We need to ask old questions in new ways so that we can think systematically about the interweaving of bodies and culture. We will not lay bare the bones of sex, but we will come to understand, instead, that our skeletons are part of a life process. If process rather than stasis becomes our intellectual goal, we will improve medical practice and have a more satisfying account of gender and sex as, to paraphrase the phenomenologists, being-in-the-world.

Notes

Thanks to the members of the Pembroke Seminar on Theories of Embodiment for a wonderful year of thinking about the process of body making and for their thoughtful response to an earlier draft of this essay. Credit for the title goes to Greg Downey. Thanks also to anonymous reviewers from *Signs* for making me sharpen some of the arguments.

1 Munro Peacock et al. write: "The pathogenesis of a fragility fracture almost always involves trauma and is not necessarily associated with reduced bone mass. Thus, fragility fracture should neither be used synonymously nor interchangeably as a phenotype for osteoporosis" (2002, 303).
2 For example, sitting height reflects trunk length (vertebral height) vs. standing height, which

reflects the length of the leg bones. These can change independently of one another. Thus height increases can result from changes in long bone length, vertebral height, or both. See Meredith 1978; Tanner et al. 1982; Malina, Brown, and Zavaleta 1987; Balthazart, Tlemçani, and Ball 1996; Seeman 1997.

3 The use of racial terms such as *Caucasian* and others in this article is fraught. But for the duration of this article I will use the terms as they appear in the sources I cite, leaving an analysis of this problematic terminology to future publications, e.g., Fausto-Sterling 2004.

4 Since a number of studies show no sex difference in hip fracture incidence between African American men and women, the "well-known" gender difference in bone fragility may really only be about white women. As so often happens, the word *gender* excludes women of color (Farmer et al. 1984).

5 Peacock et al. write, "Key bone phenotypes involved in fracture risk relate not only to bone mass but also to bone structure, bone loss, and possibly bone turnover" (2002, 306).

6 I am grateful to Peter Taylor for insisting that I read the work in life-course analysis.

7 Stepan 1982; Russett 1989; Hubbard 1990; Fausto-Sterling 1992.

8 I use the term *experience* rather than the term *environment* here to refer to functional activity. For more detail see Gottlieb, Whalen, and Lickliter 1998.

9 Perhaps because the field of archaeology is still struggling to bring gender into the fold, its practitioners often insist on the centrality of the sex/gender distinction. Yet their own conclusions undermine this dualism, precisely because they use a biological product, bone, to draw conclusions about culture and behavior (Ehrenberg 1989; Gero and Conkey 1991; Wright 1996; Armelagos 1998).

10 The validity of using bones to identify race is contested (Goodman 1997).

11 This study (cited in Melton 1988) dates from 1979, and it seems likely that subsequent cultural changes have led to different patterns of breakage; fracture incidence is a moving target.

12 Local biologies rellect local differences in biology. For example, hot flashes are far less frequent in Japan than in the United States, possibly for reasons pertaining to diet. The normalization question here is: Is it best to compare a population to its own group or some group with similar environmental and genetic histories, or to some outgroup standard?

13 These factors include: a mother having broken her hip, especially before age eighty; height at age twenty-five (taller women are more likely to break hips); extreme thinness; sedentary lifestyle; poor vision; high pulse rate; the use of certain drugs; etc.

14 One researcher states: "I think what is also of note, is that the between-center differences are greater than between-sex differences within certain centers" (Lips 1997, 95).

15 Fosamax seems to be able to prevent further bone loss in people who are losing bone and to build back lost bone at least in the hip and spine. In discussing Merck's campaign, I do not argue that the drug is useless (in fact, I am taking it!), merely (!) that drug companies play an important role in the creation of new "disease" and profit as a result.

16 "An association between the change in areal bone density and the change in fracture rates has never been documented" (Seeman 1997, 517). According to the NIH Consensus Statement Online: "Currently there is no accurate measure of overall bone strength" (2000, 5). But BMD is often used as a proxy. The National Women's Health Network cites the pitfalls of using BMD to predict future fractures (Fugh-Berman, Pearson, Allina, Zones, Worcester, Whatley, Massion, et al. 2002), but others cite a strong association between BMD and fracture rate (e.g., Melton et al. 1998; Siris et al. 2001). One overview of studies that attempted to predict osteoporosis-linked fractures with bone mineral density concluded: "Measurements of bone mineral density can predict fracture risk but cannot identify individuals who will have a fracture. We do not recommend a programme of screening menopausal women for osteoporosis by measuring bone density" (Marshall, Johnell, and Wedel 1996, 1254). See also Nelson et al. 2002.

17 For a history of the concept of osteoporosis, see Klinge 1998.

18 A secondary cause might be bone loss due to an eating disorder or a metabolic disease, or the prolonged use of a bone-leaching drug such as cortisone.

19 When I use the words *men* and *women* I refer to particular populations on which these studies were done. These are mostly Caucasian and Northern European or North American. Most of the studies have been done since the 1980s, but bone size, shape, and growth patterns would have differed at the beginning of the twentieth century compared with their appearance at the beginning of the twenty-first. I will not make these points every time I use these words.

20 So powerful is the focus on old age that the long NIH bibliography on menopause completely ignores the possible importance of pregnancy and lactation on bone development. These two processes are profoundly implicated in calcium metabolism, and if there is *no* effect on later bone

strength it would be important to find out why. What physiological mechanisms protect the bone of pregnant and lactating women? This is an example of a biological question that lies fallow because of the focus on supposed estrogen deficiency in old age.

21 The bone matrix is made up primarily of a substance called hydroxyapatite that is mostly composed of crystalline forms of the molecules calcium phosphate, calcium carbonate, and small amounts of magnesium, fluoride, and sulfate.

22 One memory device for remembering which cell is which is to think that osteoBlasts Build bone and osteoClasts Chomp on bone.

23 Stress can be from direct impact or from tension placed on the bones by attached muscles. For more details on the importance of mechanical strain on bone development, see Skerry and Lanyon 1995; Mosekilde 2000; Mosley 2000.

24 In animal models it is possible to induce new bone formation (modeling) without first having caused bone resorption (Pead, Skerry, and Lanyon 1988).

25 One review states that mechanical receptors transform signals from deforming bones into changes in the shape of DNA regions that regulate the activities of genes involved in bone formation. The authors write that "bending bone ultimately bends genes" (Pavalko et al. 2003, 104).

26 Physiological functions such as heart and breathing rate and energy metabolism are regulated through involuntary nerves belonging to the sympathetic and parasympathetic nervous systems. These systems balance each other out by stimulating or inhibiting various functions. They are controlled through brain centers without our having to think about them.

27 All cells, but especially nerve and muscle cells, need calcium. So bone is essential not only for structural support but also to maintain healthy calcium levels throughout the body.

28 The active form of vitamin D is 1,25-dihydroxycholecalciferol.

29 Parathyroid hormone also increases Ca^{++} reabsorption in the kidney and absorption in the small intestine.

30 The negative effects of estrogen treatment come from the hormone's more common mode of action—stimulating gene activities after binding to the nucleus. The researchers cited have a compound that has none of the gene-activity-stimulating actions but does behave like androgens and estrogens by preventing the death of osteoblasts. See also Moggs et al. 2003.

31 Ducy et al. 2000; Flier 2002; Takeda et al. 2002; Harada and Rodan 2003.

32 Leptin may also regulate the onset of puberty, thus linking gonadal hormones and the leptia hormone system (Chehab et al. 1997).

33 I found one eloquent but wordless example on the Web in an article on causes of vitamin D deficiency. The short segment titled "Insufficient Exposure to Sunlight" was accompanied by a photograph of two women, standing in the blazing sun, covered from head to toe in burkas, clearly insufficiently exposed to sunlight but not for want of being outdoors in the sun.

34 I use Peter Taylor's definition of systems as "units that have clearly defined boundaries, coherent internal dynamics, and simply mediated relations with their external context" (personal communication 2003).

35 This choice of systems emerges from the data presented earlier in this article. Since this is a model, others might argue for dividing the pie in a different way. To keep the diagram readable and the discussion manageable, I have not emphasized that the entire grouping of systems is embedded in a larger system I call "general health." There are many disease states that secondarily affect bone (e.g., kidney disease or endocrine disorders) by affecting calcium metabolism or preventing exercise. The relationships among the systems affecting bone strength would be shifted in dramatic ways worthy of study in their own right under such circumstances.

36 Choice of model has profound implications. For a discussion of a lifestyle model of disease that emphasizes individual choice vs. a "social production model," see Krieger and Zierler 1995. For an update on current theories of social epidemiology, see Krieger 2001.

37 To the extent that race is a legitimate category separate from class and culture, I will incorporate it into the bone systems story in pt. 2 of this work. For a model of social pathways in childhood that lead to adult health, see Kuh and Ben-Shlomo 1997.

38 Bonjour et al. 1997; Boot et al. 1997; Perry 1997; Wang et al. 2003.

39 For the effects of dietary calcium later in life, see Heaney 2000.

References

Aloia, J. F., A. N. Vaswani, J. K. Yeh, and E. Flaster. 1996. "Risk for Osteoporosis in Black Women." *Calcified Tissue International* 59(6):415–23.

Armelagos, George J. 1998. "Introduction: Sex, Gender and Health Status in Prehistoric and Contemporary Populations." In *Sex and Gender in Paleopathological Perspective*, ed. Anne L. Grauer and Patricia Stuart-Macadam, 1–10. Cambridge: Cambridge University Press.

Balthazart, Jacques, Omar Tlemçani, and Gregory F. Ball. 1996. "Do Sex Differences in the Brain Explain Sex Differences in Hormonal Induction of Reproductive Behavior? What 25 Years of Research on the Japanese Quail Tells Us." *Hormones and Behavior* 30(4):627–61.

Ben-Shlomo, Yoav, and Diana Kuh. 2002. "A Life Course Approach to Chronic Disease Epidemiology: Conceptual Models, Empirical Challenges, and Interdisciplinary Perspectives." *International Journal of Epidemiology* 31(2):285–93.

Bezkorovainy, Anatoly, and Max E. Rafelson. 1996. *Concise Biochemistry*. New York: Dekker.

Bilezikian, John P., Etah S. Kurland, and Clifford S. Rosen. 1999. "Male Skeletal Health and Osteoporosis." *Trends in Endocrinology and Metabolism* 10(6): 244–50.

Boman, U. Wide, A. Möller, and K. Albertsson-Wikland. 1998. "Psychological Aspects of Turner Syndrome." *Journal of Psychosomatic Obstetrics and Gynaecology* 19(1):1–18.

Bonjour, Jean-Phillippe, Anne-Lise Carrie, Serge Perrari, Helen Clavien, Daniel Slosman, and Gerald Theintz. 1997. "Calcium-Enriched Foods and Bone Mass Growth in Prepubertal Girls: A Randomized, Double-Blind, Placebo-Controlled Trial." *Journal of Clinical Investigation 99*(6): 1287–94.

Boot, Annemieke M., Maria A. J. de Ridder, Huibert A. P. Pols, Eric P. Krenning, and Sabine M. P. F. de Muinck Keizer-Schrama. 1997. "Bone Mineral Density in Children and Adolescents: Relation to Puberty, Calcium Intake, and Physical Activity." *Journal of Clinical Endocrinology and Metabolism* 82(1):57–62.

Boyden, Lynn M., Junhao Mao, Joseph Belsky, Lyle Mitzner, Anita Farhi, Mary A. Mitnick, Dianqing Wu, Karl Insogna, and Richard P. Lifton. 2002. "High Bone Density Due to a Mutation in LDL-Receptor-Related Protein 5." *New England Journal of Medicine* 346(20):1513–21.

Bryant, Rebecca J., Jo Cadogan, and Connie M. Weaver. 1999. "The New Dietary Reference Intakes for Calcium: Implications for Osteoporosis." *Journal of the American College of Nutrition* 18(5):S406–S412.

Chehab, Farid F., Khalid Mounzih, Ronghua Lu, and Mary E. Lim. 1997. "Early Onset of Reproductive Function in Normal Female Mice Treated with Leptin." *Science* 275 (January 3): 88–90.

Cummings, Steven R., Michael C. Nevitt, Warren S. Browner, Katie Stone, Kathleen M. Fox, Kristine E. Ensrud, Jane Cauley, Dennis Black, and Thomas M. Vogt. 1995. "Risk Factors for Hip Fracture in White Women." *New England Journal of Medicine* 332(12):767–73.

Currey, John D. 2002. *Bones: Structure and Mechanics*. Princeton, N.J.: Princeton University Press.

Dabbs, James McBride, and Mary Godwin Dabbs. 2001. *Heroes, Rogues, and Lovers: Testosterone and Behavior*. New York: McGraw-Hill.

Damien, E., J. S. Price, and L. E. Lanyon. 1998. "The Estrogen Receptor's Involvement in Osteoblasts' Adaptive Response to Mechanical Strain." *Journal of Bone and Mineral Research* 13(8):1275–82.

Dibba, Bakary, Ann Prentice, Ann Laskey, Dot Stirling, and Tim Cole. 1999. "An Investigation of Ethnic Differences in Bone Mineral, Hip Axis Length, Calcium Metabolism and Bone Turnover between West African and Caucasian Adults Living in the United Kingdom." *Annals of Human Biology* 26(3):229–42.

Ducy, Patricia, Michael Amling, Shu Takeda, Matthias Priemel, Arndt F. Schilling, Frank T. Beil, Jianhe Shen, Charles Vinson, Johannes M. Rueger, and Gerard Karsenty. 2000. "Leptin Inhibits Bone Formation through a Hypothalamic Relay: A Central Control of Bone Mass." *Cell* 100(2):197–207.

Ehrenberg, Margaret. 1989. *Women in Prehistory*. London: British Museum Press.

Ellison, Peter T. 1996. "Developmental Influences on Adult Ovarian Hormonal Function." *American Journal of Human Biology* 8(6):725–34.

Farmer, Mary E., Lon R. White, Jacob A. Brody, and Kent R. Bailey. 1984. "Race and Sex Differences in Hip Fracture Incidence." *American Journal of Public Health* 74(12):1374–80.

Fausto-Sterling, Anne. 1992. *Myths of Gender: Biological Theories about Women and Men*. 2d ed. New York: Basic Books.

——. 2000. *Sexing the Body: Gender Politics and the Construction of Sexuality*. New York: Basic Books.

——. 2004. "Refashioning Race: DNA and the Politics of Health Care." *differences: A Journal of Feminist Cultural Studies*. In press.

——. In preparation. "The Bare Bones of Sex: Part II—Race."

Flier, Jeffrey S. 2002. "Physiology: Is Brain Sympathetic to Bone?" *Nature* 420(6916):619, 620–22.

Frost, Harold M. 1986. *Intermediary Organization of the Skeleton*. Vols. 1 and 2. Boca Raton, Fla.: CRC Press.

——. 1992. "The Role of Changes in Mechanical Usage Set Points in the Pathogenesis of Osteoporosis." *Journal of Bone and Mineral Research* 7(3): 253–61.

Fugh-Berman, Adriane, C. K. Pearson, Amy Allina, Jane Zones, Nancy Worcester, and Mariamne Whatley. 2002. "Manufacturing Need, Manufacturing 'Knowledge.'" *Network News* (May/June):1, 4.

Fugh-Berman, Adriane, C. K. Pearson, Amy Allina, Jane Zones, Nancy Worcester, Mariamne Whatley, Charlea Massion, and Ellen Michaud. 2002. "Hormone Therapy and Osteoporosis: To Prevent Fractures and Falls, There Are Better Options than Hormones." *Network News* (July/August): 4–5.

Gasperino, James. 1995. "Androgenic Regulation of Bone Mass in Women." *Clinical Orthopaedics and Related Research* 311:278–86.

Gero, Joan M., and Margaret W. Conkey, eds. 1991. *Engendering Archeology: Women in Prehistory.* Oxford: Blackwell.

Gilsanz, Vicente, David L. Skaggs, Arzu Kovanlikaya, James Sayre, M. Luiza Loro, Francine Kaufman, and Stanley G. Korenman. 1998. "Differential Effect of Race on the Axial and Appendicular Skeletons of Children." *Journal of Clinical Endocrinology and Metabolism* 83(5):1420–27.

Glock, Martha, Kathleen A. Shanahan, Joan A. McGowan, and compilers, eds. 2000. "Osteoporosis [Bibliography Online]: 2,449 Citations from January 1995 through December 1999." Available online at http://www.nlm.nih.gov/pubs/cbm/osteoporosis.html. Last accessed May 5, 2004.

Goodman, Alan H. 1997. "Bred in the Bone?" *Sciences* 37(2):20–25.

Gottlieb, Gilbert, Richard E. Whalen, and Robert Lickliter. 1998. "The Significance of Biology for Human Development: A Developmental Psychobiological Systems View." In *Handbook of Child Psychology*, ed. Richard M. Lerner, 233–73. New York: Wiley.

Harada, Shun-ichi, and Gideon A. Rodan. 2003. "Control of Osteoblast Function and Regulation of Bone Mass." *Nature* 423 (May 15): 349–55.

Heaney, Robert P. 2000. "Calcium, Dairy Products, and Osteoporosis." *Journal of the American College of Nutrition* 19(2):S83–S99.

Hertzman, Clyde. 1999. "The Biological Embedding of Early Experience and Its Effects on Health in Adulthood." *Annals of the New York Academy of Science* 896:85–95.

Hu, J. F., X. H. Zhao, J. S. Chen, J. Fitzpatrick, B. Parpia, and T. C. Campbell, 1994. "Bone Density and Lifestyle Characteristics in Premenopausal and Post-menopausal Chinese Women." *Osteoporosis International* 4:288–97.

Hubbard, Ruth. 1990. *The Politics of Women's Biology.* New York: Routledge.

Ishida, Ryota, Mitsuru Emi, Yoichi Ezura, Hironori Iwasaki, Hideyo Yoshida, Takao Suzuki, Takayuki Hosoi et al. 2003. "Association of a Haplotype (196phe/532ser) in the Interleukin-1-Receptor Associated Kinase (Iraki) Gene with Low Radial Bone Mineral Density in Two Independent Populations." *Journal of Bone and Mineral Research* 18(3):419–32.

Kellie, Shirley E., and Jacob A. Brody. 1990. "Sex-Specific and Race-Specific Hip Fracture Rates." *American Journal of Public Health* 80(3):326–28.

Klinge, Ineke. 1998. "Gender and Bones: The Production of Osteoporosis, 1941–1996." Ph.D. dissertation, University of Utrecht.

Kousteni, S., T. Bellido, L. Plotkin, C. A. O'Brien, D. L. Bodenner, L. Han, G. B. DiGregorio, et al. 2001. "Nongenotropic, Sex-Nonspecific Signaling through the Estrogen or Androgen Receptors: Dissociation from Transcriptional Activity." *Cell* 104(5):719–30.

Kousteni, S., J.-R. Chen, T. Bellido, L. Han, A. A. Ali, C. A. O'Brien, L. Plotkin, et al. 2002. "Reversal of Bone Loss in Mice by Nongenotropic Signaling of Sex Steroids." *Science* 298(5594):843–46.

Krieger, Nancy. 2001. "Theories for Social Epidemiology in the Twenty-First Century: An Ecosocial Perspective." *International Journal of Epidemiology* 30(4): 668–77.

Krieger, Nancy, and Sally Zierler. 1995. "Accounting for Health of Women." *Current Issues in Public Health* 1:251–56.

Kuh, Diana, and Yoav Ben-Shlomo, eds. 1997. *A Life Course Approach to Chronic Disease Epidemiology.* Oxford: Oxford University Press.

Kuh, Diana, and Rebecca Hardy, eds. 2002. *A Life Course Approach to Women's Health.* Oxford: Oxford University Press.

Lamont, Douglas, Louise Parker, Martin White, Nigel Unwin, Stuart M. A. Bennett, Melanie Cohen, David Richardson, Heather O. Dickinson, K. G. M. M. Alberti, and Alan W. Kraft. 2000. "Risk of Cardiovascular Disease Measured by Carotid Intima-Media Thickness at Age 49–51: A Lifecourse Study." *British Medical Journal* 320 (January 29): 273–78.

Larsen, Clark Spencer, 1998. "Gender, Health, and Activity in Foragers and Farmers in the American Southeast: Implications for Social Organization in the Georgia Bight." In *Sex and Gender in*

Paleopathological Perspective, ed. Anne L. Grauer and Patricia Stuart-Macadam, 165–87. Cambridge: Cambridge University Press.

Lee, Karla, Helen Jessop, Rosemary Suswillo, Gul Zaman, and Lance E. Lanyon. 2003. "Bone Adaptation Requires Oestrogen Receptor-Alpha." *Nature* 424 (6947):389.

Lips, Paul. 1997. "Epidemiology and Predictors of Fractures Associated with Osteoporosis." *American Journal of Medicine* 103(2A):S3–S11.

Little, Randall D., John P. Carulli, Richard G. Del Mastro, Josée Dupuis, Mark Osborne, Colleen Folz, Susan P. Manning, et al. 2002. "A Mutation in the LDL Receptor-Related Protein 5 Gene Results in the Autosomal Dominant High-Bone-Mass Trait." *American Journal of Human Genetics* 70(1):11–19.

Lock, Margaret. 1998. "Anomalous Ageing: Managing the Postmenopausal Body." *Body and Society* 4(1):35–61.

Malina, Robert M., Kathryn H. Brown, and Antonio N. Zavaleta. 1987. "Relative Lower Extremity Length in Mexican American and in American Black and White Youth." *American Journal of Physical Anthropology* 72:89–94.

Marshall, Deborah, Olof Johnell, and Hans Wedel. 1996. "Meta-analysis of How Well Measures of Bone Mineral Density Predict Occurrence of Osteoporotic Fractures." *British Medical Journal* 312(7041):1254–59.

Melton, L. Joseph, III. 1988. "Epidemiology of Fractures." In *Osteoporosis: Etiology, Diagnosis, and Management*, ed. B. Lawrence Riggs and L. Joseph Melton, III, 133–54. New York: Raven.

Melton, L. Joseph, III, Elizabeth J. Atkinson, Michael K. O'Connor, W. Michael O'Fallon, and B. Lawrence Riggs. 1998. "Bone Density and Fracture Risk in Men." *Journal of Bone and Mineral Research* 13(12):1915–23.

Meredith, Howard V. 1978. "Secular Change in Sitting Height and Lower Limb Height of Children, Youths, and Young Adults of Afro-Black, European, and Japanese Ancestry." *Growth* 42(1): 37–41.

Meunier, Pierre J. 1988. "Assessment of Bone Turnover by Histormorphometry." In *Osteoporosis: Etiology, Diagnosis, and Management*, ed. B. Lawrence Riggs and L. Joseph Melton III, 317–32. New York: Raven.

Moggs, Jonathan G., Damian G. Deavall, and George Orphanides. 2003. "Sex Steroids, Angels, and Osteoporosis." *BioEssays* 25(3):195–99.

Molleson, Theya. 1994. "The Eloquent Bones of Abu Hureyra." *Scientific American* 2:70–75.

Mosekilde, Lis. 2000. "Age-Related Changes in Bone Mass, Structure, and Strength—Effects of Loading." *Zeitschrift fur Rheumatologie* 59 (Supplement 1): 1/1–1/9.

Mosley, John R. 2000. "Osteoporosis and Bone Functional Adaptation: Mechanobiological Regulation of Bone Architecture in Growing and Adult Bone." *Journal of Rehabilitation Research and Development* 37(2):189–200.

Nelson, Heidi D., Mark Helfand, Steven H. Woolf, and Janet D. Allan. 2002. "Screening for Postmenopausal Osteoporosis: A Review of the Evidence for the U.S. Preventive Services Task Force." *Annals of Internal Medicine* 137(6): 529–41.

NIH Consensus Statement Online. 2000. 17(1):1–36.

Oyama, Susan. 2000. *Evolution's Eye: A System's View of the Biology-Culture Divide.* Durham, N.C.: Duke University Press.

Pavalko, Fred M., Suzanne M. Norvell, David B. Burr, Charles H. Turner, Randall L. Duncan, and Joseph P. Bidwell. 2003. "A Model for Mechanotransduction in Bone Cells: The Load-Bearing Mechanosomes." *Journal of Cellular Biochemistry* 88(1):104–12.

Peacock, Munro, Charles H. Turner, Michael J. Econs, and Tatiana Foroud. 2002. "Genetics of Osteoporosis." *Endocrine Reviews* 23(3):303–26.

Pead, Matthew J., Timothy M. Skerry, and Lance E. Lanyon. 1988. "Direct Transformation from Quiesence to Bone Formation in the Adult Periosteum Following a Single Brief Period of Bone Loading." *Journal of Bone and Mineral Research* 3(6):647–56.

Perry, Ivan J. 1997. "Fetal Growth and Development: The Role of Nutrition and Other Factors." In Kuh and Ben-Shlomo 1997, 145–68.

Petersen, Alan. 1998. "Sexing the Body: Representations of Sex Differences in *Gray's Anatomy*, 1858 to the Present." *Body and Society* 4(1):1–15.

Rizzoli, R., and J.-P. Bonjour. 1999. "Determinants of Peak Bone Mass and Mechanisms of Bone Loss." *Osteoporosis International* 9 (Supplement 2): S17–S23.

Robinson, T. L., C. Snow-Harter, D. R. Taaffe, D. Gillis, J. Shaw, and R. Marcus. 1995. "Gymnasts Exhibit Higher Bone Mass than Runners despite Similar Prevalence of Amenorrhea and Oligomenorrhea." *Journal of Bone and Mineral Research* 10(1):26–35.

Roush, Wade. 1995. " 'Fat Hormone' Poses Hefty Problem for Journal Embargo." *Science 269* (August 4): 672.

Rubin, Gayle. 1975. "The Traffic in Women: Notes on the 'Political Economy' of Sex." In *Toward an Anthropology of Women*, ed. Rayna R. Reiter, 157–210. New York: Monthly Review Press.

Russett, Cynthia Eagle. 1989. *Sexual Science: The Victorian Construction of Womanhood.* Cambridge, Mass.: Harvard University Press.

Seeman, E. 1997. "Perspective: From Density to Structure: Growing Up and Growing Old on the Surfaces of Bone." *Journal of Bone and Mineral Research* 12(4):509–21.

——. 1998. "Editorial: Growth in Bone Mass and Size—Are Racial and Gender Differences in Bone Mineral Density More Apparent than Real?" *Journal of Clinical Endocrinology and Metabolism* 83(5):1414–19.

——. 1999. "The Structural Basis of Bone Fragility in Men." *Bone* 25(1): 143–47.

Siris, Ethel S., Paul D. Miller, Elizabeth Barrett-Connor, Kenneth G. Faulkner, Lois E. Wehren, Thomas A. Abbott, Marc L. Berger, Arthur C. Santora, and Louis M. Sherwood. 2001. "Identification and Fracture Outcomes of Undiagnosed Low Bone Mineral Density in Postmenopausal Women: Results from the National Osteoporosis Risk Assessment." *Journal of the American Medical Association* 286(22):2815–22.

Skerry, Tim. 2000. "Biomechanical Influences on Skeletal Growth and Development." In *Development, Growth, and Evolution: Implications for the Study of the Hominid Skeleton*, ed. Paul O'Higgins and Martin J. Cohn, 29–39. London: Academic Press.

Skerry, Tim, and Lance E. Lanyon. 1995. "Interruption of Disuse by Short Duration Walking Exercise Does Not Prevent Bone Loss in the Sheep Calcaneus." *Bone* 16(2):269–74.

Sowers, Maryfran. 1996. "Pregnancy and Lactation as Risk Factors for Subsequent Bone Loss and Osteoporosis." *Journal of Bone and Mineral Research* 11(8): 1052–60.

Stepan, Nancy. 1982. *The Idea of Race in Science: Great Britain, 1800–1960.* London: Macmillan.

Taha, Wael, Daisy Chin, Arnold Silverberg, Larisa Lashiker, Naila Khateeb, and Henry Anhalt. 2001. "Reduced Spinal Bone Mineral Density in Adolescents of an Ultra-Orthodox Jewish Community in Brooklyn." *Pediatrics* 107(5): c79–c85.

Takeda, Shu, Florent Elefteriou, Regis Levasseur, Xiuyun Liu, Liping Zhao, Keith L. Parker, Dawna Armstrong, Patricia Ducy, and Gerard Karsenty. 2002. "Leptin Regulates Bone Formation via the Sympathetic Nervous System." *Cell* 111(3):305–17.

Tanner, J. M., T. Hayashi, M. A. Preece, and N. Cameron. 1982. "Increase in Length of Leg Relative to Trunk in Japanese Children and Adults from 1957 to 1977: Comparison with British and Japanese Children." *Annals of Human Biology* 9(5):411–23.

Tanouye, Elyse. 1995. "Merck's Osteoporosis Warnings Pave the Way for Its New Drug." *Wall Street Journal*, June 28, B1, B4.

Thornhill, Randy, and Craig T. Palmer. 2001. *A Natural History of Rape.* Cambridge, Mass.: MIT Press.

Travis, Cheryl Brown, ed. 2003. *Evolution, Gender, and Rape.* Cambridge, Mass.: MIT Press.

Verbrugge, Martha H. 1997. "Recreating the Body: Women's Physical Education and the Science of Sex Differences in America, 1900–1940." *Bulletin of the History of Medicine* 71(2):273–304.

Walsh, Timothy B., and Michael J. Devlin. 1998. "Eating Disorders: Progress and Problems." *Science* 280(5638):1387–90.

Wang, May-Choo, Patricia B. Crawford, Mark Hudes, Marta Van Loan, Kirstin Siemering, and Laura K. Bachrach. 2003. "Diet in Midpuberty and Sedentary Activity in Prepuberty Predict Peak Bone Mass." *American Journal of Clinical Nutrition* 77(2):495–503.

Wizemann, Theresa M., and Mary-Lou Pardue, eds. 2001. *Exploring the Biological Contributions to Human Health: Does Sex Matter?* Washington, D.C.: National Academy Press.

Worthman, Carol M. 2002. "Endocrine Pathways in Differential Well-Being across the Life Course." In Kuh and Hardy 2002, 197–216.

Wright, Rita P., ed. 1996. *Gender and Archeology.* Philadelphia: University of Pennsylvania Press.

Zanchetta, J. R., H. Plotkin, and M. L. Alvarez Filgueira. 1995. "Bone Mass in Children: Normative Values for the 2–20-Year-Old Population." *Bone* 16 (Supplement 4): S393–S399.

Zelzer, Elazar, and Bjorn R. Olsen. 2003. "The Genetic Basis for Skeletal Diseases." *Nature* 423(6937):343–48.

Zuk, Marlene. 2002. *Sexual Selections: What We Can and Can't Learn about Sex from Animals.* Berkeley: University of California Press.

Section IV

The Next Generation

Bringing Feminist Perspectives into Science and Technology Studies

In Section III, we argued that scientists and engineers hold preconceptions about gender differences that have influenced their choice of subjects, research methods, standards of evidence, interpretation of data, and conclusions about the natural and physical world. When these preconceptions are directed at research on human social life, they have particularly regressive implications, as Ruth Bleier pointed out in her critique of sociobiology.

Nineteenth-century ideas about the "natural" limitations of women's intellectual capacities and biological destiny (i.e., to bear children and care for them) have been revived in the late twentieth century under the guise of evolutionary psychology. This latest incarnation of biological determinism is an attempt to explain human behaviors as "adaptations" or, in other words, as traits that are controlled by our genetic makeup and the natural selection of our genes in response to problems our ancestors faced in the distant (e.g., stone age) past. Evolutionary psychology is a controversial field, at best, and is subject to many of the same critiques as sociobiology, such as its reductionist principles and use of data from nonhuman species. In addition, critics point out (i) the tenets of evolutionary psychology are not falsifiable and cannot therefore be subjected to any rigorous testing, (ii) our knowledge about the distant past is very limited and the evolutionary "context" in which human behaviors would have evolved is not known, and (iii) the theories of evolutionary psychology lack cross-cultural support and may only be relevant to a single culture (i.e., Western, white, European/American culture).[1] Nevertheless, the tenets of sociobiology and evolutionary psychology demonstrate two principles. One is that cultural ideas about women's incapacities and men's capabilities have great staying power, despite efforts to prove them inaccurate. The other is that women of intellectual achievement make their way against a backdrop of assumptions about how exceptional they are—the exception that proves the rule that women do not belong in science.

When feminist critiques of science emerged in the 1970s, some women in science expressed fears that the feminist writings and critiques of science, in the effort to revalue women's accomplishments, implied that women scientists necessarily did science "differently," and "different" was not what they wanted to be.[2] Within a culture of male privilege and dominance, "women's science" would always be viewed as inferior science, and women in science did not want their work to be considered inferior to work done by their male colleagues. These women maintained that gender was irrelevant to the practice of science, embracing instead a belief that the scientific method removes or acts as a barrier against any bias that may arise from one's gender. However, as uncovered by the many feminist critiques of science described here and elsewhere, this belief appears to be unduly optimistic. As Evelyn Fox Keller puts it, their [the women scientists'] confidence "in the standards of scientific rigor was excessive."[3] Not only does the scientific method fail to eradicate social biases, it is sometimes used in the service of supporting them.

But is this inability to eliminate bias an intrinsic flaw in the scientific method or in those who use it? For an answer, some scholars point to the origins of modern Western science in the misogynist enclaves of Christian monasteries during the High Middle Ages, arguing that the touchstone philosophies and methods of science are deeply entangled with commitments to the superiority of men over women.[4] In this view, the scientific method is inescapably a product of its social, historical, and economic contexts, and since those reflected male-centered perspectives, so too does the scientific method. Thus Nature is imagined as a female, to be conquered and dominated by the (male) scientist.[5] Others point to the specific psychodynamics involved in practicing objectivity, and the ways that the scientific method distinguishes scientific knowledge (and scientists) from other kinds of knowledge (and scholars).[6] For example, one significant tenet of the scientific method is that the researcher is distant and separate from the subject of research, so that one may be an objective observer, presumably without influencing or being influenced by the observed phenomenon. This cognitive distance is understood to promote dispassionate, unbiased observation and control over experimental conditions; but it is also a form of rationality that is more often encouraged in men than women.[7] In addition, the very act of attempting to be distant or separate may itself be a subjective act or an act that introduces an unacknowledged bias, putting into jeopardy the assumption that distance promotes objectivity.

The pursuit of scientific knowledge conventionally requires the practice of both the scientific method and objectivity. But because both are implicated in the marginalization of women in science, the question arises, Can there be a "feminist science"? Can there be a way to retain the goals of scientific research that focuses on using systematic methods to develop reproducible, reliable knowledge about the natural world without trying to "dominate" nature and perpetuate notions of gender inequality?

Without question, there are examples of well-regarded scientific work that did not embrace this method. Barbara McClintock, a geneticist and Nobel laureate, was characterized by Evelyn Fox Keller in the biography *A Feeling for the Organism* as having a nontraditional approach to scientific study.[8] She thought it was important to work with the whole organism, to know it well enough that she did not miss the subtle changes that occurred between generations in her corn crops. She also did not throw away data that challenged prevailing views in her field of research. Though neither McClintock nor Keller may have thought of this as feminist science, it was certainly not "science as usual." Nevertheless, it was extraordinary science, and though it took some time for her work to be appreciated, Barbara McClintock's contributions, ultimately recognized with a Nobel Prize, will stand as some of the most remarkable work in the field of genetics for many years to come.

Some feminist scholars in the social sciences have embraced the need for a "feminist science" both as a matter of improving research and as an ethical imperative.[9] In their 1999 paper, Paige Smith and her colleagues described the research that they have been conducting about violence against women.[10] Specifically they wanted to find a better method for characterizing, quantifying, and understanding the battering of women. Their idea was to begin by asking women in shelters what they thought battering was. Because they began with the subjects and attended to what their subjects thought, they were able to recognize and describe battering in an entirely new way.

Prior to the work of Smith and her colleagues, descriptions and surveys of battering had a focus on acute events covering a narrow time frame, such as a single assault taking place one night, and there was an absence of any discourse on gender and on the specific personalities involved. Because the focus was so narrow, a significant percentage of battered women did not even identify themselves as battered when surveyed. Based on their interviews and documentation of the experiences of battered women, Smith and her colleagues developed a new framework by which to assess battering. This framework, called the Women's Experience with Battering (WEB) Scale, identifies ten items that are unique and consistent among women who have been victims of battering. The list (see below) is striking because of its focus on the people involved, as well as on power dynamics. It also reveals the depth to which an abusive relationship can be ongoing and have both lasting and cumulative effects. The importance of research like this cannot be understated, especially because of its focus on women and its clear goal of improving women's lives. As Smith and her colleagues say themselves, "This article (is) one way of using both qualitative methods and a feminist perspective to inform quantitative methods. A critical feminist concept that informed our research process was that battered women are the experts of their own lives."

The Women's Experience with Battering (WEB) Scale Items ask participants to respond on a scale from "strongly agree" to "strongly disagree." The items are as follows:

1. He makes me feel unsafe even in my own home.
2. I feel ashamed of the things he does to me.
3. I try not to rock the boat because I am afraid of what he might do.
4. I feel like I am programmed to react a certain way to him.
5. I feel like he keeps me prisoner.
6. He makes me feel like I have no control over my life, no power, no protection.
7. I hide the truth from others because I am afraid not to.
8. I feel owned and controlled by him.
9. He can scare me without laying a hand on me.
10. He has a look that goes straight through me and terrifies me.

The insights of Smith and her colleagues resulted in the creation of new surveys and questionnaires which resonate with battered women and provide the researcher with a much richer and more accurate set of information. This "woman-centered" approach takes into account the subjective experiences of the battered women, providing a space in which they are valued and heard. It requires that the researchers see the women as active participants in the development of the research about them. They are not objects to be observed by, and used for the benefit of, a distant researcher. Again, this is not science as usual, but it is certainly systematic and reproducible research that has as a goal the improvement of women's lives. One could argue, as such, it is a good example of feminist science.

Sandra Harding has argued that there is no unique feminist method in the social sciences, but that feminist research nonetheless shares four characteristics. (1) The origins of the questions asked emerge from woman-centered concerns. (2) The purposes of the inquiry stem from the need to foster new and more acccurate knowledge for the benefit of women and society. (3) The hypotheses and evidence are not transparently unbiased and apolitical. (4) The relationship between the researcher and the subject of study is a mutual and reflexive one.[11] Feminist research in the sciences and in engineering could involve all of these characteristics, as well. However, when the subject of study is the physical rather than the social world, new issues emerge. The sociological insight that scientific knowledge, indeed all knowledge, is socially constructed poses a major challenge to a basic assumption of science—that scientists' explanations and interpretations of natural phenomenon mirror the natural (actual, absolute) world. For many scientists, the notion that their descriptions of the natural world are subjective, even biased, models of those phenomena is counter to their professional training. It also poses a challenge to those who would bring political issues into science, since their professional communities are likely to view their political commitments as a threat to objectivity.[12]

It is an underappreciated fact that there are many feminists in science and engineering, though their philosophical positions on issues of gender in science can vary greatly—they are far from a unilateral or monolithic group. Within a commitment to gender equity, there are two camps of feminist practice, one focusing on science and one focusing on society. Some women and men are committed to gender equity in education and employment in scientific and engineering professions, but they dismiss the notion that scientific research is itself tainted by gender biases. Others feel that the meritocracy in science and engineering rewards women and men equally, most of the time, and that gender differences in professional outcomes are a result of the different choices that women and men make about their work and family responsibilities; but at the same time, they support and participate in activist initiatives for the society as a whole, such as the National Abortion Rights Action League or the National Organization for Women.

Empiricist feminists take a position that includes an analysis of the contents of science as well as the equity positions of the first two. Empiricist feminists acknowledge that gender bias occurs in the formulation of questions and the interpretation of data, but they embrace the scientific method as a corrective because they see gender bias as producing "bad science."[13] A fourth stance is taken by what we term "constructivist feminists," who accept the premise that all knowledge, including scientific knowledge, is socially constructed. For this group, the scientific method cannot eliminate bias since it is the norms and values of the scientific community that determine what is considered adequate or inadequate research. If that community sustains and promotes gender biases (however unwittingly), then the community values and research practices intertwine to sustain gender-biased science. Since social biases are unavoidable in research, scientific practice should include a recognition and evaluation of their impact as part of the research method.

One of the by-products of the concept of objectivity is that it is also employed as a way to structure the relationships among the people who practice science. It promotes a cognitive slippage that can conflate the evaluation of data with the evaluation of the researcher who produced the data. In academia, for instance, most scientists assume that the strengths and weaknesses of a data set can be determined by a set of standards that provide objective measures, just as the evaluation of a researcher's worth is measured by a set of objective standards (a "meritocracy"). The standards in both cases are established by community consensus, sometimes in unspoken ways. When the topic is an experiment or an interpretation, this seems unproblematic. When it drives the structuring of social relations within science, it means that only readily measured qualities, such as number of papers published or grants funded, are included in the evaluation. Other more subjective factors, such as the complexity of the research questions being asked, the thoroughness of the research, the establishment of successful collaborations, are rendered invisible.

Since no science is undertaken without organized social activity, a whole host of arrangements is influenced by commitments to objectivity. This includes the daily interactions among colleagues, the way that individual laboratories are structured and managed, the formation of nonoverlapping and largely noninteracting fields of research, and the unspoken but well-known ranking of these fields in terms of value and importance. Individuals are then ranked according to academic pedigree within their fields. In fact, one can view the entire process of scientific training as an exercise in learning about, or being indoctrinated in, the system that maintains hierarchy, in status and rank, in the laboratory, in the department, and in the institution vis-à-vis claims to so-called objective measures of worth.

The commitment to hierarchy as an organizing principle in professional life also has an influence on the areas of inquiry ranked as "important." This in turn influences what students learn about the natural world. From a review of research conducted in the late 1980s and 1990s, Bonnie Spanier argues, for instance, that an ideological commitment to hierarchy in science led to an emphasis on techniques from recombinant DNA technology that privilege genetics (and reductionism) over more holistic approaches to biology. Students in molecular biology are taught to understand the different subfields as hierarchically related, with molecular genetics at the top, supported by necessary but lesser fields and supplanting the wider focus of biology altogether. As a result, courses reduce life to the reproduction of genetic information, rather than the consequence of the complex interactions among physiochemical reactions and organisms and environments. The gene carries the master code for life, and all else follows and is directed by it.[14] We are left to wonder what science may look like if hierarchies could be challenged, rather than rendered invisible by their pervasiveness. How would the observations about DNA and genes be interpreted without the hierarchy embodied by the phrase "master code"? Would scientists more easily recognize the limits of genetic explanations? Does the imposition of hierarchy compromise the pursuit of scientific knowledge, and could shedding it create "better" knowledge?

If we take as given that there is an influence of practices on content, then, answering the question "Can there be a feminist science?" entails considering what elements of science and technology are available for scrutiny and revision, and how a feminist perspective might alter the activity of researchers and the nature (and interpretation) of their research results. What are the possibilities for new organizational forms of science and engineering—in the workplace, in the evaluation and assessment processes, in educational arrangements, in funding practices, and in setting priorities for research directions?

We introduce this section with excerpts from Evelyn Fox Keller's classic examination of the psychosocial dynamics that underwrite an association between masculinity and objectivity and thereby preclude the participation of women *qua* women in science and engineering. Her essay questions whether, given the importance of gender in our self-concepts, social arrangements, and professional cultures, science can be a world where "the matter of gender drops away." As Evelyn Fox Keller points out, these shared practices take place within the context of a shared language. "Sharing a language," she says, "means sharing a conceptual universe."[15] The language that scientists use thus is not transparent—it does not function simply as a passive vehicle through which scientists gather and relate knowledge of the natural world. Rather, language both represents and creates the shared meanings that a scientific community attaches to the concepts, metaphors, and images that scientists use. Specialized vocabularies develop in fields as a consequence of this process, but scientists (and all specialists) use concepts, metaphors, and images from the culture at large as well, as Carol Cohn's and Banu Subramanian's essays in Section II so persuasively document. The interpretation of any set of data thus necessarily involves tapping into the reservoir of shared meanings. A reliance on shared meanings, without which we would be unable to communicate, has important implications for understanding how science operates in our culture. As Dale Spender has observed,

> Given that language is such an influential force in shaping our world, it is obvious that those who have the power to make the symbols and their meanings are in a privileged and highly advantageous position. . . . They have, at least, the potential to order the world to suit their own ends, the potential to construct a language, a reality, a body of knowledge in which they are the central figures, the potential to legitimate their own primacy and to create a system of beliefs which is beyond challenge (so that their superiority is "natural" and "objectively" tested).[16]

Understanding the language of science as socially embedded has provoked scholars to examine the degree to which scientific research and interpretation, like most other arenas of social life, are imbued with gender biases. What they have found is that concepts, metaphors, and images that rely on beliefs about women and men—and the differences between them—have a profound influence on definitions and interpretations of "natural" phenomenon.

In some fields, where much of the early research was conducted by a fairly homogenous group of researchers (middle- to upper-class, Western and/or European white men, for example), the influence of new perspectives, in particular from women or people of color, has made clear changes to the dominant paradigms of those fields. The field of primatology is often cited as having been greatly influenced (and improved) by the presence of women. Linda Fedigan, in her article "The Paradox of Feminist Primatology," concludes "that primate studies can be called a feminist science, or that, at the very least, it has been significantly changed by the women's movement and the feminist critique of science." She notes that since the 1970s, primatologists have increasingly used so-called feminist approaches in their research, such as taking the female point of view, moving away from dualisms and reductionisms, welcoming and including researchers from formerly marginalized groups, and using science for humanitarian applications. The "paradox" in the title of this article is that primatologists themselves do not generally embrace the labeling of their field as feminist. In fact, many choose to distance themselves from this characterization. Fedigan does not believe this is merely a rejection of the label, but rather a belief among the practitioners that the basis for their research arises from "vastly different underlying assumptions than those of feminists." She also postulates that primatologists may be concerned about issues such as the undervaluing of research perceived as "feminine," a threat to the objectivity of science if there are perceived "political" concerns, and a general belief that the conduct of science is "free from sociocultural influences." Despite this apparent conflict, it is still nothing short of remarkable that in a relatively short time, primatology has responded, intentionally or otherwise, to the feminist critique of science. In this regard, it is substantially set apart from other disciplines where change, if it has occurred, is much slower. Fedigan documents in this article specific examples of how primatology as a science and how primatologists as practioners are gender sensitive and gender inclusive, despite their denial of a feminist underpinning. This documentation is structured around the eight tools of gender analysis suggested by Londa Scheibinger in her book *Has Feminism Changed Science?*[17] These themes resonate with those described throughout this section (and others), such as language use, what "counts" as science, funding priorities, institutional structures, and gender dynamics. Fedigan also describes how some primatologists, notably Jane Goodall, Dian Fossey, and Birute Galdikas, rejected the notion that a scientist needs to be distant and separate from the object of research, and instead put aside the call for "pure" research to protect the species they study.

Rachel Maines, in her article "Socially Camouflaged Technologies: The Case of the Electromechanical Vibrator," takes us back to a different social and political context to show, in historical relief, how a distorted science can affect women's lives. She examines the influence of cultural beliefs about female behavior and sexuality on the medical treatment of women for sexual desire—what the terminology of the day called "hysteria." The development of the electromechanical vibrator was in response to the needs of the medical community to reduce the cost and labor requirements of "treating" female disorders by physician-assisted masturbation. It is disquieting enough that there was a disease called "hysteria" and that this was the "cure." But that a device was developed to mechanize the treatment is even more remarkable. Indeed, this is a particularly striking example of how a new technology, ostensibly developed for the care of women, was *not* developed in response to the needs or interests of women in sexual satisfaction, nor was its development "woman-centered" in any meaningful way. Rather, this is an instance where a medical "tool" was developed in order to relieve the male-dominated medical community of a task that was at once onerous and ethically questionable, especially for "God-fearing" physicians. The sexual desire of women was "medicalized" and treated as if it were a disease. As Maines explains in the article, because of the social taboos associated with masturbation in general, and with female masturbation and sexuality in particular, physician-assisted masturbation was camouflaged, as were the associated technologies, by the authority and respectability of the medical profession. As a result, women were absent from the discourse, even as they were the topic of discussion. This theme continues in Anne Balsamo's article "On the Cutting Edge: Cosmetic Surgery and New Imaging Technologies." The practitioners of cosmetic surgery are primarily male, while the subjects operated on are primarily female. The discourse of both users and practitioners of cosmetic surgery are informed by cultural understandings of male and female bodies, whether this relates to the reason for electing such surgeries or to the feature that is to be reconstructed. Women are identified as most in need of "correction," to come closer to the prevailing "ideal." In the relatively rare cases where men are the subjects targeted, it is not to fix something that is defective, but rather to improve or enhance their appearance—not for appearance's sake, but rather to remain "competitive" in the boardroom and the bedroom.

Because ideas about gender are so dominant in the discourse and practice of cosmetic surgery, the same feature or characteristic can be viewed as an asset or a liability, depending on the gender of the body on which it appears. As Balsamo points out, the leathery, tanned faces of male farmers and construction workers connote accomplishment, experience, and purpose, but the same trait on a woman requires diagnosis and repair. Like other medical "tools" used on the body, and in particular on women's bodies, cosmetic surgery reinforces unyielding cultural stereotypes: women are valued based on their appearance, their ability to meet a feminine "ideal" set forth by pervasive media images created largely by Western, white male standards; men are valued on their individual substance, contribution, and achievement. This is not to say that there aren't unreasonable standards set for men. One could interpret the increasing use and abuse of "performance enhancers" among male athletes (professional, amateur, and recreational athletes alike), as well as the targeted advertising of erectile dysfunction medications to young, healthy men, as an indication of such standards.[18] However, even in this, the pressure borne on men is to achieve, and the standards of achievement are set by other men.

Notes

1 There are many critiques of evolutionary psychology available, but because Ruth Bleier's critiques regarding sociobiology still stand and are so clearly articulated, we are referencing some of the critiques here, rather than replacing Bleier's article in this book. For a critique from the primary literature, see L. Gannon, "A Critique of Evolutionary Psychology," *Psychology, Evolution, and Gender* 4, no. 2 (August 2002): 173–218. For an in-depth critique, see D. J. Buller, *Adapting Minds: Evolutionary Psychology and the Persistent Quest for Human Nature*, Cambridge, MA: MIT Press, 2005.

2 H. Longino and E. Hammonds, "Conflicts and Tensions in the Feminist Study of Gender and Science," in *Conflicts in Feminism*, ed. M. Hirsch and E. F. Keller, pp. 164–83. New York: Routledge, 1990.

3 For an elaboration of this debate within the history of women in science, see E. F. Keller, "The Wo/Man Scientist: Issues of Sex and Gender in the Pursuit of Science," in *The Outer Circle: Women in the Scientific Community*, ed. H. Zuckerman, J. Cole, and J. Bruer. New York: Norton, 1991.

4 D. Noble, *A World without Women: The Christian Clerical Culture of Western Science*, New York: Oxford University Press, 1992. See also L. Schiebinger, *The Mind Has No Sex? Women in the Origins of Modern Science*, Boston: Harvard University Press, 1989.

5 G. Lloyd, "Reason, Science, and the Domination of Matter," in *Feminism and Science*, ed. E. F. Keller and H. Longino. New York: Oxford University Press, 1996.

6 E. F. Keller, "Dynamic Autonomy: Objects as Subjects," in her *Reflections on Gender and Science*, New Haven, CN: Yale University Press, 1985; S. Bordo, "The Cartesian Masculinization of Thought," *Signs: Journal of Women in Culture and Society* 11, no. 3 (1986): 439–56.

7 For a discussion of object relations theory and gender in relation to scientific objectivity, see E. F. Keller, "Feminism and Science," *Signs: Journal of Women in Culture and Society* 7, no. 3 (1982): 589–602. See also N. Chodorow, *The Reproduction of Mothering*, Berkeley, CA: University of California Press, 1978; and N. Chodorow, "Gender as a Personal and Cultural Construction," *Signs: Journal of Women in Culture and Society* 20, no. 3 (1995): 516–44. For a different view that focuses on women's psychological development as driven by connections to others, see J. Jordan, M. Walker, and L. M. Hartling, eds., *The Complexity of Connection*, New York: Guilford Press, 2004.

8 E. Fox Keller, *A Feeling for the Organism: The Life and Work of Barbara McClintock*. New York: W.H. Freeman, 1983.

9 See, e.g., S. N. Hesse-Biber, ed., *Handbook of Feminist Research: Theory and Praxis*, Thousand Oaks, CA: Sage Publications, 2007. Two other standard texts are M. M. Fonow and J. A. Cook, *Beyond Methodology: Feminist Scholarship as Lived Research*, Bloomington: Indiana University Press, 1991; and Shulamit Reinharz, *Feminist Methods in Social Research*, New York: Oxford University Press, 1992.

10 P. H. Smith, J. B. Smith, and J. A. L. Earp, "Beyond the Measurement Trap: A Reconstructed Conceptualization and Measurement of Woman Battering," *Psychology of Women Quarterly* 23 (1999): 177–93, esp. pp. 189–90.

11 S. Harding and M. Hintikka, eds., *Discovering Reality: Feminist Perspectives on Epistemology, Metaphysics, Methodology, and Philosophy of Science*. Dordrecht, Holland: D. Reidel, 1983.
12 S. Harding, *Who's Science, Who's Knowledge? Thinking from Women's Lives*. Ithaca, NY: Cornell University Press, 1991, p. 79.
13 S. Harding, "The Instability of the Analytical Categories of Feminist Theory," *Signs: Journal of Women in Culture and Society* 11, no. 4 (1986): 645–64. Feminist theory includes many and varied approaches, priorities, and themes. To explore the development of feminist theory in historical context, see W. K. Kolmar and F. Bartkowski, *Feminist Theory: A Reader*, New York: McGraw-Hill, 2005. For an overview from a psychological perspective, see H. Lips, *Sex and Gender*, 6th ed., New York: McGraw-Hill, 2008. For an overview from a multicultural perspective, see G. Kirk and M. Okazawa-Rey, *Women's Lives: Multicultural Perspectives*, 3rd ed., New York: McGraw-Hill, 2004.
14 B. Spanier, *Im/Partial Science: Gender Ideology in Molecular Biology*. Bloomington: Indiana University Press, 1995.
15 E. Fox Keller, *Secrets of Life: Secrets of Death*. New York: Routledge, 1992, p. 28.
16 D. Spender, *Man Made Language*. New York: Routledge and Kegan Paul, 1980, p. 142.
17 L. Schiebinger, *Has Feminism Changed Science?* Boston: Harvard University Press, 1999.
18 National Institute on Drug Abuse, Research Report: Anabolic Steroid Abuse, NIDA NIH Publication No. 00–3721, September 2006; "Patterns of Use among Commercially Insured Adults in the United States: 1998–2002," *International Journal of Impotence Research* 16 (2004): 313–18.

17

Gender and Science

An Update

Evelyn Fox Keller

THE MEANING OF GENDER

Schemes for classifying human beings are necessarily multiple and highly variable. Different cultures identify and privilege different criteria in sorting people of their own and other cultures into groups: They may stress size, age, color, occupation, wealth, sanctity, wisdom, or a host of other demarcators. All cultures, however, sort a significant fraction of the human beings that inhabit that culture by sex. What are taken to be the principal indicators of sexual difference as well as the particular importance attributed to this difference undoubtedly vary, but, for fairly obvious reasons, people everywhere engage in the basic act of distinguishing people they call male from those they call female. For the most part, they even agree about who gets called what. Give or take a few marginal cases, these basic acts of categorization do exhibit conspicuous cross-cultural consensus: Different cultures will sort any given collection of adult human beings of reproductive age into the same two groups. For this reason, we can say that there is at least a minimal sense of the term "sex" that denotes categories given to us by nature.[1] One might even say that the universal importance of the reproductive consequences of sexual difference gives rise to as universal a preoccupation with the meaning of this difference.

But for all the cross-cultural consensus we may find around such a minimalist classification, we find equally remarkable cultural variability in what people have made and continue to make of this demarcation; in the significance to which they attribute it; in the properties it connotes; in the role it plays in ordering the human world beyond the immediate spheres of biological reproduction; even in the role it plays in ordering the nonhuman world. It was to underscore this cultural variability that American feminists of the 1970s introduced the distinction between sex and gender, assigning the term "gender" to the meanings of masculinity and femininity that a given culture attaches to the categories of male and female.[2]

The initial intent behind this distinction was to highlight the importance of non-biological (that is, social and cultural) factors shaping the development of adult men and women, to emphasize the truth of Simone de Beauvoir's famous dictum, "Women are not born, rather they are made." Its function was to shift attention away from the time-honored and perhaps even ubiquitous question of the meaning of sexual difference (that is, the meanings of masculine and feminine), *to* the question of how such meanings are constructed. In Donna Haraway's words, "Gender is a concept developed to contest the naturalization of sexual difference" (1991:131).

Very quickly, however, feminists came to see, and, as quickly, began to exploit, the considerably larger range of analytic functions that the multipotent category of gender is able to serve. From an original focus on gender as a cultural norm guiding the psychosocial development of individual men and women, the attention of feminists soon turned to gender as a cultural structure organizing social (and sexual) relations between men and women,[3] and finally, to gender as the basis of a sexual division of cognitive and emotional labor that brackets women, their work, and the values associated with that work from culturally normative delineations of categories intended as "human"—objectivity, morality, citizenship, power, often even "human nature" itself. From this perspective, gender and gender norms come to be seen as silent organizers of the mental and discursive maps of the social and natural worlds we simultaneously inhabit and construct—*even of those worlds that women never enter*. This I call the symbolic work of gender; it remains silent precisely to the extent that norms associated with masculine culture are taken as universal.

The fact that it took the efforts of contemporary feminism to bring this symbolic work of gender into recognizable view is in itself noteworthy. In these efforts, the dual focus on women as subjects and on gender as a cultural construct was crucial. Analysis of the relevance of gender structures in conventionally male worlds only makes sense once we recognize gender not only as a bimodal term, applying symmetrically to men *and* women (that is, once we see that men too are gendered, that men too are made rather than born), but also as denoting social rather than natural kinds. Until we can begin to envisage the possibility of alternative arrangements, the symbolic work of gender remains both silent and inaccessible. And as long as gender is thought to pertain only to women, any question about its role can only be understood as a question about the presence or absence of biologically female persons.

This double shift in perception—first, from sex to gender, and second, from the force of gender in shaping the development of men and women to its force in delineating the cultural maps of the social and natural worlds these adults inhabit—constitutes the hallmark of contemporary feminist theory. Beginning in the mid-1970s, feminist historians, literary critics, sociologists, political scientists, psychologists, philosophers, and soon, natural scientists as well, sought to supplement earlier feminist analyses of the contribution, treatment, and representation of men and women in these various fields with an enlarged analysis of the ways in which privately held and publicly shared ideas about gender have shaped the underlying assumptions and operant categories in the intellectual history of each of these fields. Put simply, contemporary feminist theory might be described as "a form of attention, a lens that brings into focus a particular question: What does it mean to describe one aspect of human experience as 'male' and another as 'female'? How do such labels affect the ways in which we structure the world around us, assign value to its different domains, and in turn, acculturate and value actual men and women?" (Keller 1985:6).

With such questions as these, feminist scholars launched an intensive investigation of the traces of gender labels evident in many of the fundamental assumptions underlying the traditional academic disciplines. Their earliest efforts were confined to the humanities and social sciences, but by the late 1970s, the lens of feminist inquiry had extended to the natural sciences as well. Under particular scrutiny came those assumptions that posited a dichotomous (and hierarchical) structure tacitly modeled on the prior assumption of a dichotomous (and hierarchical) relation between male and female—for example, public/private; political/personal; reason/feeling; justice/care; objective/subjective; power/love; and so on. The object of this endeavor was not to reverse the conventional ordering of these relations, but to undermine the dichotomies themselves—to expose to radical critique a worldview that deploys categories of gender to rend the fabric of human life and thought along a multiplicity of mutually sanctioning, mutually supportive, and mutually defining binary oppositions.

FEMINISM AND SCIENCE

But if the inclusion of the natural sciences under this broad analytic net posed special opportunities, it also posed special difficulties, and special dangers, each of which requires special recognition. On the one hand, the presence of gender markings in the root categories of the natural sciences and their use in the hierarchical ordering of such categories (for example, mind and nature; reason and feeling; objective and subjective) is, if anything, more conspicuous than in the humanities and social sciences. At the same time, the central claim of the natural sciences is precisely to a methodology that transcends human particularity, that bears no imprint of individual or collective authorship. To signal this dilemma, I began my first inquiry into the relations between gender and science (Keller 1978) with a quote from George Simmel, written more than sixty years ago:

> The requirements of . . . correctness in practical judgments and objectivity in theoretical knowledge . . . belong as it were in their form and their claims to humanity in general, but in their actual historical configuration they are masculine throughout. Supposing that we describe these things, viewed as absolute ideas, by the single word "objective," we then find that in the history of our race the equation objective = masculine is a valid one (cited in Keller 1978:409).

Simmel's conclusion, while surely on the mark as a description of a cultural history, alerts us to the special danger that awaits a feminist critique of the natural sciences. Indeed, Simmel himself appears to have fallen into the very trap that we are seeking to expose: In neglecting to specify the space in which he claims "validity" for this equation as a *cultural or even ideological space*, his wording invites the reading of this space as a biological one. Indeed, by referring to its history as a "history of our race" without specifying "our race" as late-modern, northern European, he tacitly elides the existence of other cultural histories (as well as other "races") and invites the same conclusion that this cultural history has sought to establish: namely, that "objectivity" is simultaneously a universal value and a privileged possession of the male of the species.

The necessary starting point for a feminist critique of the natural sciences is thus the reframing of this equation as a conundrum: How is it that the scientific mind can be *seen* at one and the same time as both male and disembodied? How is it that thinking "objectively," that is, thinking that is defined as self-detached, impersonal, and transcendent, is also understood as "thinking like a man"? From the vantage point of our newly "enlightened" perceptions of gender, we might be tempted to say that the equation "objective = masculine," harmful though it (like that other equation woman = nature) may have been for aspiring women scientists in the past, was simply a descriptive mistake, reflecting misguided views of women. But what about the views of "objectivity" (or "nature") that such an equation necessarily also reflected (or inspired)? What difference—for science, now, rather than for women—might such an equation have made? Or, more generally, what sort of work in the actual production of science has been accomplished by the association of gender with virtually all of the root categories of modern science over the three hundred odd years in which such associations prevailed? How have these associations helped to shape the criteria for "good" science? For distinguishing the values deemed "scientific" from those deemed "unscientific"? In short, what particular cultural norms and values has the language of gender carried into science, and how have these norms and values contributed to its shape and growth?

These, then, are some of the questions that feminist theory brings to the study of science, and that feminist historians and philosophers of science have been trying to answer over the last fifteen years. But, for reasons I have already briefly indicated, they are questions that are strikingly difficult to hold in clear focus (to keep distinct, for example, from questions about the presence or absence of women scientists). For many working scientists, they seem not even to "make sense."

One might suppose, for example, that once such questions were properly posed (that is, cleansed of any implication about the real abilities of actual women), they would have a special urgency for all practicing scientists who are also women. But experience suggests otherwise; even my own experience suggests otherwise. Despite repeated attempts at clarification, many scientists (especially, women scientists) persist in misreading the force that feminists attribute to gender ideology as a force being attributed to sex, that is, to the claim that women, for biological reasons, would do a different kind of science. The net effect is that, where some of us see a liberating potential (both for women *and* for science) in exhibiting the historical role of gender in science, these scientists often see only a reactionary potential, fearing its use to support the exclusion of women from science.[4]

The reasons for the divergence in perception between feminist critics and women scientists are deep and complex. Though undoubtedly fueled by political concerns, they rest finally neither on vocabulary, nor on logic, nor even on empirical evidence. Rather, they reflect a fundamental difference in mind-set between feminist critics and working scientists—a difference so radical that a "feminist scientist" appears today as much a contradiction in terms as a "woman scientist" once did.[5] . . .

THE MEANING OF SCIENCE

Although people everywhere, throughout history, have needed, desired, and sought reliable knowledge of the world around them, only certain forms of knowledge and certain procedures for acquiring such knowledge have come to count under the general rubric that we, in the late twentieth century, designate as science. Just as "masculine" and "feminine" are categories defined by a culture, and not by biological necessity, so too, "science" is the name we give to a set of practices and a body of knowledge delineated by a community. Even now, in part because of the great variety of practices that the label "science" continues to subsume, the term defies precise definition, obliging us to remain content with a conventional definition—as that which those people we call scientists do.

What has compelled recognition of the conventional (and hence social) character of modern science is the evidence provided over the last three decades by historians, philosophers, and sociologists of science who have undertaken close examination of what it is that those people we call (or have called) scientists actually do (or have done).[6] Careful attention to what questions get asked, of how research programs come to be legitimated and supported, of how theoretical disputes are resolved, of "how experiments end" reveals the working of cultural and social norms at every stage.[7] Consensus is commonly achieved, but it is rarely compelled by the forces of logic and evidence alone. On every level, choices are (must be) made that are social *even as* they are cognitive and technical. The direct implication is that not only different collections of facts, different focal points of scientific attention, but also different conceptions of explanation and proof, different representations of reality, different criteria of success, are both possible and consistent with what we call science.

But if such observations have come to seem obvious to many observers of science, they continue to seem largely absurd to the men and women actually engaged in the production of science. In order to see how cultural norms and values can, indeed have, helped define the success and shape the growth of science, it is necessary to understand how language embodies and enforces such norms and values. This need far exceeds the concerns of feminism, and the questions it gives rise to have become critical for anyone currently working in the history, philosophy, or sociology of science. That it continues to elude most working scientists is precisely a consequence of the fact that their worldviews not only lack but actually preclude recognition of the force of language on what they, in their day-to-day activity as scientists, think and do. And this, I suggest, follows as much from the nature of their activity as it does from scientific ideology.

LANGUAGE AND THE DOING OF SCIENCE[8]

The reality is that the "doing" of science is, at its best, a gripping and fully absorbing activity—so much so that it is difficult for anyone so engaged to step outside the demands of the particular problems under investigation to reflect on the assumptions underlying that investigation, much less on the language in which such assumptions can be said to "make sense." Keeping track of and following the arguments and data as they unfold, trying always to think ahead, demands total absorption; at the same time, the sense of discovering or even generating a new world yields an intoxication rarely paralleled in other academic fields. The net result is that scientists are probably less reflective of the "tacit assumptions" that guide their reasoning than any other intellectuals of the modern age.

Indeed, the success of their enterprise does not, at least in the short run, seem to require such reflectivity.[9] Some would even argue that very success demands abstaining from reflection upon matters that do not lend themselves to "clear and distinct" answers. Indeed, they might argue that what distinguishes contemporary science from the efforts of their forbears is precisely their recognition of the dual need to avoid talk *about* science, and to replace "ordinary" language by a technical discourse cleansed of the ambiguity and values that burden ordinary language, as the modern form of the scientific report requires. Let the data speak for themselves, these scientists demand. The problem is, of course, that data never do speak for themselves.

It is by now a near truism that all data presuppose interpretation. And if an interpretation is to be meaningful—if the data are to be "intelligible" to more than one person—it must be embedded in a community of common practices, shared conceptions of the meaning of terms and their relation to and interaction with the "objects" to which these terms point. In science as elsewhere, interpretation requires the sharing of a common language.

Sharing a language means sharing a conceptual universe. It means more than knowing the "right" names by which to call things; it means knowing the "right" syntax in which to pose claims and questions, and even more critically it means sharing a more or less agreed-upon understanding of what questions are legitimate to ask, and what can be accepted as meaningful answers. Every explicit question carries with it a complex of tacit (unarticulated and generally unrecognized) presuppositions and expectations that limit the range of acceptable answers in ways that only a properly versed respondent will recognize. To know what kinds of explanation will "make sense," what can be expected to count as "accounting for," is already to be a member of a particular language community.

But if there is one feature that distinguishes scientific from other communities, and that is indeed special to that particular discourse, it is precisely the assumption that the universe scientists study is directly accessible, that the "nature" they name as object of inquiry is unmediated by language and can therefore be veridically represented. On this assumption, "laws of nature" are beyond the relativity of language—indeed, they are beyond language, encoded in logical structures that require only the discernment of reason and the confirmation of experiment. Also on this assumption, the descriptive language of science is transparent and neutral; it does not require examination.

Confidence in the transparency and neutrality of scientific language is certainly useful in enabling scientists to get on with their job; it is also wondrously effective in supporting their special claims to truth. It encourages the view that their own language, because neutral, is absolute, and in so doing, helps secure their disciplinary borders against criticism. Language, assumed to be transparent, becomes impervious.

It falls to others, then, less enclosed by the demands of science's own self-understanding, to disclose the "thickness" of scientific language, to scrutinize the conventions of practice, interpretation, and shared aspirations on which the truth claims of that language depend, to expose the many forks in the road to knowledge that these very conventions have worked to obscure, and, in that process, finally, to uncover alternatives for the future. Under careful scrutiny, the hypothesized contrast between ordinary and scientific language gives way to a recognition of disconcerting similarity. Even the most purely technical discourses turn out to depend on metaphor, on ambiguity, on instabilities of meaning—indeed, on the very commonsense understanding of terms from which a technical discourse is supposed to emancipate us. Scientific arguments cannot begin to "make sense," much less be effective, without extensive recourse to shared conventions for controlling these inevitable ambiguities and instabilities. The very term "experimental control" needs to be understood in a far larger sense than has been the custom—describing not only the control of variables, but also of the ways of seeing, thinking, acting, and speaking in which an investigator must be extensively trained before he or she can become a contributing member of a discipline.

Even the conventional account scientists offer of their success has been shown by recent work in the history, philosophy, and sociology of science to be itself rooted in metaphor: The very idea, for example, of a one-to-one correspondence between theory and reality, or of scientific method as capable of revealing nature "as it is," is based on metaphors of mind or science as "mirror of nature." Simple logic, however, suggests that words are far too limited a resource, in whatever combinations, to permit a faithful representation of even our own experience, much less of the vast domain of natural phenomena. The metaphor of science as "mirror of nature" may be both psychologically and politically useful to scientists, but it is not particularly useful for a philosophical understanding of how science works; indeed, it has proven to be a positive barrier to our understanding of the development of science in its historical and social context. It is far more useful, and probably even more correct, to suppose, as Mary Hesse suggests, that "[s]cience is successful only because there are sufficient local and particular regularities between things in space-time domains where we can test them. These domains may be very large but it's an elementary piece of mathematics that there is an infinite gap between the largest conceivable number and infinity" (1989:E24).

In much the same sense, the idea of "laws of nature" can also be shown to be rooted in metaphor, a metaphor indelibly marked by its political and theological origins. Despite the insistence of philosophers that laws of nature are merely descriptive, not prescriptive, they are historically conceptualized as imposed from above and obeyed from below. "By those who first used the term, [laws of nature] were viewed as commands imposed by the deity upon matter, and even writers who do not accept this view often speak of them as 'obeyed' by the phenomena, or as agents by which the phenomena are produced."[10] In this sense, then, the metaphor of "laws of nature" carries into scientific practice the presupposition of an ontological hierarchy, ordering not only mind and matter, but theory and practice, and, of course, the normal and the aberrant. Even in the loosest (most purely descriptive) sense of the term *law*, the kinds of order in nature that laws can accommodate are restricted to those that can be expressed by the language in which laws of nature are codified. All languages are capable of describing regularity, but not all perceivable, nor even all describable, regularities can be expressed in the existing vocabularies of science. To assume, therefore, that all perceptible regularities can be represented by current (or even by future) theory is to impose a premature limit on what is "naturally" possible, as well as what is potentially understandable.

Nancy Cartwright (1990) has suggested that a better way to make sense of the theoretical successes of science (as well as its failures) would be to invoke the rather different metaphor of "Nature's Capacities." In apparent sympathy with Mary Hesse, as well as with a number of other contemporary historians and philosophers of science, she suggests that an understanding of the remarkable convergences between theory and experiment that scientists have produced requires attention not so much to the adequacy of the laws that are presumably being tested, but rather to the particular and highly local manipulation of theory and experimental procedure that is required to produce these convergences. Our usual talk of scientific laws, Cartwright suggests, belies (and elides) both the conceptual and linguistic work that is required to ground a theory, or "law," to fit a particular set of experimental circumstances and the material work required to construct an experimental apparatus to fit a theoretical claim. Scientific laws may be "true," but what they are true of is a distillation of highly contrived and exceedingly particular circumstances, as much artifact as nature.

TURNING FROM GENDER AND SCIENCE TO LANGUAGE AND SCIENCE

The questions about gender with which I began this essay can now be reformulated in terms of two separable kinds of inquiry: The first, bearing on the historical role of public and private conceptions of gender in the framing of the root metaphors of science, belongs to feminist theory proper, whereas the second, that of the role of such metaphors in the actual development of scientific theory and practice, belongs to a more general inquiry in the history and philosophy of science. By producing abundant historical evidence pertaining to the first question, and by exhibiting the in-principle possibility of alternative metaphoric options, feminist scholars have added critical incentive to the pursuit of the second question. And by undermining the realism and univocality of scientific discourse, the philosophical groundwork laid by Kuhn, Hesse, Cartwright, and many others now makes it possible to pursue this larger question in earnest, pointing the way to the kind of analysis needed to show how such basic acts of naming have helped to shape the actual course of scientific development, and, in so doing, have helped to obscure if not foreclose other possible courses.

The most critical resource available for such an inquiry is the de facto plurality of organizing metaphors, theories, and practices evident throughout the history of science. At any given moment, in any given discipline, abundant variability can be readily identified along the following four closely interdependent axes: the aims of scientific inquiry; the questions judged most significant to ask; the theoretical and experimental methodologies deemed most productive for addressing these questions; and, finally, what counts as an acceptable answer or a satisfying explanation. Different metaphors of mind, nature, and the relation between them reflect different psychological stances of observer to observed; these, in turn, give rise to different cognitive perspectives—to different aims, questions, and even to different methodological and explanatory preferences. Such variability is of course always subject to the forces of selection exerted by collective norms, yet there are many moments in scientific history in which alternative visions can survive for long enough to permit identification both of their distinctiveness and of the selective pressures against which they must struggle.

The clearest and most dramatic such instance in my own research remains that provided by the life and work of the cytogeneticist Barbara McClintock. McClintock offers a vision of science premised not on the domination of nature, but on "a feeling for the organism."[11] For her, a "feeling for the organism" is simultaneously a state of mind and a resource for knowledge: for the day-to-day work of conducting experiments, observing and interpreting their outcomes—in short, for the "doing" of science. "Nature," to McClintock, is best known for its largesse and prodigality; accordingly, her conception of the work of science is more consonant with that of exhibiting nature's "capacities" and multiple forms of order than with pursuing the "laws of nature." Her alternative view invites the perception of nature as an active partner in a more reciprocal relation to an observer, equally active, but neither omniscient nor omnipotent; the story of her life's work (especially her identification of genetic transposition) exhibits how that deviant perception bore fruit in equally dissident observations.

But history is strewn with such dissidents and deviants, often as persistent and perceptive but still less fortunate than McClintock. Normally, they are erased from the record, in a gesture readily justified by the conventional narrative of science. Without the validation of the dominant community, deviant claims, along with the deviant visions of science that had guided them, are dismissed as "mistakes," misguided and false steps in the history of science. What such a retrospective reading overlooks is that the ultimate value of any accomplishment in science—that which we all too casually call its "truth"—depends not on any special vision enabling some scientists to see directly into nature, but on the acceptance and pursuit of their work by the community around them, that is, on the prior existence or development of sufficient commonalities of language and adequate convergences between language and practice. Language not only guides how we as individuals think and act; it simultaneously provides the glue enabling others to think and act along similar lines, guaranteeing that our thoughts and actions *can* "make sense."

WHAT ABOUT "NATURE"?

Still, language does not "construct reality." Whatever force it may have, that force can, after all, only be exerted on language-speaking subjects—for our concerns here, on scientists and the people who fund their work. Though language is surely instrumental in guiding the material actions of these subjects, it would be foolhardy indeed to lose sight of the force of the material, nonlinguistic, substrata of those actions, that is, of that which we loosely call "nature." Metaphors work to focus our attention in particular ways, conceptually magnifying one set of similarities and differences while dwarfing or blurring others, guiding the construction of instruments that bring certain kinds of objects into view, and eclipsing others. Yet, for any given line of inquiry, it is conspicuously clear that not all metaphors are equally effective for the production of further knowledge. Furthermore, once these instruments and objects have come into existence, they take on a life of their own, available for appropriation to other ends, to other metaphoric schemes.

Consider, for example, the fate of genetic transposition. McClintock's search for this phenomenon was stimulated by her interest in the dynamics of kinship and interdependency; it was made visible by an analytic and interpretive system premised on "a feeling for the organism," on the integrity and internal agency of the organism. To McClintock, transposition was a wedge of resistance on behalf of the organism against control from without. But neither she herself nor her analytic and interpretive framework could prevent the ultimate appropriation of this mechanism, once exhibited, to entirely opposite aims—as an instrument for external control of organic forms by genetic engineers.

McClintock's vision of science was unarguably productive for her, and it has been seen to have great aesthetic and emotional appeal for many scientists. But it must be granted that her success pales before that of mainstream (molecular) biology. In the last few years (in part thanks to the techniques derived from genetic transposition itself), it is the successes and technological prowess of molecular biology rather than of McClintock's vision of science that have captured the scientific and popular imagination. These successes, and this prowess, cannot be ignored.

We may well be persuaded that the domain of natural phenomena is vastly larger than the domain of scientific theory as we know it, leaving ample room for alternative conceptions of science; that the accumulated body of scientific theory represents only one of the many ways in which human beings, including the human beings we call scientists, have sought to make sense of the world; even that the successes of these theories are highly local and specific. Yet, whatever philosophical accounts we might accept, the fact remains that science as we know it works exceedingly well. The question is, Can any other vision of science be reasonably expected to work as well? Just how plastic are our criteria of success?

Feminists (and others) may have irrevocably undermined our sense of innocence about the aspiration to dominate nature, but they/we have not answered the question of just what it is that is wrong with dominating nature. We know what is wrong with dominating persons—it deprives other subjects of the right to express their own subjectivities—and we may indeed worry about the extent to which the motivation to dominate nature reflects a desire for domination of other human beings.[12] But a salient point of a feminist perspective on science derives precisely from the fact that nature is not in fact a woman. A better pronoun for nature is surely "it," rather than "she." What then could be wrong with seeking, or even achieving, dominion over things per se?

Perhaps the simplest response is to point out that nature, while surely not a woman, is also not a "thing," nor is it even an "it" that can be delineated unto itself, either separate or separable from a speaking and knowing "we." What we know about nature we know only through our interactions with, or rather, our embeddedness in it. It is precisely because we ourselves are natural beings—beings *in* and *of* nature—that we *can* know. Thus, to represent nature as a "thing" or an "it" is itself a way of talking, undoubtedly convenient, but clearly more appropriate to some ends than to others. And just because there is no one else "out there" capable of choosing, we must acknowledge that these ends represent human choices, for which "we" alone are responsible. One question we need to ask is thus relatively straightforward: What are the particular ends to which the language of objectification, reification, and domination of nature is particularly appropriate, and perhaps even useful? And to what other ends might a different language—of kinship, embeddedness, and connectivity, of "feeling for the organism"—be equally appropriate and useful? But we also need to ask another, in many ways much harder, question: How do the properties of the natural world in which we are embedded constrain our social and technical ambitions? Just what is there in the practices and methods of science that permits the realization of certain hopes but not others?

Earlier in this essay, I attempted to describe the shift in mind-set from working scientist to feminist critic. But to make sense of the successes of science, however that success is measured, the traversal must also be charted in reverse: Feminist critics of science, along with other analysts of science, need to reclaim access to the mind-set of the working scientists, to what makes their descriptions seem so compelling.

For this, we need to redress an omission from many of our analyses to date that is especially conspicuous to any working scientist: attention to the material constraints on which scientific knowledge depends, and correlatively, to the undeniable record of technological success that science as we know it can boast. If we grant the force of belief, we must surely not neglect the even more dramatic force of scientific "know-how." Although beliefs, interests, and cultural norms surely can, and do, influence the definition of scientific goals, as well as prevailing criteria of success in meeting those goals, they cannot in themselves generate either epistemological or technological success. Only where they mesh with the opportunities and constraints afforded by material reality can they lead to the generation of effective knowledge. Our analyses began with the question of where, and how, does the force of beliefs, interests, and cultural norms enter into the process by which effective knowledge is generated; the question that now remains is, Where, and how, does the nonlinguistic realm we call *nature* enter into that process? How do "nature" and "culture" interact in the production of scientific knowledge? Until feminist critics of science, along with other analysts of the influence of social forces on science, address this question, our accounts of science will not be recognizable to working scientists.

The question at issue is, finally, that of the meaning of science. Although we may now recognize that science neither does nor can "mirror" nature, to imply instead that it mirrors culture (or "interests") is not only to make a mockery of the commitment to the pursuit of reliable knowledge that constitutes the core of any working scientist's self-definition, but also to ignore the causal efficacy of that commitment. In other words, it is to practice an extraordinary denial of the manifest (at times even life threatening) successes of science. Until we can articulate an adequate response to the question of how "nature" interacts with "culture" in the production of scientific knowledge, until we find an adequate way of integrating the impact of multiple social and political forces, psychological predispositions, experimental constraints, and cognitive demands on the growth of science, working scientists will continue to find their more traditional mind-sets not only more comfortable, but far more adequate. And they will continue to view a mind-set that sometimes seems to grant force to beliefs and interests but not to "nature" as fundamentally incompatible, unintegrable, and laughable.

Notes

1 A somewhat different view is given by Tom Laqueur (1990).
2 See, for example, Gayle Rubin (1975).
3 See, for example, Rubin (1975) and Catherine MacKinnon (1988).
4 Of course, scientists are not the only ones who persist in such a mistranslation; it is also made by many others, and even by some feminists who are not themselves scientists. It is routinely made by the popular press. The significant point here is that this mistranslation persists in the minds of most women scientists even after they are alerted to the (feminist) distinction between sex and gender.
5 Indeed, a striking number of those feminist critics who began as working scientists have either changed fields altogether or have felt obliged to at least temporarily interrupt their work as laboratory or "desk" scientists (I am thinking, for example, of [the late] Maggie Benston, Ruth Hubbard, Marian Lowe, Evelynn Hammonds, Anne Fausto-Sterling, and myself).
6 In large part, stimulated by the publication of Thomas S. Kuhn's *The Structure of Scientific Revolutions*, in 1962.
7 See, for example, Galison (1988); Pickering (1984); Shapin and Schaffer (1985); Smith and Wise (1989).
8 The discussion that follows begins with a recapitulation of my remarks in Keller (1985:129–32).
9 For an especially interesting discussion of this general phenomenon, see Markus (1987).
10 O.E.D., s.v. "law." The discussion here is adapted from the introduction to Part III, Keller (1985).
11 McClintock's own words, as well as the title of my book on this subject, Keller (1983).
12 See Keller (1985), Part II.

References

Cartwright, N. *Nature's Capacities*. Oxford: Oxford University Press, 1990.
Galison, P. "Between War and Peace." In *Science, Technology, and the Military*, edited by Everett Mendelsohn, Merritt Roe Smith, Peter Weingart. Dordrecht; Boston: Kluwer Academic Publishers, 1988; Mendelsohn, Smith & Weingart (1988).
Haraway, D. *Simians, Cyborgs, and Women*. New York: Routledge, 1991.
Hesse, M. "Models, Metaphors and Myths." *N.Y. Times*, October 22, 1989, p. E24.

Keller, E.F. *A Feeling for the Organism: The Life and Work of Barbara McClintock.* New York: W. H. Freeman, 1983.
Keller, E.F. "Gender and Science." *Psychoanalysis and Contemporary Thought* 1:409–33, 1978.
Keller, E.F. *Reflections on Gender and Science.* New Haven, CT: Yale Univ. Press, 1985.
Kuhn, T.S. *The Structure of Scientific Revolutions.* Chicago: Univ. of Chicago Press, 1962.
Laqueur, T. *The Making of Sexual Difference.* Cambridge: Harvard Univ. Press, 1990.
MacKinnon, C. *Feminism Unmodified.* Cambridge: Harvard Univ. Press, 1988.
Markus, G. "Why Is There No Hermeneutics of the Natural Sciences?" *Science in Context* 1(1): 5–51, 1987.
Pickering, A. *Constructing Quarks.* Chicago: Univ. of Chicago Press, 1984.
Rubin, G. "The Traffic in Women: Notes on the 'Political Economy' of Sex." In *Toward an Anthropology of Women*, ed. R.R. Reiter. New York: Monthly Review Press, 1975.
Shapin, S. and S. Schaffer. *Leviathin and the Air-Pump.* Princeton: Princeton Univ. Press, 1985.
Smith, C. and N. Wise. *Energy and Empire: A Biographical Study of Lord Kelvin.* Cambridge: Cambridge Univ. Press, 1989.

18
The Paradox of Feminist Primatology

Linda Marie Fedigan

How has primatology come to be beloved by many feminist science scholars? Has feminism truly helped to engender the field of primate studies as it exists today, or is this a case of mistaken attribution? Many science analysts have remarked favorably upon the feminist transformation of primate studies that has occurred over the past twenty-five years, and so widely accepted is the view of primatology as a feminist enterprise that Hilary Rose asks whether it has become "the goddess's discipline."[1] Donna Haraway, in her influential analysis of the history of primatology, and the authors of lesser known analyses (e.g., Rosser's application of the "six stages of feminist transformation" to primatology) have argued for a shift in primatology toward the values and practices of feminist science.[2] But what makes this case curious and paradoxical is that most primatologists vehemently deny that theirs is a feminist science. Only a small handful of primatologists are self-declared feminists, albeit scholars whose work has been very influential—Jeanne Altmann, Sarah Hrdy, Jane Lancaster, Barbara Smuts, and Meredith Small, for example. And one need only peruse a few of the strongly negative reviews by practicing scientists of Haraway's book on primatology ("infuriating" is a common reaction)[3] or attempt to casually interject the word "feminism" into a discussion with primatologists or ask them outright if they consider themselves feminists to uncover the strength and depth of their denial. In this chapter I explore how primatologists have responded to the feminist critique of science by becoming more gender inclusive and by using "tools of gender analysis,"[4] and I suggest why so many primatologists carry out work that closely adheres to the tenets of feminism while at the same time denying the appellation. I argue that this contradiction is much more than a "label problem"—rather that these primatologists see themselves as operating from vastly different underlying assumptions and models of science than those of feminists.

An attempt to understand the paradox of feminist primatology is important for several reasons: it offers insights to those who want to know how and why a science changes over time; it can help to establish better communication and working links between scientists and science analysts; it exemplifies one discipline's solution to the "women in science problem"; and it sheds some light on the central issue of this volume: What useful changes has feminism brought to science? It does take some audacity for me to even try to address the last issue since many will feel that Donna Haraway definitively answered the question of how feminism has affected primatology in her comprehensive and multifaceted analysis of the subject.[5] But Haraway herself would surely advocate multiple voices and diverse perspectives on this topic. And I have puzzled for some time over the deeply negative reactions of my fellow primatologists to Haraway's interpretation of primatology as a "genre of feminist theory," and over the extent to which some of my scientific colleagues mistrust the label of feminism even while practicing feminist virtues. I have also pondered the question of what type of evidence would convince scientists that feminism has brought useful changes to their practices. I cannot claim to have definitive answers, but perhaps I can bring to bear on this matter the multiple "situated knowledges" of a practicing primatologist, a feminist, and an interested reader of science studies.

Both Haraway and Hrdy have argued that it is not simply a coincidence that primatology in North America began to develop along feminist lines in the 1970s at the same time that the second wave of the women's movement was cresting in American society.[6] Primatology is a quantitative and empirical science, and thus, for many practitioners, the most convincing evidence of the impact of feminism on primate studies would be to demonstrate directly through experiment or observation that the feminist critique of science can cause (or has caused) primatologists to change their practices. It would also be convincing if a substantial proportion of primatologists stated that their work has been influenced by feminism. It seems to me that neither of these scenarios is likely to be enacted and that we will therefore have to rely on indirect evidence. In a previous paper, I decided to set aside the issue of which primatologists self-identify as feminists and asked instead, What would a feminist science look like? and Does primatology look like this?[7] I identified six features that many models of a feminist science hold in common (reflexivity; taking the female point of view; cooperation with, rather than domination of, nature; moving away from dualisms and reductionism; humanitarian applications of science; and greater inclusiveness of formerly marginalized groups) and argued that the field of primatology has increasingly exhibited these features over the past twenty-five years. From this circumstantial case, I concluded that primate studies can be called a feminist science or that, at the very least, it has been significantly changed by the women's movement and the feminist critique of science.

Since writing that paper, I have come to think that what is unusual and instructive about the case of primatology is not that it can (arguably) be labeled a feminist science but rather the rapidity and the extent to which it "self-corrected" in response to the feminist critique of science. By this I mean that many disciplines have been the subjects of feminist critiques (e.g., archeology, biology, cultural anthropology, history, physics), but few, if any, others moved so quickly, so extensively, and so willingly to rectify the previous androcentric aspects of their practices. Therefore, in this paper I want to bypass the probably irresolvable debate as to whether or not primatology is a feminist science and instead begin to detail some of the myriad ways in which primatologists have become increasingly gender sensitive and gender inclusive over the past twenty-five years. I will attempt this documentation through the heuristic device of what Londa Schiebinger has called "tools of gender analysis." Schiebinger has advocated a move away from prescriptive visions of feminist science and suggested that we instead highlight a set of tools of gender analysis that have been used in many different sciences for designing woman-friendly research along feminist lines—tools that do not "create some special, esoteric 'feminist' science, but rather . . . incorporate a critical awareness of gender into the basic training of young scientists and the work-a-day world of science."[8] Schiebinger cautions that these analytic devices are not peculiar to feminists, and I will provide many examples of primatologists using these devices—some of whom do, and some of whom do not, call themselves feminists. Then I will return to the paradox laid out in this introduction and suggest several reasons why many primatologists carry out research along feminist lines even while they distance themselves from any such association.

EIGHT TOOLS OF GENDER ANALYSIS

Tool 1: Scientific priorities

Schiebinger argues that one of the most important gender analytics looks at scientific priorities—given limited resources, how are choices made about what we want to know and what we will fund for researchers to study?[9]

I would argue that there are at least three ways in which primatologists over the past twenty-five years have deliberately shifted their priorities about what we study in relation to gender issues. One fundamental shift has been simply to provide more funding for research that fleshes out the picture of what female primates do, apart from bearing and rearing offspring—the patterns that are part of what Jeanne Altmann has called "dual career mothering."[10] This shift paralleled the drive in many other fields to ensure that women formed part of the study sample, for example, in psychology and medical research. A second early change in primatological priorities was to turn standard research questions around and ask them from the female's perspective. One telling example comes from the study of size and shape differences between males and females of the same species. For over a hundred years, since Darwin first posed the question in print in 1871, evolutionary biologists have puzzled over sexual dimorphism and asked themselves, Why are male mammals usually bigger than the females of the species? It was not until the late 1970s that scientists began to ask, What might be the adaptive advantages to female mammals of being smaller than the males of their species?[11] This simple inversion of the question changed our entire perspective on the conundrum—we went from talking about how males need to be large in order to protect females to examining the different bioenergetic demands on female and male mammalian bodies.

But the third, and probably the most telling, example of how primatologists have shifted their priorities about the types of research questions that are asked and funded is the increasing study in the 1980s and 1990s of gender-related issues that clearly concern women when they arise in the human context. For example, Smuts has published both empirical research and theoretical papers on topics such as male aggression and sexual coercion of females, male-female friendships, and male dominance and its repercussions for females.[12] Hrdy has studied why male primates sometimes kill infants, how females compete with other females, and why parents favor sons over daughters.[13] Lisa Rose and I embarked on a series of papers addressing the issue of why female primates live with males on a year-round basis—examining what benefits males provide and in what ways males are a liability for females.[14] Small has published books about women primatologists studying female primates, about the processes by which females choose their sexual partners, and about the evolution of female sexuality.[15] Lancaster has published extensively about parental care and its evolution in the primate order and about the evolutionary biology of women.[16] There are many more examples, but these should suffice to show that we have created a new vision of the female primate in large part by simply choosing to find out more about her and her relations with the others of her species from her own perspective.

Tool 2: Representative sampling

As pointed out by Schiebinger, one of the basic tools of gender analysis is to make scientists aware of the appropriate inclusion of females as subjects of research and of women as the conductors of research.[17] To some extent, I covered the issue of females as subjects of research above, but here I will briefly describe two ways that sampling patterns have been recognized by primatologists to have implications for the conclusions we draw about sex and gender.

The first example is the landmark and now canonical 1974 paper by the primatologist Jeanne Altmann on appropriate methods for sampling animal behavior.[18] Although she is sometimes mistakenly credited for having invented these sampling methods, what Altmann did was to codify, clarify, and label the methods used by ethologists to sample behavior up to that time and to authoritatively evaluate their relative strengths and weaknesses for different research designs. The influence of this paper is enormous: it standardized sampling practices in ethology and because it is still referenced in most methods sections of ethological publications, some credit it with being the most cited paper in the modern literature on animal behavior.[19] It also discredited for most purposes what is called "ad lib sampling"—the practice of opportunistically recording whatever strikes the observer's eye and gains one's attention. In particular, Altmann established that it is inappropriate to use such opportunistic sampling to compare rates of behavior between individual subjects or between males and females, for example. What this meant for primatologists is that we stopped watching only the larger, more swashbuckling males and started to also sample for representative periods of time the less prepossessing subordinate males, the females, and the immature individuals of the groups. Altmann made no mention in this famous publication that she was dissatisfied with the previous bias toward observing male primates more than females—rather, what she did was to convince scientists, through the use of their own methodological tools, to raise their standards of evidence, a task for which she was well qualified by her background training as a mathematician. Elsewhere, she did state that her involvement in the feminist movement contributed to her awarness of and dissatisfaction with previous androcentric sampling practices.[20]

Another way in which primatologists realized that sampling can have implications for gender is the process by which we choose which species to study. For example, out of approximately two hundred primate species, the male-dominant savannah baboon was long the favorite species for modeling the evolution of human life.[21] In the 1950s and 1960s, a large contingent of North American primatologists, influenced by Sherwood Washburn, believed that we should be able to document something they called the "primate pattern," a code term for the basic "nature" of primates.[22] Aspects of what was then presented as baboon behavior, such as male bonding and aggression and rigid male dominance hierarchies, were generalized to all nonhuman primate species as representing the primate pattern, and from there to the human species as well. The first criticisms of this "baboonization" of primatology came as early as the 1970s.[23] Initially, the baboons were replaced by other single-species models, such as the chimpanzee model for hominid evolution, but today it is realized that no one species will provide a sufficient model for early human behavior. Although primatologists are now largely aware of the biases introduced when we generalize from one (or even a few) species to the entire order, we still make many assumptions about primate nature on the basis of the species we have studied (mainly Old World monkeys and apes), and we are finding these assumptions challenged as new data roll in on the less well studied species.[24] With so many species of primates, there are examples of almost every type of social system and relationship between the sexes that one could think of, and it makes a great deal of difference to our picture of human evolution and gender relations which species we choose to study and to emphasize.

Tool 3: Dangers of extrapolating research models from one group to another

Schiebinger gives examples of the dangers of extrapolating to women research models designed for men.[25] There are some parallel examples in primatology, such as the study of sexual dimorphism mentioned earlier. One of the stumbling blocks that hindered scientists from asking why female mammals might be selected to be smaller than males was their initial neglect of the repercussions of gestation and lactation. In primatology, it was sometimes assumed that the principles that apply to body size differences between species, such as those between chimpanzees and gorillas, could be applied willy-nilly to body size differences between the males and females of a species. Thus, it was not at all uncommon in the early days of primate ecology studies to read that the larger males of dimorphic species need to eat more and use a larger share of resources than do the females. However, a female primate spends most of her adult life gestating or lactating and may have feeding and metabolic rates up to 200 percent greater than a nonreproducing female of the same size. Today, primatologists are very aware of the different physiological, reproductive, and life history processes of male and female animals and much more careful about extrapolations from one sex to the other. If anything, primatologists sometimes follow the lead of those evolutionary theorists who suggest that the males and females of a given species can be so different in their biology and behavior as to be almost "two different species."

But where I think that primatologists have become the most aware of the dangers of extrapolation from one group to another is in the projection of Western gender role stereotypes onto animal patterns and onto our human ancestors. We can see this projection most clearly in some of the Victorian scenarios that nineteenth-century Euramerican anthropologists and natural historians imagined for monkeys and for early human social life, in which females were described as domestic, reluctant, and coy, whereas males were thought to be assertive, competitive, and mobile. But even in the 1960s and 1970s, when field reports on primates were first published, we began with many assumptions specific to the gender relations of our own time and place. For example, male monkeys and apes were sometimes referred to as "owning" their females,[26] whereas we know today that males in many of these species merely attach themselves on a temporary and rotating basis to permanent kin-bonded groups of females who occupy a consistent matrilineal home range. Can it be a coincidence that a generation raised on *Father Knows Best* would have propounded the 1950s view that primate females were mothers and mates and did little else of social or ecological note? How else to explain the influence of and the vast number of papers published on the "priority of access model," which assumed that female monkeys were passive resources available to males as sexual partners simply in order of the "winner's list" from male-male competitions? One by one these assumptions have disintegrated as researchers have focused on the females and documented not only the active roles they play in primate society but the enormous variety of relations between the sexes that occur in primates, defying our attempts to simplify and extrapolate to one "human nature." And the important point for our purposes here is that ethologists and primatologists themselves became aware of the dangers of extrapolating from humans to animals and then back to humans. Several researchers published warnings about the rebounding anthropomorphism known as the "aha! reaction" (e.g., "Aha! Female animals are choosy about mates; therefore, we can refer to them as 'coy.' " "Aha! Human females are coy just like animals; therefore, female coyness must be part of our primate heritage.").[27]

Tool 4: Institutional arrangements

Schiebinger argues that institutional power structures influence the knowledge that issues from them—from formalized universities to recognized-but-informal schools of thought to nearly invisible "cliques." The examination of how such institutions structure scientific representations of gender, race, and nature was of course the primary tool used by Haraway in her extensive analysis of primate studies.[28] Use of this tool does require the assumption that "primatology is politics," and this may be one of the reasons that few, if any, primatologists themselves have commented in print on how institutional background affects what we believe to be true about primates and about gender. Furthermore, the scrutiny of institutional power structures may require a more uninvolved perspective on the science than many scientists feel they are ready or qualified to take. One exception is Hrdy's brief analysis of the relationships among several important variables in her career: her status as the sole woman in her cohort at Harvard graduate school in the 1970s, her dawning awareness of androcentric bias in animal behavior studies, and her growing ability to imagine female animals as active strategists.[29] A second exception is a recent Wenner-Gren-sponsored workshop of primatologists and science analysis in which many participants concluded that institutional background is even more important in influencing a primatologist's thinking than is theoretical affiliation or methodological preference.[30]

Tool 5: Gender dynamics in the cultures of science

I teach in a building of social science departments but attend many meetings in the neighboring biological sciences building. Perhaps because of my background training in sociocultural anthropology, I have often noticed the radically different cultures reflected in the dress codes apparent in these two buildings: suits and pantyhose in the social science building; field clothes and lab coats next door. I also note differences in the intellectual styles and acceptable behavioral patterns of the biologists and the social scientists as I commute between them daily. This has clear implications for women graduate students in primatology, who may major in either anthropology, biology, or psychology, and who then face a choice between becoming "one of the boys" in the biological field or lab situation or taking on the "woman in power suit with briefcase" role in a social science setting. Primatologists have not commented in print about the "culture of primatology," perhaps for the same reasons that they have not reflected publicly upon institutional arrangements. But its "scientific culture" is one of the several ways in which primatology is a very informative case study, because I believe that ours is by and large an androgynous culture. To some extent, this is due to critical mass—there are now nearly equal proportions of men and women primatologists in North America, and there have always been substantial proportions of women in this science.[31] The most telling example of our androgynous culture is the nature of fieldwork, which is strongly gendered masculine in most sciences (e.g., archeology, paleontology, geology, entomology) but which is not gendered masculine or feminine in primatology. For reasons which are beyond the scope of this paper (e.g., the relatively recent emergence of primate fieldwork after World War II, and the early role models of women anthropological field-workers), a macho fieldwork image never took hold in the science of primatology in North America (Japan, Latin America, and India are different stories). It has always been expected that women will be as able and as competent as men to carry out field research on primates—research that is just as physically demanding, if not more so, than it is in other sciences. In fact, the popular media's attraction to women primatologists "roughing it" has sometimes given the public the impression that only women do fieldwork in primatology, but this is not the case. There are many successful men working at primate field sites, although it is possible that the men are more associated with short-term studies and the women with longitudinal field studies (this would make a good research question). My point is that I do not think in this case that feminism changed the gendered nature of primate fieldwork; I think rather that women got in on the ground floor, proved early on that they could do good work in the field as well as in the lab, and helped to establish an androgynous culture of primate science in which feminist tenets could flourish.

Tool 6: Language use

Primatologists are caught on the horns of a powerful language dilemma in the form of anthropomorphism. One of the strongest taboos in primate studies is to attribute human characteristics to animals—before we have studied them to determine how they do, in fact, behave and think—since this implies that all organisms behave and think like ourselves. And yet, we cannot fully invent a new language to describe our observations of animals, so we must borrow terms from the human domain that seem to best capture what we observe in the behavior of our animal subjects. The more closely related the animal is to us, the more it looks like and seems to behave like us, and thus the greater the danger of anthropomorphism. Primates are our closest relatives and share many characteristics with humans. But careful observations by scientists have convinced us that the gorilla does not beat his chest because he is angry; the vervet does not present his posterior to the face of another vervet to insult him, and the macaque does not grin because she is happy. Anthropomorphic assumptions about the meaning of primate signals must be avoided at all costs in that basic tool of all primate behavior research—the ethogram, which is a list of the behavioral units to be studied. Much of the work of developing a good ethogram involves language use, and a considerable amount of time and training is invested in teaching students to use descriptors that are as value neutral as possible (e.g., "open-mouth gape" instead of "threat face").

Thus, primatologists are somewhat predisposed to be judicious in their use of language. However, this is not to say that there have not been biases in the choice of terms to describe primate behavior. Biased terminology was a sufficient problem to warrant my devoting an entire chapter to language use in my general review of primate sex roles and social bonds.[32] In the early 1980s, I identified two major forms of biases in the choice of terms to describe primate behavior: androcentrism and a preference for hostile and combative metaphors. I was certainly not the first or the only scientists to identify sexism in the language of primate behavior.[33] Perhaps a single example will suffice. One type of social organization found in some primate species consists of several adult females who rear their young together in the presence of one adult male at a time (a uni-male, multifemale, or polygynous, system). This social system had been described by Darwin (1871) as "harems" ruled by "despots" or "masters" who "possessed" the females, and some of the early primate field studies reiterated such language.[34] However, after a few years of study it became obvious that in almost all polygynous primate societies (hamadryas baboons being the exception), a network of related females forms the stable core of the group and males come and go—hardly our idea of a "harem." Once it was recognized that these sorts of terms were very inappropriate to the actual behavior of most primate species, they were largely dropped. However, the hostile and combative metaphors have been harder to discourage. I wonder what comment it makes on our scientific culture that anthropomorphic and value-laden terminology that connotes sentimentality or "a warm fuzzy feeling" (e.g., "aunting" rather than "allomothering"; "kids" rather than "juveniles"; "babies" rather than "infants") has been mainly rejected whereas equally anthropomorphic terms that imply belligerent or ethically undesirable imagery in the human context (e.g., rape, cheaters, suckers, selfish) are still widely used and accepted.

One more area of language use that has changed in primate studies concerns the active/passive connotations of how we describe the behavior of the animals. For example, there used to be two terms to describe the sexuality of female animals: attractiveness (is she attractive to the male?) and receptivity (does she accept male advances?). In other words, the female was seen as a passive resource for males. In 1976, Frank Beach pointed out that estrous female mammals often take the initiative in approaching, investigating, and soliciting sex.

He called this phenomenon "proceptivity," and it has since come to be extensively used for and documented in female primates.[35]

Tool 7: The remaking of theoretical understandings

Schiebinger notes that there has been controversy about how deep gender analysis goes and whether feminists have contributed to the remaking of the theoretical understandings of a discipline or just "added females" to the mix.[36] The idea that in scholarship women are the fact gatherers and men are the theoreticians seems to be pervasive although seldom stated in print or studied directly. In her consideration of the gender of theory, Catherine Lutz has argued that the lines separating theory from nontheory are fuzzy, that theory is intentionally or unintentionally signaled to and picked up by readers as more associated with male scholarship than female scholarship, and that feminists have continually pressed against the dualism of theory and practice.[37] This latter point was brought home to me by Naomi Quinn, who referred to some of my own work as theoretical when I thought of it as merely a critique. As she pointed out to me, "How does any new theory arise except in the context of other theory?"

One of the "signals" of theory is that it is more abstract, original, and generalized than other writings and that it articulates thoughts not spoken or published before. Bruno Latour put this last characteristic of theory somewhat differently. He argued that it is important to document the "interesting differences" made by women scientists, differences that have shown primates to us in a new light or allowed primates to "speak" to us in new ways.[38] Using this understanding of theory, I would like to address some specific cases where primatologists have been instrumental in allowing female primates to become significant social actors or where they have created new "setups" (to use another Latour term) that changed our perceptions of female primates. In the field of primatology, there have been many such important theoretical breakthroughs, brought about by the ideas of both women and men. Some of these I have briefly described already or Haraway has valorized in her widely read analysis of primatology; others are lesser known.[39] Although I recognize the many pitfalls of providing a simplified list of important theoretical insights first developed by individual primatologists, in the interest of space (and in the interest of including mention of people who are not well known outside their science) I offer here brief synopses of some of the important work by primatologists that has remade our theoretical understandings of female animals and the relations between these females and the males with which they live. This list is neither exhaustive nor representatively sampled; it is rather what comes to my mind when I think of theoretical advances that changed our understanding of female primates.

1. Several researchers contributed new insights about principles of baboon behavior that re-created the world of social baboons (and thus the generalizations about all social primates) from a male militaristic model to a representation in which females as well as males play active roles.[40] For example, Altmann conceptualized and documented female baboons as "dual career mothers" with the capacity to feed themselves and their nursing offspring, to form long-lasting kinship bonds, to act as repositories of ecological knowledge, as well as to rear their young.[41] Thelma Rowell spearheaded the critique of the "male dominance" model of baboon social life and argued that dominance may be more characteristic of human primatologists than it is of the nonhuman primates, and that its common appearance in baboons and macaques might be the result of human-induced conditions in captivity.[42] Shirley Strum documented how individuals with real power are those who can mobilize allies rather than those who can push through with brute force; they rely on systems of

social reciprocity, which they actively construct.[43] Barbara Smuts developed a model of "friendship" in baboons that showed how adult males slowly become accepted members of matrilineal societies by ingratiating themselves with individual females and their infants.[44]

2. Richard Wrangham modeled primate society as one where females first distribute themselves according to the resources (food and water) available, and males then distribute themselves according to the spatial and social pattern of females available.[45] This idea was enormously influential in moving us away from a concept of male primates as "owners" of females, because Wrangham's model causes us to think of female patterns of distribution and female sociality as being prior to that of males.

3. Karen Strier showed us that contrary to the popular image of the aggressive, competitive male primate, many New World monkey societies (primarily those of the atelines) include philopatric, affiliative, closely bonded adult males, as well as females who transfer between groups.[46] Strier has been at the forefront of challenges to the generalizations we have all developed about primates based only on the well-studied Old World primates, especially the baboons, macaques, and chimpanzees, and she has encouraged us to acknowledge and incorporate into our theories the many forms that male and female relationships may take in primate societies.[47]

4. Devra Kleiman and Patricia Wright drew our attention to the extensive parental care shown by adult males in monogamous primate societies and developed explanations for the conditions under which male care and monogamy are adaptive in primates.[48]

5. The more recent studies of prosimian primates have also challenged some of our assumptions based only on the Old World monkeys and apes. For example, Alison Jolly, Michael Pereira, et al., and Alison Richard brought dominant female lemurs to our attention and helped to develop an evolutionary and ecological model of the conditions under which it is advantageous for female primates to be dominant over the males of their groups.[49]

Although there are other examples of how our theoretical understandings of female primates and of male–female relationships have been remade, I hope that these five will suffice to show that primatologists have been very concerned to provide a more complete picture and a greater critical awareness of the roles of females in primate societies and to disassemble old sex role stereotypes.

Tool 8: Challenges to what "counts" as science

It is fairly common in the feminist literature to point out that bodies of knowledge more associated with women are often categorized as "nonscientific"—home economics, family studies, and nursing, for example. Schiebinger argues that gender analysis has challenged what "counts" as science.[50] I would like to briefly describe two areas in which some Western primatologists, mainly women, have challenged what counts as primatological science: empathy as a means of understanding the animals and "mission" science.

Empathy is the projection of one's own feelings and thoughts onto the emotions and behavior of another and thus is quite similar to anthropomorphism, the primary taboo in primate studies, which I have already briefly discussed. Although few Euramerican primatologists have dared to suggest that they employ empathy to better understand their subject matter, the exceptions that I can think of have almost all been women (several of them also on our list of self-declared feminists): Hrdy, Rowell, Sicotte and Nisan, Small, and Strum.[51] The primary exception to this gender coding of empathy as feminine in North American primatology comes from the senior and respected animal behavior field-worker George Schaller, who said that much of what we understand about primates we do through intelligent empathy.[52] Another exception comes from Japanese primatology, which developed simultaneously and independently of Western primate studies and in which most practitioners are men who use "empathetic understanding" as a scientific tool.[53] Empathy is a two-edged sword for women primatologists in North America and Europe because their hard work can easily be dismissed as some form of "female intuition," and because admitting to empathy can jeopardize one's reputation as an objective observer. It seems that not only is it taboo to employ empathy as part of one's scientific tool kit, but most scientists are even reluctant to talk about it in print or to analyze the roles that it might play in our work. At the moment, it is certainly not thought to count as science outside Japan.

Mission science in primatology mainly takes the form of conservation work. Although many men (Russ Mittermeier being the most famous)[54] and indeed most field primatologists are necessarily involved to some degree in conservation of the species they study, I would like to consider for the purposes of this chapter the example made famous by the media: the "trimates," Jane Goodall, Dian Fossey, and Birute Galdikas. There are other more in-depth analyses of these three women and their work,[55] but I want to focus here on only two aspects: their perseverance and their willingness to sacrifice scientific success in order to devote themselves to conserving and enhancing the living conditions of the great ape species that they have worked with for so long. Obviously, other primatologists have also carried out longitudinal research, but the fact that Jane Goodall is still overseeing field research on the chimpanzees of Gombe forty years after she first began is truly remarkable. Dian Fossey's research was of course cut short by her untimely death, and Birute Galdikas has run into problems renewing her permit to carry out research at her field site, but her students and field assistants continue it for her. Each of these women decided at some point in their long careers that the need to save the endangered species they study is greater than the need to save their reputation as pure and productive scientists, and each has suffered from the diminished respect that some scientists accord to popularizers of science and to those who break ranks by placing their politics on a par with, or above, their science. I do not wish to turn these women into saints or martyrs, for that they certainly are not. However, Haraway's description of Jane Goodall as the "Virgin Priestess in the Temple of Science" certainly seems apt when one sees the masses of people lined up for a simple touch of her hand.[56] Not only have these three women primatologists triggered a major tide of public goodwill and funding for primates and the scientists who study them, but their example has also encouraged many young people, particularly young women, to enter this discipline. And they have played an important role in making conservation and animal welfare a very active and recognized part of the science of primatology. Conservation symposia are now the most widely attended sessions at national and international conferences of primatology, and the *International Journal of Primatology* flags with a special symbol all articles published on endangered species. I would argue that the courage of Goodall, Fossey, and Galdikas to break ranks and their challenge to what counts as science have been crucial in legitimizing mission science as an accepted aspect of primatology.

FEMINIST PRIMATOLOGY—WHAT'S IN A NAME?

In the previous section, I have provided numerous examples of primatologists using the tools of gender analysis identified by Schiebinger as methods commonly employed by feminists to help create woman-friendly science. The literature on gender and science indicates that there are at least three fundamental ways in which feminism can influence a science: (1) it can create more opportunities for women to enter and succeed in science (which has no further implications for change in science if women "do science" just like men); (2) it can increase gender awareness or gender sensitivity in the practitioners of a science;[57] and (3) it can alter the working dynamics of a science (e.g., power relations, gender symbolism) through the practices of well-established scientists who are also feminists. I would argue that primatology has exhibited changes over the past fifty years of its existence that provide substantial evidence for all three types of influence. First of all, our discipline has included higher and higher proportions of women practitioners,[58] and these women are holding more of the important societal offices of our profession (e.g., a recent past president of the International Primatological Society was Alison Jolly, and there have also been women presidents of the American Primatological Society). Second, the many examples that I provided above of primatologists changing their minds, their descriptions, and their research foci to recognize and document the "achieving female primate" are in part the result of increasing gender awareness in primatologists. Few North American primatologists today would refer to a polygynous society as a "harem" or treat females as if they were passive resources for male competition. Third, I have given examples of the few but powerful women primatologists who identify themselves as feminists, such as Sarah Hrdy and Jeanne Altmann, and who have played significant roles in rectifying the early androcentric biases of our field.

Why then do so many primatologists distance themselves from any association with feminism and deny that they have been influenced by feminism? I will offer three possible reasons based on my knowledge of, and conversations with, primatologists. First, I think that there is an underlying concern that our science not be perceived as "feminized," because of the widespread belief that feminized disciplines become devalued.[59] Most scientists do not perceive a difference between a feminine science and a feminist science and so would confuse the influences of feminism on science with "doing science in a feminine way" (i.e., "science with pink ribbons" as my colleague Shirley Strum puts it). Primatology is a quantitative science with a vital component of evolutionary biology and a strong association with other fields coded "masculine," such as anatomy, physiology, neurology, paleontology, and quantitative ecology. Training for both lab and fieldwork is rigorous, and many primatologists feel that all the media attention to women primatologists holding baby apes will project an incorrect image of a "sissy science."

Second, like scientists everywhere, primatologists distance themselves from anything perceived as political, because "politics" implies bias and failure to adhere to the scientific credo of objectivity. As Schiebinger has pointed out, feminism is a dirty word in North America and has many negative connotations to scientists.[60] Much of the science studies literature refers back to C. P. Snow's classic description of the "two cultures" of science and humanism, two cultures that still hold sway and are dichotomized in the minds of scientists.[61] Most of the primatologists I know see themselves as scientists, whereas they classify feminists as not-scientists. They are at least vaguely aware of the "science wars" and often assume that feminists must be on the other side. An interesting point, worthy of study, is how and why scientists classify certain theories as political (e.g., any feminist theory) and other theories as nonpolitical (e.g., sociobiology). One primatologist explained this distinction by saying that he does not carry his sociobiological beliefs home to the bedroom, whereas his wife does carry her feminist theories home. I suspect that the distinction between political and apolitical theory is often made on the basis of where the theory is thought to have originated—"in" or "out" of science. Since sociobiology was developed by scientists, it is perceived as nonpolitical in spite of its strong implications for, and widespread adoption by, conservative political groups in the United States. Feminism is perceived by primatologists as having been developed outside science, in fact as a critique of science, in spite of the work of influential and established feminist primatologists such as Sarah Hrdy, who is one of the few women (and perhaps the only feminist?) to be elected to the National Academy of Sciences.

Finally, many of the primatologists that I know hold a very idealistic view of science, a view that we inculcate, both consciously and unconsciously, in our students during their graduate training. Such a view has been called the "Legend of Science" by Steven Shapin and "Science with a Capital S" or "Science-Already-Made" by Bruno Latour.[62] Holders of such an idealistic view tend to perceive science as operating in a different realm from all other human activities, a realm that is pure and objective and free from sociocultural influences. The messiness of the actual practice of primatology in the field and in the laboratory is not denied so much as it is de-emphasized or ignored. That the community of scientists insists on Science with a Capital S is exemplified in the "cleaning up" of what we report in our publications and grant proposals—our research would never be published or funded if we laid out all the influences, all the mistakes, all the blind alleys, and all the human controversy that are, in fact, an integral part of the work.

Many scientists see themselves as operating from a nonpartisan position, and they see feminists and other science analysts, who often focus on the untidiness of "science-in-the-making," as a threat to their credibility and authority. Feminist theorists of science have often been at pains to point out the social influences on science and to suggest ways to improve science by making it more inclusive and egalitarian. That scientists and science analysts adhere to very different models of science (outcome and norms vs. process and practice) was pointed out to me by Shirley Strum as we puzzled over the many difficulties that our workshop of scientists and science analysts experienced in attempting to discuss the factors that caused primatologists to change their views of primate society.[63] Often we could not even begin our discussion of the history of ideas in primatology because the participants had such different views of what science is and how it works. We finally realized that we were enacting a local battle in the larger science wars and that C. P. Snow's model of two cultures still holds sway.

Notes

My research is funded by an ongoing grant from the Natural Sciences and Engineering Research Council of Canada (NSERCC). I thank Sandra Zohar for suggestions that improved this paper, and Brian Noble, Naomi Quinn, Londa Schiebinger, Shirley Strum, Zuleyma Tang Martinez, and Alison Wylie for stimulating conversations about gender and science that have shaped my understanding of feminist primatology.

1 Hilary Rose, *Love, Power, and Knowledge: Towards a Feminist Transformation of the Sciences* (Bloomington: Indiana University Press, 1994).
2 Donna Haraway, *Primate Visions: Gender, Race, and Nature in the World of Modern Science* (New York: Routledge, 1989); Sue Rosser, "The Relationship between Women's Studies and Women in Science," in *Feminist Approaches to Science*, ed. Ruth Bleier (New York: Pergamon Press, 1986), 165–80.
3 See the following reviews of Haraway, *Primate Visions*: Susan Cachel, "Partisan Primatology," *American Journal of Primatology* 22 (1990): 139–42; Matt Cartmill, *International Journal of Primatology* 12 (1991): 67–75; Robin Dunbar, "The Apes as We Want to See Them," *New York Times Book Review*, Jan. 10, 1990, 30; Alison Jolly and Margaretta Jolly, "A View from the Other End of the Telescope," *New Scientist*, Apr. 21, 1990, 58; Peter S. Rodman, "Flawed Vision: Deconstruction of Primatology and Primatologists," *Current Anthropology* 31 (1990): 484–86; Meredith F. Small, *American Journal of Physical Anthropology* 82 (1990): 527–28; Craig B. Stanford, *American Anthropologist* 93 (1991): 1031–32.
4 Londa Schiebinger, "Creating Sustainable Science," in *Women, Gender, and Science: New Directions*, ed. Sally G. Kohlstedt and Helen E. Longino, special issue of *Osiris* 12 (1997): 201–16. See also Londa Schiebinger, *Has Feminism Changed Science?* (Cambridge, MA: Harvard University Press, 1999).
5 Haraway, *Primate Visions*.
6 Ibid., 285; Sarah B. Hrdy, introduction to *Female Primates: Studies by Women Primatologists*, ed. Meredith F. Small (New York: Alan R. Liss, 1984), 103–9; and Sarah B. Hrdy, "Empathy,

Polyandry, and the Myth of the Coy Female," in *Feminist Approaches to Science*, ed. Ruth Bleier (New York: Pergamon Press, 1986), 119–46.

7 Linda Marie Fedigan, "Is Primatology a Feminist Science?" in *Women in Human Evolution*, ed. Lori Hager (London: Routledge, 1997), 56–75.

8 Schiebinger, *Has Feminism Changed Science?* 8, 186–90.

9 Ibid.

10 Jeanne Altmann, *Baboon Mothers and Infants* (Cambridge, MA: Harvard University Press, 1980).

11 For example, Linda Marie Fedigan, *Primate Paradigms*; David G. Post, "Feeding and Ranging Behavior of the Yellow Baboon" (Ph.D. diss., Yale University, 1978); Katherine Ralls, "Mammals in Which Females Are Larger Than Males," *Quarterly Review of Biology 51* (1976): 245–76.

12 Barbara B. Smuts, "Male Aggression against Women: An Evolutionary Perspective," *Human Nature* 3 (1992): 1–44; Barbara B. Smuts, "Male Aggression and Sexual Coercion of Females in Nonhuman Primates and Other Mammals: Evidence and Theoretical Implications," in *Advances in the Study of Behavior*, vol. 22, ed. Peter J. Slater, Jay S. Rosenblatt, Charles T. Snowdon, and Manfred Milinski (New York: Academic Press, 1993), 1–63; Barbara B. Smuts, *Sex and Friendship in Baboons* (New York: Aldine, 1985); Barbara B. Smuts, "Gender, Aggression, and Influence," in *Primate Societies*, ed. Barbara B. Smuts, Dorothy L. Cheney, Robert M. Seyfarth, Richard W. Wrangham, and Thomas T. Struhsaker (Chicago: University of Chicago Press, 1987), 400–412.

13 Sarah B. Hrdy, *The Langurs of Abu: Female and Male Strategies of Reproduction* (Cambridge, MA: Harvard University Press, 1977); see also Glenn H. Hausfater and Sarah B. Hrdy, *Infanticide: Comparative and Evolutionary Perspectives* (New York: Aldine, 1984); Sarah B. Hrdy, "Care and Exploitation of Nonhuman Primate Infants by Conspecifics Other than the Mother," in *Advances in the Study of Behavior*, vol. 6, ed. Jay S. Rosenblatt, Robert A. Hinde, Evelyn Shaw, and Colin Beer (New York: Academic Press, 1976), 101–58; Sarah B. Hrdy, *The Woman That Never Evolved* (Cambridge, MA: Harvard University Press, 1981); Sarah B. Hrdy and Debra S. Judge, "Darwin and the Puzzle of Primogeniture: An Essay on Biases in Parental Investment after Death," *Human Nature* 4 (1993): 1–4S.

14 Lisa M. Rose and Linda Marie Fedigan, "Vigilance in White-Faced Capuchins, *Cebus capucinus*, in Costa Rica," *Animal Behavior* 49 (1995): 63–70; Lisa Gould, Linda Marie Fedigan, and Lisa M. Rose, "Why Be Vigilant? The Case of the Alpha Male," *International Journal of Primatology* 18 (1997): 401–14; L. Rose, "Costs and Benefits of Resident Males to Females in White-Faced Capuchins," *American Journal of Primatology* 32 (1994): 235–48.

15 Small, *Female Primates*; Small, *Female Choices*; Meredith F. Small, *What's Love Got to Do with It? The Evolution of Human Mating* (New York: Doubleday, 1995).

16 Jane Lancaster, "Carrying and Sharing in Human Evolution," *Human Nature* 1 (1978): 82–89; Jane Lancaster and Chet Lancaster, "Parental Investment: The Hominid Adaptation," in *How Humans Adapt: A Biocultural Odyssey*, ed. Donald J. Ortner (Washington, DC: Smithsonian Institution Press, 1983), 33–65; Jane Lancaster, "The Watershed: Change in Parental Investment and Family Formation Strategies in the Course of Human Evolution," in *Parenting across the Life Span: Biosocial Dimensions*, ed. Jane Lancaster, Jeanne Altmann, Alice Rossi, and Lonnie Sherrod (New York: Aldine, 1987), 187–205; Jane Lancaster, "Women in Biosocial Perspective," in *Gender and Anthropology: Critical Reviews for Research and Teaching*, ed. Sandra Morgen (Washington, DC: American Anthropological Association, 1989), 95–115; Jane Lancaster, "The Evolutionary Biology of Women," in *Milestones in Human Evolution*, ed. Alan J. Alquist and Anne Manyak (Prospect Heights, II: Waveland Press, 1993), 21–37.

17 Schiebinger, *Has Feminism Changed Science?* 187.

18 Jeanne Altmann, "Observational Study of Behavior: Sampling Methods," *Behaviour* 49 (1974): 227–67.

19 Lloyd, "Science and Anti-science."

20 Haraway, *Primate Visions*, 304–10, and Lloyd, "Science and Anti-science," 241.

21 For example, Sherwood L. Washburn and C. S. Lancaster, "The Evolution of Hunting," in *Man the Hunter*, ed. Richard B. Lee and Irven Devore (Chicago: Aldine, 1968), 293–303; see discussion in Fedigan, "The Changing Role of Women."

22 Shirley C. Strum and Linda Marie Fedigan, "Theory, Method, and Gender: What Changed Our Views of Primate Society?" in *The New Physical Anthropology*, ed. Shirley C. Strum, Don G. Lindburg, and David A. Hamburg (New York: Prentice Hall, 1999).

23 For example, Clifford J. Jolly, "The Seed-Eaters: A New Model of Hominid Differentiation Based on a Baboon Analogy," *Man 5* (1970): 5–26; Adrienne Zihlman, "Women in Human Evolution, Part II, Subsistence and Social Organization among Early Hominids," *Signs: Journal of Women in Culture and Society* 4 (1978): 4–20.

24 For example, Karen Strier, "The Myth of the Typical Primate," *Yearbook of Physical Anthropology 37* (1994): 233–71.
25 Schiebinger, *Has Feminism Changed Science?* 113–18, 189.
26 For example, John H. Crook, "Sexual Selection, Dimorphism, and Social Organization in the Primates," in *Sexual Selection and the Descent of Man, 1871–1971*, ed. Bernard Campbell (Chicago: Aldine Press, 1972), 231–81; Robin Fox, "Alliance and Constraint: Sexual Selection and the Evolution of Human Kinship Systems," in ibid., 282–331.
27 For example, R. D. Martin, "The Biological Basis of Human Behavior," in *The Biology of Brains*, ed. W. B. Broughton (London: Institute of Biology, 1974), 215–50; Hilary Callan, *Ethology and Society: Towards an Anthropological View* (Oxford: Clarendon Press, 1970); Fedigan, *Primate Paradigms.*
28 Haraway, *Primate Visions.*
29 Hrdy, "Empathy."
30 See Linda Marie Fedigan, "Gender Encounters," in Strum and Fedigan, *Primate Encounters*, 498–520.
31 Linda Marie Fedigan, "Science and the Successful Female: Why There Are So Many Women Primatologists," *American Anthropologist* 96 (1994): 529–40; and unpublished data.
32 Linda Marie Fedigan, *Primate Paradigms: Sex Roles and Social Bonds* (Montreal: Eden Press, 1982; 2d ed., with new introduction, Chicago: University of Chicago Press, 1992).
33 Anne I. Dagg, *Harems and Other Horrors: Sexual Bias in Behavioral Biology* (Waterloo, ON: Otter Press, 1983); Katherine Ralls, "Sexual Dimorphism in Mammals: Avian Models and Unanswered Questions," *American Naturalist* 111 (1977): 917–38.
34 Charles Darwin, *The Descent of Man and Selection in Relation to Sex* (London: John Murray, 1871); John H. Crook, "Gelada Baboon Herd Structure and Movement: A Comparative Report," in *Play, Exploration, and Territory in Mammals*, ed. P. A. Jewell and Caroline Loizos, Symposia of the Zoological Society of London, vol. 18 (New York: Academic Press, 1966), 237–58; Hans Kummer, "Social Organization of Hamadryas Baboons: A Field Study," *Bibliotheca Primatologica* 6 (1968): 1–189; cf. Christian Bachmann and Hans Kummer, "Male Assessment of Female Choice in Hamadryas Baboons," *Behavioral Ecology and Sociobiology* 6 (1980): 315–21.
35 Frank A. Beach, "Sexual Attractivity, Proceptivity, and Receptivity in Female Mammals," *Hormones and Behavior 7* (1976): 105–38.
36 Schiebinger, *Has Feminism Changed Science?* 189.
37 Catherine Lutz, "The Gender of Theory," in *Women Writing Culture*, ed. Ruth Behar and Deborah A. Gordon (Berkeley and Los Angeles: University of California Press, 1995), 249–66.
38 Bruno Latour, "A Well-Articulated Primatology: Reflexions of a Fellow-Traveler," in Strum and Fedigan, *Primate Encounters*, 358–81.
39 Haraway, *Primate Visions.*
40 See Linda Marie Fedigan and Laurence Fedigan, "Gender and the Study of Primates," in *Gender and Anthropology: Critical Reviews for Teaching and Research*, ed. Sandra Morgen (Washington, DC: American Anthropological Association, 1989), 41–64.
41 Altmann, *Baboon Mothers and Infants.*
42 Thelma E. Rowell, "The Concept of Dominance," *Behavioral Biology* 11 (1974): 131–54.
43 Shirley C. Strum, *Almost Human: A Journey into the World of Baboons* (New York: Random House, 1987).
44 Smuts, *Sex and Friendship in Baboons.*
45 Richard W. Wrangham, "An Ecological Model of Female-Bonded Primate Groups," *Behaviour* 75 (1980): 262–300.
46 Karen B. Strier, "Brotherhoods among Atelines: Kinship, Affiliation, and Competition," *Behaviour* 130 (1994): 151–67.
47 For example, Karen B. Strier, "The Myth of the Typical Primate," *Yearbook of Physical Anthropology* 37 (1994): 233–71.
48 Devra G. Kleiman, "Monogamy in Mammals," *Quarterly Review of Biology* 52 (1977): 39–69; Patricia C. Wright, "Biparental Care in *Aotus trivigratus* and *Callicebus moloch*," in Small, *Female Primates*, 59–75.
49 Alison Jolly, "The Puzzle of Female Feeding Priority," in Small, *Female Primates*, 197–215; Michael E. Pereira, Ruben Kaufman, Peter M. Kappeler, and Deborah J. Overdorff, "Female Dominance Does Not Characterize All of the Lemuridae," *Folia primatologica SS* (1990): 96–103; Alison Richard, "Malagasy Prosimians: Female Dominance," in Smats et al., *Primate Societies*, 25–33.
50 Schiebinger, *Has Feminism Changed Science?* 190.
51 Sarah B. Hrdy, introduction to Small, *Female Primates*, 103–9; Hrdy, "Empathy"; Rowell,

introduction to section 1 in Small, *Female Primates;* P. Sicotte and C. Nisan, "Femmes et empathie en primatologie," in *Grands singes: la fascination du double*, ed. Bertrand L. Deputte and Jacques Vauclair (Paris: Autrement, 1998), 77–102; Small, review of *Primate Visions* in *American Journal of Physical Anthropology* 82 (1990): 527–28; Strum, *Almost Human*, and unpublished manuscript.

52 George B. Schaller, *The Year of the Gorilla* (Chicago: University of Chicago Press, 1964).

53 See Pamela J. Asquith, "Primate Research Groups in Japan: Orientations and East-West Differences," in Fedigan and Asquith, *The Monkeys of Arashiyama*, 81–98; H. Takasaki, "Traditions of the Kyoto School of Field Primatology in Japan," in Strum and Fedigan, *Primate Encounters*, 151–64.

54 See R. A. Mittermeier, "A Global Overview of Primate Conservation," in *Primate Ecology and Conservation*, ed. James G. Else and Phyllis C. Lee (Cambridge: Cambridge University Press, 1986), 325–40.

55 For example, Harold T. P. Hayes, *The Dark Romance of Dian Fossey* (New York: Simon and Schuster, 1990); Marguerite Holloway, "Profile: Jane Goodall—Gombe's Famous Primate," *Scientific American*, Oct. 1997, 42–44; Sy Montgomery, *Walking with the Great Apes* (Boston: Houghton Mifflin, 1991); Virginia Morell, "Called 'Trimates,' Three Bold Women Shaped Their Field," *Science* 260 (1993): 420–25; Brian Noble, "Politics, Gender, and Worldly Primatology: The Goodall-Fossey Nexus," in Strum and Fedigan, *Primate Encounters*, 436–62; Linda Spalding, *The Follow* (Toronto, ON: Key Porter Press, 1998).

56 Haraway, *Primate Visions, 182.*

57 Alison Wylie, "Standpoint Matters, in Archaeology for Example," in Strum and Fedigan, *Primate Encounters*, 243–60.

58 Fedigan, "Science and the Successful Female"; Haraway, *Primate Visions*; Hrdy, "Empathy."

59 Margaret W. Rossiter, "Which Science? Which Women?" *Osiris* 12 (1977): 169–85.

60 Schiebinger, *Has Feminism Changed Science?*

61 C. P. Snow, *Two Cultures and the Scientific Revolution* (New York: American Library, 1959).

62 Steven Shapin, *The Social History of Truth* (Chicago: University of Chicago Press, 1994); Bruno Latour, *Science in Action: How to Follow Scientists and Engineers through Society* (Cambridge, MA: Harvard University Press, 1987); and Latour, "A Well-Articulated Primatology."

63 Shirley C. Strum, "Science Encounters," in Strum and Fedigan, *Primate Encounters*, 475–97.

19

Socially Camouflaged Technologies

The Case of the Electromechanical Vibrator

Rachel Maines

Certain commodities are sold in the legal marketplace for which the expected use is either illegal or socially unacceptable. Marketing of these goods, therefore, requires camouflaging of the design purpose in a verbal and visual rhetoric that conveys to the knowledgeable consumer the item's selling points without actually endorsing its socially prohibited uses. I refer not to goods that are actually illegal in character, such as marijuana, but to their grey-market background technologies, such as cigarette rolling papers. Marketing efforts for goods of this type have similar characteristics over time, despite the dissimilarity of the advertised commodities. I shall discuss here an electromechanical technology that addresses formerly prohibited expressions of women's sexuality—the vibrator in its earliest incarnation between 1870 and 1930. Comparisons will be drawn between marketing strategies for this electromechanical technology, introduced between 1880 and 1903, and that of emmenagogues, distilling, burglary tools, and computer software copying, as well as the paradigm example of drug paraphernalia.

I shall argue here that electromechanical massage of the female genitalia achieved acceptance during the period in question by both professionals and consumers not only because it was less cumbersome, labor-intensive, and costly than predecessor technologies, but because it maintained the social camouflage of sexual massage treatment through its associations with modern professional instrumentation and with prevailing beliefs about electricity as a healing agent.[1]

The case of the electromechanical vibrator, as a technology associated with women's sexuality, involves issues of acceptability rather than legality. The vibrator and its predecessor technologies, including the dildo, are associated with masturbation, a socially prohibited activity until well into the second half of this century.[2] Devices for mechanically assisted female masturbation, mainly vibrators and dildoes, were marketed in the popular press from the late nineteenth century through the early thirties in similarly camouflaged advertising. Such advertisements temporarily disappeared from popular literature after the vibrator began to appear in stag films, which may have rendered the camouflage inadequate, and did not resurface until social change made it unnecessary to disguise the sexual uses of the device.[3]

For purposes of this discussion, a vibrator is a mechanical or electromechanical appliance imparting rapid and rhythmic pressure through a contoured working surface usually mounted at a right angle to the handle. These points of contact generally take the form of a set of interchangeable vibratodes configured to the anatomical areas they are intended to address. Vibrators are rarely employed internally in masturbation; they thus differ from dildoes, which are generally straight-shafted and may or may not include a vibratory component. Vibrators are here distinguished also from massagers, the working surfaces of which are flat or dished.[4] It should be noted that this is a historian's distinction imposed on the primary sources; medical authors and appliance manufacturers apply a heterogeneous nomenclature to massage technologies. Vibrators and dildoes rarely appeared in household advertising between 1930 and 1955; massagers continued to be marketed, mainly through household magazines.[5]

The electromechanical vibrator, introduced as a medical instrument in the 1880s and as a home appliance between 1900 and 1903, represented the convergence of several older medical massage technologies, including manual, hydriatic, electrotherapeutic, and mechanical methods. Internal and external gynecological massage with lubricated fingers had been a standard medical treatment for hysteria, disorders of menstruation, and other female complaints at least since the time of Aretaeus Cappadox (circa 150 A.D.), and the evidence suggests that orgasmic response on the part of the patient may have been the intended therapeutic result.[6] Douche therapy, a method of directing a jet of pumped water at the pelvic area and vulva, was employed for similar purposes after hydrotherapy became popular in the eighteenth and nineteenth centuries.[7] The camouflage of the apparently sexual character of such therapy was accomplished through its medical respectability and through creative definitions both of the diseases for which massage was indicated and of the effects of treatment. In the case of the electromechanical vibrator, the use of electrical power contributed the cachet of modernity and linked the instrument to older technologies of electrotherapeutics, in which patients received low-voltage electricity through electrodes attached directly to the skin or mucous membranes, and to light-bath therapy, in which electric light was applied to the skin in a closed cabinet. The electrotherapeutic association was explicitly invoked in the original term for the vibrator's interchangeable applicators, which were known as "vibratodes." Electrical treatments were employed in hysteria as soon as they were introduced in the eighteenth century and remained in use as late as the 1920s.

Hysteria as a disease paradigm, from its origins in the Egyptian medical corpus through its conceptual eradication by American Psychological Association fiat in 1952, was so vaguely and subjectively defined that it might encompass almost any set of ambiguous symptoms that troubled a woman or her family. As its name suggests, hysteria as well as its "sister" complaint chlorosis were until the twentieth century thought to have their etiology in the female reproductive tract generally, and more particularly in the organism's response to sexual deprivation.[8] This physiological condition seems to have achieved epidemic proportions among women and girls, at least in the modern period.[9] Sydenham, writing in the seventeenth century, observed that hysteria was the most common of all diseases except fevers.[10]

In the late nineteenth century, physicians noted with alarm that from half to three-quarters of all women showed signs of hysterical affliction. Among the many symptoms listed in medical descriptions of the syndrome are anxiety, sense of heaviness in the pelvis, edema (swelling) in the lower abdomen and genital areas, wandering of attention and associated tendencies to indulge in sexual fantasy, insomnia, irritability, and "excessive" vaginal lubrication.[11]

The therapeutic objective in such cases was to produce a "crisis" of the disease in the Hippocratic sense of this expression, corresponding to the point in infectious diseases at which the fever breaks. Manual massage of the vulva by physicians or midwives, with fragrant oils as lubricants, formed part of the standard treatment repertoire for hysteria, chlorosis, and related disorders from ancient times until the post-Freudian era. The crisis induced by this procedure was usually called the "hysterical paroxysm." Treatment for hysteria might comprise up to three-quarters of a physician's practice in the nineteenth century. Doctors who employed vulvular massage treatment in hysteria thus required fast, efficient, and effective means of producing the desired crisis. Portability of the technology was also a desideratum, as physicians treated many patients in their homes, and only manual massage under these conditions was possible until the introduction of the portable battery-powered vibrator for medical use in the late 1880s.

Patients reported experiencing symptomatic relief after such treatments, and such conditions as pelvic congestion and insomnia were noticeably ameliorated, especially if therapy continued on a regular basis. A few physicians, including Nathaniel Highmore in the seventeenth century and Auguste Tripier, a nineteenth-century electrotherapist, clearly recognized the hysterical paroxysm as sexual orgasm.[12] That many of their colleagues also perceived the sexual character of hysteria treatments is suggested by the fact that, in the case of married women, one of the therapeutic options was intercourse, and in the case of single women, marriage was routinely recommended.[13] "God-fearing physicians," as Zacuto expressed it in the seventeenth century, were expected to induce the paroxysm with their own fingers only when absolutely necessary, as in the case of very young single women, widows, and nuns.[14]

Many later physicians, however, such as the nineteenth-century hydrotherapist John Harvey Kellogg, seem not to have perceived the sexual character of patient response. Kellogg wrote extensively about hydrotherapy and electrotherapeutics in gynecology. In his "Electrotherapeutics in Chronic Maladies," published in *Modern Medicine* in 1904, he describes "strong contractions of the abdominal muscles" in a female patient undergoing treatment, and similar reactions such that "the office table was made to tremble quite violently with the movement."[15] In their analysis of the situation, these physicians may have been handicapped by their failure to recognize that penetration is a successful means of producing orgasm in only a minority of women; thus treatments that did not involve significant vaginal penetration were not morally suspect. In effect, misperceptions of female sexuality formed part of the camouflage of the original manual technique that preceded the electromechanical vibrator. Insertion of the speculum, however, since it travelled the same path as the supposedly irresistible penis during intercourse, was widely criticized in the medical community for its purportedly immoral effect on patients.[16] That some questioned the ethics of the vulvular massage procedure is clear; Thomas Stretch Dowse quotes Graham as observing that "Massage of the pelvic organs should be intrusted to those alone who have 'clean hands and a pure heart'."[17] One physician, however, in an article significantly titled "Signs of Masturbation in the Female," proposed the application of an electrical charge to the clitoris as a test of salacious propensities in women. Sensitivity of the organ to this type of electrical stimulation, in his view, indicated secret indulgence in what was known in the nineteenth century as "a bad habit."[18] Ironically, such women were often treated electrically for hysteria supposedly caused by masturbation.

However they construed the benefits, physicians regarded the genital massage procedure, which could take as long as an hour of skilled therapeutic activity, as something of a chore, and made early attempts to mechanize it. Hydrotherapy, in the form of what was known as the "pelvic douche" (massage of the lower pelvis with a jet of pumped water), provided similar relief to the patient with reduced demands on the therapist. Doctors of the eighteenth and nineteenth centuries frequently recommended douche therapy for their women patients who could afford spa visits. This market was limited, however, as both treatment and travel were costly.[19] A very small minority of patients and doctors could afford to install hydrotherapeutic facilities in convenient locations; both doctor and patient usually had to travel to the spa. Electrically powered equipment, when it became available, thus had a decentralizing and cost-reducing effect on massage treatment.

In the 1860s, some spas and clinics introduced a coal-fired steam-powered device invented by a Dr. George Taylor, called the "Manipulator," which massaged the lower pelvis while the patient either stood or lay on a table.[20] This too required a considerable expenditure either by the physician who purchased the equipment or by the patient who was required to travel to a spa for treatment. Thus, when the electromechanical vibrator was invented two decades later in England by Mortimer Granville and manufactured by Weiss, a ready market already existed in the medical community.[21] Ironically, Mortimer Granville considered the use of his instrument on women, especially hysterics, a morally indefensible act, and recommended the device only for use on the male skeletal muscles.[22] Although his original battery-powered model was heavy and unreliable, it was more portable than water-powered massage and less fatiguing to the operator than manual massage (fig. 19.1).

Air-pressure models were introduced, but they required cumbersome tanks of compressed air, which needed frequent refilling. When line electricity became widely available, portable plug-in models made vibratory house calls more expeditious and cost effective for the enterprising physician. The difficulty of maintaining batteries in or out of the office was noted by several medical writers of the period predating the introduction of plug-in vibrators.[23] Batteries and small office generators were liable to fail at crucial moments during patient treatment and required more engineering expertise for their maintenance than most physicians cared to acquire. Portable models using DC or AC line electricity were available with a wide range of vibratodes, such as the twelve-inch rectal probe supplied with one of the Gorman firm's vibrators.

Despite its inventor's reservations, the Weiss instrument and later devices on the same principle were widely used by physicians for pelvic disorders in women and girls. The social camouflage applied to the older medical technology was carefully maintained in connection with the new, at least until the 1920s. The marketing of medical vibrators to physicians and the discussion of them in such works as Covey's *Profitable Office Specialties* addressed two important professional considerations: the respectability of the devices as medical instruments (including their reassuringly clinical appearance) and their utility in the fast and efficient treatment of those chronic disorders, such as pelvic complaints in women, that provided a significant portion of a physician's income.[24] The importance of a prestige image

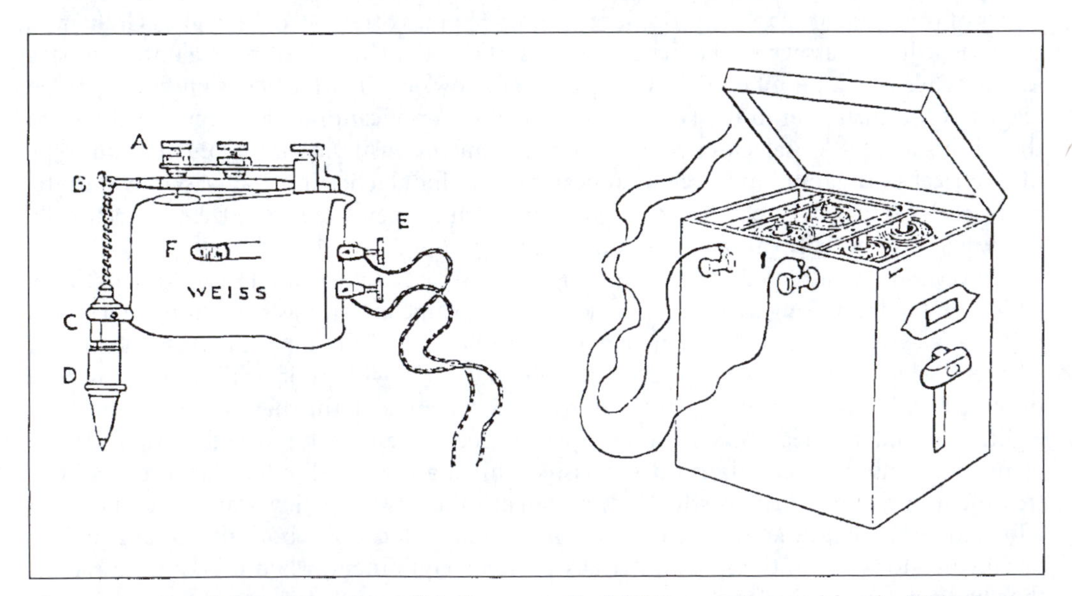

Figure 19.1 Joseph Mortimer Granville's "percuteur" of 1883, manufactured by the Weiss Instrument Company

for electromechanical instrumentation and its role in the pricing of medical vibrators is illustrated by a paragraph in the advertising brochure for the "Chattanooga" (fig. 19.2), at $200 in 1904 the most costly of the physicians' office models:

> The Physician can give with the "Chattanooga" Vibrator a thorough massage treatment in three minutes that is extremely pleasant and beneficial, but this instrument is neither designed nor sold as a "Massage Machine." It is sold only to Physicians, and constructed for the express purpose of exciting the various organs of the body into activity through their central nervous supply.[25]

I do not mean to suggest that gynecological treatments were the only uses of such devices, or that all physicians who purchased them used them for the production of orgasm in female patients, but the literature suggests that a substantial number were interested in the new technology's utility in the hysteroneurasthenic complaints. The interposition of an official-looking machine must have done much to restore clinical dignity to the massage procedure. The vibrator was introduced in 1899 as a home medical appliance and was by 1904 advertised in household magazines in suggestive terms we shall examine later on. It was important for physicians to be able to justify to patients the expense of $2–3 per treatment, as home vibrators were available for about $5.

The acceptance of the electromechanical vibrator by physicians at the turn of this century may also have been influenced by their earlier adoption of electrotherapeutics, with which vibratory treatment could be, and often was, combined.[26] Vibratory therapeutics were introduced from London and Paris, especially from the famous Hopital Salpetriere,

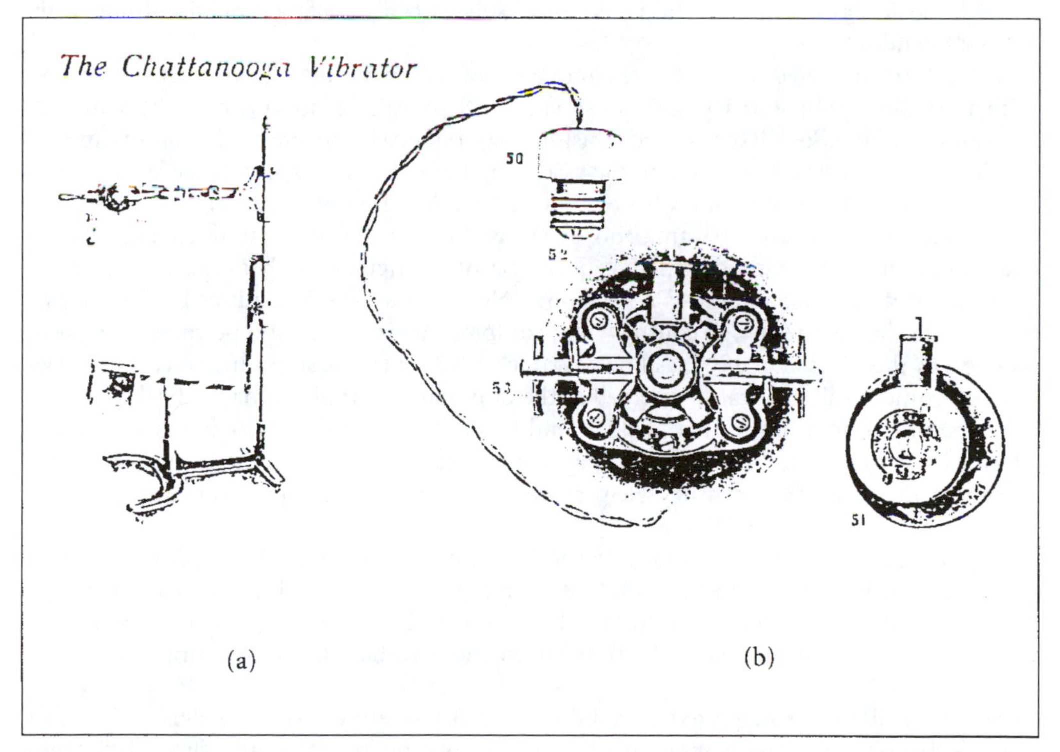

Figure 19.2 (a) The Chattanooga, at $200 the most expensive medical vibrator available in 1904, could be wheeled over the operating table and its vibrating head rotated for the physician's convenience. (b) Chattanooga Vibrator parts

which added to their respectability in the medical community.[27] It is worth noting as well that in this period electrical and other vibrations were a subject of great interest and considerable confusion, not only among doctors and the general public, but even among scientists like Tesla, who is reported to have fallen under their spell. "...[T]he Earth," he wrote, "is responsive to electrical vibrations of definite pitch just as a tuning fork to certain waves of sound. These particular electrical vibrations, capable of powerfully exciting the Globe, lend themselves to innumerable uses of great importance...."[28] In the same category of mystical reverence for vibration is Samuel Wallian's contemporaneous essay on "The Undulatory Theory in Therapeutics," in which he describes "modalities or manifestations of vibratory impulse" as the guiding principle of the universe. "Each change and gradation is not a transformation, as mollusk into mammal, or monkey into man, but an evidence of a variation in vibratory velocity. A certain rate begets a *vermis*, another and higher rate produces a *viper*, a *vertebrate*, a *vestryman*."[29]

In 1900, according to Monell, more than a dozen medical vibratory devices for physicians had been available for examination at the Paris Exposition. Of these, few were able to compete in the long term with electromechanical models. Mary L.H. Arnold Snow, writing for a medical readership in 1904, discusses in some depth more than twenty types, of which more than half are electromechanical. These models, some priced to the medical trade as low as $15, delivered vibrations from one to 7,000 pulses a minute. Some were floor-standing machines on rollers; others could be suspended from the ceiling like the modern impact wrench.[30] The more expensive models were adapted to either AC or DC currents. A few, such as those of the British firm Schall and Son, could even be ordered with motors custom-wound to a physician's specifications. Portable and battery-powered electromechanical vibrators were generally less expensive than floor models, which both looked more imposing as instruments and were less likely to transmit fatiguing vibrations to the doctor's hands.

Patients were treated in health spa complexes, in doctor's offices, or in their own homes with portable equipment. Designs consonant with prevailing notions of what a medical instrument should look like inspired consumer confidence in the physician and his apparatus, justified treatment costs, and, in the case of hysteria treatments, camouflaged the sexual character of the therapy. Hand- or foot-powered models, however, were tiring to the operator; water-powered ones became too expensive to operate when municipalities began metering water in the early twentieth century. Gasoline engines and batteries were cumbersome and difficult to maintain, as noted above. No fuel or air-tank handling by the user was required for line electricity, in contrast with compressed air, steam, and petroleum as power sources. In the years after 1900, as line electricity became the norm in urban communities, the electromechanical vibrator emerged as the dominant technology for medical massage.

Some physicians contributed to this trend by endorsing the vibrator in works like that of Monell, who had studied vibratory massage in medical practice in the United States and Europe at the turn of this century. He praises its usefulness in female complaints:

> ... pelvic massage (in gynecology) has its brilliant advocates and they report wonderful results, but when practitioners must supply the skilled technic with their own fingers the method has no value to the majority. But special applicators (motor-driven) give practical value and office convenience to what otherwise is impractical.[31]

Other medical writers suggested combining vibratory treatment of the pelvis with hydro- and electrotherapy, a refinement made possible by the ready adaptability of the new electromechanical technology.

At the same period, mechanical and electromechanical vibrators were introduced as home medical appliances. One of the earliest was the Vibratile, a battery-operated massage device advertised in 1899. Like the vibrators sold to doctors, home appliances could be hand-powered, water-driven, battery or street-current apparatus in a relatively wide range of prices from $1.50 to $28.75. This last named was the price of a Sears, Roebuck model of 1918, which could be purchased as an attachment for a separate electrical motor, drawing current through a lamp socket, which also powered a fan, buffer, grinder, mixer, and sewing machine. The complete set was marketed in the catalogue under the headline "Aids That Every Woman Appreciates" (fig. 19.3). Vibrators were mainly marketed to women, although men were sometimes exhorted to purchase the devices as gifts for their wives or to become door-to-door sales representatives for the manufacturer.[32]

The electromechanical vibrator was preceded in the home market by a variety of electro-therapeutic appliances which continued to be advertised through the twenties, often in the same publications as vibratory massage devices. Montgomery Ward; Sears, Roebuck; and the Canadian mail order department store T. Eaton and Company all sold medical batteries by direct mail by the end of the nineteenth century. These were simply batteries with electrodes that administered a mild shock. Some, like Butler's Electro-Massage Machine, produced their own electricity with friction motors. Contemporaneous and later appliances sometimes had special features, such as Dr. H. Sanche's Oxydonor, which produced ozone in addition to the current when one electrode was placed in water. "Electric" massage rollers, combs, and brushes with a supposedly permanent charge retailed at this time for prices between one and five dollars. Publications like the *Home Needlework Magazine* and *Men and Women* advertised these devices, as well as related technologies, including correspondence courses in manual massage.

Vibrators with water motors, a popular power source, as noted above, before the introduction of metered water, were advertised in such journals as *Modern Women*, which emphasized the cost savings over treatments by physicians and further emphasized the advantage of privacy offered by home treatment. Such devices were marketed through the teens in *Hearst's* and its successors, and in *Woman's Home Companion.*[33] Electromechanical vibrators were sold in the upper-middle-class market, in magazines typically retailing for between ten and fifteen cents an issue. As in the case of medical vibrators, models adapted to both AC and DC current were more expensive than those for use with DC only; all were fitted with screw-in plugs through the twenties.[34]

All types of vibrators were advertised as benefitting health and beauty by stimulating the circulation and soothing the nerves. The makers of the electromechanical American vibrator, for example, recommended their product as an ". . . alleviating, curative and beautifying agent. . . . It will increase deficient circulation—develop the muscles—remove wrinkles and facial blemishes, and beautify the complexion."[35] Advertisements directed to male purchasers similarly emphasized the machine's advantages for improving a woman's appearance and disposition. An ad in a 1921 issue of *Hearst's* urges the considerate husband to "Give 'her' a Star for Christmas" on the grounds that it would be "A Gift That Will *Keep* Her Young and Pretty." The same device was listed in another advertisement with several other electrical appliances and labelled "Such Delightful Companions!"[36] A husband, these advertisements seem to suggest, who presented his wife with these progressive and apparently respectable medical aids might leave for work in the morning secure in the knowledge that his spouse's day would be pleasantly and productively invested in self-treatment. Like other electrical appliance advertising of the time, electromechanical vibrator ads emphasized the role of the device in making a woman's home a veritable Utopia of modern technology, and its utility in reducing the number of occasions, such as visiting her physician, on which she would be required to leave her domestic paradise.[37]

Advertisements for vibrators often shared magazine pages with books on sexual matters, such as Howard's popular *Sex Problems in Worry and Work* and Walling's *Sexology*, handguns, cures for alcoholism and, occasionally, even personals, from both men and women, in which matrimony was the declared objective. Sexuality is never explicit in vibrator advertising; the tone is vague but provocative, as in the Swedish Vibrator advertisement in *Modern Priscilla* of 1913, offering "a machine that gives 30,000 thrilling, invigorating, penetrating,

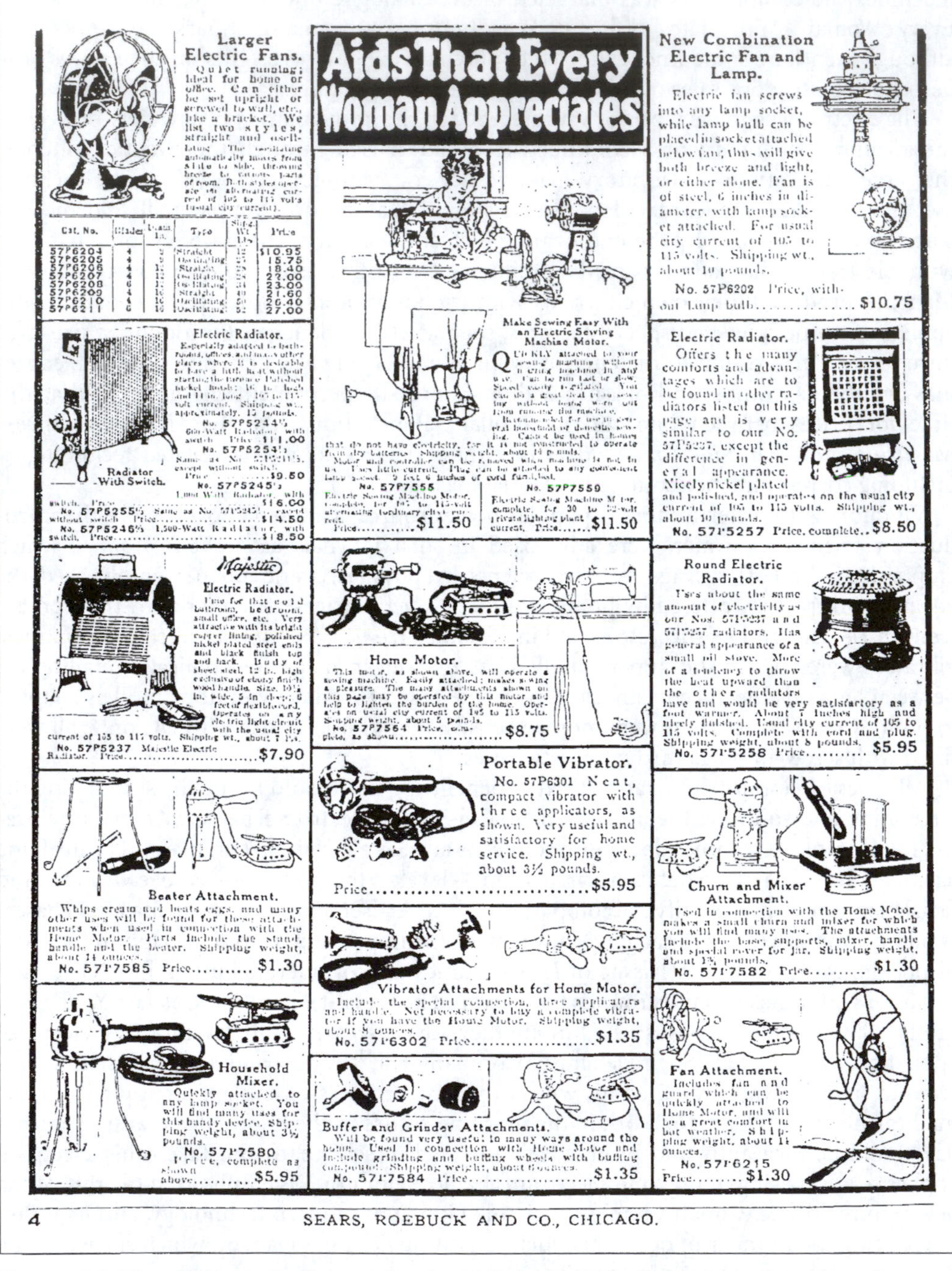

Figure 19.3 The vibratory attachments for the 1918 Sears, Roebuck home motor were only one of many electromechanical possibilities

revitalizing vibrations per minute. . . . Irresistible desire to own it, once you feel the living pulsing touch of its rhythmic vibratory motion." Illustrations in these layouts typically include voluptuously proportioned women in various states of *déshabillé*. The White Cross vibrator, made by a Chicago firm that manufactured a variety of small electrical appliances, was also advertised in *Modern Priscilla*, where the maker assured readers that "It makes you fairly tingle with the joy of living."[38] It is worth noting that the name "White Cross" was drawn from that of an international organization devoted to what was known in the early twentieth century as "social hygiene," the discovery and eradication of masturbation and prostitution wherever they appeared. The Chicago maker of White Cross appliances, in no known way affiliated with the organization, evidently hoped to trade on the name's association with decency and moral purity.[39] A 1916 advertisement from the White Cross manufacturer in *American Magazine* nevertheless makes the closest approach to explicit sexual claims when it promises that "All the keen relish, the pleasures of youth, will throb within you."[40] The utility of the product for female masturbation was thus consistently camouflaged.

Electromechanical vibrator advertising almost never appeared in magazines selling for less than 5 cents an issue (10 to 20 cents is the median range) or more than 25 cents. Readers of the former were unlikely to have access to electrical current; readers of the latter, including, for example, *Vanity Fair*, were more likely to respond to advertising for spas and private manual massage. While at least a dozen and probably more than twenty U.S. firms manufactured electromechanical vibrators before 1930, sales of these appliances were not reported in the electrical trade press. A listing from the February 1927 *NELA Bulletin* is typical; no massage equipment of any kind appears on an otherwise comprehensive list that includes violet-ray appliances.[41] A 1925 article in *Electrical World*, under the title "How Many Appliances are in Use?" lists only irons, washing machines, cleaners, ranges, water heaters, percolators, toasters, waffle irons, kitchen units, and ironers.[42] *Scientific American* listed in 1907 only the corn popper, chafing dish, milk warmer, shaving cup, percolator, and iron in a list of domestic electrical appliances.[43] References to vibrators were extremely rare even in popular discussions of electrical appliances.[44] The U.S. Bureau of the Census, which found 66 establishments manufacturing electrotherapeutic apparatus in 1908, does not disaggregate by instrument type either in this category or in "electrical household goods." The 1919 volume, showing the electromedical market at a figure well over $2 million, also omits detailed itemization. Vibrators appear by name in the 1949 *Census of Manufactures*, but it is unclear whether the listing for them, aggregated with statistics for curling irons and hair dryers, includes those sold as medical instruments to physicians.[45] This dearth of data renders sales tracking of the electromechanical vibrator extremely difficult. The omissions from engineering literature are worth noting, as the electromechanical vibrator was one of the first electrical appliances for personal care, partly because it was seen as a safe method of self-treatment.[46]

The marketing strategy for the early electromechanical vibrator was similar to that employed for contemporaneous and even modern technologies for which social camouflage is considered necessary. Technologically, the devices so marketed differ from modern vibrators sold for explicitly sexual purposes only in their greater overall weight, accounted for by the use of metal housings in the former and plastic in the latter. The basic set of vibratodes is identical, as is the mechanical action. The social context of the machine, however, has undergone profound change. Liberalized attitudes toward masturbation in both sexes and increasing understanding of women's sexuality have made social camouflage superfluous.

In the case of the vibrator, the issue is one of acceptability, but there are many examples of similarly marketed technology of which the expected use was actually illegal. One of these, which shares with the vibrator a focus on women's sexuality, was that of "emmenagogues" or abortifacient drugs sold through the mail and sometimes even off the shelf in the first few decades of this century. Emmenagogues, called in pre-FDA advertising copy "cycle restorers," were intended to bring on the menses in women who were "late." Induced abortion by any means was of course illegal, but late menses are not reliable indicators of pregnancy. Thus, women who purchased and took "cycle restorers" might or might not be in violation of antiabortion laws; they themselves might not be certain without a medical examination. The advertising of these commodities makes free use of this ambiguity in texts like the following from *Good Stories* of 1933:

> Late? End Delay—Worry. American Periodic Relief Compound double strength tablets combine Safety with Quick Action. Relieve most Stubborn cases. No Pain. New discovery. Easily taken. Solves women's most perplexing problem. RELIEVES WHEN ALL OTHERS FAIL. Don't be discouraged, end worry at once. Send $1.00 for Standard size package and full directions. Mailed same day, special delivery in plain wrapper. American Periodic Relief Compound Tablets, extra strength for stubborn cases, $2.00. Generous Size Package. New Book free.[47]

The rhetoric here does not mention the possibility of pregnancy, but the product's selling points would clearly suggest this to the informed consumer through the mentions of safety, absence of pain, and stubborn cases. The readers of the pulp tabloid *Good Stories* clearly did not require an explanation of "women's most perplexing problem."

Distilling technology raises similar issues of legality. During the Prohibition period, the classified section of a 1920 *Ainslee's* sold one- and four-gallon copper stills by mail, advising the customer that the apparatus was "Ideal for distilling water for drinking purposes, automobile batteries and industrial uses."[48] Modern advertisements for distilling equipment contain similar camouflage rhetoric, directing attention away from the likelihood that most consumers intend to employ the device in the production of beverages considerably stronger than water.[49]

Although changes in sexual mores have liberated the vibrator, social camouflage remains necessary for stills and many other modern commodities, including drug paraphernalia. The Deering Prep Kit, for example, is advertised at nearly $50 as a superlative device for grinding and preparing fine powders, "such as vitamin pills or spices."[50] Burglary tools are marketed in some popular (if lowbrow) magazines with the admonition that they are to be used only to break into one's own home or automobile, in the event of having locked oneself out. The camouflage rhetoric seems to suggest that all prudent drivers and homeowners carry such tools on their persons at all times. Most recently, we have seen the appearance of computer software for breaking copy protection, advertised in terms that explicitly prohibit its use for piracy, although surely no software publisher is so naive as to believe that all purchasers intend to break copy protection only to make backup copies of legitimately purchased programs and data.[51] As in vibrator advertising, the product's advantages are revealed to knowledgeable consumers in language that disclaims the manufacturers' responsibility for illegal or immoral uses of the product.

The marketing of socially camouflaged technologies is directed to consumers who already understand the design purpose of the product, but whose legally and/or culturally unacceptable intentions in purchasing it cannot be formally recognized by the seller. The marketing rhetoric must extoll the product's advantages for achieving the purchaser's goals—in the case of the vibrator, the production of orgasm—by indirection and innuendo, particularly with reference to the overall results, i.e., relaxation and relief from tension. The same pattern emerges in the advertisement of emmenagogues: according to the manufacturer, it is "Worry and Delay" that are ended, not pregnancy. In the case of software copyright protection programs, drug paraphernalia, and distilling equipment, the expected input and/or output are simply misrepresented, so that an expensive, finely calibrated scale with its own fitted carrying case may be pictured in use in the weighing of jelly beans. As social values and legal restrictions shift, the social camouflaging of technologies may be expected to change in response, or to be dispensed with altogether, as in the case of the vibrator.

References

1 Various versions of this paper have benefitted from comments and criticism from John Senior at the Bakken, Joel Tarr of Carnegie-Mellon University, Shere Kite of Kite Research, Karen Reeds of Rutgers University Press, my former students at Clarkson University, and participants in the Social and Economic History Seminar, Queens University (Canada), the Hannah Lecture series in the History of Medicine at the University of Ottawa, and the 1986 annual meeting of the Society for the History of Technology with the Society for the Social Study of Science. Anonymous referees of this and other journals have also provided valuable guidance in structuring the presentation of my research results.

2 Sokolow, Jayme A., *Eros and Modernization: Sylvester Graham, Health Reform and the Origins of Victorian Sexuality in America.* Rutherford, NJ: Fairleigh Dickinson University Press, 1983, pp. 77–99; Haller, John S., and Robin Haller, *The Physician and Sexuality in Victorian America.* Urbana: University of Illinois Press, 1973, pp. 184–216; Greydanus, Donald E., "Masturbation: Historic Perspective," *New York State Journal of Medicine*, November 1980, vol. 80, no. 12, pp. 1892–1896; Szasz, Thomas, *The Manufacture of Madness.* New York: Harper and Row, 1977, pp. 180–206; Hare, E.H., "Masturbatory Insanity: The History of an Idea," *Journal of Mental Sciences*, 1962, vol. 108, pp. 1–25; and Bullough, Vern, "Technology for the Prevention of 'Les Maladies Produites par la Masturbation,' " *Technology and Culture*, October 1987, vol. 28, no. 4, pp. 828–832.

3 On the vibrator in stag films, see Blake, Roger, *Sex Gadgets.* Cleveland: Century, 1968, pp. 33–46. An early postwar reference to the vibrator as an unabashedly sexual instrument is Ellis, Albert, *If This Be Sexual Heresy.* New York: Lyle Stuart, 1963, p. 136.

4 Vibrators and dildoes are illustrated in Tabori, Paul, *The Humor and Technology of Sex.* New York: Julian Press, 1969; the dildo is discussed in a clinical context in Masters, William H., *Human Sexual Response.* Boston: Little, Brown, 1966. Vibrators of the period to which I refer in this essay are illustrated in Gorman, Sam J., *Electro Therapeutic Apparatus.* 10th ed. Chicago: Sam J.Gorman, c1912; Wappler Electric Manufacturing Co. Inc., *Wappler Cautery and Light Apparatus and Accessories.* 2nd ed. New York: Wappler Electric Manufacturing, 1914, pp. 7 and 42–43; Manhattan Electrical Supply Co., *Catalogue Twenty-Six: Something Electrical for Everybody.* New York: MESCO, n.d.; and Snow, Mary Lydia Hastings Arnold, *Mechanical Vibration and Its Therapeutic Application.* New York: Scientific Author's Publication Company, 1904 and 1912. For modern vibrators, see Kaplan, Helen Singer, "The Vibrator: A Misunderstood Machine" *Redbook*, May 1984, p. 34; and Swarz, Mimi, "For the Woman Who Has Almost Everything," *Esquire*, July 1980, pp. 56–63.

5 See, for examples of such advertising, which in fact included a persistent abdominal emphasis, "Amazing New Electric Vibrating Massage Pillow," Niresk Industries (Chicago, IL) advertisement in *Workbasket*, October 1958, p. 95; "Don't be Fat," body massager (Spot Reducer) advertisement in *Workbasket*, September 1958, p. 90; and "Uvral Pneumatic Massage Pulsator," in *Electrical Age for Women*, January 1932, vol. 2, no. 7, pp. 275–276.

6 This therapy is extensively documented but rarely noted by historians. For only a few examples of medical discussions of vulvular massage in the hysteroneurasthenic disorders, see Aretaeus Cappadox, *The Extant Works of Aretaeus the Cappadocian*, ed. and transl. by Francis Adams. London: Sydenham Society, 1856, pp. 44–45, 285–287, and 449–451; Forestus, Alemarianus Petrus (Pieter van Foreest), *Observationem et Curationem Medidnalium ac Chirurgicarum Opera Omnia.* Rothomagi: Bertherlin, 1653, vol. 3, book 28, pp. 277–340; Galen of Pergamon, *De Locis Affectis*, transl. by Rudolph Siegel. Basel and New York: S. Karger, 1976, book VI, chapter II: 39; and Weber, A. Sigismond, *Traitementpar l'Electricite et le Massage.* Paris: Alex Coccoz, 1889, pp. 73–80. Of modern scholars, only Audrey Eccles discusses this therapy in detail in her

Obstetrics and Gynaecology in Tudor and Stuart England. London and Canberra: Croom Helm, 1982, pp. 76–83.

7 Baruch, Simon, *The Principles and Practice of Hydrotherapy: A Guide to the Application of Water in Disease.* New York: William Wood and Company, 1897, pp. 101, 211, 248, and 365; Dieffenbach, William H., *Hydrotherapy.* New York: Rebman, 1909, pp. 238–245; Good Health Publishing Company, *20th Century Therapeutic Appliances.* Battle Creek, MI: Good Health Publishing, 1909, pp. 20–21; Hedley, William Snowdon, *The Hydro-Electric Methods in Medicine.* London: H.K. Lewis, 1892; Hinsdale, Guy, *Hydrotherapy.* Philadelphia and London: W.B. Saunders Company, 1910, p. 224; Kellogg, John Harvey, *Rational Hydrotherapy.* Philadelphia: Davis, 1901; Irwin, J.A., *Hydrotherapy at Saratoga.* New York: Casell, 1892, pp. 85–134 and 246–248; Pope, Curran, *Practical Hydrotherapy: A Manual for Students and Practitioners.* Cincinnati, OH: Lancet-Clinic Publishing Co., 1909, pp. 181–192 and 506–538; and Trail, Russell Thacher, *The Hydropathic Encyclopedia.* New York: Fowlers and Wells, 1852, pp. 273–295. Women were reportedly in the majority as patients at spas, and some were owned by women entrepreneurs and/or physicians. See Whyman, T., "Visitors to Margate in the 1841 Census Returns," *Local Population Studies*, vol. 8, 1972, p. 23. Since at least the time of Jerome, baths and watering places have had a reputation for encouraging unacceptable expressions of sexuality. For female masturbation with water, see Aphrodite, J. [pseud.], *To Turn You On: 39 Sex Fantasies for Women.* Secaucus, NJ: Lyle Stuart, Inc., 1975, pp. 83–91; and Halpert, E., "On a Particular Form of Masturbation in Women: Masturbation with Water," *Journal of the American Psychoanalytic Association*, 1973, vol. 21, p. 526.

8 A bibliography of nineteenth-century American works on women and sexuality in relation to hysteria is available in Sahli, Nancy, *Women and Sexuality in America: A Bibliography.* Boston: Hall, 1984. See also Shorter, Edward, "Paralysis: The Rise and Fall of the 'Hysterical' Symptom," *Journal of Social History*, Summer 1986, vol. 19, no. 4, pp. 549–582; Satow, Roberta, "Where Has All the Hysteria Gone?" *Psychoanalytic Review*, 1979–80, vol. 66, pp. 463–473; Bourneville, Desire Magloire, and P. Regnard, *Iconographie Photographique de la Salpetriere.* Paris: Progres-Medical, 1878, vol. 2, pp. 97–219; Charcot, Jean-Martin, *Clinical Lectures on Certain Diseases of the Nervous System*, transl. by E.P. Hurd. Detroit: G.S. Davis, 1888, p. 141; Ellis, Havelock, *Studies in the Psychology of Sex*, vol. 1, New York: Random House, 1940, p. 270; Krohn, Alan, *Hysteria: The Elusive Neurosis.* New York: International Universities Press, 1978, pp. 46–51; McGrath, William J., *Freud's Discovery of Psychoanalysis: the Politics of Hysteria.* Ithaca, NY: Cornell University Press, 1986, pp. 152–172; Veith, Ilza, *Hysteria: The History of a Disease.* Chicago: University of Chicago Press, 1965; Wittels, Franz, *Freud and His Time.* New York: Grosset and Dunlap, 1931, pp. 215–242; and Ziegler, Dewey and Paul Norman, "On The Natural History of Hysteria in Women," *Diseases of the Nervous System*, 1967, vol. 15, pp. 301–306.

9 Bauer, Carol, "The Little Health of Ladies: An Anatomy of Female Invalidism in the Nineteenth Century," *Journal of the American Medical Woman's Association*, October 1981, vol. 36, no. 10, pp. 300–306; Ehrenreich, Barbara, and D. English, *Complaints and Disorders: The Sexual Politics of Sickness.* Old Westbury, NY: Feminist Press, 1973, pp. 15–44; and Trail, Russell Thacher, *The Health and Diseases of Women.* Battle Creek, MI: Health Reformer, 1873, pp. 7–8.

10 Sydenham, Thomas, "Epistolary Dissertation on Hysteria," in *The Works of Thomas Sydenham*, transl. by R.G. Latham. London: Printed for the Sydenham Society, 1848, vol. 2, pp. 56 and 85; and Payne, Joseph Frank, *Thomas Sydenham.* New York: Longmans, Green and Co., 1900, p. 143.

11 Only a minority of writers on hysteria associated the affliction with paralysis until Freud made this part of the canonical disease paradigm in the twentieth century.

12 Gall, Franz Josef, *Anatomie et Physiologic du Système Nerveux en Génèral.* Paris: R Schoell, 1810–1819, vol. 3, p. 86; Tripier, Auguste Elisabeth Philogene, *Lemons Cliniques sur les Maladies de Femmes.* Paris: Octave Doin, Editeur, 1883, pp. 347–351; Highmore, Nathaniel, *De Passione Hysterica et Affectione Hypochondriaca.* Oxon.: Excudebat A. Lichfield impensis R. Davis, 1660, pp. 20–35; and Ellis, *Studies in the Psychology of Sex*, vol. 1, p. 225; see also Briquet, Pierre, *Traitè Clinique et Thérapeutique de l'Hystèrie.* Paris: J.B. Baillière et Fils, 1859, pp. 137–138, 570, and 613.

13 Cullen, William, *First Lines in the Practice of Physic.* Edinburgh: Bell, Bradfute, etc., 1791, pp. 43–47; Burton, Robert, *The Anatomy of Melancholy*, Floyd Dell and Paul Jordan Smith, eds. New York: Farrar and Rinehart, 1927, pp. 353–355; Horst, Gregor, *Dissertationem . . . inauguralem De Mania. . . . Gissae: typis Viduae Friederici Kargeri*, 1677, pp. 9–18; King, A. F. A., "Hysteria," *American Journal of Obstetrics*, May 18, 1891, vol. 24, no. 5, pp. 513–532; *Medieval Woman's Guide to Health*, transl. by Beryl Rowland. Kent, OH: Kent State University Press, 1981, pp. 2, 63, and 87; Pinel, Philippe, *A Treatise on Insanity*, transl. by D.D. Davis. Facsimile edition of the London 1806 edition; New York: Hafner, 1962, pp. 229–230; and Reich, Wilhelm, *Genitality*

in the Theory and Therapy of Neurosis, transl. by Philip Schmitz. New York: Farrar, Straus and Giroux, 1980 (reprint of 1927 edition), pp. 54–55 and 93.

14 Zacuto, Abraham, *Praxis Medica Admiranda*. London: Apud loannem—Antonium Huguetan, 1637, p. 267. Zacuto is at pains to point out that some physicians regard vulvular massage as indecent: "Num autem ex hac occasione, liceat Medico timenti Deum, sopitis pariter cunctis sensibus, & una abolita respiratione in foeminis quasi animam agentibus, seu in maximo vitae periculo constitutus, veneficium illud semen, foras ab utero, titillationibus, & frictionbius partium obscoenarium elidere, different eloquenter. . . ."

15 October–November, p. 4. Kellogg's background is described in detail in Schwarz, Richard W., *John Harvey Kellogg, MD*. Nashville: Southern Publishing Association, c1970.

16 Women who regularly undergo the discomfort of gynecological examination with this instrument are justifiably amused by its nineteenth century mythology. For an example of conservative views on the speculum, see Griesinger, Wilhelm, *Mental Pathology and Therapeutics*, transl. by C. Lockhart Robinson and James Rutherford. London: New Sydenham Society, 1867, p. 202. On the inefficiency of penetration as a means to female orgasm, the standard modern work is of course Hite, Shere, *The Hite Report on Female Sexuality*. New York: MacMillan Company, 1976, but the phenomenon was widely noted by progressive physicians and others before the seventies. Most of these latter, however, regarded the failure of penetration to fully arouse about three-quarters of the female population as either a pathology on the women's part or as evidence of a natural diffidence in the female. Hite is the first to point out that the experience of the majority constitutes a norm, not a deviation. For examples of various male views on this subject, see Hollender, Marc H., "The Medical Profession and Sex in 1900," *American Journal of Obstetrics and Gynecology*, vol. 108, no. 1, 1970, pp. 139–148; Degler, Carl, "What Ought to Be and What Was," *American Historical Review*, vol. 79, 1974, pp. 1467–1490; and his *At Odds: Women and the Family in America from the Revolution to the Present*. New York: Oxford University Press, 1980, pp. 249–278; and Tourette, Gilles de la, *Traite Clinique et Therapeutique de VHysterie Paroxystique*. Paris: Plon, 1895, vol. 1, p. 46. Feminine views are seldom recorded before this century; a few examples are those reported by Katherine B. Davis, summarized in Dickson, Robert L. and Henry Pierson, "The Average Sex Life of American Women," *Journal of the American Medical Association*, vol. 85, 1925, pp. 113–117; Lazarsfeld, Sofie, *Woman's Experience of the Male*. London: Encyclopedic Press, 1967, pp. 112, 181, 271, and 308. It has also been noted that few women have difficulty achieving orgasm in masturbation, and that the median time to orgasm in masturbation is substantially the same in both sexes: Kinsey, Alfred Charles, *Sexual Behavior in the Human Female*. Philadelphia: Saunders, 1953, p. 163.

17 Dowse, Thomas Stretch, *Lectures on Massage and Electricity in the Treatment of Disease*. Bristol: John Wright and Co., 1903, p. 181.

18 Smith, E.H., in *Pacific Medical Journal*, February 1903.

19 For examples of spa expenses in the United States, see Cloyes, Samuel A., *The Healer: The Story of Dr. Samantha S. Nivison and Dry den Springs, 1820–1915*. Ithaca, NY: DeWitt Historical Society of Tompkins County, 1969, p. 24; Karsh, Estrellita, "Taking the Waters at Stafford Springs," *Harvard Library Bulletin*, July 1980, vol. 28, no. 3, pp. 264–281; McMillan, Marilyn, "An Eldorado of Ease and Elegance: Taking the Waters at White Sulphur Springs," *Montana*, vol. 35, Spring 1985, pp. 36–49; and Meeks, Harold, "Smelly, Stagnant and Successful: Vermont's Mineral Springs," *Vermont History*, 1979, vol. 47, no. 1, pp. 5–20.

20 Taylor wrote indefatiguably on the subject of physical therapies for pelvic disorders and devoted considerable effort to the invention of mechanisms for this purpose. See Taylor, George Henry, *Diseases of Women*. Philadelphia and New York: G. McClean, 1871; *Health for Women*. New York: John B. Alden, 1883 and eleven subsequent editions; "Improvements in Medical Rubbing Apparatus," U.S. Patent 175,202 dated March 21, 1876; *Mechanical Aids in the Treatment of Chronic Forms of Disease*. New York: Rodgers, 1893; *Pelvic and Hernial Therapeutics*. New York: J.B. Alden, 1885; and "Movement Cure," U.S. Patent 263,625 dated August 29, 1882.

21 An example of the early Weiss model is available for study at the Bakken (Library and Museum), Minneapolis, MN, accession number 82.100.

22 Mortimer Granville, Joseph, *Nerve-Vibration and Excitation as Agents in the Treatment of Functional Disorders and Organic Disease*. London: J.&A. Churchill, 1883, p. 57; his American colleague Noble Murray Eberhart advises against vibrating pregnant women "about the generative organs" for fear of producing contractions. See his *A Brief Guide to Vibratory Technique*. 4th ed. rev. and enl. Chicago: New Medicine Publication, c1915, p. 59. For examples of enthusiastic endorsements of the new technology, see Gottschalk, Franklin Benjamin, *Practical Electro-Therapeutics*. Hammond, IN: F.S. Betz, 1904; the same author's *Static Electricity, X-Ray and*

Electro-Vibration: Their Therapeutic Application. Chicago: Eisele, 1903; International Correspondence Schools, *A System of Electrotherapeutics.* Scranton, PA: International Textbook Company, 1903, vol. 4; Matijaca, Anthony, *Principles of Electro-Medicine, Electro-Surgery and Radiology.* Tangerine, FL, Butler, NJ, and New York, NY: Benedict Lust, 1917; Monell, Samuel Howard, *A System of Instruction in X-Ray Methods and Medical Uses of Light, Hot-Air, Vibration and High Frequency Currents.* New York: E.R. Pelton, 1902; Pilgrim, Maurice Riescher, *Mechanical Vibratory Stimulation; Its Theory and Application in the Treatment of Disease.* New York City: Lawrence Press, c1903; Rice, May Cushman, *Electricity in Gynecology.* Chicago: Laing, 1909; Rockwell, Alphonse David, *The Medical and Surgical Uses of Electricity.* New ed. New York: E.G. Treat, 1903; Snow, *Mechanical Vibration and its Therapeutic Application*; Waggoner, Melanchthon R., *The Note Book of an Electro-Therapist.* Chicago: Mcintosh Electrical Corporation, 1923; Wallian, Samuel Spencer, *Rhythmotherapy.* Chicago: Ouellette Press, 1906; and the same author's "Undulatory Theory in Therapeutics," *Medical Brief*, May and June 1905.

23 See for example, Smith, A. Lapthorn, "Disorders of Menstruation," in *An International System of Electro-Therapeutics*, Horatio Bigelow, ed. Philadelphia: F.A. Davis, 1894, p. G163.

24 Covey, Alfred Dale, *Profitable Office Specialities.* Detroit: Physicians Supply Co., 1912, pp. 16, 18, and 79–95; Bubier, Edward Trevert, *Electro-Therapeutic Hand Book.* New York: Manhattan Electric Supply Co., 1900; Duck, J.J. Co., *Anything Electrical: Catalog No. 6.* Toledo, OH: J.J. Duck, 1916, p. 162; Golden Manufacturing Co., *Vibration: Nature's Great Underlying Force for Health, Strength and Beauty.* Detroit, MI: Golden Manufacturing Co., 1914; Gorman, Sam J. Co. *Physicians Vibragenitant.* Chicago: Sam J. Gorman and Co., n.d.; Keystone Electric Co., *Illustrated Catalogue and Price List of Electro-Therapeutic Appliances . . . etc.* Philadelphia: Keystone Electric Company, c1903, pp. 63–66; Schall and Son, Ltd., *ElectroMedical Instruments and Their Management . . .* 17th ed. London and Glasgow: Schall and Son, 1925; Vibrator Instrument Co., *A Treatise on Vibration and Mechanical Stimulation.* Chattanooga, TN: Vibrator Instrument, 1902; Vibrator Instrument Co. Clinical Dept., *A Course on Mechanical Vibratory Stimulation.* New York City: Vibrator Instrument, 1903; "Vibratory Therapeutics," *Scientific American*, vol. 67, October 22, 1892, p. 265. Most of these manufacturers were quite respectable instrument firms; see Davis, Audrey B., *Medicine and Its Technology: An Introduction to the History of Medical Instrumentation.* Westport, CT: Greenwood Press, 1981, p. 22.

25 Vibrator Instrument Company, *Chattanooga Vibrator.* Chattanooga, TN: Vibrator Instrument, 1904, pp. 3 and 26.

26 Vigouroux, Auguste, *Ètude sur la Rèsistance Èlectrique chez les Melancoliques.* Paris: J. Rueff et Cie, Editeurs, 1890; Cowen, Richard J., *Electricity in Gynecology.* London: Bailière, Tindall and Cox, 1900; Engelmann, George J., "The Use of Electricity in Gynecological Practice," *Gynecological Transactions*, vol. 11, 1886; Reynolds, David V., "A Brief History of Electrotherapeutics," in *Neuroelectric Research*, D.V. Reynolds and A. Sjoberg, eds. Springfield, IL: Thomas, 1971, pp. 5–12; and Shoemaker, John V., "Electricity in the Treatment of Disease," *Scientific American Supplement*, January 5, 1907, vol. 63, pp. 25923–25924.

27 "Vibratory Therapeutics," *Scientific American*, vol. 67, October 22, 1892, p. 265.

28 O'Neill, John J., *Prodigal Genius: The Life of Nikola Tesla.* New York: Ives Washburn, Inc., 1944, p. 210.

29 *Medical Brief*, May 1905, p. 417. See also the theory of light vibrations employed in the Master Electric Company's advertising brochure *The Master Violet Ray.* Chicago: n.d.

30 Monell, *A System of Instruction . . .*, p. 595; Snow, *Mechanical Vibration.*

31 Monell, *A System of Instrumentation . . .*, p. 591.

32 See for example, "Wanted, Agents and Salesman . . ." Swedish Vibrator Company, *Modern Priscilla*, April 1913, p. 60.

33 "Agents! Drop Dead Ones!" Blackstone Water Power Vacuum Massage Machine, *Hearst's*, April 1916, p. 327; and "Hydro-Massage" Warner Motor Company, *Modern Women*, vol. 11, no. 1, December 1906, p. 190.

34 Wall receptacles are a relatively late introduction. See Schroeder, Fred E., "More 'Small Things Forgotten:' Domestic Electrical Plugs and Receptacles, 1881–1931," *Technology and Culture*, July 1986, vol. 27, no. 3, pp. 525–543.

35 "Massage is as old as the hills . . . ," American Vibrator Company, *Woman's Home Companion*, April 1906, p. 42.

36 "Such Delightful Companions!" Star Electrical Necessities, 1922, reproduced in Jones, Edgar R., *Those Were the Good Old Days.* New York: Fireside Books, 1959, unpaged; and "A Gift that Will Keep Her Young and Pretty," Star Home Electric Massage, *Hearst's International*, December 1921, p. 82.

37 See, for example, the Ediswan advertisement in *Electrical Age for Women*, January 1932, vol. 2, no. 7, p. 274, and review on page 275 of the same publication.
38 "Vibration Is Life," Lindstrom-Smith Co., *Modern Priscilla*, December 1910, p. 27.
39 Pivar, David J., *Purity Crusade: Sexual Morality and Social Control, 1868–1900.* Westport, CT: Greenwood Press, 1973, pp. 110–117.
40 See also *American Magazine*, December 1912, vol. 75, no. 2, January 1913, vol. 75, no. 3, May 1913, vol. 75, no. 7, p. 127; *Needlecraft*, September 1912, p. 23; *Home Needlework Magazine*, October 1908, p. 479, October 1915, p. 45; *Hearst's*, January 1916, p. 67, February 1916, p. 154, April 1916, p. 329, June 1916, p. 473; and *National Home Journal*, September 1908, p. 15.
41 Davidson, J.E., "Electrical Appliance Sales During 1926," *NELA Bulletin*, vol. 14, no. 2, pp. 119–120.
42 December 5,1925, vol. 86, p. 1164. See also Hughes, George A. "How the Domestic Electrical Appliances Are Serving the Country," *Electrical Review*, June 15, 1918, vol. 72, p. 983; Edkins, E.A., "Prevalent Trends of Domestic Appliance Market," *Electrical World*, March 30, 1918, pp. 670–671; and "Surveys Retail Sale of Electrical Appliances," *Printer's Ink*, vol. 159, May 19, 1932, p. 35.
43 "Electrical Devices for the Household," *Scientific American*, January 26, 1907, vol. 96, p. 95.
44 The vibrator is not included in extensive lists of appliances in Lamborn, Helen, "Electricity for Domestic Uses," *Harper's Bazaar*, April 1910, vol. 44, p. 285; and Knowlton, H.S., "Extending the Uses of Electricity," *Cassier's Magazine*, vol. 30, June 1906, pp. 99–105.
45 U.S. Bureau of the Census, *Census of Manufactures*, 1908, 1919, and 1947, pp. 216–217, 203, and 734 and 748, respectively.
46 On the early history of appliances, see Lifshey, Earl, *The Housewares Story*. Chicago: Housewares Manufacturers' Association, 1973. For the safety issue, see "Electromedical Apparatus for Domestic Use," *Electrical Review*, October 22, 1926, p. 682.
47 *Good Stories*, October 1933, p. 2; see also similar advertisement in the same issue for Dr. Roger's Relief Compound, p. 12.
48 "Water Stills," *Ainslee's Magazine*, October 1920, p. 164.
49 See, for example, Damark International, Inc., *Catalog B-330*. Minneapolis, MN: Damark International, 1988, p. 7, which emphasizes the "Alambiccus Distiller's" usefulness for distilling herbal extracts.
50 *Mellow Mail Catalogue*. Cooper Station, New York City: 1984, pp. 32–39.
51 Levy, Steven, *Hackers: Heroes of the Computer Revolution*. Garden City, NY: Anchor Press/Doubleday, 1984, p. 377.

20

On the Cutting Edge

Cosmetic Surgery and New Imaging Technologies

Anne Balsamo

COSMETIC SURGERY AND THE INSCRIPTION OF CULTURAL STANDARDS OF BEAUTY

Cosmetic surgery enacts a form of cultural signification where we can examine the literal and material reproduction of ideals of beauty. Where visualization technologies bring into focus isolated body parts and pieces, surgical procedures actually carve into the flesh to isolate parts to be manipulated and resculpted. In this way cosmetic surgery *literally* transforms the material body into a sign of culture. The *discourse* of cosmetic surgery offers provocative material for discussing the cultural construction of the gendered body because women are often the intended and preferred subjects of such discourse and men are often the agents performing the surgery. Cosmetic surgery is not simply a discursive site for the "construction of images of women," but a material site at which the physical female body is surgically dissected, stretched, carved, and reconstructed according to cultural and eminently ideological standards of physical appearance.

There are two main fields of plastic surgery. Whereas *reconstructive* surgery works to repair catastrophic, congenital, or cancer-damage deformities, *cosmetic* or aesthetic surgery is often an entirely elective endeavor. And whereas reconstructive surgery is associated with the restoration of health, normalcy, and *physical function*, cosmetic surgery is said to improve self-esteem, social status, and sometimes even professional standing.

All plastic surgery implicitly involves aesthetic judgments of proportion, harmony, and symmetry. In fact, one medical textbook strongly encourages plastic surgeons to acquire some familiarity with classical art theory so that they are better prepared to "judge human form in three dimensions, evaluate all aspects of the deformity, visualize the finished product, and plan the approach that will produce an optimal result."[1] Codifying the aspects of such an "aesthetic sense" seems counterintuitive, but in fact there is a voluminous literature that reports the scientific measurement of facial proportions in an attempt to accomplish the scientific determination of aesthetic perfection. According to one surgeon, William Bass, most cosmetic surgeons have some familiarity with the anthropological fields of anthropometry and human osteology. Anthropometry—defined in one source as "a technique for the measurement of men, whether living or dead"—is actually a critically important science used by a variety of professional engineers and designers.[2] One example of practical anthropometry is the collection of measurements of infants' and children's bodies for use in the design of automobile seat restraints.[3] Of course it makes a great deal of sense that measurement standards and scales of human proportions are a necessary resource for the design of products for human use; in order to achieve a "fit" with the range of human bodies that will eventually use and inhabit a range of products from office chairs to office buildings, designers must have access to a reliable and standardized set of body measurements.[4] But when the measurement project identifies the "object" being measured as the "American Negro" or the "ideal female face," it is less clear what practical use these measurements serve.[5]

If anthropometry is "a technique for the measurement of men," the fascination of plastic surgeons is the measurement of the ideal. One well-cited volume in a series published by the American Academy of Facial Plastic and Reconstructive Surgery, titled *Proportions of the Aesthetic Face* (by Nelson Powell and Brian Humphreys), proclaims that it is a "complete sourcebook of information on facial proportion and analysis."[6] In the preface the authors state:

> The face, by its nature, presents itself often for review. We unconsciously evaluate the overall effect each time an acquaintance is made. . . . This [impression] is generally related to some scale of beauty or balance. . . . The harmony and symmetry are compared to a mental, almost magical, ideal subject, which is our basic concept of beauty. Such a concept or complex we shall term the "ideal face."[7]

According to the authors, the purpose of their text is quite simple: to document, objectively, the guidelines for facial symmetry and proportion. Not inconsequentially, the "Ideal Face" depicted throughout this book is of a white woman whose face is perfectly symmetrical in line and profile. The authors claim that although the "male's bone structure is sterner, bolder, and more prominent . . . the ideals of facial proportion and unified interplay apply to either gender." As I discuss later, this focus on the female body is prevalent in all areas of cosmetic surgery—from the determination of ideal proportions to the marketing of specific cosmetic procedures. The source or history of these idealized drawings is never discussed. But once the facial proportions of these images are codified and measured, they are reproduced by surgeons as they make modifications to their patients' faces. Even though they work with faces that are individually distinct, surgeons use the codified measurements as guidelines for determining treatment goals in the attempt to bring the distinctive face in alignment with artistic ideals of symmetry and proportion.

The treatment of race in this book on "ideal proportions of the aesthetic face" reveals a preference for white, symmetrical faces that heal (apparently) without scarring. On the one hand, the authors acknowledge that "bone structure is different in all racial identities" and that "surgeons must acknowledge that racial qualities are appreciated differently in various cultures," but in the end they argue that "the facial form [should be] able to confer harmony and aesthetic appeal regardless of race."[8] It appears that this appreciation for the aesthetic judgment "regardless of race" is not a widely shared assumption among cosmetic surgeons. Napoleon N. Vaughn reports that many cosmetic surgeons, "mindful of keloid formation and hyperpigmented scarring, routinely reject black patients."[9] But the issue of scar tissue formation is entirely ignored in the discussion of the "proportions of the aesthetic face." Powell and Humphreys implicitly argue that black faces can be evaluated in terms of ideal proportions determined by the measurement of Caucasian faces, but they fail to address the issue of postsurgical risks that differentiate black patients from Caucasian ones.[10] Although it is true that black patients and patients with dark ruddy complexions have a greater propensity to form keloids or hypertrophic scars than do Caucasian patients, many physicians argue that black patients who are shown to be prone to keloid formation in the lower body are not necessarily prone to such formations in the facial area and upper body; therefore a racial propensity for keloid formation should not be a reason to reject a black patient's request for facial cosmetic surgery.[11] And according to Arthur Sumrall, even though "postoperative dyschromic changes and surgical incision lines are much more visible in many black patients and races of color than their Caucasian counterparts," these changes and incision lines greatly improve with time and corrective cosmetics.[12] As an abstraction, the "aesthetic face" is designed to assist surgeons in planning surgical goals; but as a cultural artifact, the "aesthetic face" symbolizes a desire for standardized ideals of Caucasian beauty.

It is clear that all plastic surgery invokes standards of physical appearance and functional definitions of the "normal" or "healthy" body. Upon closer investigation we can see how these standards and definitions are culturally determined. In the 1940s and 1950s, women reportedly wanted "pert, upturned noses," but according to one recent survey this shape has gone out of style: "the classic, more natural shape is the ultimate one with which to sniff these days."[13] The obvious question becomes, what condition does the adjective "natural" describe? In this case we can see how requests for cosmetic reconstructions show the waxing and waning of fashionable desires; in this sense, "fashion surgery" might be a more fitting label for the kind of surgery performed for nonfunctional reasons. But even as high fashion moves toward a multiculturalism in the employment of nontraditionally beautiful models,[14] it is striking to learn how great is the demand for cosmetic alterations that are based on Western markers of ideal beauty. In a *New York Times Magazine* feature, Ann Louise Bardach reports that Asian women often desire surgery to effect a more "Western"-shaped eye.[15] Indeed, in some medical articles this surgery is actually referred to as "upper lid westernization," and is reported to be "the most frequently performed cosmetic procedure in the Orient."[16] Surgeons Hall, Webster, and Dobrowski explain:

> An upper lid fold is considered a sign of sophistication and refinement to many Orientals across all social strata. It is not quite accurate to say that Orientals undergoing this surgery desire to look Western or American; rather, they desire a more refined Oriental eye. . . . An upper lid westernization blepharoplasty frequently is given to a young Korean woman on the occasion of her betrothal.[17]

Although other surgeons warn that it is "wise to discuss the Oriental and Occidental eye anatomy in terms of differences *not* defects,"[18] another medical article on this type of surgery was titled "Correction of the Oriental Eyelid."[19] In terms of eyelid shape and design, Hall and his colleagues do not comment on how the "natural" Oriental eye came to be described as having a "poorly defined orbital and periorbital appearance"; thus, when their Oriental patients request "larger, wider, less flat, more defined, more awake-appearing eyes and orbital surroundings," these surgeons offer an operative plan for the surgical achievement of what is commonly understood to be a more Westernized appearance.[20] In discussing the reasons for the increased demand for this form of blepharoplasty "among the Oriental," Marwali Harahap notes that this technique became popular after World War II; this leads some surgeons to speculate that such a desire for Westernized eyes "stem[s] from the influence of motion pictures and the increasing intermarriage of Asian women and Caucasian men."[21]

THE MARKETING OF YOUTHFULNESS

When a young girl born with "hidden eyes" was scheduled to have massive face reconstruction surgery, surgeons hoped to construct eyelids for her where there were none.[22] The key objectives for her eye surgery were "normalcy" and "functionality"; however, a review of medical literature on reconstructive surgery reveals that blepharoplasty (eyelid operations) is a common technique of "youth surgery."[23] Because body tissue loses its elasticity in the process of aging, eyelids often begin to sag when a person reaches the early fifties. Bagginess is caused by fat deposits that build up around the eye and stretch the skin, producing wrinkling and sagging, and is most likely the result of a hernia—the weakening of the tissue around the eye—in which the fat deposits push outward and downward. Although eyestrain and fatigue can result from overworking the muscles around the eyes in an effort to keep eyes looking alert and open, eyelid surgery very rarely involves a "catastrophic" or "cure-based" medical rationale. Yet it is quite common, in both the popular and professional literature, for a plastic surgeon to refer to eye bags as a "deformity." This is a simple example of the way in which "natural" characteristics of the aging body are redefined as "symptoms," with the consequence that cosmetic surgery is rhetorically constructed as a medical procedure with the power to "cure" or "correct" such physical deformities.[24]

Several types of aesthetic surgery have been marketed explicitly for an aging baby-boomer population, with the promise that external symptoms of aging can be put off, taken off, or virtually eliminated. By the end of the 1980s, the most requested techniques of cosmetic surgery included face lifts, nose reconstructions, tummy tucks, liposuction, skin peels, and hair transplants—surgical techniques that are specifically designed to counteract the effects of gravity and natural body deterioration.[25] More than a few articles have reported that baby boomers are the preferred market for these new medical procedures; as a demographic group they (1) have more money than time to spend on body maintenance, and (2) are just beginning to experience the effects of aging en masse.[26] Given the size of the baby-boomer population, it is no surprise that as the first wave of baby boomers reach their late forties we should see an increase in advertisements for services such as dental bonding and implants, requests for "revolutionary" new drugs such as Retin-A, and articles about rejuvenation drugs manufactured in Europe from dried fetal extracts.[27] Even though the size of the target market for these produces will continue to increase during the next decade, the competition among plastic surgeons has so intensified that many of them are using image consultants to design advertising campaigns to attract clients. One campaign that drew a round of criticism from other surgeons displayed a surgically sculpted shapely female body draped over an expensive car. While this is hardly a new combination for U.S. beer advertisers, many cosmetic surgeons claimed that such advertising tarnishes the dignified image of their medical profession.[28]

One of the consequences of the commodification and, correspondingly, the normalization of cosmetic surgery is that electing *not* to have cosmetic surgery is sometimes interpreted as a failure to deploy all available resources to maintain a youthful, and therefore socially acceptable and attractive, body appearance.[29] Kathryn Pauly Morgan, in an essay in a special issue of *Hypatia* on "Feminism and the Body," argues that the normalization of cosmetic surgery—"the inversion of the domains of the deviant and the pathological"—are "catalyzed by the technologizing of women's bodies."[30] From this point, Morgan goes on to discuss the more philosophical question of why "patients and cosmetic surgeons participate in committing one of the deepest of original philosophical sins, the choice of the apparent over the real." The issue I'd like to consider, drawing on Morgan's analysis of the increasing "naturalization" of cosmetic alteration, is to elaborate the mechanism whereby the apparent is transformed into the real. How are women's bodies technologized? What is the role of cosmetic surgery in the technological reproduction of gendered bodies?

COSMETIC SURGERY AS A TECHNOLOGY OF THE GENDERED BODY

In recent years, more men are electing cosmetic surgery than in the past, but often in secret. As one article reports, "previously reluctant males are among the booming number of men surreptitiously doing what women have been doing for years: having their eyelids lifted, jowls removed, ears clipped, noses reduced, and chins tightened."[31] One cosmetic surgeon elaborates the reasons why men are beginning to seek elective cosmetic surgery:

> A middle-aged male patient—we'll call him Mr. Dropout—thinks he has a problem. He doesn't think he's too old for the lovely virgins he meets, but he wants to improve things. . . . When a man consults for aging, generally he is not compulsive about looking younger but he seeks relief from one or more specific defects incidental to aging: male pattern baldness . . . forehead wrinkling . . . turkey-gobbler neck. There are many things that can be done to help the aging man look younger or more virile.[32]

According to yet another cosmetic surgeon, the reason for some men's new concern about appearance is "linked to the increasing competition for top jobs they face at the peak of their careers from women and Baby Boomers."[33] Here the increase in male cosmetic surgery is explained as a shrewd business tactic: "looking good" connotes greater intelligence, competence, and desirability as a colleague. Charges of narcissism, vanity, and self-indulgence are put aside; a man's choice to have cosmetic surgery is explained by appeal to a rhetoric of career enhancement: a better looking body is better able to be promoted. In this case, cosmetic surgery is redefined as a body management technique designed to reduce the stress of having to cope with a changing work environment, one that is being threatened by the presence of women and younger people.[34] While all of these explanations may be true—in the sense that this is how men justify their choice to elect cosmetic surgery—it is clear that other explanations are not even entertained. For example, what about the possibility that men and women are becoming more alike with respect to "the body beautiful," that men are engaging more frequently in female body activities, or even simply that a concern with appearance isn't solely a characteristic of women? What about the possibility that the boundary between genders is eroding? How is it that men avoid the pejorative labels attached to female cosmetic surgery clients?[35]

In their ethnomethodological study of cosmetic surgery, Diana Dull and Candace West examine how surgeons and patients "account" for their decisions to elect cosmetic surgery. They argue that when surgeons divide the patient's body into component parts and pieces, it enables both "surgeons and patients together [to] establish the problematic status of the part in question and its 'objective' need of 'repair.' "[36] Dull and West go on to argue that this process of fragmentation occurs "in tandem with the accomplishment of gender" which, in relying upon an essentialist view of the female body as always "needing repair," understands women's choice for cosmetic surgery as "natural" and "normal" and as a consequence of their (natural) preoccupation with appearance. However, because their "essential" natures are defined very differently, men must construct elaborate justifications for their decision to seek cosmetic alterations.

This analysis illuminates one of the possible reasons why men and women construct different accounts of their decision to elect cosmetic surgery: the cultural meaning of their gendered bodies already determines the discursive rationale they can invoke to explain bodily practices. Although the bodies and faces of male farmers and construction workers, for example, are excessively "tanned" due to their constant exposure to the sun as part of their work conditions, their ruddy, leathery skin is not considered a liability or deformity of their male bodies. In contrast, white women who display wrinkled skin due to excessive tanning are sometimes diagnosed with "The Miami Beach Syndrome"; as one surgeon claims, "we find this type of overly tanned, wrinkled skin in women who not only go to Miami every year for three or four months, but lie on the beach with a sun reflector drawing additional rays to their faces."[37] It is no surprise, then, that although any body can exhibit the "flaws" that supposedly justify cosmetic surgery, discussion and marketing of such procedures usually constructs the female as the typical patient. Such differential treatment of gendered bodies illustrates a by now familiar assertion of feminist studies of the body and appearance: the meaning of the presence or absence of any physical quality varies according to the gender of the body upon which it appears. Clearly an apparatus of gender organizes our seemingly most basic, natural interpretation of human bodies, even when those bodies are technologically refashioned. Thus it appears that although technologies such as those used in cosmetic surgery can reconstruct the "natural" identity of the material body, they do little to disrupt naturalization of feminine corporeal identity.

Wendy Chapkis amplifies this point when she writes: "however much the particulars of the beauty package may change from decade to decade—curves in or out, skin delicate or ruddy, figures fragile or fit—the basic principles remain the same. The body beautiful is woman's responsibility and authority. She will be valued and rewarded on the basis of how close she comes to embodying the ideal."[38] In the *popular media* (newspapers, magazines), advertisements for surgical services are rarely, if ever, addressed specifically to men. When a man is portrayed as a prospective patient for cosmetic surgery, he is often represented as a serious "business" person for whom a youthful appearance is a necessary business asset. And of course, many cosmetic alterations are designed especially for women: tattooed eyeliner (marketed as "the ultimate cosmetic"), electrolysis removal of superfluous hair, and face creams.[39] An advertising representative for DuraSoft explains that the company has begun marketing its colored contact lenses specifically to black women ostensibly because DuraSoft believes that "black women have fewer cosmetic alternatives," but a more likely reason is that the company wants to create new markets for its cosmetic lenses.[40] The codes that structure cosmetic surgery advertising are gendered in stereotypical ways: being male requires a concern with virility and productivity, whereas being a real woman requires buying beauty products and services.[41]

And yet women who have too many cosmetic alterations are pejoratively labeled "scalpel slaves," to identify them with their obsession for surgical fixes.[42] Women in their late thirties and forties are the most likely candidates for repeat plastic surgery. According to *Psychology Today*, the typical "plastic surgery junkie" is a woman who uses cosmetic surgery as an opportunity to "indulge in unconscious wishes."[43] *Newsweek* diagnoses the image problems of "scalpel slaves":

> Women in their 40s seem particularly vulnerable to the face-saving appeal of plastic surgery. Many scalpel slaves are older women who are recently divorced or widowed and forced to find jobs or date again. Others are suffering from the empty-nest syndrome. "They're re-entry women," says Dr. Susan Chobanian, a Beverly Hills cosmetic surgeon. "They get insecure about their appearance and show up every six months to get nips and tucks. . . . Plastic-surgery junkies are in many ways akin to the anorexic or bulimic," according to doctors. "It's a body-image disorder," says [one physician]. "Junkies don't know what they really look like." Some surgery junkies have a history of anorexia in the late teens, and now, in their late 30s and 40s, they're trying to alter their body image again.[44]

The naturalized identity of the female body as pathological and diseased is culturally reproduced in media discussions and representations of cosmetic surgery services. Moreover, the narrative obsessively recounted is that the female body is flawed in its distinctions and perfect when differences are transformed into sameness. However, in the case of cosmetic surgery the nature of the "sameness" is deceptive, because the promise is not total identity reconstruction—such that a patient could choose to look like the media star of her choice—but rather the more elusive pledge of "beauty enhancement." When cosmetic surgeons argue that the technological elimination of facial "deformities" will enhance a woman's "natural" beauty, we encounter one of the more persistent contradictions within the discourse of cosmetic surgery: namely, the use of technology to augment "nature."

MORPHING AND THE TECHNO-BODY

Surgeons are taught that the consultation process is actually an incredibly complex social exchange during which patients and surgeons must negotiate highly abstract goals. The accomplishment of goals is said to be directly related to patient satisfaction:

> [D]efining aesthetic goals with patients obviously involves the hazards of perception. . . . Any practitioner who has recommended and performed orthognathic surgery has most likely encountered patients with unrealistic aesthetic expectations. The surgical team most often accomplishes their functional and aesthetic goals, but, in this situation, the patient is disappointed. . . . Function, aesthetics, and shaping the patient's expectations into reality must all be addressed while keeping in mind the patient's best interests and desires.[45]

The most commonly used methods of patient facial analysis are radiographic and photographic analysis, where the facial profile is rendered in a two-dimensional medium.[46] The use of photographs and grease pencils is perhaps the simplest method of the surgeon-patient consultation where the task at hand is to suggest the possible benefits of cosmetic surgery at the same time that the patient must be made aware of the surgical plan. Using a Polaroid camera to produce an instantaneous photograph, surgeons often draw lines with markers to indicate the locations of incisions or stretch lines. "Photograph surgery" is a communication method to negotiate between a patient's expectations and likely surgical outcomes; the reality of those black grease-pencil lines invokes the use of surgical procedures that literally cut into the face and reconstruct it, rendering whatever features nature created obsolete and irrecoverable.[47]

The various two-dimensional consultation methods were developed to effect an "objective method of facial analysis," which is understood to be a necessary part of adequate preoperative planning and postoperative evaluation.[48] Since 1989, however, some cosmetic surgeons have been employing new visualization techniques that render the patient's face in three dimensions. The use of video imaging replaces the use of grease-pencil lines and photographic surgery, which some surgeons found to be an inadequate system of consultation because "even when adjustments have been 'drawn on' by the surgeon, it is difficult for most patients to imagine what they might look like postoperatively."[49] Using video imaging, the surgeon can manipulate an actual image of the client's face. Although the cost and skill requirements of these computerized imaging systems represents a sizable investment, using this method of consultation is promoted as a way to manage patient expectations because it provides more information about the results that surgery can accomplish. More information, in this case, is said to lead to greater patient reassurance. Indeed, one recent study reports that the use of video imaging was well accepted by patients and that most felt that "video imaging improved communication between patient and surgeon, increased confidence in surgery and surgeon, and enhanced the patient-physician relationship."[50]

The video imaging consultation begins with a series of video shots that must be taken with great precision in terms of camera angle, lighting, face position, makeup, and hair display.[51] Preoperative photograph precision is necessary to ensure that postoperative photographs will objectively record surgical results and not camera special effects. The preoperative video shots are digitally scanned into a computer and then manipulated with the use of an imaging processing system. To begin the consultation, the cosmetic surgeon displays two images of the patient's face on the computer screen. The left-hand image remains untouched and unmarked, serving as the prototypical "before" picture of the prospective cosmetic surgery client. The right-hand image is manipulated by the cosmetic surgeon, using a stylus and pressure-sensitive sketch pad. Using what is really a modified computer "painting" program, the surgeon can manipulate the image in several ways: (1) by picking up a line (a jaw line, for example) and moving it; (2) by reducing a part of the image with an eraser tool (thus eliminating a double chin, for example); or (3) by stretching a part of the face to show what heightened cheekbones might look like. Throughout the various manipulations, the right-hand image of the patient retains its visual integrity in that it continues to resemble the original, left-hand image save for the artistic manipulations performed by the surgeon. The surgeon can either display multiple procedures on one image or reproduce additional images that illustrate the effects of only one procedure at a time. With the use of a range of rendering tools, which are basically a set of artist's tools (spray can, pencil, eraser), the surgeon can redesign a client's face in the space of a 30-minute consultation.

In an interview with one surgeon who uses this method of patient consultation, he explained that when prospective patients seek surgery, they have only a layperson's understanding of facial anatomy. For example, they might believe that in order to get rid of deep lines around the nose that all they need is to stretch the cheeks and tuck the extra skin behind the ear. But what they really need, he clarified, is to heighten the cheekbones with an implant and bob the nose, which will pull the skin taut over the new cheeks; consequently the lines and folds on either side of the nose will be eliminated and the size of the nose will stay proportionate to cheek width. In this example, the imaging device would enable the surgeon to educate the patient about the different methods for accomplishing surgical goals. In fact, this surgeon emphasized that the imaging device allows him to visually demonstrate the transformation of the patient's face that he could easily accomplish in surgery, something very difficult to demonstrate in a two-dimensional format. For him, the imaging system is a mechanism whereby his artistic skill can be previewed by prospective patients.

The imaging program can also be used as a surgical planning device. The program can calculate the distance, angle, or surface of the part of the right-hand image that has been modified. In this sense, a manipulated video image is more useful than a photograph in designing the actual surgery, because the comparison between the video image and the cephalometric radiograph "allows for computerized quantification of treatment goals."[52] Thus, if a nose profile line has been redrawn, the imaging program can measure the difference between the redrawn line of the right-hand image and the original line on the left-hand image to determine the degree to which the nose needs to be modified during surgery; the surgeon can then use that measurement to plan the surgical procedure.[53]

Some physicians believe that the only way to manage patients' expectations is to assure them of the competency of the physician's skill. Traditionally, physicians have done this by showing a prospective patient photographs of previous patients' surgical results. But more recently, new high-tech imaging devices have been employed as a symbol of the quality of the physician's service.

> A computer imaging system is a wonderful educational tool in terms of marketing to patients who may not be familiar with the treatments and materials available today.... Marketing the benefits of the system to patients is easy, according to [another physician], because the "high-tech" equipment lets patients know that they can receive "high-tech" treatment. It gives you the image and identity of being on the *cutting edge* of dentistry when you can offer the newest and best materials and techniques available.[54]

So in addition to using it as a counseling and planning device, the video imaging system can also be employed as a marketing tool. In this case, the expert manipulation of a video file using a computer painting program is translated into a marker of technological expertise in the operating room. But this use of the imaging system as marketing tool is denounced by some surgeons, who believe that its use borders on the unethical because it makes it easier to manipulate patients into having procedures that they do not need or want.

During interviews with surgeons who use or have used a video imaging system, I specifically asked about the controversy surrounding the new technology. The strongest claim for the use of video imaging is that it provides a realistic image of the aesthetic treatment objective that the patient can visualize. So while some surgeons dismiss it as a possibly unethical marketing device, other physicians argue that it produces "realistic images," "realistic expectations," and a better representation of reality itself. More telling is the fact that several cosmetic surgeons in the Atlanta metro area have stopped using video imaging as a consultation method because they found that it encouraged patients to form unrealistic expectations about the kind of transformations that can be accomplished through surgical procedures. They report that patients seemed to believe that if a modification could be demonstrated on the video screen, then it could be accomplished in the operating room—that the video transformation guaranteed the physical transformation. Apparently, the digital transformation of one's own face produces a magical, liquid simulation that is difficult to reject. What some patients fail to understand is that one of the significant difficulties with any kind of cosmetic surgery is that soft tissue changes are impossible to predict accurately. A surgical incision or implanation always disrupts layers of skin, fat, and muscle. How those incised tissues heal is a very idiosyncratic matter—a matter of the irreducible distinctiveness of the material body. After hearing from a number of disappointed patients, members of the American Society of Plastic and Reconstructive Surgeons designed an official "Electronic Imaging Disclaimer" to be used by physicians who employ computerized images in preoperative consultations. Among the release statements that the patient must sign is one that reads: "I understand that because of the significant differences in how living tissue heals, there may be no relationship between the electronic images and my final surgical result."[55] Where advertising executives play with the possibilities of morphing political candidates,[56] cosmetic surgeons offer patients the promise of permanently "morphed" features. One of the key consequences that some surgeons have discovered is that witnessing video morphing dramatically undermines a patient's ability to distinguish between the real, the possible, and the likely in terms of surgical outcomes.

For some women, and for some feminist scholars, cosmetic surgery illustrates a technological colonization of women's bodies; others see it as a technology women can use for their own ends. Certainly, as I have shown here, in spite of the promise cosmetic surgery offers women for the technological reconstruction of their bodies, in actual application such technologies produce bodies that are very traditionally gendered. Yet I am reluctant to accept as a simple and obvious conclusion that cosmetic surgery is simply one more site where women are passively victimized. Whether as a form of oppression or a resource of empowerment, it is clear to me that cosmetic surgery is a practice whereby women consciously act to make their bodies mean something to themselves and to others.

Notes

1 In this article, the underlying aesthetic theory of plastic surgery is elaborared through the annotated meaning of A (awareness), R (relativity), and T (technique), where awareness means awareness of "universal qualities of form, content, lighting, color and symmetry coupled with a medical understanding of underlying anatomy and physiology"; relativity means understanding features in relation to a "norm"; and technique refers to measuring, rendering, and sculpting

techniques. Stewart D. Fordham, "Art for Head and Neck Surgeons," *Plastic and Reconstructive Surgery of the Head and Neck* (Proceedings of the Fourth International Symposium of the American Academy of Facial Plastic and Reconstructive Surgery), *Vol. I: Aesthetic Surgery*, ed. Paul H. Ward and Walter E. Berman (St. Louis, Mo.: C. V. Mosby, 1984) 5.

2 Anthropometry "can be further divided into [subfields]: somatometry—measurement of the body of the living and of cadavers; cephalometry—measurement of the head and face; osteometry—measurement of the skeleton and its parts; and craniometry—measurement of the skull." William M. Bass, *Human Osteology* (Columbia: Missouri Archaeological Society, 1971) 54. One of the well-cited texts on anthropometry is M. F. Ashley Montagu, *A Handbook of Anthropometry* (Springfield, IL: Charles C. Thomas, 1960).

3 Richard G. Snyder (and Highway Safety Research Institute, University of Michigan), "Anthropometry of Infants, Children, and Youths to Age 18 for Product Safety Design," *Final Report, Prepared for Consumer Product Safety Commission* (Warrendale, Pa.: Society for Automotive Engineers, 1977).

4 Stephen Pheasant, *Bodyspace: Anthropometry, Ergonomics, and Design* (Philadelphia: Taylor & Francis, 1986).

5 Melville Herskovits, *The Anthropometry of the American Negro* (New York: AMS Press, 1969); L. G. Farkas and J. C. Kolar, "Anthropometrics and Art in the Aesthetics of Women's Faces," *Clinics in Plastic Surgery* 14.4 (1987): 599–616.

6 Nelson Powell, DDS, MD, and Brian Humphreys, MD, *Proportions of the Aesthetic Face* (New York: Thieme-Stratton, 1984).

7 Powell and Humphreys, ix. Powell and Humphreys go on to claim that "Beauty itself is then a relative measure of balance and harmony, but most find it difficult to quantitate; however, lines, angles, and contours may be measured and gauged. Standards then can be established to evaluate the elusive goal of beauty" (ix). Thus the rest of their volume reports the geometrical constitution of the "Ideal Face." According to Powell and Humphreys, the ideal face is divided into five "major aesthetic masses," each of which is described in mathematical and geometrical detail in terms of anatomical distances, contour lines, and facial angles. The authors outline a method of analysis in which an "aesthetic triangle relates the major aesthetic masses of the face, forehead, nose, lips, chin and neck to each other" and propose that this method be used as a diagnostic tool, whereby dentofacial deformities are defined as deviation from the ideal proportions. Powell and Humphreys 51.

8 Powell and Humphreys 4.

9 Napoleon N. Vaughn, "Psychological Assessment for Patient Selection," *Cosmetic Plastic Surgery in Nonwhite Patients*, ed. Harold E. Pierce, MD (New York: Grune & Stratton, 1982) 245–251.

10 In fact, one of the most central issues discussed in Pierce's *Cosmetic Plastic Surgery in Nonwhite Patients* is that black patients, Oriental patients, and patients with dark ruddy complexions have a greater propensity to form keloids or hypertrophic scars than do Caucasian patients. Macy G. Hall Jr., MD, "Keloid-Scar Revision," Pierce 203–08.

11 Howard E. Pierce, "Ethnic Considerations," Pierce 37–49.

12 Arthur Sumrall, "An Overview of Dermatologic Rehabilitation: The Use of Corrective Cosmetics," Pierce 141–54.

13 Jackie White, "Classic Schnozz is 80s nose," *Chicago Tribune* 8 July 1988, sec. 2: I, 3.

14 The U.S. edition of *Elle* magazine offers several examples of a refashioned primitivism as high-fashion statement, where both fashion and fashion figures display lines and angles that depart significantly from the ideal (white) facial geometry discussed earlier. Two "alternative" fashion spreads feature the deconstructivist designs of Martin Margiela and provide a glimpse of the new antifashion movement: "From Our Chicest Radicals: Alternative Fashion Routes," *Elle* (Sept. 1991): 324–29; and Stephen O'Shea, "Recycling: An All-New Fabrication of Style," *Elle* (Oct. 1991): 234–39. For a discussion of the appeal of "the exotic woman" and the rise of new multicultural supermodels, see Glenn O'Brien, "Perfect Strangers: Our Love of the Exotic," *Elle* (Sept. 1991): 274–76. Two striking covers that feature both a black model and a white model—symbolizing the multicultural refashioning of ideals of feminine beauty—appear on the May 1988 and Nov. 1991 issues of *Elle*.

15 Bardach also reports that many Iranian women reportedly seek to replace their "strong arched noses" with small, pert, upturned ones. "The Dark Side of Cosmetic Surgery: Long Term Risks Are Becoming Increasingly Apparent," *New York Times Magazine* 17 Apr. 1988: 24–25, 51, 54–58.

16 Bradley Hall, Richard C. Webster, and John M. Dobrowski, "Blepharoplasty in the Oriental," Ward and Berman 210–25. Quotation is from page 210.

17 Hall et al. 210.

18 Richard T. Farrior and Robert C. Jarchow, "Surgical Principles in Face-lift," Ward and Berman 297–311.

19 J.S. Zubiri, "Correction of the Oriental Eyelid," *Clinical Plastic Surgery* 8 (1981): 725. For a discussion of the discursive strategies whereby Western scholars (anthropologists, scientists) construct "Oriental" as an ideological system of reference, see Edward Said, *Orientalism* (New York: Vintage, 1979).

20 Hall et al. 210.

21 Marwali Harahap, MD, "Oriental Cosmetic Blepharoplasty," Pierce 77–97. Quotation is from page 78.

22 The six-year-old girl was born with cryptopthalmos ("hidden eyes")—without normal eyelids or eye openings. She was treated by a University of Illinois surgeon who developed a technique for reconstructing normal openings. An ultrasound examination revealed that the girl had one eye, so the surgeon created a cavity around the eye and refashioned a "normal-appearing" set of upper and lower eyelids; hopefully, this will allow her to see. "Surgery Will Give Girl a Chance for Sight," *Chicago Tribune* 17 Jan. 1988, sec. 2: 1.

23 A 1989 article in *Longevity* magazine described in grisly detail the stitch-by-stitch procedure of a "tummy tuck"—which they identified as one technique of "youth surgery." John Camp, "Youth Surgery: A Stitch-by-Stitch Guide to Losing Your Tummy," *Longevity* June 1989: 33–35.

24 Shirley Motter Linde, ed., *Cosmetic Surgery: What It Can Do for You* (New York: Award, 1971) 7.

25 Martha Smilgis, "Snip, Suction, Stretch and Truss: America's Me Generation Signs Up for Cosmetic Surgery," *Time* 14 Sept. 1987: 70. One of the most ridiculous descriptions of cosmetic surgery applications appeared in the advertisement for a B.P.I. (Body Profile Improvement) consultation by an Atlanta plastic surgeon, who claimed to offer services for "cellulite correction" and "vertical gravity liposuction." Advertising slide show at Hoyt's Midtown Theater, opening night of *Death Becomes Her*, 31 July 1992.

26 Ruth Hamel, "Raging Against Aging," *American Demographics* 12 (Mar. 1990): 42–44. In a related article, another journalist speculated that the rising popularity of plastic surgery was evidence of the baby boomers' obsession with death and their search for some measure of control over their mortality. Debra Goldman, "In My Time of Dying: Babyboomers Experience Interest in Death," *ADWEEK* 33 (2 Mar. 1992), Eastern ed.: 18.

27 An article by Peter Jarer in *SELF* magazine describes the new tooth technology in a cavity-free era. Tooth cosmetics are a growing business and include such techniques as bonded porcelain veneers, dental implants to replace missing or decayed teeth, and ceramic braces that replace the metal ones of old. Peter Jaret, "Future Smiles," *SELF* Apr. 1989: 186–89.

Although this fear of aging might appear to be a new phenomenon, brought on by the aging baby boomers' confrontation with body deterioration, it is actually the case that from the early 1900s on, crow's-feet, the tiny wrinkles formed at the corner of the eyes, have been defined as an aging "condition" treatable through surgical methods. Sylvia Rosenthal, *Cosmetic Surgery: A Consumer's Guide* (Philadelphia: J. B. Lippincott, 1977).

Retin-A, a cream used for almost 20 years as an acne treatment, recently has been launched as a new "youth cream." Retin-A not only treats acne but also is effective in removing wrinkles and liver spots. John Voorhees, the scientist who first confirmed the ability of Retin-A to reverse skin damage is quoted as saying: "I don't want to say that this is the fountain of youth, but it's the closest thing we have today." An editorial in the *Journal of the American Medical Association*, referring to the significance of Voorhees's study, announced, "A new age has dawned." The day after the editorial appeared the stock value of Johnson & Johnson (the parent firm of Ortho Pharmaceutical) rose three points. Tim Friend, "Youth Cream: 'A New Age Has Dawned,' " *USA Today* 22 Jan. 1988, sec. 1: I.

Other rejuvenation drugs tested in Europe but not available in the United States include: Gerovital, a mixture of procaine and stabilizers that seems to improve memory, muscular strength, and skin texture; Centrophenoxine, a compound that slows the skin aging process; DHEA (dehydroepiandrosterone), a naturally occurring hormone found in young adults that has been found to increase survival and improve immune function in animals; Piracetam, a nootropic which shows some signs of improving memory function; and cerebral vasodilators, a category of drug that improves blood circulation to the brain, which in turn is supposed to improve mental ability. Lynn Payer, "Rejuvenation Drugs," *Longevity* June 1989: 25.

28 The news item read: "Dr Charles D. Smithdeal's ad in *Los Angeles* magazine is a definite eye-catcher. In full color, on a full page, model Rebecca Ferratti leans her nearly bare body on a red Ferrari. Her flawless proportions are credited to Smithdeal, a Los Angeles cosmetic

surgeon." Donna Kato, "A Shot of Glitz for Medical Marketing." *Chicago Tribune* 30 Jan. 1989, sec. 2: I, 3.

29 In discussing the nonwhite patient's motivation for rhinoplasty, Harold Pierce rejects the argument that "the non-white patient who seeks rhinoplasty is attempting, symbolically to deny his heritage"; rather, Pierce asserts that these patients want "a nose that is smaller, more symmetrical and pleasing in three dimensional contour—a desire shared by patients requiring rhinoplasty in all racial groups" ("Ethnic" 48). He notes the irony that in an era marked by increased displays of ethnic pride, the number of black and Asian cosmetic surgery patients is increasing. He explains this paradox by suggesting that economic forces demand an "attractive appearance" as a professional attribute.

30 Kathryn Pauly Morgan, "Women and the Knife: Cosmetic Surgery and the Colonization of Women's Bodies," *Hypatia* 6.3 (Fall 1991): 25–53. Quotation is from page 28.

31 Suzanne Dolezal, "More Men Are Seeing Their Future in Plastic—the Surgical Kind," *Chicago Tribune* 4 Dec. 1988, sec. 5, p. 13. Quote from page 13.

32 Michael M. Gurdin, MD, "Cosmetic Problems of the Male," Linde 105–14. Quotation is from page 107.

33 Dolezal, "More Men Are Seeing Their Future in Plastic," 13. A 1989 Ann Landers column reported that Texas prisons often provide free cosmetic surgery as therapy for convicts: "A convicted rapist serving time in Louisiana received an implanted testicle at Charity Hospital in New Orleans that cost the state an estimated $5,000. The implanted testicle replaced one that was diseased and had been surgically removed in 1987. The rationale offered by the Texas prison system suggests that cosmetic surgical procedures performed on inmates provide practice for plastic surgeons and that cosmetic surgery makes a person feel better about himself. Studies were cited to prove that inmates were less likely to return to prison if they had a higher level of self-esteem." Ann Landers column, 13 July 1989.

34 Numerous articles on "the cost of beauty" suggest that as women earn more money they will demand better cosmetic services and conveniences. *Vogue* reports that many companies are responding by offering convenient maintenance programs, which for "the new breed of executive woman" can become a substantial investment and part of her business style; for some executives, in fact, an important perk is a contract that covers the cost of image upkeep and exercise. This would suggest that the differences between men's and women's rationalizations for cosmetic surgery are eroding: women, too, are beginning to justify cosmetic alterations within a logic of the workplace. Dorothy Schefer, "The Real Cost of Looking Good," *Vogue* Nov. 1988: 157–68.

35 The horror stories of women who justify cosmetic surgery for business-related reasons are often reported with an exceedingly critical edge. A female real estate agent in Beverly Hills felt pretty enough in her own way, but totally inadequate when compared to the glamorous female clients she worked with. After three years of silicone treatments to produce artificial "high cheekbones," her face began changing grotesquely; relentless calls to her plastic surgeon went unanswered, and two years later he committed suicide. She is still plagued by shifting silicone lumps under her face skin, and though she has undergone surgery several times to repair the damage, she will never regain her previous unconstructed features. She is described as a woman who just wanted to get "an edge" on the competition, but ended up getting more than she bargained for. Bardach, "The Dark Side of Cosmetic Surgery" 24–25, 51, 54–58.

36 Dull and West offer an interesting analysis of the social process whereby gender is constructed. They label this process "the accomplishment of gender," which they describe as "an ethnomethodological view of gender as an accomplishment, that is, an achieved property of situated social action" (64). Building on their work, my essay is concerned with the elaboration of how gender is also a fully *cultural* accomplishment. Diana Dull and Candace West, "Accounting for Cosmetic Surgery: The Accomplishment of Gender," *Social Problems* 38.1 (Feb. 1991): 54–70. Quotation is from page 67.

37 Blair O. Rogers, MD, "Management after Surgery in Facial and Eyelid Patients," Linde, 53–61. Quotation is from page 57.

38 Wendy Chapkis, *Beauty Secrets: Women and the Politics of Appearance* (Boston: South End, 1986) 14.

39 Eye surgeon Giora Angres of Las Vegas implants a permanent eyeliner just under the skin, so it is always there. The most popular colors are earth-tone shades of gray and brown. Implanted pigments look very natural and last about 10 years. It takes 20 minutes to complete the tattooing effect, and costs from $800 to $1,000. "It's probably one of the most effective tattooing methods yet developed," says a spokesman for the American Academy of Ophthalmology. *American Health* Dec. 1984: 33. Removing tattooed eyeliner is becoming a common "spin-off" surgery. As

reported in the *Chicago Tribune*, two Chicago surgeons describe the difficulties of surgery performed on a woman unhappy with the appearance of her tattooed eyeliner. "Medical Notes," *Chicago Tribune* 21 Aug. 1988, sec. 2: 5.

40 In a 1990 survey, DuraSoft found that 43 percent of black women were interested in hazel lenses, 26 percent in blue, and 14 percent in green. Leslie Savon, "Green looks very natural on Black women; but in blue, they look possessed," *Village Voice* 2 May 1988: 5z.

41 Carol Lynn Mithers, "The High Cost of Being a Woman," *Village Voice* 24 Mar. 1987: 31.

42 In the medical literature, patients who show an insatiable desire or addiction to surgery are said to display a "Polysurgical Syndrome." In her article, "How to Recognize and Control the Problem Patient," Mary Ruth Wright argues that "surgical addiction reflects deep psychological conflicts" (532). Wright goes on to report that her research on the psychological profile of the cosmetic surgery patient supports the argument that "all cosmetic surgery patients are psychiatric patients" and that all are "potential problem patients" (530). According to Wright, paranoid schizophrenics are the most dangerous. She notes that "homicides involving elective surgeons are increasing as elective surgery increases" (532). She refers to the case of Dr. Vasquez Anon, who was assassinated after he refused to see a patient who wanted more surgery. Mary Ruth Wright, "How to Recognize and Control the Problem Patient," Ward and Berman 530–35. For more information on Dr. Anon, see U. T. Hinderer, "Dr. Vasquez Anon's Last Lesson," *Aesthetic Plastic Surgery* 2 (1978): 375.

43 Annette C. Hamburger, "Beauty Quest," *Psychology Today* May 1988: 28–32.

44 "Scalpel Slaves Just Can't Quit," *Newsweek* 11 Jan. 1988: 58–59.

45 David M. Sarver, DMD, Mark W. Johnston, DMD, and Victor J. Marukas, DDS, MD, "Video Imaging for Planning and Counselling in Orthognatic Surgery," *Journal Oral and Maxillofacial Surgery* 46 (1988): 939–45. Quotation is from page 939. The authors point out that mismatched goals are a common occurrence: "what surgeons or orthodonrists consider ideal may not be the same as the patient's desires" (939).

46 Wayne F. Larrabee Jr., John Sidles, and Dwight Sutton describe the traditional two-dimensional methods of facial analysis and the new three-dimensional digitizers that offer a new approach to facial analysis. Wayne F. Larrabee Jr., MD, John Sidles, and Dwight Sutton, "Facial Analysis," *Laryngoscope* 98 (Nov. 1988): 1273–75.

47 Kathryn Pauly Morgan does an excellent job of uncovering the "idea of pain" associated with the surgical instruments used by cosmetic surgeons: "Now look at the needles and at the knives. Look at them carefully. Look at them for a long time. Imagine them cutting into your skin. Imagine that you have been given this surgery as a gift from your loved one." Morgan, "Women and the Knife" 26.

48 Larrabee et al. 1274.

49 J. Regan Thomas, MD, M. Sean Freeman, MD, Daniel J. Remmler, MD, and Tamara K. Ehlert, MD, "Analysis of Patient Response to Preoperative Computerized Video Imaging," *Archives of Otolaryngol Head and Neck Surgery* 115 (July 1989): 793–96.

50 Thomas et al., "Facial Analysis" 793.

51 One Atlanta cosmetic surgeon uses a proprietary image-processing system designed by Truevision, Inc. (Indianapolis), which includes an IBM computer (with peripherals), mouse and tablet, analog RGB monitor and video camera, and Truevision's TARGA+ board and Imager-I software (by Cosmetic Imaging Systems, Inc., Santa Monica, Calif.).

52 Sarver et al., "Video Imaging" 940.

53 The use of new medical imaging devices is well documented through the 1980s. A. Favre, Hj. Keller, and A. Comazzi, "Construction of VAP, A Video Array Processor Guided by Some Applications of Biomedical Image Analysis," *Proceedings of the First International Symposium on Medical Imaging and Image Interpretation*, Vol. 375 (Berlin, FRG, 26–28 Oct. 1982).

Computer imaging is also being tested in instructional uses where, in one report, resident plastic surgeons are taught how to conduct a patient planning session—normally a skill that is considered very difficult to teach. Ira D. Papel, MD, and Robert I. Park, MD, "Computer Imaging for Instruction in Facial Plastic Surgery in a Residency Program," *Archives of Otolaryngol Head and Neck Surgery* 114 (Dec. 1988): 1454–60.

In another article, the use of video imaging as a "means of predicting results of orthognatic surgery" is said to increase a surgeon's treatment-planning skills. Sarver et al., "Video Imaging" 939.

54 Dido Franceschi, MD, Robert L. Gerding, MD, and Richard B. Fratianne, MD, "Micro-computer Image Processing for Burn Patients," *Journal of Burn Care and Rehabilitation* 10.6 (Nov.–Dec. 1989): 546–49.

55 The release form includes five statements that must be signed by the patient, the physician, and a witness. A copy of the disclaimer appears in William B. Webber, MD, "A More Cost-Effective Method of Preoperative Computerized Imaging," *Plastic and Reconstructive Surgery* 84.1 (July 1989): 149.

56 In a recent newspaper article, one New York ad man claimed that he showed "a client how to use such in-motion retouching techniques in political advertising [by showing] how we could take Michael Dukakis and make him as tall as Bill Bradley. . . . We also made Bush look drunk. That's possible." "Image 'Morphing' Changes What We See—and Believe," *Atlanta Journal-Constitution* 29 June 1992: sec. 4, p. 6.

Section V

Reproducible Insights

Women Creating Knowledge, Social Policy, and Change (with Elizabeth Adams and Jennifer Schneider)

In the introduction to Section IV, we reviewed feminist theory that extends and redefines scientific theories and methods so that women's lives and experiences are fully included. The absence of women, as embodied subjects and as objective researchers, has fundamentally limited the scope and accuracy of much of what is known from Western science. Some argue that the development of modern science depended on and reinscribed notions of women's intellectual and physical inferiority to men.[1] Contemporary scientists, too, sometimes sift their ideas through a sociopolitical filter that reinforces, rather than challenges, assumptions about fundamental physiological and psychological differences between the sexes.[2]

In the last decade, some feminist scientists have argued that the root of women's subordination is located in broad acceptance of biological explanations for gender-based social arrangements.[3] Their efforts seek to develop a fuller account, a better explanation of (1) if, (2) how, and (3) under what conditions there are demonstrably meaningful biological and behavioral differences between women and men. But feminist scientists cannot, by themselves, undo the centuries of misinformation that informs ideologies of sex differences. We can critique canonical knowledge; we can offer alternative interpretations and hypotheses of existing data; we can generate new research that challenges old ideas; we can develop new models of social organization and interpersonal relations. Unless those critiques, new ideas, and new social organizations are woven into the scientific community and society in general, they will have little impact.

Thus, no effort in feminist education would be complete without a focus on the process of implementing the new perspectives offered by feminist studies of science. In feminist education, the bridge between theory and practice is a thoroughgoing concern. It is a central goal of feminist initiatives to improve and enrich the health, well-being, and happiness of all women (and men) through fighting the oppressions that privilege the few at the expense of many.[4] Scientific research has sometimes played a role in legitimating the oppressions.[5] Nevertheless, precisely because scientific research cannot be neutral, because it is informed by and informs contemporary social, historical, and economic academic and public policy debates, and because careful research can be used persuasively, those who are scientifically literate emerge as key players in public debates about the policy issues that shape the conditions of women's (indeed everyone's) daily lives.

We have, in multiple ways, made the point that it matters that women have been excluded and/or marginalized from the mainstream of scientific and technological education, research, and development. It matters for a number of reasons, the most practical of which is that, by exclusion and marginalization, women are denied equal access to well-paying and satisfying careers in the fields of their choice. It matters because the contents of scientific research, where inquiry begins with assumptions of "male as norm," present an incomplete and inaccurate picture of the natural world. It also matters because where women are present, although women's presence alone is no guarantee, there is a higher probability that someone will notice and question research assumptions and directions that reinforce cultural biases against women. It matters, too, because scientific and technological research and innovations make critical contributions to contemporary social and economic life.

Wherever women are absent from decision-making processes about the research focus, funding, and promoting of new developments, and wherever women's lives and experiences are absent from this picture, there is lost opportunity for innovative thinking that improves everyone's lives. Instead we confront replaced, revised, and reinforced versions of nineteenth-century ideas that celebrated the great wonders of *man*kind's intellect.[6] So it does matter that the science and technologies of today's world have been constructed by and for the benefit of (with important exceptions) a small group of men trained in the Western scientific tradition. To reverse this continuing epidemic, it is important to increase the prominence of women and feminist perspectives across all sectors of society.

In the governmental arena, where policy-making power resides from the local to the state and federal levels, women are only just beginning to gain a foothold. Despite the fact that most of the world's governments claim to be formally committed to full equality for women, in no country of the world are women represented in government in proportion to the population.[7] In the United States, recent elections have brought the percentage of women holding seats in state houses to an average of about 23%. In a few states, voters have succeeded in electing higher percentages of women to their state legislatures, but only thirteen (Vermont, New Hampshire, Colorado, Minnesota, Arizona, Maryland, Hawaii, Washington, Oregon, Delaware, Maine, Nevada, and New Mexico) have over 30% and none have over 40% women among their legislators.[8] While the percent of women elected to office has increased steadily in the last ten years, women are still drastically underrepresented in the political sector.

Women's participation in political activity can have a profound impact on public policy creation. Research shows that women legislators are more committed to women's issues (such as reproductive issues, women's health concerns, and protection against violent crime) and are more likely than their male colleagues to vote for women's issue bills.[9] However, when the policies at hand are less directly related to women, gender-based differences in voting practices lessen significantly. Nonpartisan organizations, such as Women's Policy, Inc., bring women's policy issues to the forefront of political discussion so that policymakers are equipped to make informed decisions.[10] Still, among state and congressional legislators, both men and women, those who seek to develop public policy about science and technology from a woman-centered and feminist perspective are few and far between.[11] Among those few, there is but a handful, nationally, who have scientific expertise. This should be no surprise. How could it be any different, with the history of exclusion from both the institutions of scientific/technological expertise *and* the institutions of political power and practice?

If change is our goal but we have limited access to the institutions and leadership positions through which change is promoted, then how can we accomplish anything? We can teach one another and ourselves. We can pay attention. We can learn how to speak out, how to make the argument, how to think critically about the issues at hand. We can take comfort in the fact that the process of moving new ideas into the mainstream is neither orderly nor predictable. Unintended consequences can work to our benefit, just as they may work to our disadvantage. The need is to pay attention, to monitor, to stay in the struggle, to understand what is happening, and to always keep women at the center of the frame.

The process of innovation and technology development is one area where feminist perspectives can have a profound multilevel influence. Unfortunately, women and constructivist feminist methodology are not always incorporated into the process and are, often, overlooked in favor of traditional linear processes. For example, in the conventional paradigm for the emergence and development of new technologies, researchers develop ideas about potentially important processes or products to address an identified need. These processes and products are subjected to a series of experiments, tested, elaborated, re-tested, and eventually shaped into a marketable process or material product. One may assume that this process is directed by apparently unbiased decisions dictated not by the researchers but by the outcome of each experiment, and where the success of a technological innovation is determined not by human actors but by a faceless "marketplace" where competitive forces promote the survival of the products that best fit the market's needs. In this model, social changes necessarily follow as the new product, if it is an important one such as the computer, is incorporated into social life.[12]

But as historical and social studies of "technoscience" have amply demonstrated, this conventional linear model for the development of innovations captures neither the more dynamic processes that influence their emergence and use nor the social changes that influence and transpire from the innovations.[13] By abandoning an understanding of innovation and social change as driven by a linear, rational, and inevitable scientific process, we acquire a more accurate vision of social change and can reject powerlessness and marginalization. Opportunities for promoting or, alternatively, discouraging a technological innovation become visible at many, sometimes simultaneous, points within systems of research and development that do not conform to simplistic linear models.

For example, in the past decade, consumers have become increasingly aware of the social implications and health risks surrounding agricultural production technologies. Agricultural production technologies are processes and products used by farmers to grow, harvest, and store crops. To meet market demands for uniform, low-cost agricultural products farmers often use technologies, e.g., large equipment, synthetic pesticides, fertilizers, and genetically modified crops, that favor reduced direct costs associated with production, human labor, and management, but that can result in negative consequences for local communities and economies, ecosystems and biodiversity, and health of farm families, farm labor, and, potentially, consumers.

At the end of World War II, wartime technological advances in chemicals brought about widespread use of synthetic fertilizers, pesticides, and antibiotics in agriculture. Ammonium nitrate, a chemical used to create munitions during the war, became an inexpensive source of nitrogen for fertilizers, while pesticides developed from nerve poisons became widely used on American farms.[14] Postwar agricultural production continued to become highly industrialized and reliant on synthetic inputs to replace "inefficient" biological cycles. Petroleum-derived synthetic fertilizers have largely replaced animal and green manures to manage soil fertility, herbicides have replaced crop rotation and cultivation to control weeds, and insecticides have replaced biological control by beneficial predatory and parasitic organisms to control insect pests. More than 16,000 pesticide products are marketed in the U.S., and approximately one billion pounds of pesticide-active ingredient are used annually in the U.S. The U.S. Environmental Protection Agency estimates that 10,000–20,000 physician-diagnosed pesticide poisonings occur each year among the approximately 3,380,000 U.S. agricultural workers.[15]

Women and children often suffer more from exposure to environmental contamination than men.[16] Yet data about pesticides on women, children, and human pregnancy is sparse. Research by the California Birth Defects Monitoring Program provides a framework for clarifying issues surrounding human pregnancy, pesticide exposure, and birth defects.[17] Over 2,000 women were questioned about a wide range of potential pesticide exposures and other pregnancy factors. Half of these women worked in agriculture; others had jobs such as florist or animal handler, where potential for pesticide exposure is high. The authors of the study observed elevated risks, at least 1.5 times greater among exposed women, for these birth defects and exposures—pesticide use in household gardening and certain types of oral clefts, neural tube defects, heart defects, and limb defects; and living within 1/4 mile of agricultural crops and neural tube defects multiple exposure sources.[18] New research suggests that children whose mothers lived near applications of specific organochlorine pesticides during the first trimester of pregnancy are at significantly greater risk for developing Autism Spectrum Disorders (ASD). Researchers found the incidence of ASD among children born to mothers living near the locations of highest use of organochlorine insecticides during the first trimester of pregnancy is 6.1 times the incidence for mothers not exposed to these chemicals.[19]

Postwar changes in animal production from pasture- or range-based to industrial, high density confinement production necessitated the nontherapeutic use of antibiotics and synthetic hormones. An estimated 70% of antibiotics and related drugs produced in the U.S. are used for nontherapeutic purposes. These purposes include accelerating animal growth and preventing infection in compensation for overcrowded, unsanitary conditions on large-scale confinement facilities known as "factory farms." This is equivalent to about 25 million pounds of antibiotics and related drugs fed every year to livestock for nontherapeutic purposes, almost eight times the amount given to humans to treat disease.[20] These nontherapeutic uses of antibiotics in the agricultural industry promote the selection of antibiotic resistance in bacterial populations, and microbial resistance to antibiotics is increasing. The resistant bacteria from agricultural environments may be transmitted to humans, in whom they cause disease that cannot be treated by conventional antibiotics.[21] Even though the European Union has banned the use of antibiotics of human importance in farm animals for nontreatment purposes since 1998, producers in the United States continue to use this unsustainable practice. The Centers for Disease Control and the World Health Organization have stated that antibiotics used for human medicine should no longer be used as growth promoters in agriculture.[22]

The use of industrial crop production technologies has had a profound impact on agricultural trade and the local economies.[23] Since the 1970s the increased number of multinational food companies has increased the size of both farms and the overall food system. Local farmers are no longer able to compete with multinational, industrial-scale agricultural production, processing, and distribution by corporations that are able to seek and source the lowest-cost agricultural products from around the world. Women, in particular, were displaced from their role in commercial farming, as they were less likely to maintain high-level or decision-making positions on the farm.[24]

More recently, a social movement of farmers and consumers has focused on building relationships and changing purchasing habits to include less industrial, more locally produced foods.[25] The local food movement considers that the way food is produced and marketed has a great effect on human and animal health, the ecosystem, local economies, and cultural diversity. Consumption decisions favoring locally produced food directly affect the well-being of people and improve local economies.[26] The recent proliferation of local, organic food movements provides a clear example of the nonlinear, iterative, and holistic innovation process offered within feminist perspectives. National USDA organic standards prohibit the use of toxic and persistent chemicals, irradiation, and genetic modification as well as the use of sewage sludge as fertilizer, a common practice in conventional agriculture. Organic standards for meat and poultry prohibit growth hormones and antibiotics; instead, sick animals are removed from the herd and treated. Additionally, organic animal husbandry requires humane treatment of animals and their access to the outdoors.[27]

In general, women have faced difficulty in trying to establish themselves in conventional, industrialized agriculture. Women who are Hispanic, Black, and Native American have been especially disadvantaged in commercial agriculture because of the historical and structural racism in farm organizations and federal and state laws in the United States.[28] Women farmers are increasingly turning to alternative and sustainable agriculture, and they are at the forefront of this movement to produce and access health-focused, locally and sustainably produced foods. Although research on women on organic farms is limited, it indicates that there are a greater percentage of women among organic farmers than among all farmers (conventional plus organic). The biannual survey of organic farmers by the Organic Farming Research Foundation has shown for the past decade that about 22% of its respondents are women.[29] A 2001 study by the Women on U.S. Farms Research Initiative at Pennsylvania State University concluded that women farmers are less likely than men to use chemical-intensive production and instead use "sustainable" agriculture practices—those that are ecologically and socially responsible as well as profitable. Those practices include certified organic production methods.[30]

Analyses of organic farming in the context of industrial agriculture point to the complexities of democratic practices within the innovation process. Donna Haraway argues that activist responses are necessary for a critical, multisided, public and democratic science. Her vision of what she calls "technoscientific democracy"

> does not necessarily mean an anti-market politics, and certainly not an anti-science politics. But such a democracy does require engaging in critical science politics at the national, as well as at local, state, and regional levels. "Critical" means evaluative, public, multi-actor, multi-agenda, oriented to equality and heterogeneous well-being.[31]

She argues that "situated knowledge" is an untapped and yet essential component of democratic science, one that recognizes the limitations of any individual's capacity to "know reality" and offers an antidote—a conception of diverse, informed participants engaged in a fully democratic process of creating public policy that serves the public interest. Yet, implementing such a vision will not be easy. Even the definition of "public interest" is contested territory. The United States is a culturally heterogeneous country with a capitalist economy. Which public? Whose interests?

Nowhere are these issues more in evidence than in the development and delivery of medical knowledge and health care specific to women. In her book *Women's Health: Missing from U.S. Medicine*, Sue Rosser catalogues the ways in which research about women's bodies suffers from the biases and distortions identified by feminist studies of science and technology. Research directions, standards of evidence, interpretation of data, and the implementation of research findings—all of these have been influenced by the inadequate incorporation of women into all aspects of scientific research and development. Women have been absent from studies that were expressly focused on issues in women's health, so that, for instance, drug trials on the role of estrogen in protecting against heart attacks originally included only male participants.[32]

Alternatively, women have been included in high-risk or poorly researched studies when there was potential for large profits, for instance, in the case of contraceptives. The Dalkon Shield is a notorious example. It was an intrauterine contraceptive device developed by Hugh Davis, M.D., a faculty member of the Johns Hopkins University Medical School who stood to profit from its manufacture and sale. Based on one small study of 600 women for one year, and ignoring the advice of scientists who had concerns about the shortage of reliable data, A. H. Robins purchased the device and introduced it to the market in 1970. They stopped sales of the Dalkon Shield just four years later but did not recall those already sold and in use. About 235,000 American women suffered injuries, most of which involved life-threatening pelvic infections, many of which left women infertile, and twenty women died. A. H. Robins filed for bankruptcy in reaction to a class-action suit. The litigation ended when a trust fund was set up to compensate the injured women, but the majority received only $725 or less. Final payments from the company were not completed until 1996.[33] The controversy surrounding the Dalkon Shield case led to legislation that required companies to show their medical products to be safe and effective before they could be approved by the Food and Drug Administration for marketing.[34]

Yet controversies continue to surround currently available medical products for women's bodies. For example, in 2005 over 40 women brought a lawsuit against manufacturers of the Ortho-Evra birth control patch for failing to properly warn against the heightened risk of blood clots for patients using the patch. In response to the lawsuit, the Associated Press conducted research investigating the safety of the Ortho-Evra patch, finding that there were over 16,000 reported adverse side effects, 23 of which resulted in death and one of which resulted in paralysis. Since the investigation and lawsuit, the FDA revised the label of Ortho-Evra to reflect the increased risk of blood clots and possible fatality. The manufacturers of Ortho-Evra continue to make financial settlements with patch users who have experienced life-threatening blood clots.[35]

Recent embryonic stem-cell research provides a prime example of the dialectic relationship between medical technoscience and international policy, which can result in powerful, often negligent, outcomes for women. Public controversy over embryonic stem-cell research reached new heights in 2005 when prominent South Korean researchers (subsidized by both the U.S. and U.K.) were sanctioned for fabricating data and ethically mistreating women participants. Subsequent investigation found that, despite publicly reporting that women participants voluntarily donated eggs, researchers actually paid for egg procurement.[36] Given the economic inequity facing South Korean women, bartering payment in exchange for eggs is exploitive. Moreover, the informed consent form provided to the participants did not include any information about the potentially fatal side effects of egg procurement, such as severe ovarian hyper-stimulation syndrome. Indeed, two junior researchers were prodded into donating eggs, and because they were subordinates, their consent could not be given without at least the implied threat of coercion. In the race to develop new stem-cell technology and gain notoriety, stem-cell researchers in this case valued research objectives as paramount to the well-being of women participants.

Due to the high profile of embryonic stem-cell research and efforts of feminist groups across the world, the question of research ethics and gender-based discrimination became a topic of international discussion. In response to an international outcry, university and governmental bodies in South Korea and the United States enacted task forces to investigate their own research practices.[37] Sanctions and regulations were created to ensure that women participants and researchers in high-tech fields are treated ethically in the future.[38] With focus, attention, and expertise, it is possible to create discussion, affect policy, involve feminist critique in scientific and technological innovation, and enact change on a global level.

Because the physiological processes of women's bodies are underresearched relative to men, women are underdiagnosed for some diseases (such as AIDS) and overdiagnosed for others relative to men (such as mental illnesses). For example, despite the fact that heart disease is the leading cause of death in women worldwide, women are regularly underdiagnosed. In addition, women with heart disease are subjected to surgical and drug regimens that may not be as effective for women as men (i.e., women are less likely than men to survive coronary bypass surgery).[39] In an effort to better understand this incidence, NIH funded a research study, known as WISE, investigating heart disease in women. The WISE study found that men and women often manifest very different patterns of symptoms and artery blockages, and that the traditional examination procedures, surgeries, and drug regimens are not effective for diagnosing or counteracting heart disease in women.[40] It is unclear whether medical practitioners have incorporated findings from the WISE study into their examination practices. What remains clear, however, is that women are underresearched, underdiagnosed, and undertreated relative to men.

To complicate the picture still further, women have different health concerns depending on their racial/ethnic families of origin, including sickle-cell anemia and systemic lupus erythematosus among African Americans, alcoholism among Native Americans, and hemoglobin E disease and hepatitis B among Asian Americans.[41] Cervical cancer rates are twice as high among Hispanic women as non-Hispanic women, and African American women have higher rates of hypertension than European American women. Japanese- and Chinese American women are less likely than Filipina women to suffer from hypertension.[42] Research on and treatment for diseases as linked to racial/ethnic groups are particularly underdeveloped, often focusing primarily on behavioral determinants, as opposed to the social and cultural context, of chronic illnesses.[43]

Indeed, even the components of women's physiology that are reputed to be at the heart of femininity itself, our "hormonal hurricanes," as Anne Fausto-Sterling dubbed them, are sometimes wildly misunderstood.[44] Yet, understanding how hormone cycles work is critical to understanding how the immune system works, which is critical to women's health in particular since an estimated 75% of autoimmune disease patients are women. It is also critical to the general science of medicine, since the relationship between hormones and the immune system has implications for areas as diverse as drug testing, the timing of vaccinations, infertility problems, and administering chemotherapy.[45]

Critiques of the limitations of conventional medical research related to women emerged in the scholarly literature in the 1980s, along with feminist studies of science more generally, and the women's health movement. However, it was women in Congress, sparked by activist communities, who worked together across conventional political lines to begin efforts toward a more inclusive national research agenda. Among these were Democrats Patricia Schroeder and Barbara Mikulski and Republicans Constance Morella and Olympia Snowe, whose lobbying efforts eventually led to the establishment of the Office of Research on Women's Health within NIH, the Women's Health Initiative (WHI), and the NIH Revitalization Act of 1993.[46] The NIH Revitalization Act requires that women and people of color be explicitly included (or explain why not) in all grant proposals using human subjects. The Women's Health Initiative funds a national network of Centers of Excellence in Women's Health through which federally sponsored research is organized. The WHI is a 15-year project designed to collect longitudinal data on cardiovascular diseases, cancer, and osteoporosis, the three leading health threats among women over forty-five. To date, the study includes over 160,000 postmenopausal women and has produced over 160 papers on women's health outcomes. The research studies conducted thus far have been used to inform FDA recommendations for treatment regimens, labeling, and dosage recommendations, most notably for hormone replacement therapy drug regimens. Yet, the extent to which such efforts on the part of mainstream institutions will lead to the re-centering of our knowledge base, so that it fully includes women's lives (in all their variety) to the same degree that it includes men's lives, remains to be seen.[47] It will depend on how much public support there is, and how much support in Congress.

Grassroots women's health initiatives also have contributed to the currents of change. Their focuses most often emphasize community-based empowerment, health delivery, and disease prevention, within the context of the biological or genetic elements of health. These organizations sponsor educational efforts to support a wide variety of health care issues for women, including parent and child health care, breast cancer and cervical cancer examinations, blood pressure monitoring, nutrition needs, and childbirth.[48] Though each has its own goals, projects, and community strategies, they share a focus on how social factors influence patterns of risk and access to adequate treatment. Among the most well-known national efforts are the National Black Women's Health Project (NBWHP), founded by Byllye Avery in 1983 and now known as the Black Women's Health Imperative; the Boston Women's Health Book Collective, which has published *Our Bodies Our Selves* since 1970; and the National Women's Health Network, which is an advocacy organization that lobbies Congress and other government agencies, as well as a clearinghouse of women's health information. These organizations and initiatives are committed to both critiquing the status quo of medical knowledge and treatment and developing an understanding of the political and social processes involved in using the knowledge and accessing the treatment. All are aimed at improving, enriching, and expanding the conditions of women's lives within the larger social, political, and economic context.

Similarly, the readings we feature in this section of the book explore the ways in which science and politics are complicit and implicit partners in controlling the conditions in which women live. Ruth Perry, in her essay "Engendering Environmental Thinking: A Feminist Analysis of the Present Crisis," points out the ways in which multinational corporations, increasingly empowered by state and national legislation, pose the greatest threat to our environment. Because women more often than men take seriously "the labor of reproducing the conditions for life" so critical to the survival of the earth, women make up the majority of unpaid environmental activists around the world. If public policy-makers likewise took this labor seriously, environmental pollution would be inconceivable. Instead, policy-makers and environmental engineers can argue that "dilution is the solution" when there is potential for environmental degradation from all sorts of landfills, factories, power plants, agrichemicals, or even home cleaning products.

Shruti Rana, in the article "Fulfilling Technology's Promise: Enforcing the Rights of Women Caught in the Global High-Tech Underclass," illustrates the relationship between public policy and gender-based racial discrimination in the high-tech industry in both Malaysia and the United States. According to Rana, multinational companies in the high-tech industry exploit the stereotype of Asian women as dexterous and obedient in making decisions about where to locate their manufacturing plants. Men are placed in managerial positions while Asian women are placed in low-paying, low-skilled, expendable, and often hazardous assembly-line positions. Countries ignore unfair labor practices, create tax breaks for companies, and lift restraints on multinational investments in order to attract the commerce, but the resulting public policies do little to protect women laborers. The significance of Rana's article is that this blend of gender and racial exploitation is not limited to Third World countries, as commonly assumed. Instead, workforce stratification of the high-tech industry by race and gender can be found in both the United States and Malaysia, illustrating the global and yet local scale of the problem.

Deeply embedded in the long history of gender-based workforce discrimination is the notion that women are weak-bodied, frail, and incapable. Emily Martin, in her essay "Premenstrual Syndrome, Work Discipline, and Anger," looks at nineteenth- and twentieth-century debates about white middle- and upper-class women in the paid workforce. She points to the history of medical ideas about menstruation as a case in point, where male physicians' beliefs about women's mental and physical abilities lent the weight of scientific authority to arguments against women in the workplace. Though prototypical feminist research by psychologist Leta Hollingworth in 1914 documented that women were no less capable of manual and mental work when they were menstruating than when they were not, her findings were swamped by those more conveniently aligned with conventional attitudes that saw women as perpetually diseased and disabled. Martin does not dismiss the experiences of menstruation as unimportant, but she argues that instead of understanding our bodies in negative terms we should challenge the conditions and descriptions that mark us as, by nature, unhealthy.

The perception of women's bodies as unnatural and unhealthy is not limited to beliefs about menstruation. As Laura Woliver points out in her essay "Reproductive Technologies, Surrogacy Arrangements, and the Politics of Motherhood," women continue to be treated according to a male medical model. Even a process as intimately linked to being a woman as is childbirth has been reinvented to perpetuate a patriarchal conception of the birthing experience. The gestation process is segmented into phases, and women are treated as birthing machines, whose only value is in the end "product" they manufacture. In effect, reproductive technologies are used as a means to assert male dominance and control over women's bodies, allowing medical practitioners to set precedents for what is "normal." Women who do not conform to the medical model and prefer a more holistic, natural view of their bodies are considered abhorrent and irresponsible. Public policies often do little to protect the rights of women, emphasizing instead the sociocultural and economic context within which men operate.

* * *

We have tried throughout this book to present feminist science studies in such a way as to make the general point that scientists and engineers, and a science- and technology-literate public, can enrich and expand their outlook on the social and natural world by including women. This is not a new idea, but it bears repeating because the mother field to feminist science studies, women's studies, is one of the great educational innovations of the twentieth century. In less than thirty-five years, interest in scholarship by, about, and for women has grown from a few isolated courses offered informally and off-campus to over 500 women's studies programs enrolling over a million students in courses in the United States alone. Internationally, the story is even more dramatic, since scholarship on women has emerged within universities and research centers all over the world, bringing with it a concurrent commitment to international themes.[49] This is a profound accomplishment, and we tip our bonnets to the early pioneers, especially including those brave women who first critiqued science. As new struggles emerge in the effort to institutionalize and implement what has been learned about the social and political processes that can either encourage or dismiss women as educators, researchers, and subjects of study, it is easy to forget how much ground has been gained.

Notes

1 D. Noble, *A World without Women: The Clerical Culture of Western Science*, New York: A.A. Knopf, 1992; L. Schiebinger, *The Mind Has No Sex? Women in the Origins of Modern Science*, Cambridge, MA: Harvard University Press, 1989.
2 For an overview, see H. Longino, *Science as Social Knowledge*, Princeton, N.J.: Princeton University Press, 1990, in particular chapter 6, "Research on Sex Differences," pp. 103–32. See also A. Eagly, "Sex Differences in Social Behavior: Comparing Social Role Theory and Evolutionary Psychology," *American Psychologist* 52, no. 12 (1997): 1380–83.
3 This is not a new critique by feminists. Mary Putnam Jacobi in her 1877 book, *The Question of Rest for Women during Menstruation,* argued that biological sex differences were a myth. Leta Stetter Hollingworth, in 1916, directly challenged assertions by social Darwinists that women were innately inferior to men, pointing to the inadequacy of the evidence upon which their interpretations relied. For a discussion of these foremothers, see F. Denmark and L. Fernandez, "Historical Development of the Psychology of Women," in *Psychology of Women: A Handbook of Issues and Theories*, ed. F. Denmark and M. Paludi, Westport, CN: Greenwood Press, 1993. For recent critiques of biological arguments, see Longino, *Science as Social Knowledge*; A. Fausto-Sterling, *Myths of Gender*, New York: Basic Books, 1992; L. Schiebinger, *Has Feminism Changed Science?*, Cambridge, MA: Harvard University Press, 1999.
4 bell hooks defines feminism as the struggle to end sexist oppression by opposing all forms of domination. See *Feminist Theory: From Margin to Center*, Boston: South End Press, 1984. In this definition, there is no room for "natural" forms of human domination.
5 S. Harding, ed., *The "Racial" Economy of Science*, Bloomington: Indiana University Press, 1993; C. F. Epstein, *Deceptive Distinctions: Sex, Gender, and the Social Order*, New Haven, CN: Yale University Press, 1988.
6 For a discussion of this phenomenon in terms of innovative teaching techniques such as "collaborative learning," see M. Mayberry, "Reproductive and Resisant Pedagogies: The Comparative Roles of Collaborative Learning and Feminist Pedagogy in Science Education," in *Meeting the Challenge: Innovative Feminist Pedagogies in Action*, ed. M. Mayberry and E. C. Rose, New York: Routledge, 1999.
7 J. Seager, *The State of Women in the World Atlas*. New York: Penguin Books, 1997, pp. 14, 90.
8 Center for American Woman and Politics, New Brunswick, NJ, Rutgers University, 2007.
9 M. Swer, *The Difference Women Make: The Policy Impact of Women in Congress*. Chicago: University of Chicago Press, 2002.
10 Retrieved online from http://www.womenspolicy.org/2007.
11 Women have made important gains in many other areas of public policy and politics. For a review of these gains and the organizational efforts behind them, see S. Hartmann, *From Margin to Mainstream: American Women and Politics since 1960*, Philadelphia: Temple University Press, 1989.
12 For a discussion of the limitations of this model in relation to computers, see P. N. Edwards, "From 'Impact' to Social Process: Computers in Society and Culture," in *Handbook of Science and Technology Studies*, ed. S. Jasanoff *et al.*, Thousand Oaks, Sage Publications, 1995. For a discussion of the conventional model, see E. M. Rogers, *The Diffusion of Innovations*, 4th ed., New York: The Free Press, pp. 132–60, esp. fig. 4–1.
13 D. Haraway, *Modest_Witness@Second_Millennium*, New York: Routledge, 1997, p. 95; Edwards, "From 'Impact' to Social Process."
14 J. Malcolm, "The Use of Chemicals in Agriculture," *Water and Environment Journal* 3, no. 5 (1989): 522–25.
15 http://www.cdc.gov/niosh/topics/pesticides/.
16 M. Jacobs and B. Dinham, eds., *The Silent Invaders: Pesticides, Livelihoods and Women's Health*. New York: Zed Books, 2002.
17 http://www.cbdmp.org/pdf/pesticides.pdf.

18 G. M. Shaw, C. R. Wasserman, C. D. O'Malley, V. Nelson, and R. J. Jackson, "Maternal Pesticide Exposure from Multiple Sources and Selected Congenital Anomalies," *Epidemiology* 10, no. 1 (1999): 60–66.
19 E. M. Roberts, P. B. English, J. K. Grether, G. C. Windham, L. Somberg, and C. Wolff, "Maternal Residence Near Agricultural Pesticide Applications and Autism Spectrum Disorders among Children in the California Central Valley," *Environmental Health Perspectives* 115, no. 10 (2007).
20 Office of Technology Assessment, *Drugs in Livestock Feed: Volume 1, Technical Report.* Washington, DC: U.S. Government Printing Office, 1979. govinfo.library.unt.edu/ota/Ota_5/DATA/1979/7905.pdf.
21 G. G. Khachatourians, "Agricultural Use of Antibiotics and the Evolution and Transfer of Antibiotic-Resistant Bacteria," *Canadian Medical Association Journal* 159, no. 9 (1998): 1129–36.
22 W. Boyd, "Making Meat: Science, Technology, and American Poultry Production," *Technology and Culture* 42 (2001): 631–64; M. G. Mellon, C. Benbrook, and K. L. Benbrook, *Hogging It! Estimates of Antimicrobial Abuse in Livestock*, Cambridge, MA: Union of Concerned Scientists, 2001; C. Viola and S. J. DeVincent, "Overview of Issues Pertaining to the Manufacture, Distribution, and Use of Antimicrobials in Animals and Other Information Relevant to Animal Antimicrobial Use Data Collection in the United States," *Preventive Veterinary Medicine* 73, nos. 2–3 (2006): 111–31.
23 C. B. Flora, "Social Capital and Sustainability: Agriculture and Communities in the Great Plains and Corn Belt," Journal Paper No. J16309, Iowa Agriculture and Home Economics Experiment Station, Ames, Iowa (Project No. 3281), 1995.
24 D. Barndt, *Women Working the NAFTA Food Chain: Women, Food and Globalization.* Toronto: Second Story Press, 1999.
25 http://www.slowfood.com/; http://www.foodroutes.org/.
26 For example, Organic Consumers Association (www.organicconsumers.org), Womens Agricultural Network (http://www.uvm.edu/~wagn/), and Women, Food and Agriculture Network (http://www.wfan.org/), among others.
27 http://www.ams.usda.gov/nop/indexNet.htm.
28 Berit Brandth, "On the Relationship Between Feminism and Farm Women," *Agriculture and Human Values* 19 (2004): 107–17; Anne Effland, Robert Hoppe and Peggy Cook, "Special Outlook Report: Minority and Women Farmers in the U.S.," May 1998, http://www.ers.usda.gov/publications/agoutlook/may1998/ao251d.pdf.
29 E. Walz, OFRF Fourth National Organic Farmers' Survey: Sustaining Organic Farms in a Changing Organic Marketplace, 2004.
30 http://agwomen.aers.psu.edu/index.htm; http://wagn.cas.psu.edu/.
31 Haraway, *Modest Witness*, p. 95. The current practice of publishing new rules and regulations in the *Federal Register* and inviting comment is a limited attempt to encourage public participation in decision-making.
32 T. Gura, "Estrogen: Key Player in Heart Disease among Women," *Science* 269 (1995): 771–73; S. Rosser, *Women's Health: Missing from U.S. Medicine*, Bloomington: Indiana University Press, 1994.
33 R. Sobol, *Bending the Law: The Story of the Dalkon Shield Bankruptcy*, Chicago: University of Chicago Press, 1991. See also L. Laurence and B. Weinhouse, *Outrageous Practices: How Gender Bias Threatens Women's Health*, New Brunswick, NJ: Rutgers University Press, 1994. For a general discussion of the ethics of research in universities, see D. Shenk, "Money + Science = Ethical Problems on Campus," *The Nation*, March 22, 1999, pp. 11–18.
34 For a discussion of FDA guidelines for the participation of women in clinical trials, see L. A. Sherman, R. Temple, and R. B. Merkatz, "Women in Clinical Trials: An FDA Perspective," *Science* 269 (1995): 793–94; C. Meinert, "The Inclusion of Women in Clinical Trials," *Science* 269 (1995): 795–96; and S. J. Heymann, "Patients in Research: Not Just Subjects, but Partners," *Science* 269 (1995): 797–98.
35 E. Pringle, "Ortho-McNeil Knew Ortho-Evra Patch was Lethal," Media Moniters Network, 2006. Retrieved online July 1, 2007, from http://usa.mediamonitors.net/headlines/ortho_mcneil_knew_ortho_evra_patch_was_lethal.
36 C. Sang-Hun and N. Wade, Cloning fabricated, Seoul panel concludes International Herald Tribune Asia-Pacific, 2006. Retrieved online July 20, 2007, from http://www.iht.com/articles/2006/01/10/news/clone.php.
37 MS NBC South Korea widens stem cell probe, Associated Press. Retrieved online July 20, 2007, from http://www.msnbc.msn.com/id/10940784/from/RSS/. The President's Council on Bioethics, Monitoring Stem Cell Research. Retrieved online July 20, 2007, from http://www.bioethics.gov/reports/stemcell/chapter2.html.

38 National Institute of Health Stem Cell Task Force, Policy and Guidelines Documents, 2006. Retrieved online July 20, 2007, from http://stemcells.nih.gov/policy/guidelines.asp.
39 "Women: Absent Term in the AIDS Research Equation," *Science* 269 (1995): 777–80; Rosser, *Women's Health*, pp. 37, 81; B. Healy, "Women's Health, Public Welfare," *Journal of the American Medical Association* 266, no. 4 (July 24/31, 1991): 566–68.
40 National Heart Lung and Blood Institute. National Institute of Health Women and Ischemia Syndrome Evaluation (WISE), 2002, http://www.nhlbi.nih.gov/meetings/workshops/wise/session05_kelsey.pdf.
41 Rosser, *Women's Health*, pp. 90–91.
42 L. Schiebinger, *Has Feminism Changed Science?* Boston: Harvard University Press, 1999, p. 119.
43 A. Krumeich, W. Weijts, P. Reddy, and A. Meijer-Weiz, "The Benefits of Anthropological Approaches for Health Promotion Research and Practice," *Health Education Research* 16, no. 2 (2001).
44 A. Fausto-Sterling, *Myths of Gender: Biological Theories about Women and Men*. New York: Basic Books, 1992.
45 V. Morell, "Zeroing in on How Hormones Affect the Immune System," *Science* 269 (1995): 773–75.
46 For a discussion of the debates surrounding these initiatives, see Schiebinger, *Has Feminism Changed Science?* and Healy, "Women's Health, Public Welfare."
47 For a review of the controversies surrounding these initiatives, see A. Clarke and V. Olesen, eds., *Revisioning Women, Health, and Healing*, New York: Routledge, 1999, pp. 14–17.
48 For an overview of the growth of the women's health movement in the United States, see S. Morgen, *Into Our Own Hands: The Women's Health Movement in the United States, 1969–1990*, Piscataway, NJ: Rutgers University Press, 2002.
49 For a history of the development of women's studies in the United States, see M. Boxer, *When Women Ask the Questions: Creating Women's Studies in America*, Baltimore: Johns Hopkins University Press, 1998. For discussions of the challenges that international themes bring to women's studies, see M. M. Ley, J. Monk, and D. D. Rosenfelt, eds., *Encompassing Gender: Integrating International Studies and Women's Studies*, New York: Feminist Press at the City University of New York, 2002.

21

Engendering Environmental Thinking

A Feminist Analysis of the Present Crisis

Ruth Perry

UNPAID REPRODUCTIVE LABOR

Once, at a time of great stress in my life, I bought a cottage on a salt marsh south of Boston. I found the tidal rhythms infinitely soothing, a reminder that life was not structured by semesters or fiscal years. Twice every day the tides flush the channels, making silvery little waterways in what otherwise looks like a meadow. Seabirds—especially seagulls but also egrets, herons, cormorants, and ducks—swim in it, rest on it, or circle around the marsh, fishing. When an especially high tide comes in, the channels fill to overflowing; twenty-five times or so in the course of the year, particularly in the winter when the tides are deepest, the marsh floods into a little lake.

Wetlands are the most productive ecosystems in the world. They produce more organic matter than tropical rain forests. They control flooding by storing tidal overflows, thus limiting erosion and buffering uplands from the effects of storms. They clean and filter water by trapping sediment in the tangle of marsh plants, and quite literally slow down the flow of water long enough to permit plant roots to absorb this sediment even when it is toxic. A study of a Pennsylvania marsh showed "significant reductions in BOD (biochemical oxygen demand), phosphorous, and nitrogen within three to five hours in samples taken from heavily polluted waters flowing through a 512-acre marsh."[1] Ironically, given the destruction of so much natural marshland, experiments are now being conducted using man-made wetlands as tertiary treatment facilities for domestic and industrial waste. Many wetland plant species are peculiarly adapted to fix nitrogen, and marsh plants figure centrally in the recycling of carbon and methane as well. The constant tidal influx keeps these lands abundantly supplied with plankton and algae and other nutrients so that they are invaluable as spawning and breeding grounds. Thus, although wetlands comprise only 5% of the landscape, they contribute disproportionately to the sustenance of the environment—and hence to the perpetuation of the natural world and the health of its citizens.

By the mid 1980s, 58% of the wetlands in the United States had been filled in for real estate developments, agriculture, highways; used as landfills or garbage pits; or polluted by drilling for fossil fuels. Most of this appropriation of wetlands occurred between 1950 and 1980.

The work that wetlands do to maintain the environmental balance—to sweeten the water and the air, to provide suitable breeding grounds for a great variety of plant and animal species, to fix nitrogen and recycle carbon, to absorb toxicity and purify the land—this work, the work of reproducing the world, is not valued, not even noticed, and certainly not counted in any measure of national wealth. There is no way to add in the work that wetlands do given the way we now calculate the GNP (Gross National Product), a figure based on the aggregate cash transactions of the nation—but not on the health of its natural resources or even its people. Given this definition of wealth, any financial profit, however insignificant, is thought more valuable when weighed in the balance than these irreplaceable amphibious centers of life; they are traded away casually for the kind of profit that *can* be counted in the GNP. On August 9, 1991, in order to stimulate the flagging money-based economy, the United States government redrafted the official definition of a wetland so as to remove millions of acres of wetlands from federal protection and permit their sale to businessmen in the private sector with interests in real estate, transportation, oil, timber, and finance. Short-term gains for individuals—so long as they are the kind of profits that show up on the balance sheets of the present accounting system—matter more than infinitely renewable gains in the quality of air and water and regional biodiversity. The value that wetlands create accrues to everyone and therefore is not to the advantage of anyone in particular. The service of cleaning and renewing common resources can be sold for the temporary profit of a handful of businessmen.

The analogy to housekeeping is obvious: in the culture of commodity capitalism, no one values the labor of maintenance, of subsistence, the labor that makes possible domestic life as opposed to the labor that produces commodities or commodified services. In our society the labor of maintenance—reproductive labor in the largest sense of the phrase—is unappreciated because unremunerated. What is not paid for is not valued in our culture; and the reproductive labor of women, whether as unpaid housewives or "unskilled" cleaning women or babysitters, is notoriously undervalued by our society. By reproductive labor I mean such activities as raising children, preparing food, tending the old and sick, cleaning and mending, what Adrienne Rich calls "the activity of world-protection, world preservation, world-repair—the million tiny stitches, the friction of the scrubbing brush, the scouring cloth, the iron across the shirt, the rubbing of cloth against itself to exorcise the stain, the renewal of the scorched pot, the rusted knife-blade, the invisible weaving of a frayed and threadbare family life, the cleaning up of the soil and waste left behind by men and children. . . ."[2]

Such labors that reproduce the conditions for life have always been understood and valued in subsistence cultures. They were probably not even conceived of as separable from labor that produced goods until industrialization, with the shift to a cash economy, commodification of labor, and production for a market. Prior to this, class determined who would perform these maintenance functions; perhaps power is always evidenced by who, symbolically speaking, cleans up and takes out the garbage. But the commodification of labor and alienation from subsistence that industrialization brought with it restructured the economies of everyday life in ways that were distinctly gendered as well as class based. Increasingly, tasks associated with reproduction of the conditions for life were understood to be women's work, the responsibility of women of all classes, while the work of production was man's work, performed and remunerated in the formal economy whose meaning came precisely from its being disconnected from private domestic life.

This sexual division of labor in the public and private spheres evolved first in eighteenth-century English society as a cultural marker of the middle class. While middle-class men pursued profit in the marketplace, their women were creating new spaces for living insulated from the pressures of that world, as geographically separate from the public world of finance and government as the suburbs, where a man kept his wife and family, were distant from the city where he did his business. Henry Thrale, the brewer, was thought odd because he settled his wife and children in a house next door to his brewery in the heart of industrial

London rather than outside of the city, as was becoming more usual for eighteenth-century industrialists.[3] This physical separation of functions simultaneously granted the importance of women's reproductive labor in the private sphere while setting the terms for its devaluation. These newly feminized functions—tending and educating children, arranging basic subsistence for one's family in the form of clean and mended clothes, warm and comfortable shelter, nourishing cooked food, and most importantly unconditional emotional support to offset the calculations of the market—these were distinguished from the money-making business of life in spaces increasingly thought of as women's domain. Middle-class women performed these unremunerated tasks and superintended female servants who earned scarcely more than their board, as proof of their difference, their "softer" instincts, and their unsuitability for public affairs or high wages.

While these cultural redefinitions were transforming English urban society—the rise of the middle class, the development of the suburbs, the evolution of new arenas of domestic life presided over by women—rural life was also being transformed forever by the commercialization of agriculture. During the second half of the eighteenth century, lands that from time immemorial had belonged to no one in particular were seized as private property in an unprecedented sequence of Parliamentary Enclosure Acts proposed and partially financed by large landowners and representatives of the Church of England. Although there had been enclosure of common lands in England since the sixteenth century, this appropriation was on an entirely new scale. In the first sixty years of the eighteenth century, from the reign of Queen Anne through that of George II, 244 such Acts were passed, amounting to the removal of 337,876 acres from public holdings. But under George III, 3,554 Acts were passed and 5,686,400 acres were "improved"—fenced, removed from common usage, and put into production.[4] In the last half of the eighteenth century, one-quarter of England's arable land was taken out of common usage and assigned to individual property owners. This privatization of what had hitherto been common land, available to all for fodder, fuel, herbs, berries, barks, and kitchen gardens, was justified as being a more efficient use of land, a more complete exploitation of the country's natural resources, contributing to the general prosperity measured as productions for domestic and foreign markets. Landowners with capital could invest in the latest technologies (irrigation ditches, breeding stock, fertilizers) and apply these new techniques of agriculture to reclaiming "waste" lands, as they were called, for large-scale production. Thus, in the same period as the distinction deepened between unremunerated reproductive labor increasingly done by women and productive labor for wages increasingly done by men, common lands that had provided subsistences and recreation to communities for centuries were being fenced by individuals for private gain. Both women's reproductive labor and the free resources of nature were being appropriated to stoke the market economy—a pattern repeated, as we shall see, in twentieth-century examples of capital-driven development.

The immediate result of transferring these fallow lands from those to whom they had use value to those with the capital and technology to extract maximum exchange value from them, was increased production by private owners on an unprecedented scale. Manufacture as well as agriculture thrived in England as never before during this period. The newly mechanized textile industry produced Lancashire cottons and Yorkshire woolens in astonishing profusion, as Adam Smith, Patrick Colquhoun, and other economic commentators never tired of pointing out. Grain poured in from Warwickshire, Leicestershire, the Midlands, Oxfordshire, Gloucestershire, and Worcestershire—the soil made more profitable by irrigation, rotating crops, and new fertilizers. The mania for production spread to breeding livestock, and sheep and cows too began to be grown on a grander and grander scale.[5]

It has been estimated that the rents landlords extracted from their land doubled in those parts of England enclosed in this period (the Midlands, Northamptonshire, Warwickshire,

Leicestershire, Lincolnshire, East Riding). Arable land went from renting at 14 shillings an acre to 28 shillings an acre and grasslands for pasture from 40 shillings to 3 pounds an acre. Even counting in the cost of fencing and the legal fees for the paperwork, large-scale agriculture was one of the best investments of the age.[6]

But there were social costs to this increased productivity. Food prices began their long upward climb, encouraged by the growth of the market and encouraging in turn the enclosure of more land for profit. Thousands of cottagers squatting on common lands were displaced and their centuries-old reciprocal, ecological relationships to forests, fens, and swamps were abruptly terminated. Successive waves of historians have argued that enclosure and industrialization ultimately raised the standard of living for all Englishmen and produced whole new categories of jobs for those displaced from their land. Progress entails disruption, so the argument goes, but in the long run, the whole nation benefits. Others argue that only a fraction profited from the enclosures: only those with education, capital, or connections benefited from this re-allocation of land. Although enormous gains in productivity are undeniable, there was tremendous turnover in the landholders of enclosed areas during this period—as much as 60% in some places—and at least one-third of the existing farms disappeared in the process.[7]

Recently it has also been argued that if the enclosures of the late eighteenth century affected the entire cottager class in England, pauperizing many by reducing their access to subsistence and forcing them off the land altogether, women may have suffered especially from this movement from subsistence to waged labor. For one thing, women are always paid less than men when they enter the labor market, even when they put in the same hours doing the same work. But more importantly, the enclosures made serious inroads on women's ability to contribute to the subsistence of their households. Women had generally seen to the chores involved in keeping chickens or a cow, in maintaining kitchen gardens, or in gathering wild herbs in season, as extensions of their other forms of reproductive labor such as caring for children, cooking, preparing medicines, and the like. These labors which, because they depended on the natural world, included habits that reproduced the resource base, were made more difficult or even altogether impossible by the enclosures. Women, in particular, were thus impoverished and made vulnerable by the enclosure of common lands, because their tasks in a subsistence household economy were worth more to their families than their labor in a waged economy.[8] Without access to the commons, a woman could not earn a little extra with her butter and eggs at the local Saturday market; and without access to domestic animals and a kitchen garden, the lives of her family were impoverished of contact with the natural world. Thus, these two factors combined in eighteenth-century England to impoverish women in particular and to interrupt traditional ecological habits of life: (1) the gendering and devaluing of reproductive labor in a market economy and (2) the enclosure of common land.

As I have said, the work that wetlands do is central to reproducing the conditions for life, but we have no mechanism in our postindustrial world for assessing its value or for recognizing its contribution to our collective well-being. True cost-benefit analysis of the natural resource base has made as much headway as the "wages for housework" campaign of the late 1970s—which is to say almost none. This is true despite the paradoxical fact that all people, men and women alike, have to live on the earth and share its resources.

In our own time, as in the eighteenth century, there is a dramatic correlation between increased production of commodities and the privatization of public resources. Barry Commoner noted in 1971 in *The Closing Circle: Nature, Man & Technology*, that the economy of the previous thirty years literally fed off the environment. The relationship between high profits and damage to the environment was not accidental, he argued. In the post-World War II era, what extended margins of profit were techniques of manufacture and delivery that made use of the free natural resource base. In his analysis, the most profitable new enterprises were based on innovative technologies that replaced older, less polluting manufacturing processes. He instanced the production and use of detergents rather than soap, transportation by truck instead of by train, and the manufacture of synthetic materials rather than the growth and extraction of natural plant fibers as examples of new industries whose profit margins were assured only by positing an infinite world in which to freely dispose of polluting by-products. In other words, manufacturers exercised their citizenly rights to use their country's natural resources at a rate that far exceeded what they were entitled to as individuals. The excess, which constituted their profit, could be said to be subsidized by the publicly owned resource base. After demonstrating that farmers' profits were directly correlated to their use of high nitrogen fertilizers, Commoner concluded that there is "evidence that a high rate of profit is associated with practices that are particularly stressful toward the environment and that when these practices are restricted, profits decline."[9] Use, not ecological maintenance, he tells us, is rewarded by our present economic system.

NO MAN'S LAND

The degradation of the environment in the United States, the pollution of our air and water, is the twentieth-century equivalent of the enclosure of the commons. Just as in the eighteenth century enclosing landlords appropriated for private profit what had hitherto belonged in common to those living in a given locale, the use of public-access land, water, and air for industrial processes uses up what belongs to all for the profit of few. Moreover, as in the eighteenth century, the loss of the commons has been accompanied by a distinctive feminization of poverty, a phenomenon at least in part owing to the same configuration of cultural forces that contributed to the feminization of poverty in eighteenth-century England: undervaluing women's labor in the public sector and assuming their unpaid reproductive labor in the private sphere. The degradation of natural resource bases from which women are expected to reproduce the conditions for life—clean water and air for washing and drying clothes and household articles, safe land for children to roam in, and wild plants and a patch of garden to supplement their family's diet—is seen as lamentable but unavoidable. So long as no one's livelihood is threatened—defined as profit in the commercial sector—environmental degradation is seen as a necessary evil, trivial to the extent that its immediate visible impact is on women and children.

Perhaps for this reason, because environmental degradation impinges most directly on women's work of subsistence and maintenance—given our current sexual division of labor—or because the health and safety of families is still seen to be women's responsibility, women have responded most actively to the present ecological crisis. Grassroots activists at the local level, internationally, tend to be women. (Inevitably, as these movements become bureaucratized, men often move into leadership positions.) At every stage, from discovering the problem—whether dangerous industrial by-products, contaminated ground water, hazardous waste dumps, or inadequate sewage treatment—to organizing the community and lobbying the powers that be to clean it up, it is women who act for the safety of their neighborhoods. At home for longer stretches of the day, doing domestic chores and caring for children, comparing stories with neighbors, women often are the first to notice mysterious chemical barrels in vacant lots, odd smells or stinging eyes, or the sudden coincidence of too many deformed births in an area.

In Las Playas de Tijuana, a Mexican border town, a group of women calling themselves The Association of Las Playas Housewives managed to stop the opening of Mexico's first chemical incinerator. Despite the assurances of Chemical Waste Management of Mexico,

Inc., to the contrary, these women were convinced that the fumes from incinerated toxic waste would be a health hazard to everyone in the vicinity who had to breathe the air. Chemical Waste Management, a Chicago-based company, built the facility to dispose of some of the toxic waste building up from foreign manufacturers along the 2,000-mile United States–Mexico border. The Las Playas Housewives, who pressured the government to revoke the plant's license after investigating the health and safety record of similar toxic incineration facilities, are now afraid that a North American Free Trade Agreement will reverse their temporary victory and authorize the relicensing of the plant. Ultimately, they fear that plants like this, on the Mexican side of the border, will be used to process chemical waste from United States manufacturing operations.[10]

The Las Playas Housewives, operating outside formal institutional channels, are only one example among thousands of women around the world mobilizing against local environmental disasters. Possibly as an extension of their responsibility for children, possibly because they are underpaid and undervalued by the commercial sector and hence not invested in protecting local industries or defending business interests, women are often the ones who begin the unremunerated, supererogatory labor of organizing to protect their environments. I am not claiming that it is in women's nature to protect the earth, nor even that women are particularly suited to take care of others. I am simply observing that the majority of unpaid environmental activists around the world *are* women. . . .

I have argued that the labor of reproducing the conditions for life, scorned in the past as "women's work," is crucial to the survival of the earth. I have sketched a situation in which nature's mechanisms of sustainability are not being preserved and in which business profits are tied to extraordinary consumption of natural resources. We need to reeducate our society's estimation of reproductive labor, and what is implicated in that estimation, the status of women. Some feminist theorists have framed this issue as a problem in the definition of national wealth and national security. The GNP, after all, is comprised of all the goods and services produced in the money economy but not the labor that reproduces the conditions for life.[11] The GNP does not measure economies of subsistence and needs satisfaction—such as gardening, fishing, barter, cleaning, raising children—even though these activities often underwrite and make possible the production of commodities for profit.

As the production of commodities and commodified services rises in so-called developing countries, calculated by such measures as a rising GNP, women in these countries are increasingly impoverished, as measured by health, nutrition, life span, and access to cash. Development always seems to entail lengthening the working day of women, as if the costs of industrialization are borne especially by women. One explanation for this is that capitalism erodes the material base of women's subsistence production by commodifying more and more of that material base. Bina Agarwal describes this alienation of natural resources in India from women who rely on them for their livelihood. "Because women are the main gatherers of fuel, fodder, and water," she says, "it is primarily their working day (already averaging ten to twelve hours) that is lengthened with the depletion of and reduced access to forests, waters, soils." She reports that the hardship caused by scarcity of these natural resources in Uttar Pradesh has caused a rise in young women's suicide rates in recent years.[12] Vandana Shiva's example of the women of Garhwal also shows the way women bear the ecological costs of modernization in unremunerated labor in the private sphere. Because industrial needs for wood in that region exceed the regenerative power of the forest ecosystem, women who could once gather all the fuel and fodder they needed in a few hours must now travel by truck up to two days to collect what they need.[13] The unanimous consensus about the decade 1975–1985, named, ironically, the United Nations decade for women, is that "with a few exceptions, women's relative access to economic resources, incomes and employment has worsened, their burden of work has increased, and their relative and even absolute health, nutritional and educational status has declined."[14] Overvaluing the production of commodities while undervaluing the functions that reproduce life leads inescapably to the impoverishment of women.

It has been suggested that one corrective would be to compute the GNP differently and to make the unit of measurement the household and not the enterprise.[15] This would have the salutary effect of putting women back into the equation. Arguably, a nation's wealth is comprised of the net worth of each collective household; that might be a better measure of the well-being of its citizens than its profit-making businesses. As if to illustrate the principle that women's access to public resources is through their households rather than through new business enterprises, a group of women in a village in India recently opposed a scheme to cut down a tract of oak forest in order to establish a potato-seed farm. They argued that the project would take away their only local source of fuel and fodder, thus adding five kilometers to their daily collection journey, and that the cash that would accrue to their men from the project would not necessarily benefit themselves or their children.[16]

Nor do these forces operate only in the countries of the southern hemisphere. In the United States the devaluation of the labor of social reproduction—in this case raising children—together with disproportionate power given to those interested solely in profitable enterprise (rather than in citizens' health and safety) has created a particularly ugly situation in the controversy over lead paint. Although the harmful effects of even low doses of lead paint to children are so well documented that its use was outlawed in most West European countries and Australia in the early twentieth century, it was legal in the United States until 1978, when it was banned from use in homes, on furniture, and on toys. Today, lead paint is still legally manufactured and sold in the United States for painting outdoor structures such as bridges and water towers. But the primary health hazard posed by lead paint is what remains on the walls and woodwork of most older housing in the northeastern United States insofar as lead is released into the air every time anyone opens a window or door, or brushes against a painted surface. Lead dust taken in through the nose or mouth ends up inert in the bones of most people over seven, but in children under that age it circulates in the blood until it lodges in the soft tissue of the brain, where it is highly neurotoxic. It is estimated that 10–12% of the children under six in the United States "lead belt" have neurotoxic levels of lead in their blood sufficient to cause reduced I.Q., attention deficit disorder, hyperactivity, learning disability and impaired growth.[17] Notwithstanding the thorough understanding of the neurotoxicity of lead to growing children, public health departments do not, as a rule, subsidize lead paint removal. Moreover, mothers rather than landlords continue to be blamed for the damage it does to their children's health.

[It has been] suggested that what keeps this problem intractable is that male policymakers simply do not take seriously their responsibility for the safety of homes, which are considered women's domain. It is a notorious fact that the most polluted air in the United States is indoor air, the air of the home, the air in women's space, saturated with the gasses emitted from plastics and cleaning products. As far-fetched as it sounds, the sexual division of labor in our culture that makes women responsible for the health and safety of their families apparently disconnects men from responsibility for these issues. . . .

SUSTAINING KNOWLEDGE

Feminists have also observed that women's special knowledge about the environment is often ignored or passed over as old wives' tales or superstition. As the primary agricultural workers in many societies, not to mention the gatherers of fuel and water, women's traditional knowledge about sustainable resource use, accrued slowly over generations, is often overlooked by those who make policy about economic development projects. Just as the labor that women do is not valued—women's unpaid labor of love, to use Hilary Rose's phrase—so women's non-commodified knowledge about the reproductive processes of the natural world is insufficiently valued. Bina Agarwal notes that poor rural women in India have "an elaborate knowledge of the nutritional and medicinal properties of plants, roots, and trees," which can be called upon during drought or famine conditions. She also notes that hill women usually manage seed selection for their communities, having learned more about nature and agriculture "in the process of their everyday contact with and dependence on nature's resources," a function of the gendered division of labor in that society.[18] This knowledge about environmental processes must also exist among poor rural women in our own country who similarly depend on a reciprocal relationship with their local natural environment and thus have a stake in maintaining it. Yet I cannot imagine the EPA seriously entertaining a proposal to collect, codify, and put to use this lore.

Although we need to develop cleaner industrial processes, and to test the effects of chemical by-products on living organisms, much environmental thinking is not a matter of technical training so much as a matter of common sense. At a recent conference on women and the environment, two intelligent women without technical training told their stories about how they fought to end the dumping of hazardous substances in their home towns. Each demonstrated how quickly they could learn everything technical that they needed to know about the particular toxins being produced and buried in their neighborhoods, despite the deliberate mystification of some public health officials concerned to keep the public ignorant and to protect commercial interests. One described how she had slowly and painstakingly, over years, collected stories of cancer cases in her neighborhood, putting pins on a map every time she heard of a new case. She told us that these instances of cancer had clustered just where the water flowed and where the wind tended to blow. There was a simple elegance to her description that was like the best scientific observation. Such direct information about natural processes is too often passed over in preference for technically sophisticated but locally uninformed conjectures about the environmental effects of industrial processes.

Let me note in passing, too, the ecofeminist contribution to a gendered analysis of environmental issues. Both in foregrounding the metaphorical connections between the body of the earth and the bodies of women and in explaining how social justice and environmental justice are inextricably linked, ecofeminists insist on the continuities between how we treat the natural world and how we treat each other. According to an ecofeminist analysis, pollution reveals social dominance whereas sustainable production, that is, the reproduction of subsistence, is the visible sign of well-balanced interactions between humans and the natural world. Ecofeminists reject the view that the planet's processes can be understood merely mechanistically, its bio-masses reduced to "natural resources" available for exploitation by humans. Instead, they see the whole as a set of interlinked processes involving earth, air, water, and organic life. "Life on earth is an interconnected web, not a hierarchy," as Ynestra King puts it. "Human hierarchy is projected onto nature and then used to justify social domination. Therefore, ecofeminist theory seeks to show the connections between all forms of domination, including the domination of nonhuman nature, and ecofeminist practice is necessarily antihierarchical."[19]

Ecofeminists sometimes argue that women themselves have been thought of as if they were "natural resources," and thus "know" what it feels like to be viewed that way. As evidence for this way of thinking, one might begin by observing that everyone's first environment is a woman. Gerda Lerner's claim that the earliest form of slavery was the reification and appropriation by men of women's reproductive capacity is a historical argument based upon this difference.[20] Sherry Ortner's important essay "Is Female to Male as Nature Is to Culture?" was an early meditation on the social meanings of this cultural analogue between attitudes towards nature and attitudes towards women.[21] Thus men's disrespectful attitudes towards women and nature could be said to reflect and reinforce one another.

In closing, let me observe that in our own day the most intractable opposition to environmental legislation comes from multinational corporations, the latest avatar of the absentee landlord enclosing publicly held resources for private profit. These multinational corporations represent a further attenuation in the journey away from subsistence, the ultimate triumph of productive values over reproductive values. Accountable neither to families, neighborhoods, municipalities, or even nations, stripped of any organic relation to place, these multinationals pose the greatest threat of all to our environment. Increasingly empowered to invalidate the environmental agendas set by our own elected legislative bodies, they represent the disembodied privilege of profit. For example, although DDT has been banned for more than twenty years in the United States following the publication of Rachel Carson's *The Silent Spring*, the General Agreement on Trade and Tariffs (GATT) could force United States citizens to import DDT-laden produce again on the grounds that our pesticide laws are a barrier to trade.[22] A law passed in Nassau County, Long Island, requiring that newspapers published there be printed on 50% recycled paper was challenged by Canada as a "barrier to trade," spurred on by the Canadian Pulp and Paper Workers Union. Our ban on importing asbestos was recently challenged by the terms of the United States–Canada trade agreement. Although upheld in the end, our law was pronounced extreme, and it was recommended that a less "trade distorting" alternative would be to require people working with asbestos to wear gloves and goggles. In the summer of 1992, George Bush refused to sign the biodiversity treaty in Rio on the grounds that it posed a danger to biotechnology industries.

Empirically and theoretically, the greatest threat to our world lies in permitting the transfer of political authority from citizens' groups to multinational corporations. Such a move will completely disempower those whose labor reproduces our social world, and definitively locate it with those who produce commodities and commodified services. The implications for women's status, women's knowledge, and the reproductive labor of subsistence and maintenance within such a system are appalling. As political authority for environmental decisions is transferred from elected officials responsive to people to appointed boards adjudicating conflicts between multinational corporations, we will witness the final disappearance of the concept of the commons and a publicly owned resource base.

Entities larger than nations, regional blocs like the EEC, are likely to be the powers negotiating future world trade agreements. How will citizens' groups be able to influence the priorities of these regional blocs? Who will speak up for the rights of neighborhoods, communities, families, individuals? How can we retain control of the environments in which we live in the face of such international pressures? We need to reestablish the fundamental claims of ordinary people to clean air, water, and safe soil as part of a basic economic bill of rights. We need to affirm the importance of small-scale invisible economies that create subsistence even when they are weighed against corporate profits. The quality of our lives and the health of our world depends on our ability to reauthorize a society that gives full value to the labor of reproducing the conditions for life.

Notes

1 Jon A. Kusler, *Our National Wetland Heritage* (An Environmental Law Institute Publication), 1.
2 Adrienne Rich, "Conditions for Work: The Common World of Women," Foreword to *Working It Out*, ed. Sara Ruddick and Pamela Daniels (New York: Pantheon, 1977), xvi.

3 Mary Nash, Hester Thrale's biographer, conjectures that urban pollution may have accounted for Hester Thrale's many miscarriages. For information on the suburban estates of eighteenth-century men of fortune, see Catherine Hall and Lenore Davidoff, *Family Fortunes: Men and Women of the English Middle Class, 1780–1850* (Chicago: University of Chicago Press, 1987), especially chapter 8.
4 W. Hasbach, *A History of the English Agricultural Labourer*, trans. Ruth Kenyon (London: P.S. King & Son, 1908), 57–58.
5 Harriet Ritvo, "Possessing Mother Nature: Genetic Capital in Eighteenth-Century Britain," in *Early Modern Conceptions of Property*, ed. John Brewer and Susan Staves (New York: Routledge, forthcoming).
6 Michael Turner, *Enclosures in Britain, 1750–1830* (London: Macmillan, 1984), 39, 41.
7 Turner, *Enclosures*, 72–75.
8 Bridget Hill, *Women, Work, and Sexual Politics in Eighteenth-Century England* (Oxford: Basil Blackwell, 1989), especially chapter 3. After the first family wage law in 1795, the infamous Speenhamland Act, women's wages in England were sometimes fixed below subsistence, which made them unable any longer even to earn enough to support themselves independently.
9 Barry Commoner, *The Closing Circle: Nature, Man & Technology* (New York: Knopf, 1971), 262.
10 "Women Blocked Incinerator," *The Boston Globe*, June 15, 1992, 33.
11 Marilyn Waring, *If Women Counted: A New Feminist Economics* (New York: HarperCollins, 1988).
12 Bina Agarwal, "The Gender and Environment Debate: Lessons from India," *Feminist Studies* 18 (Spring 1992):138.
13 Vandana Shiva, *Staying Alive: Women, Ecology, and Development* (London: Zed Books, 1989), 8.
14 DAWN, *Development Crisis and Alternative Visions: Third World Women's Perspectives* (Bergen: Christian Michelsen Institute, 1985), 21, quoted in Vandana Shiva, *Staying Alive*, 3.
15 At the feminist environmental conference held at MIT in May 1992, both Joni Seager and Gita Sen suggested that computing a nation's wealth by households rather than by businesses would reorient the national accounting system to include women's work.
16 Bina Agarwal, "The Gender and Environment Debate: Lessons from India," 147. In *Women Counted* Marilyn Waring asks: "Why do nutritional deficiencies result, when family food availability declines as subsistence (nonmonetary) farmland is taken for cash crops and men get paid an income?" (19).
17 Alliance to End Childhood Lead Poisoning, 1992, quoted in newsletter of the Citizen's Clearinghouse for Hazardous Waste, *Everybody's Backyard*, June 1992.
18 Bina Agarwal, "The Gender and Environment Debate: Lessons from India," 142. Parallel arguments about women's knowledge of sustainable processes elsewhere in the world can be found in Jodi L. Jacobson's 1992 Worldwatch Paper 110, "Gender Bias: Roadblock to Sustainable Development."
19 Ynestra King, "The Ecology of Feminism and the Feminism of Ecology," in *Healing the Wounds: The Promise of Ecofeminism*, ed. Judith Plant (Philadelphia and Santa Cruz: New Society Publishers, 1989), 19.
20 Gerda Lerner; *The Creation of Patriarchy* (New York: Oxford University Press, 1986), especially chapter 2.
21 Sherry B. Ortner, "Is Female to Male as Nature Is to Culture?" in *Women, Culture and Society*, eds. Michelle Zimbalist Rosaldo and Louise Lamphere (Stanford: Stanford University Press, 1974), 67–89.
22 Kristin Dawkins, Institute for Agriculture and Trade Policy, February 1993.

22

Fulfilling Technology's Promise

Enforcing the Rights of Women Caught in the Global High-Tech Underclass

Shruti Rana

I. INTRODUCTION

This article will focus on two groups of women within the high-tech underclass: factory workers in the free trade zones of Malaysia[1] and Asian immigrant assembly workers in the United States. Comparing the Malaysian electronics industry to its U.S. counterpart is instructive because both utilize a significant number of assembly workers, have developed highly stratified workforces in the midst of heterogeneous populations, and have been shaped by similar external economic and social forces.

My analysis has two primary goals. First, in analyzing and comparing the experiences of women assembly workers in Malaysia and the United States, this article will explain not only why companies seek out these women, but also how the creation of this workforce has in turn influenced, and become embedded in, state policies. Second, by identifying the common struggles and parallels of two groups of women seemingly worlds apart, this article aims to show that this modern version of a gender-based, immobile industrial underclass can develop anywhere. These labor structures are not confined to developing countries, but can flourish in the most "advanced" industries and most protective legal regimes in the world. This article is not intended to focus simply on the problems these women face, but also to inquire into ways that these jobs and this industry can improve to allow these women to truly benefit from the technology that they help build. . . .

II. WOMEN IN THE MALAYSIAN ELECTRONICS INDUSTRY

In many ways the electronics assembly system in Malaysia, and the role of women within it, provides a microcosmic setting for exploring the roles of gender and race in the global electronics industry. The electronics industry has been in Malaysia almost since the beginning of the industry's shift overseas, and has had time to mature and develop along with the country itself.[2] In fact, this growth and development has been so successful that today Malaysia is the world's largest exporter of electronic chips.[3] The same stereotypes, economic forces, and policy issues that have shaped links along the assembly line elsewhere are present in Malaysia.[4] In addition, the heterogeneous population of Malaysia makes it more comparable to the United States than other countries with growing high-tech sectors. For these reasons, the Malaysian electronics assembly experience can provide important lessons and historical perspectives for countries at various stages of development.

Malaysia is a multicultural nation with a workforce that has historically been divided along racial lines.[5] Malaysia has a current population of about 22.23 million, of which 5.52 million are ethnic Chinese, 1.56 million are Indian, and 12.84 million are "Bumiputra" (translated as "sons of the soil," this category encompasses the ethnic Malay people).[6] Historically, people of Chinese origin held much of the country's capital and wealth, while the people of Indian origin occupied a middle status, and people of Malay origin held the least wealth and education.[7] This hierarchy, along with a gender hierarchy, is replicated in the electronics industry. People of Chinese origin are by far the largest group of managers and professionals (69%; 52% male and 17% female).[8] People of Indian and Malay origin are concentrated in the lower ranks: Indians make up 13%, and people of Malay origin comprise 70% of skilled and semiskilled labor (Indian and Malay women comprise 12.5% and 61% of the Indian and Malay totals, respectively).[9] . . .

A. Penang—Asia's "Silicon Island"

The current centerpiece of Malaysian high-tech development is the idyllic island of Penang in Western Malaysia. Penang is often called "Silicon Island," reflecting both its aims and origins.[10] Since the 1970s, Penang's transformation from an agricultural and service-based economy to a primarily manufacturing-based economy has been fueled by foreign investment.[11]

Electronics companies make up by far the largest employers in Penang, employing 200,000 people, a little over sixty percent of the island's workforce.[12] As of 1998, there were 152 electronics companies in Penang, manufacturing computers, microchips, data storage products, wireless communication products, automotive electronic parts, and other electronic parts used in the manufacturing and assembly centers.[13] Japanese, U.S., and Taiwanese companies are the biggest investors in the electronics industry in Penang, followed by various European companies.[14] In 1998, there were 62 Japanese companies in Penang, 36 U.S. companies, and 68 Taiwanese companies.[15] Penang contains two Free Industrial Zones (FIZs), which are centralized zones for export-oriented industries, six industrial parks, and licensed manufacturing warehouses.[16] Most electronics companies are located in the Bayan Lepas Free Trade Zone, where nearly 55,000 people worked in 1998.[17] . . .

B. The development of a gender-based labor force

As in other free trade zones,[18] women have formed the backbone of the labor force since the beginning of Penang's development.[19] At one point, women workers comprised nearly eighty percent of the workforce in Penang's electronic industry.[20] By the mid-1980s this proportion declined to roughly two-thirds of the total workforce as the percentage of low-level jobs dropped in proportion to the number of highly skilled or managerial positions, which are mainly held by men.[21] The continuing stratification of the labor force along gender lines is evident. In 1998, a survey of sixteen large multinational electronics companies in Penang showed that 77% of the jobs described as managerial, professional, supervisory, or technical were held by men, whereas women held 87% of the lower-ranked clerical, general, skilled, and semiskilled jobs.[22] Assembly workers fall into the semiskilled labor category.[23]

The development of this highly divided labor force can be traced to the views shared by both multinational companies and the Malaysian government on race and gender.[24] The reasons why multinational companies have specifically targeted certain groups of workers can be roughly divided into three categories: physical stereotypes, social stereotypes, and economic positions.[25] In each category, gender and racial images are employed to create the image of the "ideal" assembly worker as a young, single, and meek Asian female.[26]

Employers have maintained that the highly precise nature of electronics assembly required workers with "nimble fingers, agile hands and keen eyesight."[27] Women, as evidenced by their skill at tasks like knitting and needlework, were ideal candidates for such labor. Adding a layer of racial imagery, it was felt that Asian women had "the smallest hands in the world."[28] Thus, the "fast-fingered Malaysian" became a competitive advantage for electronics firms.[29]

In addition to their physical qualifications, Asian women were thought to have the ideal social background for electronics assembly jobs.[30] The painstaking and tedious nature of assembly jobs requires employees with patience who would obey the demands of their supervisors and work accurately and efficiently for long periods of time.[31] As one employer put it, "[y]ou cannot expect a man to do very fine work for eight hours. . . . Our work is designed for females . . . if we employ men, within one or two months they'd run away. . . . Girls [*sic*] under thirty are easier to train and easier to adapt to the job function."[32] Asian women were supposed to be accustomed to life in a traditional patriarchal atmosphere and would have already learned to be respectful of authority.[33] Young Malay women from villages were considered more likely to have these qualities and were lured to factories with promises of work, adventure, company benefits, and cash incentives.[34] In this manner, employers' perceptions about the patriarchal structures and the role of women in Asian families and societies enhanced the attractiveness of Asian women as the prime labor source for electronics companies.[35]

A third advantage of employing women was their economic position.[36] From the 1970s through the mid-1980s, while wages in Malaysia were already lower than wages in the U.S. or Europe, women's wages were even lower than men's, and their position in the labor force more precarious.[37] Women were also supposed to be temporary workers, who would leave outside employment when they married or had children.[38] In essence, they were perfect candidates for an industry that needed intense work for short periods of time and required a flexible labor force that could easily be expanded or contracted.

Such a manipulation of physical, social, and economic characteristics to define and seek out the ideal assembly worker has been replicated around the world.[39] However, what is significant about this occurrence in Malaysia is the way in which the formation of this gender-based labor force in turn influenced, and ultimately became an integral part of, government policy.[40] In fact, it was the Malaysian government that issued the following lines in a now infamous investment brochure:

> The manual dexterity of the oriental female is famous the world over. Her hands are small and she works fast with extreme care. . . . Who, therefore, could be better qualified by nature and inheritance, to contribute to the efficiency of a bench-assembly production line than the oriental girl? No need for a Zero Defects program here! By nature, they "quality control" themselves.[41]

The following section analyzes the ways in which gender and race can become embedded in the very laws that are supposed to protect and enhance workers' rights.

C. Gender, race, and law in Malaysian government policy

Clearly, the qualities of Malaysia's female labor force were not the only factors involved in attracting foreign investment. Nevertheless, this labor force became an important part of a government-designed package for foreign investment that included tax incentives, increased bureaucratic efficiency, and fewer restraints on capital and profits.[42] Foreign companies demanded well-trained and docile labor, and this is what the government began to provide.

Moreover, the government did not feel the need to accord to women electronics workers the "respect, status and material well-being commensurate with the industry's indispensability to economic policy" because of the pervading gender ideology of women as secondary and temporary workers.[43] Rather than protect its female workers, the government itself began to manipulate and exploit the gendered nature of the electronics labor force.[44] In this manner, the Malaysian government and legal system began to perpetuate the conditions that made women an attractive low-cost labor force.[45]

Government industrial relations and labor legislation deliberately favored the needs of multinational companies over workers.[46] The government did not set a minimum wage, nor did it try to rectify the considerable wage gap between men and women.[47] In order to appease foreign companies' demands for round-the-clock labor, the government exempted electronics companies from laws that forbade employing women to work at night.[48] Also, the government did not enforce many existing labor laws, including those regulating shift work.[49] As young, single women were thought to be the best workers, companies manipulated laws which were designed to protect pregnant workers, to discourage married women from applying for jobs.[50] In addition, the government enacted labor laws, or selectively applied existing ones, to effectively prohibit the formation of most unions in electronics factories, limiting an important means to recognize the employee's rights.[51] In these ways, the government helped Malaysian women become the efficient, low-cost, and available workforce that foreign investors sought.

The effects of this gendered and racial ideology, in both business and government, continue to have important consequences today. In analyzing policies affecting women electronics workers, or in designing strategies to overcome their problems, these underlying constructions of gender and race cannot be ignored. The next section will discuss the current problems facing women electronics assembly workers in Malaysia and explore the most recent manifestations and consequences of this ideology of gender and race in the electronics industry.

D. Current issues facing women assembly workers in the Malaysian electronics industry

Rapid industrialization and technological advances have tended to exacerbate, rather than improve, the stratification of labor in the Malaysian electronics industry. Increasing globalization has made Malaysian workers more vulnerable to capital flight, market dynamics, and ever-growing cost pressures.[52] The competitive threat from other countries seeking to benefit from the high-tech industry is rising.[53]

Several economic and technological trends in particular have increased the vulnerability of assembly workers in Malaysia. In line with the increasing decentralization of multinational firms,[54] more and more electronics companies are subcontracting work to smaller firms.[55] As seen in the apparel industry, this allows companies to shift responsibility and liability for employment conditions and practices.[56] Moreover, in Malaysia, smaller, national firms are not subject to the higher employment, health, and safety standards that multinational companies have to meet, leading to reduced protection for these employees.[57] Workers in smaller firms are paid less,[58] and smaller firms can more easily exploit legal loopholes. For example, it has been reported that in some cases small firms have shut down after nearly five years of operation and then reopened in order to evade the national laws which allow the formation of unions after the first five years of a firm's existence.[59]

Also, due to market pressures, electronics companies are increasingly relying on temporary or "casual" workers rather than full-time, salaried employees.[60] Temporary workers are severely disadvantaged as compared to full-time workers in the kinds of benefits they receive.[61] "Casual" workers are mostly women who do piecework or other home-based assembly work.[62] They lack the protection of the Malaysian Employment Act and social security laws and may be working in inadequate environments lacking proper lighting or work conditions.[63] Due to the nature of their employment contracts, they are often the first to find themselves unemployed during market downturns.[64] The traditional image of women as secondary earners allows companies and governments to justify the lack of benefits and security they receive.[65]

For women working as regular employees in established electronics firms, employment laws are frequently violated.[66] Not only are labor laws, such as those regulating maximum overtime, frequently disregarded and unenforced,[67] but companies can even obtain exemptions or waivers from the government from certain provisions of labor laws.[68]

Another important problem is the lack of advancement opportunities for women employees despite increasing numbers of high-level positions. As career track and more highly skilled jobs open up, they are primarily filled by men.[69] Female assembly workers have little opportunity to enhance their skills and receive promotions.[70]

Perhaps the most pressing problem that these women face is in the area of health and safety. Although Malaysia passed a comprehensive Occupational Health and Safety Act in 1994,[71] the government is already facing charges that it is not being adequately followed or enforced.[72] While U.S. and Japanese companies appear to have strong health and safety practices, it is unclear what level of monitoring is involved and what measures, if any, smaller Malaysian subcontractors follow.[73]

The health issues that assembly workers face are numerous. During the course of their work, they are exposed to many hazardous solvents and chemicals, which have reportedly caused respiratory illnesses, skin problems, and eyesight degeneration.[74] Exposure to these chemicals has also been linked to reproductive problems such as miscarriage.[75] Moreover, many of the effects of these chemicals on humans are unknown.[76] The particular demands of shift work and long work hours have also been connected to various stress-related illnesses and problems related to non-ergonomically designed assembly lines.[77] These hazards make electronics assembly work a highly dangerous job in need of strict regulation.

In view of the severe health and safety risks, economic conditions, and hurdles to legal protection that Malaysian female assembly workers face, the task of identifying legal remedies and protections takes on added urgency. To this end, the experiences of women in other countries who occupy similar positions along the electronics assembly chain must also be recognized. The next section will analyze female electronics assembly workers in the United States and examine the emergence of a parallel workforce worlds away from Malaysia.

III. WOMEN ELECTRONICS ASSEMBLY WORKERS IN THE UNITED STATES

Women assembly workers in the United States are also subject to a subordinating ideology predicated on gender and race. Their experiences mirror those of Malaysian women situated in similar racial and gender systems, providing a revealing look at the way women in the United States can also be trapped in the high-tech underclass.

Currently in Silicon Valley women are thought to comprise between eighty and ninety percent of the workforce in low-skilled assembly jobs and are greatly underrepresented in management positions.[78] While in many ways these numbers reflect the status of women in the American workforce as a whole, the composition of the electronics industry is also significant because of the large numbers of immigrants who have entered this industry, particularly in Silicon Valley.[79] Many of these immigrants have been able to parlay their previous education or skills into high-paying and high-status jobs.[80] Others, however, have become trapped in jobs similar in nature, prospects, and hazards to assembly lines in the countries they left behind.[81] In an area fueled by the wealth of high technology, these jobs are part of high-tech's "hidden" labor[82] that most Americans never see.

Today, Silicon Valley contains several hundred electronics assembly houses, which constitute the largest concentration of assembly houses in the United States.[83] The workers of choice on these assembly lines are immigrants. In addition, "[i]mmigrant women constitute about seventy percent of the entry-level labor force in Silicon Valley."[84] Still others work at home, doing contract-based piecework.[85] "I call them Vietnamese mother-in-laws, people who can't get into the general workforce," says one employer, describing his home-based employees.[86] The origins of this hidden workforce lie again in the economic and political forces that have shaped the current structure of the electronics industry.

A. The development of a divided workforce

Although, as noted earlier, the electronics industry began to shift much of its production work overseas in the 1960s, a significant portion of this work was kept close to home. Faced with intense competition, electronics companies needed a quick and convenient source of labor for short-term projects or prototypes, so some assembly jobs remained in Silicon Valley to fill this role.[87] As the electronics industry expanded, the percentage of these assembly jobs in the industry declined drastically, but the actual number of these jobs has remained relatively constant, leaving a significant number of women in the United States doing production work that "closely resembles the same 'low-tech' labor done by their 'sisters' overseas."[88]

This "low-tech" labor force is highly stratified along racial and gender lines. Women and minorities are concentrated in the jobs with the least opportunities and benefits, while white males hold most of the high-level jobs.[89] Immigrant women are preferred over immigrant men for the lowest-level jobs.[90] The most common countries of origin are Mexico, Vietnam, the Philippines, and Korea, though immigrants from many developing countries are represented.[91] Also, while electronics assembly workers in Malaysia are predominantly young and single,[92] Silicon Valley assembly workers have historically included large numbers of married women across a wider age range.[93]

The reasons behind these characteristics are quite telling and reveal much about the status and structure of assembly work in the economy of Silicon Valley. The preference for immigrants, for example, is a reflection of the low status of many immigrants in the U.S. economy, much like the female labor force in Malaysia. Because of their precarious financial positions, their lack of knowledge about wage rates and U.S. labor protections, and the comparison of their present wages to the lower ones in their homelands, immigrants are often willing to accept lower wages, allowing electronics companies to keep wages and benefits low.[94]

Stereotypes also play a large role. Electronics companies claim that immigrants are the only ones willing to take assembly jobs, but studies have shown that white American applicants are often deliberately discouraged from taking such jobs.[95] Hiring personnel claim that most men and white women are not suited for tedious assembly line work and encourage such applicants to apply for management or professional positions.[96] Companies also use language barriers to control immigrant employees.[97] Managers also believe that immigrants unfamiliar with the English language and the American legal system will be less familiar with occupational health and safety laws and will be less likely to organize or protest.[98] Employers thus not only spend less on wages for immigrant workers but are also able to exert greater control over them.[99]

Among these immigrant workers, women are preferred for some of the same reasons that companies seek out Malaysian women. As in Malaysia, employers in the United States can capitalize on the secondary status of women and immigrants in the American workforce.[100] While immigrants in general accept lower pay, immigrant women also fit employers' notions of women as a flexible, reserve source of labor who can justifiably be paid lower wages or be hired only "temporarily."[101] The declining supply of unmarried women workers in the United States has forced employers since the 1970s to hire married women and women with children,[102] while in Malaysia the number of unmarried women workers is increasing.[103] These trends are exacerbated by the highly competitive nature of the electronics industry, which drives employers to seek out the cheapest and most flexible labor.[104] Immigrant women, in other words, are a cheap and expendable workforce, suited to the demands of the industry.[105]

Unfortunately, like their counterparts in Malaysia, employers in the United States often view Asian immigrant women through the lens of racial prejudice. The stereotypes of "nimble fingers" and passive personalities operate in the United States as well,[106] though perhaps not as blatantly. Stereotypes of Asian women are employed in ways that devalue the work and lifestyles of the workers. For instance, a study by Karen Hossfeld showed that white managers consistently referred to Asian workers as "girls," while non-Asian workers were "women."[107] One manager explained,

> Asian women are more subservient than American females: if I refer to them as "girls" it's because to me, they act like girls: They only speak when spoken to, do exactly as they are told, and so forth. So I play into it—I treat them firmly like a father figure. . . .[108]

Such "docility" and fear of reprisals appears attractive to employers in an industry where unions and activism have been shunned.[109]

Through such stereotyping, images of Asian immigrant women are molded into an American version of the ideal employee. Perhaps even more disturbing is the way these images of Asian immigrant women have become twisted and adopted into corporate and government policy, thus intensifying the denial of their rights. These processes will be discussed below.

B. Gender and race in law and corporate policy

At the corporate level, images of Asian immigrant women have allowed employers and managers to create manipulative strategies to control the women.[110] Managers have used perceptions of the worker's femininity against them.[111] For example, employers have been known to characterize union action or assertiveness as inappropriately unfeminine behavior.[112] Also, employers are able to call the precise and detailed work involved in electronics assembly "unskilled" work because they are able to successfully equate it with skills that these women are already supposed to have—assembly work is compared to "following a recipe" or knitting, and the tolerance for tedious work that assembly workers need to have is something that Asian women are assumed to have as an ingrained skill.[113] Asian immigrant women are supposedly "suited" by temperament and economic position for the jobs that are perceived to be too low-paid and boring for white native-born women.[114] Once a task is classified as unskilled, the workers who do it can be paid less than they would for comparable "skilled" work.

Other tactics devalue these tasks as women's work and operate to deny Asian immigrant women opportunities for advancement or skill acquisition. Most assembly jobs are viewed by employers, and often employees, as "temporary" or supplemental jobs, though they often do not turn out to be.[115] The devaluation of immigrant women's work itself seems to underlie employers' view of assembly work as temporary. As one electronics factory owner stated, "Let's face it, when you have to expand and contract all the time, you need people who are more expendable. When I lay off immigrant housewives, people don't get as upset as if you were laying off regular [*sic*] workers."[116]

Even when assembly jobs become long-term positions,[117] these images allow employers to reduce the opportunities available to immigrant women. By portraying assembly work as temporary and as women's work, employers can justify the lack of training and advancement opportunities that assembly workers receive.[118] Male workers are promoted faster, and men, especially white men, are often hired to supervise the very female workers who will then have to train them, underscoring the fact that female assembly workers could have just as easily been trained to be supervisors.[119] One employee described the limiting stereotypes at work:

> Most managers automatically feel that the assembly job is the highest aspiration Asian American women have. When [managers] see a white women [*sic*] as an assembler, they feel that she looks out of place—she should be in the office. When they see an Asian American man, they feel he looks out of place—he should be an engineer. But for Asian American women—it seems logical she should only be an assembler. This perception contributes to limiting the mobility of Asian American women in the industry.[120]

Thus, by assuming that Asian immigrant women are best "suited" to assembly line jobs, employers reinforce racial and gender stereotypes and limit these women's opportunities by making the lowest-level jobs the only jobs that are available to them.[121]

Not only are Asian women's race and gender used to control and deny privileges to them as workers, but the precarious position of many undocumented immigrant women in the legal system is also manipulated to further marginalize their position in the electronics industry. Most undocumented workers labor at the margins of the Silicon Valley assembly line.[122] Both legal and undocumented immigrants have been employed by the electronics industry to assemble transistors, circuitry, or other electronic devices at home, a system called piecework or "home work."[123] This type of "home work" has been traditionally associated with women's work and the devaluation associated with this characterization.[124] The extent of this work by undocumented workers is unknown, though the majority are said to be Asian immigrants.[125] The undocumented workers often keep silent in the face of abuses and are not able to benefit from U.S. laws in many situations.[126] For example, sending piecework to the homes of immigrants allows companies to evade overtime and minimum wage laws, since no one is monitoring who is working and for how long.[127] Also, there is no one to monitor health and safety or provide safety precautions, even though dangerous chemicals may be required.[128] Finally, home workers do not get the benefits that salaried factory workers do.[129] These violations are a concern for legal immigrants who do such home work as well.

C. Current issues affecting the rights of electronics assembly workers

In addition to the concerns undocumented workers face, other recent developments effectively reduce assembly workers' rights and legal recourses. One is the industry's increasing reliance on subcontractors and contract workers.[130] Women are favored as temporary contract workers and, as such, usually do not get the medical or other benefits that regular workers do.[131] Also, if the experience of Malaysian women offers any examples, it is that the use of subcontractors can allow a company to shift responsibility and liability for its actions, no small fear in an industry known for its use of toxic chemicals.

Further, in an industry that changes as rapidly as the electronics industry, it is difficult for government regulators to enforce current laws, and even more difficult for them to modify existing regulations to cover ever-changing processes. In other words, the pace of technological innovation is so rapid that traditional regulatory methods, such as those of government administrative agencies or legislative bodies, simply cannot keep up.[132] Thus, environmental and health laws may not provide adequate protection for many assembly workers. Indeed, many of the problems that assembly workers face can be traced to these inadequate protections.[133]

Like women in Malaysia, some of the most pressing problems that assembly workers in Silicon Valley face are related to health and safety. In some ways, work in U.S. factories may be even more hazardous because of the more chemical-intensive processes in use.[134] Assembly workers may be exposed to dangerous organic solvents, such as xylene and methylene chloride, as well as other chemicals which may cause neurological, vision, or reproductive damage.[135]

Immigrant women in the factories of Silicon Valley echo the same complaints that Malaysian assembly workers describe. For example, the women complain of noise pollution, eye strain, and injuries from repetitive motions.[136] They also cite respiratory and allergy problems from exposure to chemicals and fumes.[137] Workers are sometimes not told what chemicals they are working with, making self-protection measures difficult, and are often not able to prove that their illnesses are work-related because doctors do not have experience in assessing the effects of the chemicals they have been exposed to.[138] Recent studies have also shown that women working in some types of electronics factories have a higher risk of miscarriage.[139]

However, unlike women in Malaysia, assembly workers in the United States have access to a more comprehensive and more readily enforced body of legal protection. These provisions, along with the relevant Malaysian legal structure, will be discussed in the following section.

IV. AN INTERNATIONAL LEGAL FRAMEWORK FOR RECOGNIZING THE RIGHTS OF WOMEN ON THE GLOBAL HIGH-TECH ASSEMBLY LINE

The previous sections have surveyed the origins, development, and maintenance of the stratified assembly workforces that have developed in the United States and in Malaysia. Expanding on those analyses, this section will examine the Malaysian and U.S. regulatory systems intended to protect the rights of assembly workers and compare their strengths and weaknesses. . . .

A. Domestic laws

1. Malaysian Laws

In Malaysia, perhaps the biggest legal and regulatory hurdle is enforcement. The Malaysian government is trying to balance rapid economic development with the task of building a legal structure to accommodate these changes and does not have the resources to enforce all laws or punish most violations.[140]

The primary law setting out the rights of workers in Malaysia is the Employment Act of 1955 (with subsequent amendments).[141] Of great significance to women assembly workers is the fact that there is no provision mandating equal pay for equal work, so under this Act, it would not be unlawful for an employer to pay female employees less than male employees doing the same job.[142] Moreover, there is no minimum wage law for the manufacturing sector,[143] and the role of unions is greatly circumscribed.[144] Unions may be disallowed for the first five years of a company's operations, and trade unions are not allowed to be affiliated with national unions.[145] Furthermore, the government has waived many of the Employment Act's provisions for electronics companies, and many of the remaining provisions are not enforced.[146] Also, since many assembly workers, like home-based or contract workers, are not covered by the Act, it is of limited use for the protection of women assembly workers.[147]

In the area of health and safety, Malaysia passed an Occupational Safety and Health Act in 1994.[148] The Act is intended to provide a "legislative framework to promote, stimulate and encourage high standards of safety and health at work," and supersedes existing laws pertaining to health and safety.[149] Since the Act is relatively recent, many of the provisions have not yet been implemented, and enforcement is a problem.[150] However, because it is relatively comprehensive on paper,[151] the Act does have great future potential.

Even with these laws on the books, there are formidable barriers towards recognizing and enforcing the rights of assembly workers. Activists recognize a reluctance among women workers to complain, as well as a lack of acknowledgment of the potential hazards of assembly work.[152] Moreover, many workers are not aware of their rights or how to enforce them.[153] In addition, workers and nongovernmental organizations (NGOs) do not have access to data or information about safety and health problems in the factories.[154] There are no public disclosure mechanisms, and any disclosure of chemicals used is usually accomplished through voluntary corporate action.[155] European, Japanese, and U.S. companies maintain that they voluntarily exceed local regulations in applying the same standards in Malaysia as those required in their own countries.[156]

Consequently, in Malaysia, the primary means of enforcing the rights of women assembly workers is through educational campaigns by NGOs[157] and government departments or ad-hoc protests and complaints, which are highly important strategies since public awareness is very low.[158] Again, it is difficult for the groups themselves to obtain the necessary information, and they are constrained by a lack of resources.[159] Thus, while the Malaysian electronics industry is rapidly approaching the world's highest technological standards, its legal standards and enforcement levels lag far behind.

2. U.S. Laws

The situation in the United States is very different, though still inadequate in certain areas. Assembly workers in Silicon Valley can turn to a comprehensive system of federal, state, and administrative regulations. These include the California Injury and Illness Prevention Program, CAL/OSHA standards, the Hazard Communication Standard, and a variety of state and federal laws governing wages, hours, and other occupational health and safety issues.

The cornerstone of workers' rights and protection in California is the Injury and Illness Prevention Program (IIPP), implemented in 1991.[160] Under the California Occupational Safety and Health Act of 1973, every employer has a legal obligation to provide and maintain a safe and healthful workplace.[161] The IIPP expands this obligation for employers like electronics companies who deal with hazardous substances, requiring such employers to design and maintain a written and effective Injury and Illness Prevention Program.[162] Under this program, employers must provide health and safety training to employees, inform them of hazardous situations, act promptly to rectify problems, and keep adequate records of hazardous substances and problems.[163]

CAL/OSHA standards are intended to provide another level of protection.[164] These standards implement at the state level the provisions of the federal Hazard Communication

Standard (HAZCOM).[165] Adopted in 1983, HAZCOM requires chemical manufacturers and employers who use dangerous chemicals to assess their hazards and disclose this information to their employees and/or customers through the use of warning labels, information sheets, and other methods.[166] CAL/OSHA provisions regulate workers' exposure levels to dangerous chemicals[167] and cover other standards related to work environments such as ventilation[168] and temperature.[169]

To enforce these standards, OSHA's compliance officers can inspect workplaces and issue citations and fines for violations of these standards.[170] While this appears to be an effective system, it is often rendered ineffective because of the small numbers of chemicals and hazards that OSHA has identified as needing such regulation.[171] For electronics companies, especially semiconductor manufacturers, this problem is particularly acute because of the high rate of technological change in their manufacturing processes.[172] New processes supplant old ones before there is adequate time to study the effects of exposure to the chemicals used in the old ones, and with workers exposed to so many processes and chemicals during their careers, it is difficult to isolate the effects of any one of them.[173] In effect, OSHA's regulatory system, further constrained by the rate of bureaucratic change, lags so far behind the current practices of semiconductor companies that OSHA itself has admitted that its standards are inadequate to regulate semiconductor companies.[174] This situation has left a critical accountability gap.

For immigrant assembly workers, these provisions have not been sufficient to guarantee their rights or protect their health and safety. Workers have complained of inadequate ventilation or malfunctioning systems, forcing them to inhale chemical fumes.[175] Others have said that they were never told what chemicals they were using or what their effects were.[176] Their complaints do not appear to be unfounded; it has been reported that about ten percent of chipmaking plants inspected since 1992 provided inadequate training, and that nearly one-third of Silicon Valley firms inspected were cited for shoftfalls in meeting health and safety standards.[177] However, it has been difficult for women suffering from exposure-related problems to prove that their symptoms were work related, which means they cannot receive worker's compensation.[178] Activists and labor organizations are trying to improve this situation by conducting training programs and disseminating health, safety, and legal information to assembly workers, empowering them to monitor work conditions.[179]

Companies must also comply with applicable federal environmental regulations that protect workers in industries where hazardous substances are used. The Environmental Protection Agency (EPA) implements the Toxic Release Inventory (TRI),[180] which essentially implements "right-to-know" provisions for the public about companies' use of chemicals.[181] Manufacturing firms must report to the EPA what chemicals they are releasing, as well as the amounts and types of releases, and the EPA compiles this data into an annual national report and a computer database.[182] Environmental organizations, communities, and the media have used this easily accessible information to exert legal and political pressure on companies and to publicize potentially dangerous corporate practices.[183]

However, while TRI is a very valuable instrument, the set of chemicals covered is not nearly comprehensive enough to adequately regulate the conduct of microchip manufacturers.[184] Environmentalists have estimated that nearly ninety-five percent of all toxic chemicals released by corporations have escaped reporting, and that company non-compliance rates are very high, possibly up to thirty percent.[185] Many toxic chemicals are simply not covered under TRI, including many of those used by semiconductor companies.[186] Also, since TRI standards only apply in the United States, companies can easily escape their provisions by locating manufacturing operations in other countries (although the U.S. government has mandated its application in certain isolated international contexts, such as U.S. companies operating maquiladoras in Mexico).[187]

This analysis demonstrates that there are important gaps in the current domestic legal regimes for the protection of assembly workers. Workers in Malaysia have minimal legal protection and must rely on the self-policing and initiatives of electronics companies for protection. In the United States, even a comprehensive system of health and safety laws falls short in an industry where rapid changes outpace regulators. In this light, it is important to assess the potential applicability of international norms and standards. . . .

V. CONCLUSION

In analyzing the same industry in different cultures, this article aimed to show how gender and race have been manipulated in the electronics industry to control women and deny them benefits or opportunities that could have been available. Women in both countries describe similar dangerous working conditions, health problems, and a lack of advancement opportunities. Their prospects for improvement are also limited by the inadequacies of the current domestic and international legal systems. These problems cannot be ignored, and the electronics industry should be held accountable for its role in creating them. With public awareness of these problems increasing, this industry is poised at a critical point in its history. Electronics companies can continue to expand current opportunities for constructive dialogue with workers, activists, and regulators, or they can ignore these issues until legal action is brought against them. As the experience of the garment industry shows, constructive engagement and pre-emptive voluntary corporate actions can help ameliorate rights violations before damaging public condemnation and lawsuits begin. The electronics industry has just begun to initiate voluntary actions and acknowledge problems, and should heed this warning for the future:

> The electronics industry is in an analogous position to Tobacco back in the 1960s. . . . Huge damage awards would wreak havoc in what could become this country's most important economic sector if it stays healthy. It has the opportunity to face up to potential problems squarely now, or ignore it and hope it goes away.[188]

Notes

This work was supported by a Selected Professions Fellowship from the American Association of University Women, I would like to thank Dr Lyuba Zarsky, the California Global Accountability Project, and Kai Lintumaa for their support and guidance during this project, and Sandra Santos and Maki Arakawa at the *Berkeley Women's Law Journal* for their help in the editing process

1 Malaysia is a major assembly and manufacturing site for American electronics companies, and the factory workers in this industry are almost all female. *See* Kristal E. Alley, American Malaysian Chamber of Commerce (AMCHAM), *Malaysian-American Electronics Industry Annual Survey 1998/1999*, 27 (1999); Sha'ban Muftah Ismail, *Women, Economic Growth & Development in Malaysia* 72 (1997). Furthermore, Malaysian laws predicated on stereotypical roles for women have reduced the rights of women electronics workers. *See* Elizabeth Grace, *Shortcircuiting Labour, Unionizing Electronic Workers in Malaysia* 15–17 (1991) (discussing employment and "family development" policies in Malaysia that restrict women's opportunities).

2 Malaysia has been a major exporter of electronics products for the past two decades. *See* Ismail, *supra* note 1, at 72 (citing Fong Chan Onn, "Industrialization in Malaysia: Role of Small and Medium Scale Industries," in *The Malaysian Economy in Transition* 71 [Ambrin Buang, ed., 1990]).

3 *See* Michael Yeoh, "Preface," in *Penang into the 21st Century* (Asian Strategy & Leadership Institute, 1995) (stating that Malaysia is currently the world's largest exporter of electronic chips because of the success of the electronics industry in the state of Penang).

4 *See* Karen F. Travis, "Women in Global Production and Worker Rights Provisions in U.S. Trade

Laws," 17 *Yale Journal of International Law* 173 (1992), at 189–94 (describing common characteristics of EPZs such as employers' preferences for young female workers, low wage rates, gender-based wage differentials, and government restrictions on bargaining rights, and noting that these characteristics are present in Malaysian EPZs as well).

5 *See* Grace, *supra* note 1, at 6–7.

6 *See* AMCHAM, 1999 *Malaysian Business* 6 (1998).

7 *See*, e.g., Grace, *supra* note 1, at 6–7 (describing the hierarchical social and political structure of Malaysian society). *See also* Raja Rohana Raja Mamat, *The Role and Status of Malay Women in Malaysia: Social and Legal Perspectives* 10–11 (1991) (noting that the historical and continuing socioeconomic differences between Chinese, Indians, and Malays are among the most significant features of Malaysian society, that the Malays have always lagged behind the Chinese in these areas, and that this has led to social and political partisanship in Malaysia).

8 *See* Alley, *supra* note 1, at 27 (providing the numerical data that the author used to compute the above percentages).

9 Ibid.

10 *See* Joseph Chin & Lee Min Keong, "Silicon Island," *Asia Inc.*, Sept. 1999, at 17.

11 Ibid. at 17–19. As a sign of the maturity of Penang's electronics industry, among Association of South East Asian member countries, Malaysian wages are now the second-highest, below only Singapore's wage levels. Ibid. at 19.

12 Ibid., pull-out section

13 Ibid.

14 Ibid.

15 Ibid. While Taiwanese companies have the most factories in Penang, American companies employ more workers and are more technology intensive. *See also* Ibrahim Saad, "Industrial Development Strategy," in *Penang into the 21st Century*, *supra* note 3, at 37–38 (noting that in 1993, American companies in Penang had over 44,000 employees, while Taiwanese firms had less than 19,000).

16 *See* "Silicon Island," *supra* note 12.

17 Ibid. (figure included in a table of "Penang's Industrial Spread," 1998).

18 *See* International Labour Organization (ILO), *Labour and Social Issues Relating to Export Processing Zones* 3, 14 (1998), at 20–21 (reporting that women make up the majority of workers in EPZs around the world; for example, in 1995, women workers comprised 80% of the employees in the free trade zones of Guatemala and Nicaragua).

19 *See* Linda Lim, "*Women Workers in Multinational Corporations: The Case of the Electronics Industry in Malaysia and Singapore*," 9 *MICH. Occasional Papers in Women's Studies* 1, 7 (1978), at 6–8 (stating that women have constituted a significant part of the Malaysian electronics industry's labor force since electronics companies first began to locate in Malaysia).

20 *See* Travis, *supra* note 4, at 190.

21 *See* Shanti Diariam, "New Technologies and the Future of Women's Work in Asia" 7 (1994) (unpublished report from the workshop New Technologies and the Future of Women's Work in Asia) (on file with author). Positions requiring higher education levels are usually offered to recent male graduates, while women assembly workers or operators are usually not given a chance to upgrade their skills and apply for these higher-level jobs. *See* Jamilah Ariffin, "Economic Development and Women in the Manufacturing Sector," in *Readings on Women and Development in Malaysia* 53, 66–67 (Jamilah Ariffin, ed., 1994), at 205, 210 (hereinafter Ariffin, "Economic Development") (noting that by the 1970s at least 75% of employees in the Malaysian electronics industry were women).

22 *See* Alley, *supra* note 1, at 27 (providing the numerical data that the author used to compute the above percentages).

23 Ibid. *See also* Aihwa Ong, "Global Industries and Malay Peasants in Peninsular Malaysia," in *Women, Men, and the International Division of Labor* 426, 434–35 (June Nash & Maria Patricia Fernandez-Kelly eds., 1983), at 429–30 (hereinafter Ong, "Global Industries") (illustrating the types of semiskilled jobs available to Malaysian women in EPZs, including electronics assembly work).

24 *See* Victor G. Devinatz, *High-Tech Betrayal: Working and Organizing on the Shop Floor* 9–10 (1999), at 9–10 (describing multinational electronics companies' preferences for nonwhite women assembly workers based on sexual stereotypes of women as more patient and dexterous workers who can be paid less than men). *See also* Grace, *supra* note 1, at 15 (noting the large amount of literature on the electronics industries' preference for female workers).

25 *See* Vivian Lin, *Health, Women's Work, and Industrialization: Semiconductor Workers in Singapore*

and Malaysia 5 (1991), at 7 (hereinafter Lin, *Women's Work*) (examining, in selected countries, the concentration of women workers in the electronics industry).

26 *See* Lim, *supra* note 19, at 7; Devinatz, *supra* note 24, at 9–10.

27 Lin, *Women's Work, supra* note 25, at 7.

28 Lim, *supra* note 19, at 7 (explaining that small hands are thought to be an advantage for assembly workers, whose jobs require manual dexterity and involve delicate, precise operations with extremely small electronics components).

29 Ibid.

30 *See* Cynthia H. Enloe, "Women Textile Workers in the Militarization of Southeast Asia," in *Women, Men, and the International Division of Labor, supra* note 23, at 407, 412–13 (discussing the supposed docility of Asian women as an attractive factor for multinational garment and electronic manufacturers and noting that this stereotype reflected the belief that the influence of patriarchal cultures made Asian women obedient to both their parents and factory managers).

31 *See* Aihwa Ong, *Spirits of Resistance and Capitalist Discipline: Factory Women in Malaysia* 204–10 (1987) (hereinafter Ong, *Spirits of Resistance*), at 152.

32 Ibid. (quoting an engineer on corporate policy, from the author's survey of comments from corporate personnel in Malaysian electronics factories).

33 *See* Lin, *Women's Work, supra* note 25, at 7 (examining the concentration, in selected countries, of women workers in electronics).

34 *See* Lim, *supra* note 19, at 13.

35 *See* Enloe, *supra* note 30, at 412–13; Lim, *supra* note 19, at 12–14 (demonstrating how managers of multinational corporations often viewed Asian women as docile and subservient to patriarchal authority, stereotypes which supported their belief that Asian women would make ideal employees).

36 *See* Enloe, *supra* note 30, at 412–13 (describing how multinational manufacturing companies have been able to tap into women as a cheap and available source of labor).

37 *See* Lin, *Women's Work, supra* note 25, at 6–7. Lin also cites statistics showing that in 1980, Malaysian electronics workers were paid in U.S. dollars an average of 42 cents an hour, as compared to $6.96 an hour in the U.S., and that in 1985, average wages were 75 cents in Malaysia and $12.59 in the United States. *See* ibid. at 7 (Table 1.2). Women were, and still are, also seen as temporary or supplemental workers who did not need to be paid as well, nor given the same advancement opportunities as men. *See* ibid. at 8. *See* generally Diariam, *supra* note 21, at 2–3 (noting that women in the electronics industry tend to enter the market unskilled, are more likely to face employment risks, and receive lower wages than men).

38 *See* Lim, *supra* note 19, at 11–12.

39 *See* Enloe, *supra* note 30, at 412–13 (describing how women have been similarly targeted as a cheap and malleable workforce by multinational manufacturing companies in the Philippines, South Korea, Taiwan, Hong Kong, and Singapore).

40 *See* Grace, *supra* note 1, at 15–16 (explaining ways that gender biases influenced government policies concerning the Malaysian electronics industry; for example, when foreign companies' preference for female labor became apparent, the government passed laws or removed restrictions which had the effect of making female labor more accessible to the companies).

41 Lim, *supra* note 19, at 7. The brochure quoted by Lira was issued abroad by the Malaysian government to entice multinational electronic companies to locate in Malaysia.

42 *See* Ong, *Spirits of Resistance, supra* note 31, at 146. *See also* Jean Larson Pyle & Leslie Dawson, "The Impact of Multinational Technological Transfer on Female Workforces in Asia," 25 *Colum. Journal of World Business* 40, 42 (1990) (citing a researcher who concluded that Asian governments have viewed the female labor supply as their principal resource for industrial development).

43 Grace, *supra* note 1, at 15–16.

44 *See* Lance Compa, "International Labor Standards and Instruments of Recourse For Working Women, 17 *Yale Journal of International Law* 151, 159–161 (1992) (illustrating the ways in which female workers are exploited in the electronics industry, and explaining how even in the face of international criticism the Malaysian government refused to enhance worker protections, maintaining that it had a right to put the preferences of the electronics industry above worker protection). *See also* Jamilah Ariffin, "Women Workers in the Manufacturing Industries," in *Malaysian Women: Problems and Issues* 49, 52 (Evelyn Hong, ed., 1983) (hereinafter Ariffin, "Women Workers") (noting that the Malaysian government relaxed or failed to enforce regulations regarding women workers in order to protect the interests of foreign investors).

45 *See* Ariffin, "Women Workers," *supra* note 44, at 52, 55 (providing examples of cases where the government made it even easier for foreign companies to exploit low-cost female workers; for

instance, it allowed firms to send recruiting agents to rural areas, thus encouraging women to migrate to cities, where they then enlarged the pool of low-cost labor available to electronics companies).

46 *See* Ong, *Spirits of Resistance, supra* note 31, at 146–47.
47 Ibid. at 147.
48 *See* Grace, *supra* note 1, at 16.
49 *See* Diariam, *supra* note 21, at 3.
50 *See* Ong, *Spirits of Resistance, supra* note 31, at 147–48.
51 *See* Grace, *supra* note 1, at 22–25; Ong, *Spirits of Resistance, supra* note 31, at 148.
52 *See* P. Y. Lai, "Value-added Manufacturing: A New Paradigm," in *Penang into the 21st Century, supra* note 43, at 52–53 (describing current and projected increases in Malaysian wage rates which, combined with the emergence of lower cost economies in Vietnam, China, and India, threaten to make Malaysia a less attractive location for labor-intensive manufacturing industries). *See also* Chin & Keong, *supra* note 10, at 19 (citing the growing competition that Penang's electronics industry is facing from countries like China with lower labor costs, at the same time that the costs of production are rising in Penang). *See* generally ILO, *supra* note 18, at 14–15 (describing how the pressures of globalization and a corresponding rise in international competition are forcing multinational companies to become increasingly decentralized and ready to move or expand according to market changes).
53 *See* generally ILO, *supra* note 18, at 3 (noting that more and more countries are seeking to attract foreign investment and are trying to expand their export processing zones).
54 Ibid.
55 *See also* M. Patricia Fernandez-Kelly, "Labor Force Recomposition and Industrial Restructuring in Electronics: Implications for Free Trade, 10 *Hofstra Lab. L.J.* 623, 634 (1993), at 650 (calling subcontracting "the backbone of the electronics industry," and noting that more than half of the electronics companies in her study relied heavily on various forms of domestic and international subcontracting). *See also* Diariam, *supra* note 21, at 2 (noting the increase in the number of electronics firms subcontracting work in Malaysia).
56 *See* Laura Ho *et al.*, "(Dis)assembling Rights of Women Workers Along the Global Assembly Line: Human Rights and the Garment Industry," 31 *Harv. C.R.-C.L.L. Rev.* 383, at 391.
57 *See* generally Federal Dep't of Town and Country Planning Ministry of Hous. & Local Gov't Malaysia and Universiti Putra Malaysia, *Environmental Guidelines for the Multimedia Super Corridor (MSC): Investor's Guide* 2 (undated) (describing how foreign corporations will be held to "world-class" standards, such as U.S., European, or Japanese standards, while Malaysian firms will be subject to Malaysian standards).
58 *See* Diariam, *supra* note 21, at 2.
59 Ibid. at 3.
60 Ibid. at 2–3.
61 *See* generally ibid. at 2–3.
62 Ibid. at 5–6.
63 Ibid. at 6.
64 *See* Ariffin, "Economic Development," *supra* note 21, at 212 (discussing the sharply fluctuating demands of the electronics industry in Malaysia for cheap, disposable labor, and how female workers are preferred for these jobs where they are promptly laid off when the market declines).
65 *See* Grace, *supra* note 1, at 16 (describing how the perception of women as secondary or temporary workers allows the government to justify the lack of protection of their jobs or work conditions); Ariffin, "Economic Development," *supra* note 21, at 212 (stating that in electronics companies, "[t]he management's view and assumption that women workers are secondary income earners provide rationality for their policy of paying lower wages to women workers and retrenching them during periods of slack market demand for their products"). *See* generally "Working Without a Net: Women and the Asian Financial Crisis," *Gender Matters Q.* (USAID Office of Women in Development, GenderReach Project, Washington, DC), Jan. 2000, at 6 (stating that in countries following the "Asian Model of Development," including Malaysia, "[w]omen are the last hired, the first fired, and the least likely to qualify for benefits offered by their employers or provided by their governments").
66 *See* Tan Pek Leng, "Women Factory Workers and the Law," in *Malaysian Women: Problems and Issues, supra* note 44, at 64–65 (describing the eagerness of the Malaysian government and legal system to misinterpret and disregard the law in order to promote the interests of foreign investors at the expense of female electronics workers).
67 *See* Amarjit Kaur, "An Historical Analysis of Women's Economic Participation in Development,"

in *Readings on Women and Development in Malaysia*, *supra* note 21, at 3, 18 (discussing the lack of enforcement of Malaysian labor laws). *See* Leng, "Women Factory Workers and the Law," in *Malaysian Women: Problems and Issues*, *supra* note 44, at 64, 73–74 (listing instances of particularly egregious violations of Malaysian labor laws by electronics companies, such as forcing employees to work overtime against their will, and retaliating against those who refuse to work overtime; mislabeling overtime work as "overlap" shifts so companies would not have to pay overtime rates or conform to laws regulating maximum overtime; and forcing workers to take leave days during company shutdowns), and at 73 (noting that the government stood by idly while these violations occurred).

68 *See* Diariam, *supra* note 21, at 3. *See also* Grace, *supra* note 1, at 14–15 (explaining how the Malaysian government has prevented workers in the electronics industry from unionizing by inconsistently applying labor laws for the benefit of companies).

69 *See* Diariam *supra* note 21, at 7.

70 Ibid.

71 For the full text of the act, see National Institute of Occupational Safety and Health (NIOSH) (Malay.), *Occupational Safety and Health Act 1994* (visited Jan. 28, 2000) <http://www2.jaring.my/niosh/law/act/osh%5fact.htm>.

72 *See* A. Kathirasen, "Flexibility Vital to Promote Conducive Occupational Environment," *New Straits Times*, Aug. 8, 1998, at 2. The article describes a 1998 dispute between electronics workers and their employers, in which the employees charged that they were being forced to stand for over eight hours a day, straining their health and possibly violating the Occupational Safety and Health Act. *See* note 71. However, when the workers complained to the government, the Department of Occupational Safety and Health (DOSH) sided with the companies and excluded workers from their inquiry, generating criticism from affected workers, their representatives, and workers' organizations about the government's willingness to adequately enforce the law. *See* note 71. Anecdotal evidence also suggests that the government lacks the will and resources to effectively enforce the law. *See* Malaysian Science and Tech. Info. Centre (MASTIC), 1994 National Survey of Innovation in Industry 64 (1996), at 65 (describing the "manpower" (*sic*) shortage in all sectors that may hinder the government's productivity goals).

73 *See* Ariffin, "Economic Development," *supra* note 21, at 217 (describing the need for monitoring and research in the electronics industry). *See*, for example, *AMD Environmental, Health and Safety Report 1997* (visited Jan. 28, 2000) <http://wwwamd.com/about/about.html> (hereinafter *AMD*), for a statement by a U.S. company maintaining that the company follows the same environmental and safety standards abroad as in the United States.

74 *See* Khoo Hoon Eng, "Hazards Faced by Women at Work," in *Malaysian Women: Problems and Issues*, *supra* note 44, at 80, 82–83 (describing how the industrial and organic solvents used in the electronics industry, such as isopropyl alcohol and chloroform, can cause dermatitis, headaches, nausea, weakness, and perhaps even cancer, along with other symptoms and explaining that hand soldering work often required in electronics assembly may expose workers to poisonous fumes from metals like cadmium or zinc).

75 *See* Lin, *Women's Work*, *supra* note 25, at 16; Marc Schenker, "Epidemiologic Study of Reproductive and Other Health Effects Among Workers Employed in the Manufacture of Semiconductors, Final Report to the Semiconductor Industry Association," 1992 (on file with author).

76 *See* Leslie Byster & Ted Smith, "High-Tech and Toxic," F. for Applied Res. & Pub. Pol'y, Apr. 1, 1999 (noting that toxicity information is unavailable for more than half of the chemical substances used in the electronics industry). *See also* Judy Mann, "A Hard Look at the Health of Working Women," *Washington Post*, Sept. 18, 1998, at D20 (quoting Sheila Hoar Zahm of the National Cancer Institute, who notes that in new industries, such as the semiconductor industry, there are no studies yet available on the effects on women of exposure to workplace carcinogens).

77 *See* Lin, *Women's Work*, *supra* note 25, at 10 (citing studies by Grossman [1978], Lira [1978], Woon [1982], and Paglaban [1978]), and at 15 (discussing the repetitious and detailed manual work required on electronics assembly lines that can lead to ergonomic problems).

78 *See* Devinatz, *supra* note 24, at 9 (citing studies estimating that "approximately 83 percent of all professional workers and technicians employed in Silicon Valley semiconductor companies are Anglo-American males" and studies estimating that the lowest skilled production jobs are composed of 80–90% women).

79 *See* Tom Abate, "Laboring in the Silicon Jungle Activists Charge High-Tech Industry With Exploiting Mainly Immigrant, Female Employees," *SF Examiner*, Apr. 25, 1993, at E1 (hereinafter Abate, "Laboring in the Silicon Jungle") (quoting a researcher who states that immigrant women constitute about 70% of the entry-level labor force in Silicon Valley); Ken McLaughlin & Ariana

Eunjung Cha, "Power: Politics and the Workplace Divisions: Segregation Trends Emerge in High-Tech Industry, Experts Say," *San Jose Mercury News*, Apr. 16, 1999, at 1A (suggesting that large numbers of immigrants have entered the high-tech industry in California).

80 *See* McLaughlin & Cha, *supra* note 79 (citing a survey showing that 10% of the CEOs of Silicon Valley's top 150 companies are Asian, and the white-collar workforce in Silicon Valley is 31% Asian, although citizens and immigrants of Asian ancestry have been lumped together). *See also* Paul Ong & Tania Azores, "Asian Immigrants in Los Angeles: Diversity and Divisions," in *The New Asian Immigration in Los Angeles and Global Restructuring* 100, 111 (Paul Ong *et al.*, eds., 1994) (describing how many highly skilled Asian immigrants, especially those entering the United States after 1965, have been able to find high-level managerial or professional jobs commensurate with their skills in the United States).

81 *See* generally Eric Lai & Theresa C. Viloria, "Down & Out in Silicon Valley," *A. Mag.*, Nov. 30, 1995, at 32 (describing how Vietnamese and Filipino immigrants came to dominate the lowest-paying and most dangerous high-tech jobs in Silicon Valley). They note that "Vietnamese and Filipino women working in electronics and high-tech jobs earn the lowest average salaries in [Silicon] Valley." *Id.*

82 *See* Miranda Ewell & K. Oanh Ha, "High-Tech's Hidden Labor Outside the Eyes of the Law: Silicon Valley Companies Pay Asian Immigrants by the Piece to Assemble Parts at Home," *San Jose Mercury News*, June 27, 1999, at A1 (hereinafter Ewell & Ha, "High-Tech's Hidden Labor") (describing predominantly Asian immigrant piecework home assembly workers as a "hidden" labor force).

83 Ibid.

84 *See* Abate, "Laboring in the Silicon Jungle," *supra* note 79 (quoting Karen Hossfeld).

85 *See* Ewell & Ha, "High-Tech's Hidden Labor," *supra* note 82.

86 Ibid.

87 *See* Karen J. Hossfeld, "Their Logic Against Them: Contradictions in Sex, Race, and Class in Silicon Valley," in *Women Workers and Global Restructuring* 149, 155, 167 (Kathryn Ward, ed., 1990), at 153 [hereinafter Hossfeld, "Their Logic Against Them"]; Ewell & Ha, "High-Tech's Hidden Labor," *supra* note 82.

88 *See* Hossfeld, "Their Logic Against Them," *supra* note 87, at 153. As discussed below, male immigrant employees generally have greater career prospects, are paid more, and are promoted more than female immigrant employees. *See id.* at 167–68; Naomi Katz & David S. Kemnitzer, "Women and Work in Silicon Valley," in *My Troubles Are Going to Have Trouble With Me* 209, 210 (Karen Brodkin Sacks & Dorothy Remy, eds., 1984) (noting that immigrant women make up most of the low-level production workers in the Silicon Valley electronics industry).

89 *See* Devinatz, *supra* note 24, at 9 (noting that the jobs in Silicon Valley are largely segregated by race and gender; white men occupy more than 80% of the managerial and technical positions with the greatest power and incomes, while nonwhite men, mostly immigrants, occupy the middle rung of this wage structure, and minority women, again mostly immigrants, occupy most of the least-skilled and lowest-paid jobs). *See also* Karen J. Hossfeld, "Hiring Immigrant Women: Silicon Valley's 'Simple Formula,' " in *Women of Color in U.S. Society* 65, 72 (Maxine Baca Zinn & Bonnie Thornton Dill, eds., 1994) (hereinafter Hossfeld, "Hiring Immigrant Women") (describing the proportion and distribution of racial minorities in the electronics industry).

90 *See* Hossfeld, "Hiring Immigrant Women," *supra* note 89, at 74 (reporting that employers in her study indicated that they preferred immigrant women over immigrant men for assembly work, due to the beliefs that women could afford to work for less pay and that their physical characteristics, such as smaller size, made them better suited to assembly work than men).

91 *See id.* at 66; Sucheta Nazumdar, "General Introduction: A Woman-Centered Perspective on Asian American History," in *Making Waves: An Anthology of Writings By and About Asian American Women* 1, 19 (Asian Women United of Cal., ed., 1989) (noting that about half of the women electronics workers in Silicon Valley are Filipinas, Vietnamese, Koreans, and South Asians).

92 *See* Ong, *Spirits of Resistance*, *supra* note 31, at 148.

93 *See* Susan S. Green, "Silicon Valley's Women Workers: A Theoretical Analysis of Sex-Segregation in the Electronics Industry Labor Market," in *Women, Men, and the International Division of Labor*, *supra* note 23, at 273.

94 *See* Rebecca Villones, "Women in the Silicon Valley," in *Making Waves: An Anthology of Writings By and About Asian American Women*, *supra* note 91, at 172, 173. *See also* Ewell & Ha, "High Tech's Hidden Labor," *supra* note 82 (stating that immigrants are often unaware of the protections offered by U.S. laws and prevailing wage rates, and thus are sometimes "grateful" for such work).

95 *See* Hossfeld, "Hiring Immigrant Women," *supra* note 89, at 77 (describing the author's experience as a white, North American woman applying for assembly jobs, reporting that she was repeatedly told by personnel directors that the work would not suit her because she was "an American"; another white, North American researcher faced similar reactions when she applied for assembly jobs).

96 Ibid. The author describes the experiences of white North American women and men who applied for assembly jobs who were told that they were better suited to professional positions. When she applied, the author was told that assembly work would not suit her and that she would be much happier at a professional job, because she was an American. Also, female students with foreign accents who called to inquire about entry-level positions at production plants were told that there might be jobs available three times more often than male students with Anglo accents.

97 *See* Hossfeld, "Their Logic Against Them," *supra* note 87, at 173–74 (noting that electronics companies have been known to separate workers who speak the same language to minimize socializing and solidarity). Other companies employ language as a method of control by hiring supervisors who speak the same language as assembly workers, thus reinforcing the supervisors' power over the workers. *See* McLaughlin & Cha, *supra* note 79 (quoting a Vietnamese émigré who runs an assembly contract firm, "I speak Vietnamese and my workers speak Vietnamese. I have control. . . . [i]f I also hired Hispanics, it would be very difficult to control them").

98 *See* Hossfeld, "Hiring Immigrant Women," *supra* note 89, at 78 (describing how immigrants often have little experience with organizing, and are also seen as "unlikely to 'make waves' against any part of the American system for fear of jeopardizing their welcome").

99 *See* Villones, *supra* note 94, at 173 (noting that employers often pay immigrants lower wages). *See also* McLaughlin & Cha, *supra* note 79 (describing ways electronics employers exert control over their employees).

100 *See* Green, *supra* note 93, at 281–86 (discussing possible explanations for the predominately female workforce in the electronics industry).

101 *See* Hossfeld, "Their Logic Against Them," *supra* note 87, at 163.

102 *See* Green, *supra* note 93, at 287.

103 *See* Tiun Ling Ta & Marsitah Mohd Radzi, *Demography and the Development of Penang Island: Survey on Family, Women and Work* 28 (1994) (noting that more and more Malaysian women are choosing to work and postpone marriage). *See also* Mamat, *supra* note 7, at 37 (noting that the rate of participation and numbers of women in the labor force in Malaysia are increasing).

104 *See* Fernandez-Kelly, *supra* note 55, at 634–35 (describing the increasingly competitive nature of the electronics industry, and the ability of electronics companies to relocate in search of cheaper labor).

105 *See* Hossfeld, "Their Logic Against Them," *supra* note 87, at 157. *See also* Fernandez-Kelly, *supra* note 55, at 665 (reporting findings that electronics companies are increasingly tapping into a labor force of immigrant and minority women).

106 *See* Green, *supra* note 93, at 292; Hossfeld, "Hiring Immigrant Women," *supra* note 89, at 74 (citing managers and employers who stated that immigrant women are particularly suited to tiring and detailed electronics assembly work because of their patience, coordination, and small size).

107 *See* Edward Jang-Woo Park, "Asian Americans in Silicon Valley: Race and Ethnicity in the Postindustrial Economy" 118 (1992) (unpublished Ph.D. dissertation, University of California, Berkeley) (on file with the U.C. Berkeley Library).

108 Ibid.

109 *See* Villones, *supra* note 94, at 173–74. *See also* Park, *supra* note 107, at 122 (stating that the image of Asian women as docile and silent "has made them indispensable in an industry that is constantly worried about workers organizing").

110 *See* Hossfeld, "Their Logic Against Them," *supra* note 87, at 156–57.

111 Ibid. at 158–62.

112 Ibid. at 161.

113 *See* Katz & Kemnitzer, *supra* note 88, at 210.

114 *See* Hossfeld, "Hiring Immigrant Women," *supra* note 89, at 77. There is also evidence that employers do not distinguish between Asian American and Asian immigrant women, often perceiving both as "foreign." *See id.* at 78.

115 *See* Katz & Kemnitzer, *supra* note 88, at 211. *See also* Hossfeld, "Their Logic Against Them," *supra* note 87, at 163 (pointing out that nearly 80% of the female assembly workers in her study were the primary wage earners in their families).

116 Hossfeld, "Hiring Immigrant Women," *supra* note 89, at 78.
117 *See* Eric Lai & Theresa C. Viloria, *supra* note 81, at 32–33 (giving examples of Vietnamese and Filipina women immigrants who have worked on the electronics assembly lines in Silicon Valley for decades).
118 Ibid.
119 *See* Katz & Kemnitzer, *supra* note 88, at 211.
120 Park, *supra* note 107, at 122 (quoting a Korean American supervisor in the electronics industry). The terminology used here suggests that similar stereotypes operate against Asian American women as well as Asian immigrant women.
121 *See* Hossfeld, "Hiring Immigrant Women," *supra* note 89, at 79.
122 *See* Miranda Ewell & K. Oanh Ha, "High-Tech Firms Reap Benefits of Products' Home Assembly," *San Jose Mercury News*, June 28, 1999, at A1 (hereinafter Ewell & Ha, "High-Tech Firms").
123 *See* Katz & Kemnitzer, *supra* note 88, at 213.
124 *See* Eileen Boris, *Home to Work: Motherhood and the Politics of Industrial Homework in the United States* 1–2 (1994).
125 *See* Ewell & Ha, "High-Tech Firms," *supra* note 122.
126 Ibid. *See also* Hiroshi Motomura, "Federalism, International Human Rights, and Immigration Exceptionalism," 70 *University of Colorado Law Review* 1361, 1385–92 (describing some of the very limited constitutional rights that undocumented immigrants have in comparison to citizens or documented immigrants).
127 *See* Ewell & Ha, "High-Tech's Hidden Labor," *supra* note 82.
128 Ibid. Note that a 1980 government investigation revealed that many Silicon Valley electronics companies have discontinued "home work."
129 Ibid.
130 *See* Onus Benner *et al.*, *Walking the Lifelong Tightrope: Negotiating Work in the New Economy* at iii–iv (1999) (describing the growing trend of technology companies in California using contract or temporary workers for whom they are far less likely to provide health care, pension, or other benefits).
131 *See* Ewell & Ha, "High-Tech's Hidden Labor," *supra* note 82; *see also* Benner *et al.*, *supra* note 130.
132 For example, the Occupational Safety and Health Administration (OSHA) has admitted that its regulations are inadequate to protect workers in the semiconductor industry from dangerous chemicals. Among other problems, most of OSHA's regulations were designed over thirty years ago, nearly a decade before the semiconductor industry began its rapid growth, and long before the processes in use today were invented. *See* Smith & Schmit, *supra* note 76.
133 *See* Byster & Smith, *supra* note 76 (explaining the difficulty of instituting effective health and safety measures because of the rapid technological changes in the electronics industry, and how this inadequacy leaves many workers unprotected from the risks of exposure to hazardous chemicals, gases, and metal fumes).
134 *See* Judy Mann, *supra* note 76, at D20; *see also* Byster & Smith, *supra* note 76 (describing the chemicals that electronics industry employees may be exposed to in the workplace).
135 *See* Donna Mergler *et al.*, "Visual Dysfunction Among Former Microelectronics Assembly Workers," 46 *Archives of Environmental Health* 326, (1991); Byster & Smith, *supra* note 76 (listing the various materials to which electronics workers may be exposed and some of the health problems that have been linked to those materials).
136 *See* Sonja Kim & Helen Kim, "AIWA San Jose Members Step Up Environmental Health and Safety Project," *AIWA News* (Asian Immigrant Women Advoc., Oakland/San Jose, Cal.), Nov. 1996, at 2, 2–3 (findings from surveys of women participating in AIWA's Environmental Health and Safety Project).
137 Ibid.
138 *See* "Peer Health Promoter Network," *AIWA News* (Asian Immigrant Women Advoc., Oakland/San Jose, Cal.), Sept. 1998, at 4, 4 (noting some of the problems that the women in the AIWA Peer Health Promoter Network have reported). *See also* Byster & Smith, *supra* note 76 (noting that electronics companies often withhold information on the chemicals and materials in use for proprietary reasons, and also that the health effects of many chemicals in commercial use in the industry are unknown).
139 *See* Schenker, *supra* note 75, at ix (describing the risk of miscarriage for women working in semiconductor fabrication plants).
140 *See* MASTIC, *supra* note 72, at 65 (1996) (explaining how the rapid growth of the Malaysian economy has contributed to a shortage of "manpower" in all sectors).

141 The full text of the Act is available at: <http://natlex.ilo.org/txt/E55MYS01.htm>.
142 *See* Mamat, *supra* note 7, at 47 (noting that there is no provision in the Act making it illegal for employers to pay women employees less than men doing the same type of work, so the principle of equal pay for equal work has not been legally adopted in Malaysia).
143 *See* ILO, *supra* note 18, at 26.
144 Ibid. (noting Malaysian trade unions "may not negotiate on matters relating to promotion, transfer, recruitment, retrenchment, dismissal, reinstatement, and allocation of duties").
145 Ibid.
146 *See* Kaur, *supra* note 67, at 3 (noting that weak enforcement of Malaysian labor laws facilitates the exploitation of women workers by multinational companies); Diariam, *supra* note 21 (noting that the Minister of Human Resources has the power to exempt firms from complying with provisions of the labor laws).
147 *See* Diariam, *supra* note 21.
148 *See* NIOSH, *supra* note 71.
149 *See* NIOSH, *Contents of Occupational Safety and Health Act* (visited Mar. 31, 2000) <http://www2.jaring.my/niosh/lawlact/act%5find.htm>.
150 *See* Diariam, *supra* note 21.
151 It is at least as comprehensive as European and U.S. occupational health and safety laws, since it was modeled after them. *See* NIOSH, *supra* note 71. It remains to be seen, however, how effective it will be in practice. *See*, e.g., Kathirasen, *supra* note 72, at 2 (reporting a recent incident where company compliance with and government enforcement of the Act faced criticism).
152 *See* Davaki Arumugam, "Persatuan Sahabat Wanita Selangor," in *New Technologies and the Future of Women's Work in Asia: Workshop Report* 38, 38–39 (Cecilia Ng Choon Sim & Anne Munro Kua, eds., 1995) (describing employers' lack of desire to recognize potential health risks).
153 Ibid. at 39 (describing the necessity of educating women about their rights and about the ways they can control the effects of technology); Kaur, *supra* note 67, at 19. *See also* Leng, *supra* note 66, at 68–69 ("[M]ost workers [in the Malaysian electronics industry] are not informed of the labour laws. Often, they do not even realize that their rights have been violated. Even given that they know they have been wronged and want to lodge a complain[*sic*], they have no idea how to go about it.").
154 *See* Arumugam, *supra* note 152, at 39.
155 Ibid.
156 *See*, e.g., *AMD, supra* note 73, at I (indicating in its annual report that U.S. semiconductor company AMD uses "Best Practices" standards in the U.S. and abroad).
157 *See*, e.g., "Sahabat Alam Malaysia (SAM)," Preface to *Hazardous Industries and Workers Health: Proceedings of the National Seminar on Workers Health and Safety Problems in Malaysia* (1986) (suggesting that public interest organizations such as SAM can play a role in encouraging health and safety for workers).
158 *See* Ismail, *supra* note 1, at 120–21. *See also* Grace, *supra* note 1, at 20–21 (describing a protest carried out by laid-off electronic factory workers in the Bayan Lepas Free Trade Zone during 1985 and 1986 when workers conducted a 32-day picket along with sit-ins to protest their treatment and draw attention to their plight); Leng, *supra* note 66, at 68–69 (discussing electronics industry workers' frequent lack of awareness of their legal rights).
159 *See* Arumugam, *supra* note 152, at 39.
160 *See Cal. Code Regs.* tit. viii, Section 3203 (1995) (effective 1991).
161 *See Cal. Lab. Codes* Section 6300, 6307 (West 1929) (authorizing health and safety standards and giving the state Division of Occupational Safety and Health the power to enforce them).
162 *See Cal. Code Regs.* tit. viii, Section 3203(a) (1995). *See also* CAL/OSHA Consultation Service, *Guide to Developing Your Workplace Injury and Illness Prevention Program*, About This Guide (1998).
163 *See Cal. Code Regs.* tit. viii, Section 3203 (a)(2), (3) and (6) (1995).
164 *See Cal. Code Regs.* tit. viii (1990).
165 *See* 29 C.F.R. Section 1910.1200(1999).
166 *See* 29 C.F.R. Section 1910.200 (b). *See also Cal. Code Regs.* tit. viii, Section 339 (outlining hazardous chemicals regulated in California in the Hazardous Substances List).
167 *See Cal. Code Regs.* tit. viii, Section 3463 (1990) (outlining regulations for hazardous atmospheres and substances).
168 *See Cal. Code Regs.* tit. viii, Section 1230 (1999) (outlining temperature, illumination, sanitation, and ventilation).
169 Ibid.

170 *See* Mark A. Rothstein, *Occupational Safety and Health Law* 8–10 (2d ed. 1983).
171 *See* John M. Mendeloff, *The Dilemma of Toxic Substance Regulation: How Overregulations Cause Underregulations at OSHA* 2, 7 (1988).
172 *See* Jan Mazurek, *Making Microchips: Policy, Globalization, and Economic Restructuring in the Semiconductor Industry* 22 (1999), at 201.
173 *See* Bruce T. Rubenstein, "For Electronics Industry, Short History is Still History," *Corp. Legal Times*, Mar. 12, 1999, at 4.
174 *See* Smith & Schmit, *supra* note 76.
175 *See* Kim & Kim, *supra* note 136.
176 *See* International Campaign for Responsible Technology, *Global Semiconductor Health Hazards Exposed* (visited Jan. 25, 2000) <http://www.svtc.org/natsem/press.htm>.
177 *See* Julie Schmit, "Dirty Secrets, Exposing the Dark Side of a 'Clean' Industry," *USA Today*, Jan. 12, 1998, at B1 (citing OSHA and fire department records).
178 *See* "Peer Health Promoter Network," *supra* note 138.
179 Ibid. *See also* "A Successful Collaboration, Immigrant Women Trained to Become Peer Trainers," *AIWA News* (Asian Immigrant Women Advoc., Oakland/San Jose, Cal.), June 1999, at 3, 3.
180 TRI was created in 1986 under the Emergency Planning and Community Right to Know Act (EPCRA), Section 313. *See* Mazurek, *supra* note 172, at 58.
181 Ibid.
182 Ibid. at 58–59.
183 *See* Sidney M. Wolf, "Fear and Loathing About the Public Right to Know: The Surprising Success of the Emergency Planning and Community Right-to-Know Act," 11 *Journal of Land Use and Environmentel Law* 217, 281 (1996) ("[T]he TRI has spawned extensive grassroots agitation, numerous governmental and environmental organization reports, significant regulatory and legislative actions, major industry initiatives, and wider public consciousness about massive toxic releases and the need to reduce them. The TRI confirms the observation that more information on an important public issue tends to lead to public pressure which can lead to reform."). Commentators have described these significant results as "regulation by information" or "regulation by embarrassment."
184 *See* Mazurek, *supra* note 172, at 59.
185 *See* Wolf, *supra* note 183, at 267.
186 *See* Mazurek, *supra* note 172, at 59.
187 Ibid.
188 Rubenstein, *supra* note 173, at 4.

23

Premenstrual Syndrome, Work Discipline, and Anger

Emily Martin

Looming over the whole current scene in England and the United States is the enormous outpouring of interest—the publishing of magazine and newspaper articles, popular books, and pamphlets; the opening of clinics; the marketing of remedies—devoted to premenstrual syndrome.

The dominant model for premenstrual syndrome (PMS) is the physiological/medical model. In this model, PMS manifests itself as a variety of physical, emotional, and behavioral "symptoms" which women "suffer." The list of such symptoms varies but is uniformly negative, and indeed worthy of the term "suffer." Judy Lever's list in her popular handbook serves as an example (Table 23.1).[1]

The syndrome of which this list is a manifestation is a "genuine illness,"[2] a "real physical problem"[3] whose cause is at base a physical one.[4] Although psychological factors may be involved as a symptom, or even as one cause, "the root cause of PMT [short for premenstrual tension, another term for PMS], no matter how it was originally triggered, is physical and can be treated."[5] This physical cause comes from "a malfunction in the production of hormones during the menstrual cycle, in particular the female hormone, progesterone. This upsets the normal working of the menstrual cycle and produces the unpleasant symptoms of PMT." Astonishingly, we are told that "more than three quarters of all women suffer from symptoms of PMT."[6] In other words, a clear majority of all women are afflicted with a physically abnormal hormonal cycle.

Various "treatments" are described that can compensate a women for her lack of progesterone or her excess of estrogen or prolactin. Among the benefits of this approach are that psychological and physical states that many women experience as extremely distressing or painful can be alleviated, a problem that had no name or known cause can be named and grasped, and some of the blaming of women for their premenstrual condition by both doctors and family members can be stopped. It seems probable that this view of PMS has led to an improvement from the common dismissals "it's all in your mind," "grin and bear it," or "pull yourself together." Yet, entailed also in this view of PMS are a series of assumptions about the nature of time and of society and about the necessary roles of women and men.

... [In] the nineteenth century ... menstruation began to be regarded as a pathological process. Because of ideas prevailing among doctors that a woman's reproductive organs held complete sway over her between puberty and menopause, women were warned not to divert needed energy away from the uterus and ovaries. In puberty, especially, the limited

Table 23.1 A List of the Symptoms of Premenstrual Syndrome from a Popular Handbook

Complete Checklist of Symptoms		
Physical Changes		
Weight gain	Epilepsy	Spontaneous bruising
Skin disorders	Dizziness, faintness	Headache, migraine
Painful breasts	Cold sweats	Backache
Swelling	Nausea, sickness	General aches and pains
Eye diseases	Hot flashes	
Asthma	Blurring vision	
Concentration		
Sleeplessness	Lowered judgment	Lack of coordination
Forgetfulness	Difficulty concentrating	
Confusion	Accidents	
Behavior Changes		
Lowered school or work performance	Avoid social activities	Drinking too much alcohol
Decreased efficiency	Taking too many pills	
Lethargy	Food cravings	
Mood Changes		
Mood swings	Restlessness	Tension
Crying, depression	Irritability	Loss of sex drive
Anxiety	Aggression	

Source: Judy Lever and Michael G. Brush, *Pre-menstrual Tension*, © 1981 by Bantam Books; reproduced by permission.

amount of energy in a woman's body was essential for the proper development of her female organs.

> Indeed physicians routinely used this energy theory to sanction attacks upon any behavior they considered unfeminine; education, factory work, religious or charitable activities, indeed virtually any interests outside the home during puberty were deplored.[7] . . .

This view of women's limited energies ran very quickly up against one of the realities of nineteenth-century America: many young girls and women worked exceedingly long and arduous hours in factories, shops, and other people's homes. The "cult of invalidism" with its months and even years of inactivity and bed rest, which was urged on upper-class women, was manifestly not possible for the poor. This contradiction was resolved in numerous ways: by detailing the "weakness, degeneration, and disease" suffered by female clerks and operatives who "strive to emulate the males by unremitting labor"[8] while callously disregarding the very poor health conditions of those workers,[9] or by focusing on the greater toll that brain work, as opposed to manual work, was thought to take on female bodies. According to

Edward Clarke's influential *Sex in Education* (1873), female operatives suffer less than schoolgirls because they "work their brain less. . . . Hence they have stronger bodies, a reproductive apparatus more normally constructed, and a catamenial function less readily disturbed by effort, than their student sisters."[10]

If men like Clarke were trying to argue that women (except working-class women) should stay home because of their bodily functions, feminists were trying to show how women could function in the world outside the home in spite of their bodily functions. Indeed, it is conceivable that the opinions of Clarke and others were in the first place a response to the threat posed by the first wave of feminism. Feminists intended to prove that the disciplined, efficient tasks required in the workplace in industrial society could be done by women when they were menstruating as well as when they were not.

In *Functional Periodicity: An Experimental Study of the Mental and Motor Abilities of Women During Menstruation* (1914), Leta Hollingworth had 24 women (who ironically held research and academic jobs) perform various tests of motor efficiency and controlled association both when they were and when they were not menstruating. These included tapping on a brass plate as many times as possible within a brief time; holding a 2.5-mm rod as steady as possible within a 6-mm hole while trying not to let it touch the edges; naming a series of colors as quickly as possible; and naming a series of opposites as quickly as possible. Ability to learn a new skill was also tested by teaching the subjects to *type* and recording their progress while menstruating and not. The findings: "the records of all the women here studied *agree* in supporting the negative conclusion here presented. None of them shows a characteristic inefficiency in the traits here tested at menstrual periods."[11]

Similarly, in *The Question of Rest for Women During Menstruation* (1877), Mary Putnam Jacobi showed that "women do work better, and with much greater safety to health when their work is frequently intermitted; but that those intermittences should be at short intervals and lasting a short time, not at long intervals and lasting longer. Finally that they are required at all times, and have no special reference to the period of the menstrual flow."[12] Given the nature of the organization of work, men would probably also work better if they had frequent short breaks. What is being exposed in these early studies, in addition to the nature of women's capacities, is the nature of the work process they are subjected to.

It is obvious that the relationship between menstruation and women's capacity to work was a central issue in the nineteenth century. When the focus shifted from menstruation itself to include the few days before menstruation, whether women could work outside the home was still a key issue. It is generally acknowledged that the first person to name and describe the symptoms of premenstrual syndrome was Robert T. Frank in 1931.[13] Two aspects of Frank's discussion of what he called "premenstrual tension" deserve careful attention. The first is that he carried forward the idea, which flourished in the nineteenth century, that women were swayed by the tides of their ovaries. A woman's ovaries were known to produce female sex hormones, and these were the culprit behind premenstrual tension. His remedy was simple and to the point: "It was decided to tone down the ovarian activity by roentgen [x-ray] treatment directed against the ovaries."[14]

Frank reserved x-ray treatment for the most severe cases, but it was not long before the influence of female hormones on a woman was extended to include her emotional states all month long. In an extraordinary study in 1939, Benedek and Rubenstein analyzed psychoanalytic material from therapy sessions and dreams, on the one hand, and basal body temperature and vaginal smears, on the other, from nineteen patients under treatment for various neurotic disturbances. They were able to predict when the patients had ovulated and menstruated from the physiological record as well as from the psychoanalytic record. . . . But Benedek and Rubenstein went far beyond showing simple correlations between hormones and emotions. Without evidence of a causal link one way or the other, they concluded that "human [they meant 'adult female'] instinctual drives are *controlled by* gonadal hormone production" (emphasis added).[15]

Benedek and Rubenstein may have been unusually avid in their attempt to derive women's emotional states from hormones. But their study was still being quoted approvingly and elaborated on in the late 1960s. One later study (1968) concluded that "the menstrual cycle exercises gross influences on female behavior."[16] It was not until the 1970s that some researchers began to insist that women's moods had important social, cultural, and symbolic components and that even though *correlation* between biochemical substances and emotional changes can be observed, "the direction of causality is still unclear. Indeed, there is abundant evidence to suggest that biochemical changes occur in *response* to socially mediated emotional changes."[17]

The second aspect of Frank's study that deserves attention is his immediate interest in the effect of premenstrual tension on a woman's ability to work, such that in mild cases employers have to make provision for an employee's temporary care and in severe ones to allow her to rest in bed for one or two days.[18] It strikes me as exceedingly significant that Frank was writing immediately after the Depression, at a time when the gains women had made in the paid labor market because of World War I were slipping away. Pressure was placed on women from many sides to give up waged work and allow men to take the jobs.[19]

Can it be accidental that many other studies were published during the interwar years that showed the debilitating effects of menstruation on women?[20] Given this pattern of research finding women debilitated by menstruation when they pose an obstacle to full employment for men, it is hardly surprising that after the start of World War II a rash of studies found that menstruation was not a liability after all.[21] "Any activity that may be performed with impunity at other times may be performed with equal impunity during menstruation" wrote Seward in 1944, reversing her own earlier finding in 1934 that menstruation was a debility. Some of the evidence amassed for this conclusion seems astoundingly ad hoc: Seward argues that when women miss work because of menstrual complaints, they are indulging in "a bit of socially acceptable malingering by taking advantage of the popular stereotype of menstrual incapacitation." Her evidence? That when a large life insurance company discontinued pay for menstrual absenteeism after a limited time allowance, menstrual absenteeism markedly declined![22] She missed the point that if people need their wages and have used up their sick leave, they will go to work, even in considerable discomfort.

After World War II, just as after World War I, women were displaced from many of the paid jobs they had taken on.[23] The pattern seems almost too obvious to have been overlooked so long, but as we know, there was a spate of menstrual research after the Second World War that found, just as after the first, that women were indeed disabled by their hormones. Research done by Katherina Dalton in the 1940s was published in the *British Medical Journal* in 1953,[24] marking the beginning of her push to promote information about the seriousness of premenstrual syndrome. As we will see, one of her overriding concerns was the effect on women's performance at school and work and the cost to national economies of women's inability to work premenstrually.

Although Dalton's research fit in nicely with the postwar edging of women out of the paid workforce, it was not until the mid- to late 1970s that the most dramatic explosion of interest in PMS took place. This time there were no returning veterans to demand jobs for which women were suddenly "unqualified"; instead, women had made greater incursions into the paid workforce for the first time without the aid of a major war.

> First single women, then wives, and then mothers of school-aged children were, in a sense, freed from social constraints against work outside the home. For each of these

> groups, wage labor was at one time controversial and debatable, but eventually employment became a socially acceptable—and even expected—act.[25]

Many factors were responsible for women's emergence in the paid workforce: the second wave of feminism and stronger convictions about women's right to work, a lower birth rate, legislative support barring sex discrimination, increasing urbanization, and growth in educational opportunities for women.[26] It goes without saying that women's move toward center stage in the paid workforce (as far away as equality still remains) is threatening to some women and men and has given rise to a variety of maneuvers designed to return women to their homes.[27] Laws has suggested that the recent burgeoning of emphasis on PMS is a "response to the second wave of feminism." I think this is a plausible suggestion, made even more convincing by the conjunction between periods of our recent history when women's participation in the labor force was seen as a threat and, simultaneously, menstruation was seen as a liability. . . .[28]

Many PMS symptoms seem to focus on intolerance for the kind of work discipline required by late industrial societies. But what about women who find that they become clumsy? Surely this experience would be a liability in any kind of social setting. Perhaps so, and yet it is interesting that most complaints about clumsiness seem to focus on difficulty carrying out the mundane tasks of keeping house: "You may find you suddenly seem to drop things more often or bump into furniture around the house. Many women find they tend to burn themselves while cooking or cut themselves more frequently."[29] "It's almost funny. I'll be washing the dishes or putting them away and suddenly a glass will just jump out of my hands. I must break a glass every month. But that's when I know I'm entering my premenstrual phase."[30] Is there something about housework that makes it problematic if one's usual capacity for discipline relaxes?

On the one hand, for the numbers of women who work a double day (hold down a regular job in the paid workforce and come home to do most of the cooking, cleaning, and child care), such juggling of diverse responsibilities can only come at the cost of supreme and unremitting effort. On the other hand, for the full-time homemaker, recent changes in the organization of housework must be taken into account. Despite the introduction of "labor-saving" machines, time required by the job has increased as a result of decline in the availability of servants, rise in the standards of household cleanliness, and elaboration of the enterprise of childrearing. . . .[31]

Perhaps the need for discipline in housework comes from a combination of the desire for efficiency and a sense of its endlessness, a sense described by Simone de Beauvoir as "like the torture of Sisyphus . . . with its endless repetition: the clean becomes soiled, the soiled is made clean, over and over, day after day. The housewife wears herself out marking time: she makes nothing, simply perpetuates the present."[32] Not only sociological studies[33] but also novels by women attest to this aspect of housework:

> First thing in the morning you started with the diapers. After you changed them, if enough had collected in the pail, you washed them. If they had ammonia which was causing diaper rash, you boiled them in a large kettle on top of the stove for half an hour. While the diapers were boiling, you fed the children, if you could stand preparing food on the same stove with urine-soaked diapers. After breakfast, you took the children for a walk along deserted streets, noting flowers, ladybugs, jet trails. Sometimes a motorcycle would go by, scaring the shit out of the children. Sometimes a dog followed you. After the walk, you went back to the house. There were many choices before nap time: making grocery lists; doing the wash; making the beds; crawling around on the floor with the children; weeding the garden; scraping last night's dinner

> off the pots and pans with steel wool; refinishing furniture; vacuuming; sewing buttons on; letting down hems; mending tears; hemming curtains. During naps, assuming you could get the children to sleep simultaneously (which was an art in itself), you could flip through *Family Circle* to find out what creative decorating you could do in the home, or what new meals you could spring on your husband.[34]

Here is Katharina Dalton's example of how a premenstrual woman reacts to this routine:

> Then quite suddenly you feel as if you can't cope anymore—everything seems too much trouble, the endless household chores, the everlasting planning of meals. For no apparent reason you rebel: "Why should I do everything?" you ask yourself defiantly. "I didn't have to do this before I was married. Why should I do it now?" . . . As on other mornings you get up and cook breakfast while your husband is in the bathroom. You climb wearily out of bed and trudge down the stairs, a vague feeling of resentment growing within you. The sound of cheerful whistling from upstairs only makes you feel a little more cross. Without any warning the toast starts to scorch and the sausages instead of happily sizzling in the pan start spitting and spluttering furiously. Aghast you rescue the toast which by this time is beyond resurrection and fit only for the trash. The sausages are charred relics of their former selves and you throw those out too. Your unsuspecting husband opens the kitchen door expecting to find his breakfast ready and waiting, only to see a smoky atmosphere and a thoroughly overwrought wife. You are so dismayed at him finding you in such chaos that you just burst helplessly into tears.[35]

Needless to say, by the terms of the medical in which Dalton operates, the solution for this situation is to seek medical advice and obtain treatment (usually progesterone).[36] The content of the woman's remarks, the substance of what she objects to, escape notice.

A woman who drops things, cuts or burns herself or the food in this kind of environment has to adjust to an altogether different level of demand on her time and energy than—say —Beng women in the Ivory Coast. There, albeit menstrually instead of premenstrually, women specifically must not enter the forest and do the usual work of their days—farming, chopping wood, and carrying water. Instead, keeping to the village, they are free to indulge in things they usually have no time for, such as cooking a special dish made of palm nuts. This dish, highly prized for its taste, takes hours of slow tending and cooking and is normally eaten only by menstruating women and their close friends and kinswomen.[37] Whatever the differing demands on Beng as opposed to Western women, Beng social convention requires a cyclic change in women's usual activities. Perhaps Beng women have fewer burned fingers.

For the most part, women quoted in the popular health literature do not treat the cyclic change they experience as legitimate enough to alter the structure of work time. However, several of the women I interviewed did have this thought. One woman expressed this as a wish, while reinterpreting what she had heard about menstrual huts (places of seclusion used by women in some societies when they are menstruating):

> [Does menstruation have any spiritual or religious significance for you?] I like the idea of menstrual huts a great deal. They intrigue me. My understanding is it's a mysterious thing in some ways. It infuriates me that we don't know more about it. Here are all these women—apparently when you get your period you go off to this hut and you hang around. [Because you're unclean?] That's what I feel is probably bull, that's the masculine interpretation of what's going on passed on generally by men to male

> anthropologists, whereas the women probably say, "Oh, yeah, we're unclean, we're unclean, see ya later." And then they race off to the menstrual hut and have a good time. (Meg O'Hara)

Another got right to the heart of the matter with simplicity:

> Some women have cramps so severe that their whole attitude changes; maybe they need time to themselves and maybe if people would understand that they need time off, not the whole time, maybe a couple of days. When I first come on I sleep in bed a lot. I don't feel like doing anything. Maybe if people could understand more. Women's bodies change. (Linda Matthews)

. . . These women are carrying on what amounts to a twin resistance: to science and the way it is used in our society to reduce discontent to biological malfunction and to the integrity of separate spheres which are maintained to keep women in one while ruling them out of the other.

Given that periodic changes in activity in accord with the menstrual cycle are not built into the structure of work in our society, what does happen to women's work during their periods? Much recent research has attempted to discover whether women's actual performance declines premenstrually. The overwhelming impression one gets from reading the popular literature on the subject is that performance in almost every respect does decline. According to Dalton's influential account, women's grades drop, they are more likely to commit crimes and suicide, and they "cost British industry 3% of its total wage bill, which may be compared with 3% in Italy, 5% in Sweden and 8% in America."[38] Yet other accounts make powerful criticisms of the research on which these conclusions are based: they lack adequate controls, fail to report negative findings, and fail to report overall levels of women's performance in comparison to men's.[39] Still other studies find either increased performance or no difference in performance at all.[40]

Some women we interviewed expressed unforgettably the double message that women workers receive about PMS:

> Something I hear a lot that really amazes me is that women are discriminated against because they get their period. It makes them less capable to do certain kinds of work. It makes me angry. I never faced it in terms of my own personal experience, but it's something I've heard. I grew up thinking you shouldn't draw attention to your period, it makes you seem less capable than a man. I always tried to be kind of a martyr, and then all of a sudden recently I started hearing all this scientific information that shows that women really do have a cycle that affects their mood, and they really do get into bad moods when they have their periods. I don't know whether all of a sudden it gives legitimacy to start complaining that it's okay. I think I have a hard time figuring out what that's supposed to do. Then again you can look at that as being a really negative thing, medical proof that women are less reliable. It's proven now that they're going to have bad moods once a month and not be as productive. (Shelly Levinson)

I think the way out of this bind is to focus on women's experiential statements—that they function differently during certain days, in ways that make it harder for them to tolerate the discipline required by work in our society. We could then perhaps hear these statements not as warnings of the flaws inside women that need to be fixed but as insights into flaws in society that need to be addressed.

What we see from the list of PMS symptoms in Table 23.1 is not so much a list of traits that would be unfortunate in any circumstance but traits that happen to be unfortunate in our particular social and economic system, with the kind of work it requires. This consideration gives rise to the question of whether the decreases reported by women in their ability to concentrate or discipline their attention are accompanied by gains in complementary areas. Does loss of ability to concentrate mean a greater ability to free-associate? Loss of muscle control, a gain in ability to relax? Decreased efficiency, increased attention to a smaller number of tasks?

Here and there in the literature on PMS one can find hints of such increased abilities. Women report:

> No real distress except melancholy which I actually enjoy. It's a quiet reflective time for me.
>
> My skin breaks out around both ovulation and my period. My temper is short; I am near tears, I am depressed. One fantastic thing—I have just discovered that I write poetry just before my period is due. I feel very creative at that time.[41]

Others find they "dream more than usual, and may feel sexier than at other times of the cycle."[42]

> A sculptor described her special abilities when she is premenstrual. "There is a quality to my work and to my visions which just isn't there the rest of the month. I look forward to being premenstrual for its effect on my creativity, although some of the other symptoms create strains with my family." Another woman, prone to depression, described in the journal she kept, "When I am premenstrual I can write with such clarity and depth that after I get my period I don't recognize that those were my thoughts or that I could have written anything so profound."[43]
>
> I don't know what it is, but I'll wake up one morning with an urge to bake bread. I can hardly wait to get home from work and start mixing the flour, kneading the dough, smelling the yeast. It's almost sensual and very satisfying. Maybe it's the earth-mother in me coming out. I don't know. But I do enjoy my premenstrual time.[44]
>
> I have heard that many women cry before their period. Well, I do too. Sometimes I'll cry at the drop of a hat, but it's a good crying. I'll be watching something tender on TV or my children will do something dear, and my eyes fill up. My heart is flooded with feelings of love for them or for my husband, for the world, for humanity, all the joy and all the suffering. Sometimes I could just cry and cry. But it strengthens me. It makes me feel a part of the earth, of the life-giving force.[45]

And from my interviews:

> I dream very differently during my period; my dreams are very, very vivid and sometimes it seems that I hear voices and conversations. My dreams are very vivid and the colors are not brighter but bolder, like blues and reds and that's also very interesting. The last three days I feel more creative. Things seem a little more colorful, it's just that feeling of exhilaration during the last few days. I feel really great. (Alice Larrick)
>
> I like being by myself, it gives me time to forget about what people are thinking. I like the time I don't have to worry about talking to anybody or being around anybody. It's nice to be by yourself. Time alone. (Kristin Lassiter)

Amid the losses on which most accounts of PMS focus, these women seem to be glimpsing increased capacities of other kinds. If these capacities are there, they are certainly not ones that would be given a chance to flourish or would even be an advantage in the ordinary dual workday of most women. Only the exception—a sculptor or writer—would be able to put these greater emotional and associative capacities to work in her regular environment. Perhaps it is the creative writing tasks present in most academic jobs that lead to the result researchers find puzzling: if premenstrual women cannot concentrate as well, then why are women academics' work performance and concentration better than usual during the premenstrual phase?[46] The answer may be that there are different kinds of concentration; some requiring discipline inimical to body and soul that women reject premenstrually and some allowing expression of the depth within oneself that women have greater access to premenstrually.

We can gain some insight into how women's premenstrual and menstrual capacities can be seen as powers, not liabilities, by looking at the ethnographic case of the Yurok.[47] Thomas Buckley has shown how the Yurok view of menstruation (lost in ethnographic accounts until his writing) held that

> a menstruating woman should isolate herself because this is the time when she is at the height of her powers. Thus, the time should not be wasted in mundane tasks and social distractions, nor should one's concentration be broken by concerns with the opposite sex. Rather, all of one's energies should be applied in concentrated meditation on the nature of one's life, "to find out the purpose of your life," and toward the "accumulation" of spiritual energy.[48]

Michelle Harrison, with a sense of the appropriate setting for premenstrual women, says poignantly, "Women who are premenstrual often have a need for time alone, time to themselves, and yet few women actually have that time in their lives. One woman wrote, 'When I listen to music I feel better. If I can just be by myself and listen quietly, then the irritability disappears and I actually feel good. I never do it, though, or rarely so. I feel guilty for taking that time for myself, so I just go on being angry or depressed.' "[49] What might in the right context be released as powerful creativity or deep self-knowledge becomes, in the context of women's everyday lives in our societies, maladaptive discontent.

A common premenstrual feeling women describe is anger, and the way this anger is felt by women and described by the medical profession tells a lot about the niche women are expected to occupy in society. An ad in a local paper for psychotherapeutic support groups asks: "Do you have PMS?—Depression—irritability—panic attacks—food cravings —lethargy—dizziness—headache—backache—anger. How are other women coping with this syndrome? Learn new coping mechanisms; get support from others who are managing their lives."[50] Anger is listed as a symptom in a syndrome, or illness, that afflicts only women. In fuller accounts we find that the reason anger expressed by women is problematic in our society is that anger (and allied feelings such as irritability) makes it hard for a woman to carry out her expected role of maintaining harmonious relationships within the family.

> Serious problems can arise—a woman might become excessively irritable with her children (for which she may feel guilty afterwards), she may be unable to cope with her work, or she may spend days crying for no apparent reason. Life, in other words, becomes intolerable for a short while, both for the sufferer and for those people with whom she lives. . . . PMT is often referred to as a potential disrupter of family life. Women suffering from premenstrual irritability often take it out on children, sometimes violently. . . . Obviously an anxious and irritable mother is not likely to promote harmony within the family.[51]

This entire account is premised on the unexamined cultural assumption that it is primarily a woman's job to see that social relationships work smoothly in the family. Her own anger, however substantial the basis for it, must not be allowed to make life hard on those around her. If she has an anger she cannot control, she is considered hormonally unbalanced and should seek medical treatment for her malfunction. If she goes on subjecting her family to such feelings, disastrous consequences—construed as a woman's *fault* in the PMS literature—may follow. For example, "Doctor Dalton tells the story of a salesman whose commissions dropped severely once a month, putting a financial strain on the family and worrying him a great deal. Doctor Dalton charted his wife's menstrual cycle and found that she suffered from severe PMT. This affected her husband, who became anxious and distracted and so less efficient at his job. The drop in his commissions coincided with her premenstrual days. Doctor Dalton treated his wife and cured the salesman!"[52]

Not only can a man's failure at work be laid at the doorstep of a woman's PMS, so also can a man's violence. Although the PMS literature acknowledges that many battered women do nothing to provoke the violence they suffer from men, it is at times prone to suggest that women may themselves bring on battering if the man has a "short fuse": "[The woman's] own violent feelings and actions while suffering from PMT could supply the spark that causes him to blow up."[53] Or consider this account, in which the woman is truly seen as a mere spark to the man's blaze:

> One night she was screaming at him, pounding his chest with her fists, when in her hysteria she grabbed the collar of his shirt and ripped so hard that the buttons flew, pinging the toaster and the microwave oven. But before Susan could understand what she had done, she was knocked against the kitchen wall. Richard had smacked her across the face with the back of his hand. It was a forceful blow that cracked two teeth and dislocated her jaw. She had also bitten her tongue and blood was flowing from her mouth. . . . [Richard took her to the emergency room that night and moved out the next morning.] He was afraid he might hit her again because *she was so uncontrollable* when she was in a rage. [Emphasis added.][54]

In this incident, who was most uncontrollable when in a rage—Richard or Susan? Without condoning Susan's actions, we must see that her violence was not likely to damage her husband bodily. A woman's fists usually do not do great harm when pounding a man's chest, and in this case they evidently did not. Ripping his clothes, however unfortunate, is not on the same scale as his inflicting multiple (some of them irreversible!) bodily injuries that required her to be treated in a hospital emergency room. The point is not that she was unable to injure him because of her (presumed) smaller size and lesser strength. After all she could have kicked him in the groin or stabbed him with a knife. The point is, she chose relatively symbolic means of expressing her anger and he did not. Yet in the PMS literature *she* is the one cited as uncontrollable, and responsible for his actions. The problems of men in these accounts are caused by outside circumstances and other people (women). The problems of women are caused by their own internal failure, a biological "malfunction." What is missing in these accounts is any consideration of why, in Anglo and American societies, women might feel extreme rage at a time when their usual emotional controls are reduced.

That their rage is extreme cannot be doubted. Many women in fact describe their premenstrual selves as being "possessed." One's self-image as a woman (and behind this the cultural construction of what it is to be a woman) simply does not allow a woman to recognize herself in the angry, loud, sometimes violent "creature" she becomes once a month.

I feel it is not me that is in possession of my body. My whole personality changes, making it very difficult for the people I live and work with. I've tried. Every month I say, "This month it's going to be different, I'm not going to let it get hold of me." But when it actually comes to it, something chemical happens to me. I can't control it, it just happens.[55]

Something seems to snap in my head. I go from a normal state of mind to anger, when I'm really nasty. Usually I'm very even tempered, but in these times it is as if someone else, not me, is doing all this, and it is very frightening.[56]

It is something that is wound up inside, you know, like a great spring. And as soon as anything triggers it off, I'm away. It is very frightening. Like being possessed, I suppose.[57]

I try so hard to be a good mother. But when I feel this way, it's as if there's a monster inside me that I can't control.[58]

I just get enraged and sometimes I would like to throw bookshelves through windows, barely feeling that I have control . . . these feelings of fury when there's nothing around that would make that necessary. Life is basically going on as before, but suddenly I'm furious about it. (Meg O'Hara)

I was verbally abusive toward my husband, but I would really thrash out at the kids. When I had these outbursts I tended to observe myself. I felt like a third party, looking at what I was doing. There was nothing I could do about it. I was not in control of my actions. It's like somebody else is taking over.[59]

Once a month for the last 25 years this wonderful woman (my wife) has turned into a .[60]

. . . Women say they feel "possessed," but what the society sees behind their trouble is really their own malfunctioning *bodies*. Redress for women may mean attention focused on the symptoms but not on the social environment in which the "possession" arose. The anger was not really the woman's fault, but neither was it to be taken seriously. Indeed, one of women's common complaints is that men treat their moods casually:

Sometimes if I am in a bad mood, my husband will not take me seriously if I am close to my period. He felt if it was "that time of the month" any complaints I had were only periodic. A few weeks ago I told him that until I am fifty-five he will have taken me seriously only half the time. After that he will blame it on menopause.[61]

Or if husbands cannot ignore moods, perhaps the moods, instead of whatever concrete circumstances from which they arise, can be treated.

The husband of a woman who came for help described their problem as follows, "My wife is fine for two weeks out of the month. She's friendly and a good wife. The house is clean. Then she ovulates and suddenly she's not happy about her life. She wants a job. Then her period comes and she is all right again." He wanted her to be medicated so she would be a "good wife" all month.[62]

Marilyn Frye, in discussing the range of territory a woman's anger can claim, suggests: "So long as a woman is operating squarely within a realm which is generally recognized as a woman's realm, labeled as such by stereotypes of women and of certain activities, her anger will quite likely be tolerated, at least not thought crazy." And she adds, in a note that applies precisely to the anger of PMS, "If the woman insists persistently enough on her anger being taken seriously, she may begin to seem mad, for she will seem to have her values all mixed up and distorted."[63]

What are the sources of women's anger, so powerful that women think of it as a kind of possessing spirit? A common characteristic of premenstrual anger is that women often feel it has no immediate identifiable cause: "It never occurred to me or my husband that my totally unreasonable behavior toward my husband and family over the years could have been caused by anything but basic viciousness in me."[64]

Women often experience the depression or anger of premenstrual syndrome as quite different from the depression or anger of other life situations. As one woman described this difference: "Being angry when I know I'm right makes me feel good, but being angry when I know it's just me makes me feel sick inside."[65]

Anger experienced in this way (as a result solely of a woman's intrinsic badness) cannot help but lead to guilt. And it seems possible that the sources of this diffuse anger could well come from women's perception, however inarticulate, of their oppression in society—of their lower wage scales, lesser opportunities for advancement into high ranks, tacit omission from the language, coercion into roles inside the family and out that demand constant nurturance and self-denial, and many other ills. Adrienne Rich asks:

> What woman, in the solitary confinement[66] of a life at home enclosed with young children, or in the struggle to mother them while providing for them single-handedly, or in the conflict of weighing her own personhood against the dogma that says she is a mother, first, last, and always—what woman has not dreamed of "going over the edge," of simply letting go, relinquishing what is termed her sanity, so that she can be taken care of for once, or can simply find a way to take care of herself? The mothers: collecting their children at school; sitting in rows at the parent-teacher meeting; placating weary infants in supermarket carriages; straggling home to make dinner, do laundry, and tend children after a day at work; fighting to get decent care and livable schoolrooms for their children; waiting for child-support checks while the landlord threatens eviction . . . —the mothers, if we could look into their fantasies—their daydreams and imaginary experiences—we would see the embodiment of rage, of tragedy, of the overcharged energy of love, of inventive desperation, we would see the machinery of institutional violence wrenching at the experience of motherhood.[67]

Rich acknowledges the "embodiment of rage" in women's fantasies and daydreams. Perhaps premenstrually many women's fantasies become reality, as they experience their own violence wrenching at all of society's institutions, not just motherhood as in Rich's discussion.

Coming out of a tradition of psychoanalysis, Shuttle and Redgrove suggest that a woman's period maybe "a moment of truth which will not sustain lies." Whereas during most of the month a woman may keep quiet about things that bother her, "maybe at the paramenstruum, the truth flares into her consciousness: this is an intolerable habit, she is discriminated against as a woman, she is forced to underachieve if she wants love, this examination question set by male teachers is unintelligently phrased, I will not be a punch-ball to my loved ones, this child must learn that I am not the super-natural never-failing source of maternal sympathy."[68] In this rare analysis, some of the systematic social causes of women's second-class status, instead of the usual biological causes, are being named and identified as possible sources of suppressed anger.

If *these* kinds of causes are at the root of the unnamed anger that seems to afflict women, and if they could be named and known, maybe a cleaner, more productive anger would arise from within women, tying them together as a common oppressed group instead of sending them individually to the doctor as patients to be fixed.

> And so her anger grew. It swept through her like a fire. She was more than shaken. She thought she was consumed. But she was illuminated with her rage; she was bright with fury. And though she still trembled, one day she saw she had survived this blaze. And after a time she came to see this anger-that-was-so-long-denied as a blessing.[69]

To see anger as a blessing instead of as an illness, it may be necessary for women to feel that their rage is legitimate. To feel that their rage is legitimate, it may be necessary for women to understand their structural position in society, and this in turn may entail consciousness of themselves as members of a group that is denied full membership in society simply on the basis of gender. Many have tried to describe under what conditions groups of oppressed people will become conscious of their oppressed condition. Gramsci wrote of a dual "contradictory consciousness, . . . one which is implicit in his [humans'] activity and which in reality unites him with all his fellow-workers in the practical transformation of the world; and one, superficially explicit or verbal, which he has inherited from the past and uncritically absorbed."[70] Perhaps the rage women express premenstrually could be seen as an example of consciousness implicit in activity, which in reality unites all women, a consciousness that is combined in a contradictory way with an explicit verbal consciousness, inherited from the past and constantly reinforced in the present, which denies women's rage its truth.

It is well known that the oppression resulting from racism and colonialism engenders a diffused and steady rage in the oppressed population.[71] Audre Lorde expresses this with power: "My response to racism is anger. That anger has eaten clefts into my living only when it remained unspoken, useless to anyone. It has also served me in classrooms without light or learning, where the work and history of Black women was less than a vapor. It has served me as fire in the ice zone of uncomprehending eyes of white women who see in my experience and the experience of my people only new reasons for fear or guilt." Alongside anger from the injustice of racism is anger from the injustice of sexism: "Every woman has a well-stocked arsenal of anger potentially useful against those oppressions, personal and institutional, which brought that anger into being."[72]

Can it be accidental that women describing their premenstrual moods often speak of rebelling, resisting, or even feeling "at war"?[73] It is important not to miss the imagery of rebellion and resistance, even when the women themselves excuse their feelings by saying the rebellion is "for no apparent reason"[74] or that the war is with their own bodies! ("Each month I wage a successful battle with my body. But I'm tired of going to war.")[75]

Elizabeth Fox-Genovese writes of the factors that lead women to accept their own oppression: "Women's unequal access to political life and economic participation provided firm foundations for the ideology of gender difference. The dominant representations of gender relations stressed the naturalness and legitimacy of male authority and minimized the role of coercion. Yet coercion, and frequently its violent manifestation, regularly encouraged women to accept their subordinate status."[76] Looking at what has been written about PMS is certainly one way of seeing the "naturalness" of male authority in our society, its invisibility and unexamined, unquestioned nature. Coercion in this context need not consist in the violence of rape or beating: sometimes women's violence is believed to trigger these acts, as we have seen, but other times it is the women who become violent. In a best-selling novel, a psychopathic killer's brutal murders are triggered by her premenstruum.[77] In either case, physical coercion consists in focusing on women's bodies as the locus of the operation of power and insisting that rage and rebellion, as well as physical pain, will be cured by the administration of drugs, many of which have known tranquilizing properties.[78]

Credence for the medical tactic of treating women's bodies with drugs comes, of course, out of the finding that premenstrual moods and discomfort are regular, predictable, and in accord with a woman's menstrual cycle. Therefore, it is supposed, they must be at least partially caused by the changing hormonal levels known to be a part of the cycle. The next step, according to the logic of scientific medicine, is to try to find a drug that alleviates the unpleasant aspects of premenstrual syndrome for the millions of women that suffer them.

Yet if this were to happen, if women's monthly cycle were to be smoothed out, so to speak, we would do well to at least notice what would have been lost. Men and women alike in our society are familiar with one cycle, dictated by a complex interaction of biological and psychological factors, that happens in accord with cycles in the natural world: we all need to sleep part of every solar revolution, and we all recognize the disastrous consequences of being unable to sleep as well as the rejuvenating results of being able to do so. We also recognize and behave in accord with the socially determined cycle of the week, constructed around the demands of work-discipline in industrial capitalism.[79] It has even been found that men structure their moods more strongly in accord with the week than women.[80] And absenteeism in accord with the weekly cycle (reaching as high as 10 percent at General Motors on Mondays and Fridays)[81] is a cause of dismay in American industry but does not lead anyone to think that workers need medication for this problem.

Gloria Steinem wonders sardonically,

> What would happen if suddenly, magically, men could menstruate and women could not?
>
> Clearly, menstruation would become an enviable, boast-worthy, masculine event:
>
> Men would brag about how long and how much.
>
> Young boys would talk about it as the envied beginning of manhood. Gifts, religious ceremonies, family dinners, and stag parties would mark the day.
>
> To prevent monthly work loss among the powerful, Congress would fund a National Institute of Dysmenorrhea.[82]

Perhaps we might add to her list that if men menstruated, we would all be expected to alter our activities monthly as well as daily and weekly and enter a time and space organized to maximize the special powers released around the time of menstruation while minimizing the discomforts.

PMS adds another facet to the complex round of women's consciousness. Here we find some explicit challenge to the existing structure of work and time, based on women's own experience and awareness of capacities that are stifled by the way work is organized. Here we also find a kind of inchoate rage which women, because of the power of the argument that reduces this rage to biological malfunction, often do not allow to become wrath. In the whole history of PMS there are the makings of a debate whose questions have not been recognized for what they are: Are women, as in the terms of our cultural ideology, relegated by the functions of their bodies to home and family, except when, as second best, they struggle into wartime vacancies? Or are women, drawing on the different concepts of time and human capacities they experience, not only able to function in the world of work but able to mount a challenge that will transform it?

Notes

1 Lever 1981:108. Other examples of uniformly negative symptomatology are Halbreich and Endicott 1982 and Dalton 1983.
2 Lever 1981:1
3 Ad for a drug in *Dance* magazine, Jan. 1984:55
4 Few accounts reject the description of PMS as a disease. One that does is Witt, who prefers the more neutral term "condition" (1984:11–12). It is also relevant to note that I do not experience

severe manifestastions of PMS, and there is a possibility for that reason that I do not give sufficient credit to the medical model of PMS. I have, however, experienced many similar manifestations during the first three months of each of my pregnancies, so I have some sense that I know what women with PMS are talking about.

5 Lever 1981:47
6 Lever 1981:2, 1. Other estimates are "up to 75%" (Southam and Gonzaga 1965:154) and 40% (Robinson, Huntington and Wallace 1977:784)
7 Smith-Rosenberg 1974:27
8 Clarke 1873:130
9 Ehrenreich and English 1973:48
10 Clarke 1873:133–34
11 Hollingworth 1914:93
12 Jacobi 1877:232
13 Frank 1931
14 Frank 1931:1054
15 Benedek and Rubenstein 1939, II:461
16 Ivey and Bardwick 1968:344
17 Paige 1971:533–34
18 Frank 1931:1053
19 Kessler-Harris 1982:219, 259, 254–55
20 Seward 1934; McCance 1937; Billings 1934; Brush 1938
21 Altmann 1941; Anderson 1941; Novak 1941; Brinton 1943; Percival 1943
22 Seward 1944:95
23 Kessler-Harris 1982:295
24 Dalton and Greene 1953
25 Weiner 1985:118
26 Kessler-Harris 1982:311–16; Weiner 1985:89–97, 112–18
27 Kessler-Harris 1982:316–18
28 Laws 1983:25. I am working on a more extensive study of the literature on menstruation and industrial work from the late nineteenth century to the present, including publications in industrial hygiene as well as medicine and public health.
29 Lever 1981:20
30 Witt 1984:129
31 Vanek 1974; Scott 1980; Cowan 1983:208
32 de Beauvoir 1952:425
33 Oakley 1974:45
34 Ballantyne 1975:114
35 Dalton 1983:80
36 Dalton 1983:82
37 Gottlieb 1982:44
38 Dalton 1983:100. These figures are still being cited in major newspapers (Watkins 1986).
39 Parlee 1973:461–62
40 Golub 1976; Sommer 1973; Witt 1984:160–62
41 Both quoted in Weideger 1977:48
42 Birke and Gardner 1982:23
43 Harrison 1984:16–17
44 Witt 1984:150
45 Witt 1984:151
46 Birke and Gardner 1982:25; Bernstein 1977
47 Powers (1980) suggests that the association generally made between menstruation and negative conditions such as defilement may be a result of *a priori* Western notions held by the investigator. She argues that the Oglala, Plains Indians, have no such association. This does not mean that menstruation is *never* regarded negatively, of course (see Price 1984:21–22).
48 Buckley 1982:49
49 Harrison 1984:44
50 *Baltimore City Paper* 20 April 1984:39
51 Birke and Gardner 1982:25
52 Lever 1981:61
53 Lever 1981:63
54 Lauersen and Stukane 1983:18

55 Lever 1981:25
56 Lever 1981:28
57 Lever 1981:68
58 Angier 1983:119
59 Lauersen and Stukane 1983:80
60 Letter in *PMS Connection* 1982, 1:3. Reprinted by permission of PMS Action, Inc., Irvine, CA.
61 Weideger 1977:10
62 Harrison 1984:50
63 Frye 1983:91
64 Lever 1981:61
65 Harrison 1984:36
66 On isolation of housewives, see Gilman 1903:92.
67 Rich 1976:285
68 Shuttle and Redgrove 1978:58, 59
69 Griffin 1978:185
70 Gramsci 1971:333
71 Fanon 1963; Genovese 1976:647
72 Lorde 1981:9,
73 Dalton 1983:80; Harrison 1984:17; Halbreich and Endicott 1982:251, 255, 256
74 Dalton 1983:80
75 *PMS Connection* 1984, 3:4
76 Fox-Genovese 1984:272–73
77 Sanders 1981
78 Witt 1984:205, 208; Herrmann and Beach 1978
79 Thompson 1967
80 Rossi and Rossi 1974:32
81 Braverman 1974:32
82 Steinem 1981:338

References

Altmann, M. 1941 "A Psychosomatic Study of the Sex Cycle in Women." *Psychosomatic Medicine* 3:199–225.

Anderson, M. 1941 "Some Health Aspects of Putting Women to Work in War Industries." *Industrial Hygiene Foundation 7th Annual Meeting*, 165–69.

Ballantyne, Sheila 1975 *Normajean the Termite Queen*. New York: Penguin.

Benedek, Therese and Boris B. Rubenstein 1939 "The Correlations Between Ovarian Activity and Psychodynamic Processes. I. The Ovulative Phase; II. The Menstrual Phase." *Psychosomatic Medicine* 1(2):245–70; 1(4):461–85.

Bernstein, Barbara Elaine 1977 "Effect of Menstruation on Academic Women." *Archives of Sexual Behavior* 6(4):289–96.

Billings, Edward G. 1933 "The Occurrence of Cyclic Variations in Motor Activity in Relation to the Menstrual Cycle in the Human Female." *Bulletin of Johns Hopkins Hospital* 54:440–54.

Birke, Lynda and Katy Gardner 1982 *Why Suffer? Periods and Their Problems*. London: Virago.

Brinton, Hugh P. 1943 "Women in Industry," pp. 395–419 in *Manual of Industrial Hygiene and Medical Service in War Industries*, National Institutes of Health, Division of Industrial Hygiene, Philadelphia: W.B. Saunders.

Brush, A.L. 1938 "Attitudes, Emotional and Physical Symptoms Commonly Associated with Menstruation in 100 Women." *American Journal of Orthopsychiatry* 8:286–301.

Buckley, Thomas 1982 "Menstruation and the Power of Yurok Women: Methods in Cultural Reconstruction." *American Ethnologist* 9(1):47–60.

Clarke, Edward H. 1873 *Sex in Education; or a Fair Chance for the Girls*. Boston: James R. Osgood and Co.

Cowan, Ruth Schwartz 1983 *More Work for Mother: The Ironies of Household Technologies from the Open Hearth to the Microwave*. New York: Basic Books.

Dalton, Katharina 1983 *Once a Month*. Claremont, CA: Hunter House.

Dalton, Katharina and Raymond Greene 1983 "The Premenstrual Syndrome." *British Medical Journal* May:1016–17

de Beauvoir, Simone 1952 *The Second Sex*. New York: Knopf.

Ehrenreich, Barbara and Deirdre English 1973 *Complaints and Disorders: The Sexual Politics of Sickness.* Old Westbury, New York: The Feminist Press.
Fanon, Frantz 1963 *The Wretched of the Earth.* New York: Grove Press.
Fox-Genovese, Elizabeth 1982 "Gender, Class and Power: Some Theoretical Considerations." *The History Teacher* 15(2):255–76.
Frank, Robert T. 1931 "The Hormonal Causes of Premenstrual Tension." *Archives of Neurology and Psychiatry* 26:1053–57.
Frye, Marilyn 1983 *The Politics of Reality: Essays in Feminist Theory.* Trumansburg, NY: The Crossing Press.
Genovese, Eugene D. 1974 *Roll: Jordan, Roll: The World the Slaves Made.* New York: Vintage.
Gilman, Charlotte Perkins 1903 *The Home: Its Work and Influence.* Urbana: University of Illinois Press (1972 reprint).
Golub, Sharon 1976 "The Effect of Premenstrual Anxiety and Depression on Cognitive Function." *Journal of Personality and Social Psychology* 34(1):99–104.
Gottlieb, Alma 1982 "Sex, Fertility and Menstruation among the Beng of the Ivory Coast: A Symbolic Analysis." *Africa* 52(4):34–47.
Gramsci, Antonio 1971 *Prison Notebooks.* New York: International Publishers.
Griffin, Susan 1978 *Woman and Nature: The Roaring Inside Her.* New York: Harper and Row.
Halbreich, Uriel and Jean Endicott 1982 "Classification of Premenstrual Syndromes," pp. 243–65 in *Behavior and the Menstrual Cycle*, Richard C. Friedman, ed. New York: Marcel Dekker.
Harrison, Michelle 1984 *Self-Help for Premenstrual Syndrome.* Cambridge, MA: Matrix Press.
Herrmann, W.M. and R.C. Beach 1978 "Experimental and Clinical Data Indicating the Psychotropic Properties of Progestogens." *Postgraduate Medical Journal* 54:82–87.
Hollingworth, Leta 1914 *Functional Periodicity: An Experimental Study of the Mental and Motor Abilities of Women During Menstruation.* New York: Teachers College Press.
Ivey, Melville E. and Judith M. Bardwick 1968 "Patterns of Affective Fluctuation in the Menstrual Cycle." *Psychosomatic Medicine* 30(3):336–45.
Jacobi, Mary Putnam 1877 *The Question of Rest for Women During Menstruation.* New York: Putnam's Sons.
Kessler-Harris, Alice 1982 *Out to Work: A History of Wage-Earning Women in the United States.* New York: Oxford University Press.
Lauersen, Niels H. and Eileen Stukane 1983 *PMS Premenstrual Syndrome and You: Next Month Can Be Different.* New York: Simon and Schuster.
Laws, Sophie 1983 "The Sexual Politics of Pre-Menstrual Tension." *Women's Studies International Forum* 6(1):19–31
Lever, Judy with Dr. Michael G. Brush 1981 *Pre-menstrual Tension.* New York: Bantam.
Lorde, Audre 1981 "The Uses of Anger." *Women's Studies Quarterly* 9(3):7–10.
——. 1982 *Chosen Poems: Old and New.* New York: W.W. Norton.
McCance, R.A., M.C. Luff, and E.E. Widdowson 1937 "Physical and Emotional Periodicity in Women." *Journal of Hygiene* 37:571–614.
Novak, Emil 1941 "Gynecologic Problems of Adolescence." *Journal of the American Medical Association* 117:1950–53.
Oakley, Ann 1974 *The Sociology of Housework.* New York: Pantheon.
Paige, Karen E. 1971 "Effects of Oral Contraceptives on Affective Fluctuations Associated with the Menstrual Cycle." *Psychosomatic Medicine* 33(6):515–37.
Parlee, Mary 1973 "The Premenstrual Syndrome." *Psychological Bulletin* 80(6):454–65.
Percival, Eleanor 1943 "Menstrual Disturbances as They May Affect Women in Industry." *The Canadian Nurse* 39:335–37.
Powers, Maria N. 1980 "Menstruation and Reproduction: An Oglala Case." *Signs* 6:54–65.
Price, Sally 1984 *Co-wives and Calabashes.* Ann Arbor: University of Michigan Press.
Rich, Adrienne 1976 *Of Woman Born.* New York: Bantam.
Robinson, Kathleen, Kathleen M. Huntington, and M.G. Wallace 1977 "Treatment of the Premenstrual Syndrome." *British Journal of Obstetrics and Gynaecology* 84:784–88.
Rossi, Alice S. and Peter E. Rossi 1977 "Body Time and Social Time: Mood Patterns by Menstrual Cycle Phase and Day of the Week." *Social Science Research* 6:273–308.
Sanders, Lawrence 1981 *The Third Deadly Sin.* New York: Berkley Books.
Scott, Joan Wallach 1980 "The Mechanization of Women's Work." *Scientific American*, March: 167–85.
Seward, G.H. 1934 "The Female Sex Rhythm." *Psychological Bulletin* 31:153–192.
Shuttle, Penelope and Peter Redgrove 1978 *The Wise Wound: Eve's Curse and Everywoman.* New York: Richard Marek.

Smith-Rosenberg, Carroll 1974 "Puberty to Menopause: The Cycle of Femininity in Nineteenth-century America," pp. 23–37 in *Clio's Consciousness Raised*, Mary Hartman and Lois W. Banner, eds. New York: Harper and Row.

Sommer, Barbara 1973 "The Effect of Menstruation on Cognitive and Perceptual-Motor Behavior: A Review." *Psychosomatic Medicine* 35(6):515–34.

Steinem, Gloria 1981 *Outrageous Acts and Everyday Rebellions*. New York: Holt, Rinehart and Winston.

Thompson, E.P. 1967 "Time, Work-Discipline, and Industrial Capitalism." *Past and Present*, 38:56–97.

Vanek, Joan 1974 "Time Spent in Housework." *Scientific American* 231(5):116–20.

Watkins, Linda M. 1986 "Premenstrual Distress Gains Notice as a Chronic Issue in the Workplace." *Wall Street Journal*, 22 Jan.

Weideger, Paula 1977 *Menstruation and Menopause: The Physiology and Psychology, the Myth and the Reality*. New York: Delta.

Weiner, Lynn Y. 1985 *From Working Girl to Working Mother: The Female Labor Force in the United States, 1820–1980*. Chapel Hill: University of North Carolina Press.

Witt, Reni L. 1984 *PMS: What Every Woman Should Know about Premenstrual Syndrome*. New York: Stein and Day.

24

Reproductive Technologies, Surrogacy Arrangements, and the Politics of Motherhood

Laura Woliver

New reproductive technologies and surrogacy arrangements are subtly altering women's lives by making conception, gestation, and birth something that predominately male authorities increasingly monitor, examine, and control. Motherhood is powerfully shaped by culture (Firestone 1970; Bernard 1974; Rich 1976; Chodorow 1978; Ruddick 1980; O'Brien 1981; Dworkin 1983:173–88; Sevenhuijsen and Vries 1984; Belenky, Clinchy, Goldberger, and Tarule 1986; Martin 1987; Tong 1989: chap. 3; Duden 1993). Even though motherhood has never had much power and prestige in many societies (despite the sweet talk about "motherhood and apple pie"), one power women did have was their ability to gestate and give birth to babies. Many feminists are very wary of these new reproductive arrangements, given the disappointing track record of the medical profession's treatment of women (Rich 1976; Daly 1978; Ehrenreich and English 1978; S. Rothman 1978: 142–53; Graham and Oakley 1981; B. K. Rothman 1982; Edwards and Waldorf 1984; Oakley 1984; Pollock 1984; Corea 1985a, 1985b; Farrant 1985:103; Fisher 1986; Martin 1987; Dutton, Preston, and Pfund 1988, especially the case study on DES; Rowland 1992). Feminist scholars are also justly skeptical about modern science given its history and ethic of dominance, control, and insensitivity to women's lives (Griffin 1978; Merchant 1980; Gould 1981; Rothschild 1983; Bleier 1984; Arnold and Faulkner 1985; Keller 1985; Harding 1986; Rosser 1986, 1989).

Defining human sexuality and reproduction issues within a medical and scientific model also has the subtle power of displacing an alternative focus on the social, economic, and environmental issues of reproduction (Elkington 1985; Diamond 1990). A biological paradigm frames the issue medically and individually while distracting from the political and economic context of reproductive decisions. Medical technology, rather than social change, therefore is offered as the solution to reproductive problems and concerns. The medical profession's gate-keeping role and its monopoly over birth control information and services already display these tendencies to control and medicalization (Petchesky 1984; Jaquette and Staudt 1985; Hartmann 1987).

Additionally, many legal systems have already displayed a marked tendency to devalue and belittle the experience and desires of women (Finley 1986; Estrich 1987; MacKinnon 1987; Burniller 1988; Eisenstein 1988; Fineman 1988; Pateman 1988; Woliver 1988; Rhode 1993–94). Changes in American divorce and child custody policies have actually harmed many women, for example, in part because reformers did not consider women's economic and emotional circumstances, but instead adopted a neutrality stance that was favorable to men (Fineman 1983, 1988). Medical technologies have engendered much debate and concern about their economic, legal, and ethical impact (see, for example, Hanmer 1984; Hubbard 1984; Raymond 1984; Saxton 1984; Wikler 1986; Kishwar 1987; Spallone and Steinberg 1987; Stanworth 1987; Diamond 1988; Elshtain 1989; Glover 1989; St. Peter 1989; Taub and Cohen 1989; Teich 1990; to name just a few). In addition, surrogacy arrangements and reproductive technologies, which are presented as seemingly "neutral" procedures that result in "fair" policies, in fact gloss over women's relative poverty and powerlessness in society, thus skewing these arrangements toward abuse of women. These technologies, then, might actually decrease women's power over their bodies.

"As with most technologies," Rosser points out, "intrinsically the new reproductive technologies are neither good nor bad; it is the way they are used that determines their potential for benefit or harm" (1986:40). Introduced into sexist and class-based cultures, though, the new reproductive technologies raise serious issues of eugenics, the specter of abortions for sex selection, and the diminution of women's relationship to reproduction (Hanmer 1981: 176–77; Powledge 1981:196–97; Hoskins and Holmes 1984; Roggenkamp 1984:266–77; Corea 1985b:188–212; Warren 1985; Holmes and Hoskins 1987; Steinbacher and Holmes 1987; Weisman 1988:1, 6; Rowland 1992).

SURROGACY

Surrogacy illustrates the marginalization of the woman involved, while highlighting fetuses as distinct legal entities, children as commodities, and the primacy of father-rights. In fact, the term *surrogate* itself is a misleading and insensitive term for these arrangements, since the so-called surrogate is the biological mother, and the contracting or adopting wife of the genetic father is the actual surrogate. The highly publicized 1987 "Baby M" custody trial in the United States displayed many of the issues in surrogacy arrangements and, possibly, future innovative reproductive arrangements. The Baby M case illustrated that women's experience of maternity tends to be belittled even in the least technologically dependent situation of surrogacy (Harrison 1987; Ketchum 1987; Pollitt 1987; Rothman 1989; Woliver 1989a:27–33).

In Baby M, a woman was artificially inseminated with the contracting man's sperm. The contract specified, among many things, that the birthmother would relinquish the baby to the contracting man and his wife at birth. Instead, she tried to keep her baby. In Baby M, the genetic father was the contracting man, and the genetic mother was the surrogate mother. When she tried to keep her baby, she at least was recognized as having a genetic tie to the child.

In the future, there will be many arrangements where the contracting man and woman are the genetic parents while the birthmother, who is not genetically related to the baby, has gestated and birthed the child. Here matters become very complicated and the powerlessness and vulnerability of these so-called full surrogates is acute. Already at least one case has highlighted many of these issues.

Anna M. Johnson contracted to gestate and birth a child conceived from the ovum and sperm of the contracting couple. When Johnson sued in a California court to have her contract invalidated, the court upheld the contract. Johnson appealed to the California Supreme Court, which upheld the contact and awarded custody of the child to the genetic (contracting) parents (*Johnson v. Calvert*, 1993 WL 167739, May 20, 1993 [Cal.]). In part, Johnson had no claim to the child, the court said, because she had no genetic link. At the time the contract was signed, Johnson was a single mother and the contracting couple were middle class.

Many feminist scholars, after studying the Baby M trial and other testimonials of surrogate mothers, oppose these arrangements. Pateman writes, "The surrogacy contract also indicates that a further transformation of modern partiarchy may be under way. Father-right is reappearing in a new, contractual form" (1988:209). A legal system based on the male standard as the norm rationalizes surrogate contacts as "fair" and impartial arrangements. This logic means that sexual difference becomes irrelevant to physical reproduction: "The former status of 'mother' and 'father' is thus rendered inoperative by contract and must be replaced by the (ostensibly sex-neutral) 'parent' " (Pateman 1988:216). But, as Pateman explains, this is dangerous to women:

> In classic patriarchalism, the father is *the* parent. When the property of the "surrogate" mother, her empty vessel, is filled with the seed of the man who has contracted with her, he, too, becomes the parent, the creative force that brings new life (property) into the world. Men have denied significance to women's unique bodily capacity, have appropriated it and transmuted it into masculine political genesis. The story of the social contract is the greatest story of men giving political birth, but, with the surrogacy contract, modern pariarchy has taken a new turn. Thanks to the power of the creative political medium of contract, men can appropriate physical genesis too. The creative force of the male seed turns the empty property contracted out by an "individual" into new human life. Patriarchy in its literal meaning has returned in a new guise. (1988:216–17; emphasis in original)

Similarly, Mary Shanley analogizes surrogacy contracts to aspects of slavery: "In both contract pregnancy and consensual slavery, fulfilling the agreement, even if it appears to be freely undertaken, violates the ongoing freedom of the individual in a way that does not simply restrict future options (such as whether I may leave my employer) but does violence to the self (my understanding of who I am)" (1993:629; see also P. Williams 1988 and 1991:224–26).

Some scholars are concerned about the seemingly patronizing ideology behind prohibitions on surrogacy contracts. They believe it is based on the idea that women, unlike men, cannot make rational decisions about emotional issues like genetic parenthood, signing a contract, and then living up to the terms of the contract. One voice for this point of view is that of Lori Andrews:

> My personal opinion is that it would be a step backward for women to embrace any policy argument based on a presumed incapacity of women to make decisions. That, after all, was the rationale for so many legal principles oppressing women for so long, such as the rationale behind the laws not allowing women to hold property. (1989:369–70; see also Andrews 1988 and 1989a)

On the other hand, some scholars argue that surrogacy, if seen as a contract for reproductive services rather than as baby-selling, can potentially empower women: "The idea of personal agency in contracting to become a parent seeks to empower women to reclaim the power of their wombs," Shalev writes, "and to wield it responsibly with due respect for the biological vulnerability of men who must be able to trust and depend on women if they are to become fathers" (1989:17). The ability of women to contract the sale of their reproductive services, Shalev believes, would diminish the prevalent view of women as passive in reproductive matters. These contracts would enable women to exercise "birth power" (Shalev 1989) and fulfill the procreative desires of contracting couples (Robertson 1988).

Taken as a given by pro-surrogacy scholars is the right to sell human beings, the government's lack of jurisdiction to regulate such personal freedoms and choices, and the belittling of those who doubt the wisdom of such arrangements as merely expressing "personal opinions." Governments have a legitimate role to play in determining the permissibility of these arrangements in part because of the effect on all people. Surrogacy challenges society to assert fundamental principles regarding human dignity. "What is probably most remarkable about the debate over surrogate motherhood is that it has necessitated defending a claim that was previously taken as self-evident: namely, that society has an interest in people being regarded as intrinsically valuable, not as monetized units in a marketplace" (Capron and Radin 1988:63; see also Holder 1988).

Surrogacy debates must be informed by what motherhood means to women and by the socioeconomic conditions that pressure women into "choosing" these arrangements. Careful reading of surrogate contract stories reveals the heartaches these women endure and the sexist and classist nature of these arrangements (see, for example: Ketchum 1987:5–6; Pollitt 1987; Annas 1988; Coles 1988; Field 1989; B. Rothman 1989; Rowland 1992:177–80). "We are asked not to look behind the resulting children to see their lower-middle-class and lower-class mothers. But the core reality of surrogate motherhood is that it is both classist and sexist: a method to obtain children genetically related to white males by exploiting poor women" (Annas 1988:43). Arguments by feminists who defend surrogacy "attribute freedom to the person only as an isolated individual and fail to recognize that individuals are also ineluctably social creatures" (Shanley 1993:636). Radin, for example, argues for some exchanges to be "market-inalienable," because they are "grounded in noncommodification of things important to personhood" (1987:1903). Radin continues,

> In an ideal world markets would not necessarily be abolished, but market-inalienability would protect all things important to personhood. But we do not live in an ideal world. In the nonideal world we do live in, market-inalienability must be judged against a background of unequal power. In that world it may sometimes be better to commodify incompletely than not to commodify at all. Market-inalienability may be ideally justified in light of an appropriate conception of human flourishing, and yet sometimes be unjustifiable because of our nonideal circumstances. (1987:1903)

In our culture, "the rhetoric of commodification has led us into an unreflective use of market characterizations and comparisons for almost everything people may value," Radin points out, "and hence into an inferior conception of personhood" (1987:1936). With regard to paid surrogacy arrangements, therefore, "the role of paid breeder is incompatible with a society in which individuals are valued for themselves and are aided in achieving a full sense of human well-being and potentiality" (Capron and Radin 1988:62).

Discussions of surrogacy sometimes emphasize the sanctity of contracts ("She was a competent adult when she freely chose to sign this contract") and analogies to wet nurses and babysitters. Structuring discussions in those terms ignores gestation and birth, exclusively female experiences. Women's services are not merely rented by surrogacy; a part of the woman is taken away. The gestalt of maternity—women's psychological as well as physical experiences—is compartmentalized and devalued by defining childbirth in discontinuous terms (a "tented womb") rather than as the woman's complete experience of maternity (O'Brien 1981; B. Rothman 1986, 1989).

Equating the male experience of genetic parenthood to the woman's, comparing men's abilities to separate their sperm from their bodies, sell it (as in artificial insemination), or pass it on to a "surrogate" mother to women's "equal opportunity" to do the same through surrogacy, imposes male experience as the universal. As Pateman notes, "women's equal standing must be accepted as an expression of the freedom of women *as women*, and not treated as an indication that women can be just like men" (1988:231, emphasis in original; see also Woliver 1988). Similarly, Eisenstein argues, "because law is engendered, that is, *structured through the multiple oppositional layerings embedded in the dualism of man/woman*, it is not able to move beyond the male referent as the standard for sex equality" (1988:42, emphasis in original).

Feminist concern about surrogacy arrangements is grounded in what Binion calls "progressive feminism," which argues for the inclusion of women's values and experiences in our legal system (Binion 1991). Mary Shanley, for instance, points out that the woman's experiences of her pregnancy "trump" prior agreements in a contractual pregnancy because "enforcement of a pregnancy contract against the gestational mother's wishes would constitute a legal refusal to recognize the reality of the woman and fetus as beings-in-relationship, which the law should protect as it does many other personal relationships" (1993:632). Similarly, other progressive feminist scholars explain the need to incorporate the experiences and aspirations of teenage girls into public policy-making regarding teenage pregnancy issues (Gordon 1976: chap. 15; Rhode 1993–94).

In addition, surrogacy could shift the cultural meaning of "to mother," making it more biological, discontinuous, and distant. Presently, "to mother" means, among other things, to nurture, a long-term emotional commitment. "To father," in contrast, means the biological provision of sperm. Surrogacy and justifications voiced in its favor cloud the present meanings of these terms and might limit "mothering" to mere genetic connection, comparable to "fathering."

NEW REPRODUCTIVE TECHNOLOGIES

Feminist scholars are concerned about the impact reproductive technologies will have on women as a group, possibly restricting the group's choices rather than modestly increasing individual choices (Corea 1985b; Rowland 1987; Merrick and Blank 1993). Women's physical and emotional experiences of childbirth are being drastically altered at the same time as we are told that these technologies are for our own good and are responding to our demands.

Behind seemingly benign, neutral, and objective scientific practices and research often lie subtle systems of power. Murray Edelman (1977: chap. 4), for example, displays the way phrases used by mental health professionals that apparently imply progress, therapy, and empathy toward patients in fact disguise and justify systems of control and dominance. Modern feminists also view medical and legal power over issues of reproduction skeptically.

Expanding indicators for these new technologies mean that they are becoming part of more and more women's experience of motherhood. Where once they were used for emergencies and special cases, now many of them (ultrasound and fetal monitors, for example) are used routinely on healthy women. In these and many other ways, the naturalness of pregnancy and childbirth has been transformed into an illness, an unnatural condition, with an assumption of risk to fetal and maternal health that only the medical profession can rectify. Eventually, it is feared, the definition of a "good mother" will include women whose pregnancies and births are "managed" by the medical establishment using these technologies. At the same time, an "abusive" mother might include one who refuses to utilize the available technologies. In 1979 the president of the American College of Obstetricians and Gynecologists, for example, referred to home births as "the earliest form of child abuse" (quoted in Oakley 1984:219). Indeed, there have already been cases where mothers have been reprimanded for staying away from doctors' prenatal care and having home births (Oakley 1984:219, 246; Sherman 1988a, 1988b; Maier 1989; Rowland 1992:129).

The modern women's movement includes a women's health component, disenchanted with many medical practices and seeking to empower women to be better informed and more assertive consumers of health care (see, for example, Gardner 1981; Boston Women's

Health Book Collective 1971, 1984; Edwards and Waldorf 1984). The potential force of the women's health movement to critique and change the medicalization of pregnancy and birth, though, has been partially co-opted by the doctors and clinics themselves. What started out, for example, as exercise, nutrition, natural childbirth, and similar classes organized and staffed by community women have slowly been absorbed by the birthing centers of hospitals and the outpatient services of large ob-gyn clinics. The original classes were alternatives to, if not in opposition to, the traditional doctor's treatment of pregnancy and childbirth; their goals included encouraging mothers to keep healthy, strong, and fit, in body and mind, for the upcoming birth. Mothers were educated about their bodies, what to expect from doctors and hospital staff, what their rights as patients were, and how best to resist and avoid the unnecessary and demeaning aspects of the usual medicalized childbirth. In contrast, hospital- and clinic-based classes, while trying to educate women a little and encourage them to keep healthy, are not training them to challenge doctors or hospital routines but rather to be prepared and compliant future patients (see also Oakley 1984:238, 248–49). In addition, some health insurance policies might pay for doctor- and clinic-based prenatal classes, but not for those based in community organizations that, for instance, might tend to track women into the doctor-organized classes.

Oakley, for example, found from her study of the history of the medical treatment of pregnant women in Britain that "if one single message emerged, it was that pregnant women were themselves deficient: they lacked the necessary intelligence, foresight, education or responsibility to see that the only proper pathway to successful motherhood was the one repeatedly surveyed by medical expertise" (1984:72). Within this context, new reproductive technologies, such as ultrasound, seem "revolutionary" to Oakley because, "for the first time, they enable obstetricians to dispense with mothers as intermediaries, as necessary informants on fetal status and lifestyle" (1984:155). Ultrasound is like "a window on the womb," a long-desired goal for the professional providers of maternity care (Oakley 1984:156).

Researchers have noted that many women going through ultrasound are excited to see their fetus, relieved to learn it appears normal, and lovingly affectionate toward their future baby. "Ultrasound must, therefore," Oakley writes, "take its place in a long line of other well-used strategies for educating women to be good mothers. . . . Antenatal care has finally discovered mother love. Along with postnatal bonding, prenatal bonding will now in future be added to the repertoire of reproductive activities named and controlled by obstetricians" (1984:185; see also B. Rothman 1986:78–85, 114; Martin 1987:145–48).

THE CHANGING NATURE OF ABORTION POLITICS

New reproductive technologies add at least three new issues to political debates about abortion in the United States. First, new reproductive technologies are predicated on the availability of legal abortion, which is constantly under threat in this country. Amniocentesis, ultrasound, genetic screening, embryo transfer, test-tube babies, to name but a few, are procedures in which the option of abortion or destruction of excess embryos are inherent. Any changes in abortion laws will directly affect the use of these technologies. These technologies bring into abortion debates a new group of potential abortion users, who desire the abortion option for slightly different reasons than the original position on reproductive choice taken by the early women's movements. In addition, new medical knowledge is beginning to blur the boundaries between contraception and abortion. The controversy over introducing the French "abortion pill," RU-486, into the United States is one current example (Richard 1989:947–48).

Second, reproductive technologies such as ultrasound and fetal medicine are playing a significant, if subtle, role in the efforts to restrict abortion rights by furthering the construction of the image of the fetus as an entity distinct from the woman who carries and delivers it. These images are powerful symbols in United States prolife, antiabortion politics. "The idea that knowledge of fetal life, and especially confrontation with the visual image of the fetus, will 'convert' a woman to the pro-life position has been a central theme in both local and national right-to-life activism," Ginsburg found. "A popular quip summarizes this position: 'If there were a window on a pregnant woman's stomach, there would be no more abortions' " (1989:104). Images of the fetus used by antiabortion activists in the United States, it should be emphasized, usually exclude visualization of the woman the fetus is within (Petchesky 1984:353; F. Ginsburg 1989:107–9; Condit 1990:79–95; Duden 1993). These carefully constructed fetal images are powerful aspects of the antiabortion discourse.

> When pro-Life rhetors talk about why they believe as they do, the role of the photographs and films becomes quite clear. Without these pictures, pro-Life advocates would have only an abstract argument about the importance of chromosomes in determining human life or a religious argument about the "soul" . . . neither of those options could sustain the righteous fire of the public movement. (Condit 1990:80)

Fetal images, combined with artful political commentary and voice-overs by antiabortion spokespeople, help the antichoice movement in its attempt to define the fetus as a human person (Petchesky 1984; Condit 1990:86–89). In ultrasound images, or microscopic views of zygotes, the viewer's eye must be trained to "see" the "life," since it is not clear to the uninitiated (Duden 1993:73–78; see also Rowland 1992); this training, as well as the viewing of women's innards, is an integral part of the politics and the cultural shift of re-visioning (Duden 1993). Technologies, therefore, that allow viewing, studying, and possibly treating fetuses medically "are likely to elevate the moral status of the fetus" (Blank 1988:148).

Fetal medicine and the technologies that allow the visualization of the fetus push women out of the center of medical attention in gestation and birth, except for efforts to control female behavior for the well-being of the fetus. Discussions and lawsuits about what pregnant women eat and drink (alcohol, tobacoo, drugs) or how they live (working in unsafe occupations, engaging in unsafe sex, or practicing risky sports) have troubling implications in the United States for controlling women's behavior. Much of the emphasis in these discussions is on what the women do, not what the men in their lives are doing, or on the socioeconomic conditions that might influence women's habits, employment patterns, and prenatal care. Particularly chilling is the fact that some earlier medically recommended treatments and advice for pregnant women have subsequently proved very harmful (the use of X-rays in prenatal care is just one example). How do we know that the prevailing medical wisdoms of today will not be proven unsafe in the future? The inadequate, short-term testing of ultrasound is a troubling case in point (Oakley 1984:155–86; 1987:44–48).

Envisioning the fetus as a separate patient, combined with more detailed and heightened monitoring of the behavior of pregnant women, might have an impact on abortion choice. After all, if society oversees the nutrition and lifestyle of pregnant women in order to ensure the health of the fetal "patient," how would the woman be allowed to abort this "patient"? The elevation of the fetus as a patient in medicine and politics, though, marginalizes the women involved. As Gallagher reminds us, "given the very geography of pregnancy, questions as to the status of the fetus must follow, not precede, an examination of the rights of the woman within whose body and life the fetus exists" (1989:187–98; see also Wingerter 1987).

Third, neonatal technologies are altering perceptions of fetal viability, and thus undermining one of the premises of *Roe v. Wade* (the 1973 U.S. precedent for legal abortion, based partly on the viability of fetuses). As U.S. Supreme Court Justice Sandra Day O'Connor noted, *Roe* is on "a collision course with itself" because of these changes (O'Connor, quoted from *City of Akron v. Akron Center for Reproductive Health, Inc.*:458). Neonatal technologies are pushing back the gestational age when fetuses might be viable outside the womb (Blank 1984b:584–602; Blank 1988:64–65). The result might be to pit the woman's rights against those of her fetus. Indeed, recent U.S. cases indicate that "when maternal actions are judged detrimental to the health or life of the potential child, the court has shown little hesitancy to constrain the liberty of the mother" (Blank 1984a:150). In reality only a very small number of women and babies have access to these neonatal technologies, but the experience of a privileged few is being generalized into the abortion debate as a whole. In addition to the attack by conservatives and antiabortion activists on abortion choice (see, for example, Steiner 1983; Luker 1984; Cohan 1986; Glendon 1987; F. Ginsburg 1989; Condit 1990; Himmelstein 1990:89–90; to name but a few), feminists now must respond to the pressure the new technologies exert on abortion politics.

INCREASED MARGINALIZATION OF WOMEN

The values inherent in these technologies are making women, as Rothman states it, "transparent" (1986). A cultural shift has occurred where women have been "skinned" and authorities are permitted to examine and monitor their innards (Duden 1993:7). The role of the woman is becoming secondary to that of the medical profession and those who can broker surrogacy and adoption contracts.

When they are used for contraception, sex-selective abortions, and sterilization, reproductive technologies have already seriously violated the rights, dignity, and indigenous cultures of poor women the world over while failing to address the underlying poverty and inequalities in these women's societies (Shapiro 1985; Hartmann 1987; Tobin 1990). The coercive "choices" presented to impoverished third-world women to "select" sterilization or dangerous contraceptives such as Depo-Provera in exchange for food, clothing, or other benefits show how professional medical technologies can be used to harm women, yet be justified with the language of individual choice (see, for example, Bunkle 1984; Clarke 1984; Akhtar 1987; Hartmann 1987; Kamal 1987). "We will have to lift our eyes from the choices of the individual woman," writes Rothman, "and focus on the control of the social system that structures her choices, which rewards some choices and punishes others, which distributes the rewards and punishments for reproductive choices along class and race lines" (1984:33; see also Dworkin 1983:182; Corea 1985:2–3, 27; Rowland 1987:84; Gordon 1976, especially chap. 15; Rhode 1993–94).

In population control politics, the desires of women are largely ignored, their reproductive choices are narrowed, their knowledge of their own bodies is belittled, and traditional female and community-based healers are driven out of business. Contraceptive technologies emphasized by international aid agencies for less-developed countries, for example, sacrifice "women's health and safety in the indiscriminate promotion of hormonal contraception, the IUD, and sterilization, at the same time as they have neglected barrier methods and natural family planning" (Hartmann 1987:xv). The problem with barrier methods, from the point of view of many population planners, is that they are not effective enough, since they are under women's control and discretion. Hartmann found: "The thrust of contraceptive research in fact has been to remove control of contraception from women, in the same way that women are being increasingly alienated from the birth process itself" (1987:32).

Similarly, studies of the United States show that minority and poor women receive less quality health care, are subject to more intrusive medical procedures, and have limited "choice" to use expensive reproductive technologies (Martin 1987; Nsiah-Jefferson 1989; Davis 1990:53–65). Medical advice to poor and minority women is often uninformed regarding the language skills, cultural background, and desires of these patients. In these settings, therefore, "the graph she is asked to look at during her visit to the clinic only serves to mystify her experience. In ways that she cannot fathom, expert professionals claim to know something about her future child, much more, in fact, than she could ever find out by herself. Long before she actually becomes a mother she is habituated to the idea that others know better and that she is dependent on being told" (Duden 1993:29).

The marginalization of women in these new reproductive arrangements is shockingly clear in the language that is used to describe the women involved. Women are discussed in the new reproductive literature by bodily parts: "maternal environment" replaces a woman's womb, a pregnant woman becomes "an embryo carrier" (see, for example, Klein 1987:66). Surrogate mothers are likened to reproductive machines and are described as inanimate objects: "rented wombs," "incubators," "receptacles," "a kind of hatchery," "gestators," "a uterine hostess," or a "surrogate uterus" (see, for example, Ince 1984:99–116; Corea 1985:222; Hollinger 1985:901, 903). Fetal well-being appears more important than "invasions" of mother's bodies (Laborie 1987; Burfoot 1988:108, 110).

In vitro fertilization is sometimes described without once intimating that a human woman is involved. No woman, for example, is mentioned in the following overview of the 1978 birth of Louise Brown, the first "test-tube" baby:

> After many years of frustrating research, Drs. Edwards and Steptoe had succeeded in removing an egg from an ovarian follicle, fertilizing it in a dish, and transferring the developing zygote to a uterus where it implanted and was brought to term. (Robertson 1986:943)

The women involved in these procedures are truly marginal. It is important to recall that "test-tube" babies are born from mothers who carried them during pregnancy. In addition, these women were sometimes experimented on without their full consent, or they participated based on misleading information concerning the probability of actually having a baby (Corea 1985:112–17, 166–85; Corea and Ince 1987; Lasker and Borg 1987:53–55).

The future role of mothers in a medical system oriented toward technological intervention and control in conception, gestation, and birth, where life itself is just another commodity and women's bodies producers of quality products, is very troubling. "To use the law for these complicated moral decisions," writes Rothman, "is to lose the nuances, the idiosyncrasies, and the individuality that protect us from fundamentally untrustworthy political institutions" (1989:87). In addition to the marginalization of women in these reproductive arrangements, these technologies deflect pressures for social reforms by promising technological fixes for reproductive difficulties. Some women delay motherhood and possibly increase reproductive risks, for example, to conform to male career timetables (such as the tenure system in universities or the partner process in law firms, to name but two). The new reproductive technologies allow this to continue by implying to women that they will suffer few consequences by delaying motherhood. Women's delay of motherhood is also used to justify the "demands" for the technologies by women. Potential pressures to change corporations, universities, and other employers, then, are blunted by this technological turn (Woliver 1989a:39; see also Woliver 1989b and 1990). New reproductive technologies and surrogacy arrangements are increasingly making women marginal in the new politics of motherhood.

Note

An earlier version of this paper was presented at the 1990 Feminism and Legal Theory Conference on Motherhood, University of Wisconsin-Madison Law School.

References

Akhtar, Farida. 1987. "Wheat for Statistics: A Case Study of Relief Wheat for Attaining Sterilization Targer in Bangladesh." In Patricia Spallone and Deborah Lynn Steinberg, eds., *Made to Order: The Myth of Reproductive and Genetic Progress*, pp. 154–60. New York: Pergamon.

Andrews, Lori B. 1988. "Surrogate Motherhood: The Challenge for Feminists." *Law, Medicine, & Health Care* 16, nos. 1–2: 72–80. Also published in Larry Gostin, ed., *Surrogate Motherhood: Politics and Privacy*, pp. 167–82. Bloomington: Indiana University Press, 1988.

——. 1989. *Between Strangers: Surrogate Mothers, Expectant Fathers, and Brave New Babies.* New York: Harper and Row.

Annas, George J. 1988. "Fairy Tales Surrogate Mothers Tell." In Larry Gostin, ed., *Surrogate Motherhood: Politics and Privacy*, pp. 43–55. Bloomington: Indiana University Press.

Arnold, Erick and Wendy Faulkner. 1985. "Smothered by Invention: The Masculinity of Technology." In Erik Arnold and Wendy Faulkner, eds., *Smothered by Invention: Technology in Women's Lives*, pp. 18–50. London: Pluto.

Belenky, Mary Field, Blythe McVicker Clinchy, Nancy Rule Goldberger, and Jill Martuck Tarule. 1986. *Women's Ways of Knowing: The Development of Self, Voice, and Mind.* New York: Basic Books.

Bernard, Jessie. 1974. *The Future of Motherhood.* New York: Dial.

Binion, Gayle. 1991. "Toward a Feminist Regrounding of Constitutional Law." *Social Science Quarterly* 72, no. 2 (June): 207–20.

Blank, Robert H. 1984a. *Redefining Human Life: Reproductive Technologies and Social Policy.* Boulder: Westview.

——. 1984b. "Judicial Decision Making and Biological Fact: *Roe v. Wade* and the Unresolved Question of Fetal Viability." *Western Political Quarterly* 37:584–602.

——. 1988. *Rationing Medicine.* New York: Columbia University Press.

Bleier, Ruth. 1984. *Science and Gender: A Critique of Biology and Its Theories on Women.* New York: Pergamon.

Boston Women's Health Book Collective. 1971. *Our Bodies, Our Selves.* Boston: New England Free Press. 1973–92. Reprint, New York: Simon and Schuster.

——. 1984. *The New Our Bodies, Our Selves.* New York: Simon and Schuster.

Burniller, Kristen. 1988. *The Civil Rights Society: The Social Construction of Victims.* Baltimore: Johns Hopkins University Press.

Bunkle, Phillida. 1984. "Calling the Shots? The International Politics of Depo-Provera." In Arditti, Klein, and Minden, eds., *Test-Tube Women: What Future for Motherhood?*, pp. 165–87. London: Pandora.

Burfoot, Annette. 1988. "A Review of the Third Annual Meeting of the European Society of Human Reproduction and Embryology." *Reproductive and Genetic Engineering: Journal of International Feminist Analysis* 1, no. 1: 107–11.

Capron, Alexander M. and Margaret J. Radin. 1988. "Choosing Family Law Over Contract Law as a Paradigm for Surrogate Motherhood." In Larry Gostin, ed., *Surrogate Motherhood: Politics and Privacy*, pp. 59–76. Bloomington: Indiana University Press.

Chodorow, Nancy. 1978. *The Reproduction of Mothering: Psychoanalysis and the Sociology of Gender.* Berkeley: University of California Press.

Clarke, Adele. 1984. "Subtle Forms of Sterilization Abuse: A Reproductive Rights Analysis." In Arditti, Klein, and Minden, eds., *Test-Tube Women: What Future for Motherhood?*, pp. 188–212.

Cohan, Alvin. 1986. "Abortion as a Marginal Issue: The Use of Peripheral Mechanisms in Britain and the United States." In Joni Lovenduski and Joyce Outshoorn, eds., *The New Politics of Abortion*, pp. 27–48. Newbury Park, Calif.: Sage.

Coles, Robert. 1988. " 'So, You Fell in Love with Your Baby.' " *New York Times Book Review* (June 26): 1, 34–35.

Condit, Celeste M. 1990. *Decoding Abortion Rhetoric: The Communication of Social Change.* Urbana: University of Illinois Press.

Corea, Gena. 1985a, Updated Edition. *The Hidden Malpractice: How American Medicine Mistreats Women.* New York: Harper Colophon.

——. 1985b. *The Mother Machine: Reproductive Technologies from Artificial Insemination to Artificial Wombs.* New York: Harper and Row.
Corea, Gena and Susan Ince. 1987. "Report of a Survey of IVF Clinics in the USA." In Spallone and Steinberg, eds., *Made to Order: The Myth of Reproductive and Genetic Progress*, pp. 133–45.
Daly, Mary. 1978. *Gyn/Ecology: The Metaethics of Radical Feminism.* Boston: Beacon.
Davis, Angela Y. 1990. *Women, Culture, and Politics.* New York: Vintage.
Diamond, Irene. 1988. "Medical Science and the Transformation of Motherhood: The Promise of Reproductive Technologies." In Ellen Boneparth and Emily Stoper, eds., *Women, Power, and Policy: Toward the Year 2000*, pp. 155–67. New York: Pergamon.
——. 1990. "Babies, Heroic Experts, and a Poisoned Earth." In Irene Diamond and Gloria Feman Orenstein, eds., *Reweaving the World: The Emergence of Ecofeminism*, pp. 201–10. San Francisco: Sierra Club Books.
Duden, Barbara. 1993. *Disembodying Women: Perspectives on Pregnancy and the Unborn.* Cambridge: Harvard University Press.
Dutton, Diana B., Thomas A. Preston, and Nancy E. Pfund. 1988. *Worse Than the Disease: Pitfalls of Medical Progress.* New York: Cambridge University Press.
Dworkin, Andrea. 1983. *Right-Wing Women.* New York: Perigee.
——. 1991. *Ideology: An Introduction.* London: Verso.
Edelman, M. W. 1987. *Families in Peril: An Agenda For Social Change.* Cambridge: Harvard University Press.
Edwards, Margot and Mary Waldorf. 1984. *Reclaiming Birth: History and Heroines of American Childbirth Reform.* Trumansburg, N.Y.: Crossing.
Ehrenreich, Barbara and Deirdre English. 1978. *For Her Own Good: 150 Years of the Experts' Advice to Women.* Garden City, N.Y.: Anchor.
Eisenstein, Zillah R. 1988. *The Female Body and the Law.* Berkeley: University of California Press.
Elkington, John. 1985. *The Poisoned Womb: Human Reproduction in a Polluted World.* New York: Penguin.
Elshtain, Jean Bethke. 1989. "Technology as Destiny: The New Eugenics Challenges Feminism." *The Progressive* 53, no. 6 (June): 19–23.
Estrich, Susan. 1987. *Real Rape.* Cambridge: Harvard University Press.
Farrant, Wendy. 1985. "Who's for Amniocentesis? The Politics of Prenatal Screening." In H. Homans, ed., *The Sexual Politics of Reproduction*, pp. 96–122. Brookfield, Vt.: Gower.
Field, Martha A. 1988. *Surrogate Motherhood.* Cambridge: Harvard University Press.
Fineman, Martha A. 1983. "Implementing Equality: Ideology, Contradiction, and Social Change; A Study of Rhetoric and Results in the Regulation of the Consequences of Divorce." *Wisconsin Law Review*, pp. 789–886.
——. 1988. "Dominant Discourse, Professional Language, and Legal Change in Child Custody Decision Making." *Harvard Law Review* 101:727.
Finley, Lucinda M. 1986. "Transcending Equality Theory: A Way Out of the Maternity and Workplace Debate." *Columbia Law Review* 86:1118–82.
Firestone, Shumalith. 1970. *The Dialectic of Sex: The Case for Feminist Revolution.* New York: Morrow.
Fisher, Sue. 1986. *In the Patient's Best Interest: Women and the Politics of Medical Decisions.* New Brunswick: Rutgers University Press.
Gallagher, Janet. 1989. "Fetus as Patient." In Nadine Taub and Sherrill Cohen, eds., *Reproductive Laws for the 1990s*, pp. 185–235. Clifton, N.J.: Humana.
Gardner, Katy. 1981. "Well Woman Clinics: A Positive Approach to Women's Health." In Helen Roberts, ed., *Women, Health, and Reproduction*, pp. 129–43. London: Routledge and Kegan Paul.
Ginsburg, Faye D. 1989. *Contested Lives: The Abortion Debate in an American Community.* Berkeley: University of California Press.
Ginsburg, Ruth Bader. 1985. "Some Thoughts on Autonomy and Equality in Relation to *Roe v. Wade.*" *North Carolina Law Review* 63, no. 2: 375–86.
——. 1992. "Speaking in a Judicial Voice." *New York University Law Review* 67, no. 6: 1185–209.
Glendon, Mary Ann. 1987. *Abortion and Divorce in Western Law.* Cambridge: Harvard University Press.
Glover, Jonathan et al. 1989. *Ethics of New Reproductive Technologies: The Glover Report to the European Commission.* De Kalb: Northern Illinois University Press.
Gordon, Linda. 1976. *Woman's Body, Woman's Right: A Social History of Birth Control in America.* New York: Grossman.
Gould, Stephen Jay. 1981. *The Mismeasure of Man.* New York: Norton.

Graham, Hillary and Ann Oakley. 1981. "Competing Ideologies of Reproduction: Medical and Maternal Perspectives on Pregnancy." In Helen Roberts, ed., *Women, Health, and Reproduction*, pp. 50–74. London: Routledge and Kegan Paul.

Griffin, Susan. 1978. *Woman and Nature: The Roaring Inside Her*. New York: Harper Colophon.

Hanmer, Jalna. 1981. "Sex Predetermination, Artificial Insemination, and the Maintenance of Male-Dominated Culture." In Helen Roberts, ed., *Women, Health, and Reproduction*, pp. 163–90. London: Routledge and Kegan Paul.

——. 1984. "A Womb of One's Own." In Arditti, Klein, and Minden, eds., *Test-Tube Women: What Future for Motherhood?*, pp. 438–48.

Harding, Sandra. 1986. *The Science Question in Feminism*. Ithaca: Cornell University Press.

Harrison, Michelle. 1987. "Social Construction of Mary Beth Whitehead." *Gender & Society* 1, no. 3: 300–11.

Hartmann, Betsy. 1987. *Reproductive Rights and Wrongs: The Global Politics of Population Control and Contraceptive Choice*. New York: Harper and Row.

Himmelstein, Jerome L. 1990. *To the Right: The Transformation of American Conservatism*. Berkeley: University of California Press.

Holder, Angela R. 1988. "Surrogate Motherhood and the Best Interest of Children." In Larry Gostin, ed., *Surrogate Motherhood: Politics and Privacy*, pp. 77–87. Bloomington: Indiana University Press.

Hollinger, J. H. 1985. "From Coitus to Commerce: Legal and Social Consequences of Noncoital Reproduction." *University of Michigan Journal of Law Reform* 18 (Summer): 865–932.

Holmes, Helen B. and Betty B. Hoskins. 1987. "Prenatal and Preconception Sex Choice Technologies: A Path to Femicide?" In Gena Corea et al., eds., *Man-Made Women: How the New Reproductive Technologies Affect Women*, pp. 15–29. Bloomington: Indiana University Press.

Hoskins, Betty B. and Helen B. Holmes. 1984. "Technology and Prenatal Femicide." In Arditti, Klein, and Minden, eds., *Test-Tube Women: What Future for Motherhood?*, pp. 237–55.

Hubbard, Ruth. 1984. "Personal Courage Is Not Enough: Some Hazards of Child-bearing in the 1980s." In Arditri, Klein, and Minden, eds., *Test-Tube Women: What Future for Motherhood?*, pp. 331–55.

Ince, Susan. 1984. "Inside the Surrogate Industry." In Arditti, Klein, and Minden, eds., *Test-Tube Women: What Future for Motherhood?*, pp. 99–116.

Jaquette, Jane S. and Kathleen A. Staudt. 1985. "Women as 'at Risk' Reproducers: Biology, Science, and Population in U.S. Foreign Policy." In Virginia Sapiro, ed., *Women, Biology, and Public Policy*, pp. 235–68. Newbury Park, Calif.: Sage.

Kamal, Sultana. 1987. "Seizure of Reproductive Rights? A Discussion on Population Control in the Third World and the Emergence of the New Reproductive Technologies in the West." In Spallone and Steinberg, eds., *Made to Order: The Myth of Reproductive and Genetic Progress*, pp. 146–53.

Keller, Evelyn Fox. 1985. *Reflections on Gender and Science*. New Haven: Yale University Press.

Ketchum, Sara Anne. 1987. "New Reproductive Technologies and the Definition of Parenthood: A Feminist Perspective." Paper presented at the June Feminism and Legal Theory Conference on Intimacy, University of Wisconsin Law School, Madison, Wisc.

Kishwar, Madhu. 1987. "The Continuing Deficit of Women in India and the Impact of Amniocentesis." In Gena Corea et al., eds., *Man-Made Women: How New Reproductive Technologies Affect Women*, pp. 30–37. Bloomington: Indiana University Press.

Kline, Marlee. 1989. "Race, Racism, and Feminist Legal Theory." *Harvard Women's Law Journal* 12:115.

Laborie, Françoise. 1987. "Looking for Mothers, You Only Find Fetuses." In Spallone and Steinberg, eds., *Made to Order: The Myth of Reproductive and Genetic Progress*, pp. 48–57.

Lasker, Judith and Susan Borg. 1987. *In Search of Parenthood: Coping with Infertility and High-Tech Conception*. Boston: Beacon.

Luker, Kristin. 1984. *Abortion and the Politics of Motherhood*. Berkeley: University of California Press.

MacKinnon, Catharine A. 1987. *Feminism Unmodified: Discourses on Life and Law*. Cambridge: Harvard University Press.

Maier, Kelly E. 1989. "Pregnant Women: Fetal Containers or People with Rights?" *Affilia: Journal of Women and Social Work* 4, no. 2 (Summer): 8–20.

Martin, Emily. 1987. *The Woman in the Body: A Cultural Analysis of Reproduction*. Boston: Beacon.

Merchant, Carolyn. 1980. *The Death of Nature: Women, Ecology and the Scientific Revolution*. San Francisco: Harper and Row.

Merrick, Janna C. and Robert H. Blank, guest eds. 1993. "The Politics of Pregnancy: Policy Dilemmas in the Maternal-Fetal Relationship." *Women & Politics*, special issue 13, nos. 3/4.

Nsiah-Jefferson, Laurie. 1989. "Reproductive Laws, Women of Color, and Low-Income Women." In Nadine Taub and Sherrill Cohen, eds., *Reproductive Laws for the 1990s*, pp. 23–67. Clifton, N.J.: Humana.

Oakley, Ann. 1984a. *Taking It Like a Woman*. London: Cape.

——. 1984b. *The Captured Womb: A History of the Medical Care of Pregnant Women*. Oxford: Basil Blackwell.

——. 1987. "From Walking Wombs to Test-Tube Babies." In Michelle Stanworth, ed., *Reproductive Technologies: Gender, Motherhood, and Medicine*, pp. 36–56. Minneapolis: University of Minnesota Press.

O'Brien, Mary. 1981. *The Politics of Reproduction*. Boston: Routledge and Kegan Paul.

Pateman, Carole. 1988. *The Sexual Contract*. Stanford: Stanford University Press.

Petchesky, Rosalind P. 1984. *Abortion and Woman's Choice: The State, Sexuality, and Reproductive Freedom*. New York: Longman, and Boston: Northeastern University Press.

Pollitt, Katha. 1987. "Contracts and Apple Pie: The Strange Case of Baby M." *The Nation* 244 (May 23): 667, 682–688.

Pollock, Scarlet. 1984. "Refusing to Take Women Seriously: 'Side Effects' and the Politics of Contraception." In Arditti, Klein, and Minden, eds., *Test-Tube Women: What Future for Motherhood?*, pp. 138–52.

Powledge, Tabitha M. 1981. "Unnatural Selection: On Choosing Children's Sex." In Helen B. Holmes, Betty B. Hoskins, and Michael Gross, eds., *The Custom-Made Child? Women-Centered Perspectives*, pp. 193–99. Clifton, N.J.: Humana.

Radin, Margaret Jane. 1987. "Market-Inalienability." *Harvard Law Review* 100: 1849–937.

Raymond, Janice. 1984. "Feminist Ethics, Ecology, and Vision." In Arditti, Klein, and Minden, eds., *Test-Tube Women: What Future for Motherhood?*, pp. 427–37.

——. 1993–94. "Adolescent Pregnancy and Public Policy." *Political Science Quarterly* 108, no. 4: 635–69.

Rich, Adrienne. 1976. *Of Woman Born: Motherhood as Experience and Institution*. New York: Norton.

Richard, Patricia Bayer. 1989. "Alternative Abortion Policies: What Are the Health Consequences?" *Social Science Quarterly* 70, no. 4: 941–55.

Robertson, John A. 1986. "Embryos, Families, and Procreative Liberty: The Legal Structure of the New Reproduction." *Southern California Law Review* 59: 939–1041.

——. 1988. "Procreative Liberty and the State's Burden of Proof in Regulating Noncoital Reproduction." In Larry Gostin, ed., *Surrogate Motherhood: Politics and Privacy*, pp. 24–42. Bloomington: Indiana University Press.

Roggenkamp, Viola. 1984. "Abortion of a Special Kind: Male Sex Selection in India." In Arditti, Klein, and Minden, eds., *Test-Tube Women: What Future for Motherhood?*, pp. 266–77.

Rosser, Sue. 1986. *Teaching Science and Health from a Feminist Perspective*. New York: Pergamon.

——, ed. 1989. "Feminism and Science: In Memory of Ruth Bleier." *Women's Studies International Forum*. Special issue 12.

Rothman, Barbara Karz. 1982. *In Labor: Women and Power in the Birthplace*. New York: Norton.

——. 1986. *The Tentative Pregnancy: Prenatal Diagnosis and the Future of Motherhood*. New York: Viking.

——. 1989. *Recreating Motherhood: Ideology and Technology in a Patriarchal Society*. New York: Norton.

Rothman, Sheila M. 1978. *Woman's Proper Place: A History of Changing Ideals and Practices, 1870 to the Present*. New York: Basic Books.

Rothschild, Joan. 1983. "Introduction: Why Machina ex Dea?" In Joan Rothschild, ed., *Machina ex Dea: Feminist Perspectives on Technology*, pp. ix–xxix. New York: Pergamon.

Rowland, Robyn. 1987. "Technology and Motherhood: Reproductive Choice Reconsidered," *Signs* 12:512–28.

——. 1992. *Living Laboratories: Women and Reproductive Technologies*. Bloomington: Indiana University Press.

Ruddick, Sara. 1980. "Maternal Thinking," *Feminist Studies* 6:432–67.

St. Peter, Christine. 1989. "Feminist Discourse, Infertility, and Reproductive Technologies." *NWSA Journal* 1, no. 3 (Spring): 353–67.

Saxton, Marsha. 1984. "Born and Unborn: The Implications of Reproductive Technologies for People with Disabilities." In Arditti, Klein, and Minden, eds., *Test-Tube Women: What Future for Motherhood?*, pp. 298–312.

Sevenhuijsen, Selma and Petra de Vries. 1984. "The Women's Movement and Motherhood." In Anja

Meulenbelt et al., eds., *A Creative Tension: Key Issues in Socialist-Feminism*, pp. 9–25. Boston: South End.

Shalev, Carmel. 1989. *Birth Power: The Case for Surrogacy*. New Haven: Yale University Press.

Shanley, Mary Lyndon. 1993. " 'Surrogate Mothering' and Women's Freedom: A Critique of Contracts for Human Reproduction." *Signs* 18: 618–39.

Shapiro, Thomas. 1985. *Population Control Politics: Women, Sterilization, and Reproductive Choice*. Philadelphia: Temple University Press.

Sherman, Rorie. 1988a. " 'Fetal Rights' Cases Draw Little Attention." *National Law Journal* 11, no. 4: 25.

——. 1988b. "Keeping Baby Safe from Mom." *National Law Journal* 11, no. 4: 1, 24, 26.

Spallone, Patricia and Deborah Lynn Steinberg. 1987. "Introduction." In Spallone and Steinberg, eds., *Made to Order: The Myth of Reproductive and Genetic Progress*, pp. 13–17. New York: Pergamon Press.

Stanworth, Michelle. 1987. *Reproductive Technologies: Gender, Motherhood, and Medicine*. Minneapolis: University of Minnesota Press.

Steinbacher, Roberta and Helen B. Holmes. 1987. "Sex Choice: Survival and Sisterhood." In Gena Corea et al., eds., *Man-Made Women: How the New Reproductive Technologies Affect Women*, pp. 52–63.

Steiner, Gilbert Y., ed. 1983. *The Abortion Dispute and the American System*. Washington, D.C.: The Brookings Institution.

Taub, Nadine and Sherrill Cohen, eds. 1989. *Reproductive Laws for the 1990s*. Clifton, N.J.: Humana.

Teich, Albert H., ed. 1990. *Technology and the Future*. 5th ed. New York: St. Martin's.

Tobin, Richard J. 1990. "Environment, Population, and Development in the Third World." In Norman J. Vig and Michael E. Kraft, eds., *Environmental Policy in the 1990s*, pp. 279–300. Washington, D.C.: Congressional Quarterly Press.

Tong, Rosemarie. 1989. *Feminist Thought: A Comprehensive Introduction*. Boulder: Westview.

Warren, Mary Anne. 1985. *Gendercide: The Implications of Sex Selection*. Totowa, N.J.: Rowman & Allanheld.

Weisman, S. R. 1988. "State in India Bars Fetus Sex-Testing." *New York Times* (July 20): 1, 6.

Wikler, Norma J. 1986. "Society's Response to the New Reproductive Technologies: The Feminist Perspectives." *Southern California Law Review* 59, no. 5: 1043–57.

Williams, Patricia. 1988. "On Being the Object of Property." *Signs* 14:5–24.

——. 1991. *The Alchemy of Race and Rights: Diary of a Law Professor*. Cambridge: Harvard University Press.

Wingetter, Rex B. 1987. "Fetal Protection Becomes Assault on Motherhood." *In These Times* (June 10–23): 3, 8.

Woliver, Laura R. 1988. "Review Essay: The Equal Rights Amendment and the Limits of Liberal Legal Reform." *Polity* 21, no. 1: 183–200.

——. 1989a. "The Deflective Power of Reproductive Technologies: The Impact on Women." *Women & Politics* 9, no. 3 (November): 17–47.

——. 1989b. "New Reproductive Technologies: Challenges to Women's Control of Gestation and Birth." In Robert Blank and Miriam K. Mills, eds., *Biomedical Technology and Public Policy*, pp. 43–56. Westport, Conn.: Greenwood.

——. 1990. "Reproductive Technologies and Surrogacy: Policy Concerns for Women." *Politics and the Life Sciences* 8, no. 2 (February): 185–93.

Contributors

Elizabeth Adams is a doctoral student in the Psychology in the Public Interest Program, Department of Psychology, North Carolina State University, Raleigh, North Carolina.

Anne Balsamo is a Professor in the Interactive Media Division and Gender Studies at the University of Southern California in Los Angeles, California.

Mary Barbercheck is a Professor of Entomology at Pennsylvania State University in University Park, Pennsylvania.

Ruth Bleier was a Professor in the Department of Neurophysiology at the University of Wisconsin, Madison, Wisconsin, and a founding member of the Women's Studies Program there. She died January 4, 1988. She is missed.

Carol Cohn is the executive director of the Boston Consortium on Gender, Security and Human Rights. She also is a program director for the Fletcher School of Law and Diplomacy at Tufts University in Medford, Massachusetts.

Wendy Faulkner is a Reader in Technology Studies at the University of Edinburgh in Edinburgh, Scotland.

Anne Fausto-Sterling is a Professor of Biology and Gender Studies at Brown University in Providence, Rhode Island.

Linda Marie Fedigan is a Professor and Canada Research Chair in Primatology and Bioanthropology at the University of Calgary in Calgary, Alberta, Canada.

Dara Horn received her doctorate in comparative literature from Harvard University. She is an award-winning author and currently resides in New York City.

Evelyn Fox Keller is a Professor in the Program in Science, Technology and Society at Massachusetts Institute of Technology in Cambridge, Massachusetts.

Suzanne Kessler is a Professor of Psychology and Dean of the School of Natural and Social Sciences at the State University of New York-Purchase in Purchase, New York.

Jaekyung Lee is an Associate Professor of Counseling, School, and Educational Psychology at the University of Buffalo in Buffalo, New York.

Rachel Maines is a Visiting Scholar in the Department of Science & Technology Studies at Cornell University in Ithaca, New York.

Emily Martin is a Professor of Anthropology at New York University in New York.

Ruth Perry is a Professor of Literature at the Massachusetts Institute of Technology in Cambridge, Massachusetts.

Hilary Rose is a Visiting Professor of Sociology at City University in London, England.

Aimée Sands is a Department Instructor of English at Clark University in Worcester,

Massachusetts. She is also the co-director of the Brookline Poetry Series and an award-winning independent documentary producer.

Jennifer Schneider is a doctoral student in the Psychology in the Public Interest Program, Department of Psychology, North Carolina State University.

Shruti Rana is an Assistant Professor of Law at the University of Maryland School of Law in Baltimore, Maryland.

Banu Subramaniam is an Associate Professor of Women's Studies at University of Massachusetts-Amherst in Amherst, Massachusetts.

Andrea Tone is a Professor of History and Canada Research Chair in the Social History of Medicine at McGill University in Montreal, Quebec, Canada.

Marta Wayne is an Associate Professor in Zoology at the University of Florida in Gainesville, Florida.

Christine Wennerás is a Professor and Chief Physician in the Department of Clinical Bacteriology at Göteborg University, Sweden.

Agnes Wold is a Professor and Chief Physician in the Department of Clinical Bacteriology at Göteborg University, Sweden.

Laura Woliver is a Professor of Political Science and Women's Studies at the University of South Carolina in Columbia, SC.

Permissions

Section Ia. Education: Out of the Frying Pan . . .

Keller, Evelyn Fox, "The Anomaly of a Woman in Physics," in *Working It Out: Twenty-three Writers, Scientists and Scholars Talk about Their Lives*, ed. Sara Ruddick and Pamela Daniels, pp. 77–91. New York: Pantheon. Copyright © 1977 by Sara Ruddick and Pamela Daniels. Reprinted by permission of Pantheon Books, a division of Random House, Inc.

Sands, Aimée, "Never Meant to Survive, a Black Woman's Journey: An Interview with Evelynn Hammonds," in *The Racial Economy of Science*, ed. Sandra Harding, pp. 239–48. Bloomington, Indiana University Press. Copyright © 1993 by Radical Teacher Press. Reprinted by permission of Radical Teacher Press.

Subramaniam, Banu, "Snow Brown and the Seven Detergents: A Metanarrative on Science and the Scientific Method," *Women's Studies Quarterly* 28, nos. 1 & 2 (spring/summer 2000): 296–304, special issue, *Building Inclusive Science: Connecting Women's Studies and Women in Science and Engineering*, edited by Sue V. Rosser. Copyright © 2000 by Banu Subramaniam. Reprinted by permission of The Feminist Press at the City University of New York.

Horn, Dara, "The Shoulders of Giants," *Science* 280 (1998): 1354–55. Copyright © 2000 by American Association for the Advancement of Science. Reprinted with permission from AAAS.

Section Ib. Careers: . . . And into the Fire

Wennerás, Christine, and Wold, Agnes, "Nepotism and Sexism in Peer Review." Reprinted by permission from *Nature* vol. 387 (1997): 341–43. Copyright © 1997 by Macmillan Publishers Ltd. Reprinted by permission from Macmillan Publishers Ltd.

Rose, Hilary, "Nine Decades, Nine Women, Ten Nobel Prizes: Gender Politics at the Apex of Science," in *Love, Power, and Knowledge: Toward a Feminist Transformation of the Sciences*, pp. 136–70. Bloomington: Indiana University Press. Copyright © 1994 by Indiana University Press. Reprinted by permission of Indiana University Press.

Lee, Jaekyung (2005), "Asian Americans and the Gender Gap in Science and Technology," in *"Strangers" of the Academy: Asian Female Scholars in Higher Education*, edited by Guofang

Li and Gulbahar H. Beckett, pp. 37–55. Sterling, VA: Stylus Publishing, LLC. Copyright © 2006, Stylus Publishing, LLC. Reprinted by permission of the publisher.

Wayne, Marta, "Walking a Tightrope: The Feminist Life of a *Drosophila* Biologist," *NWSA Journal* 12, no. A3 (2000): 139–50. Copyright © 2000 by *National Women's Studies Association Journal.* Reprinted with permission of The Johns Hopkins University Press.

Section II. Stereotypes, Rationality, and Masculinity in Science and Engineering

Cohn, Carol, "Sex and Death in the Rational World of Defense Intellectuals," *Signs: Journal of Women in Culture and Society* 12, no. 4 (1987): 687–718. Copyright © 2000 by the University of Chicago Press. Reprinted with permission from the University of Chicago Press.

Barbercheck, Mary, "Science, Sex, and Stereotypical Images in Scientific Advertising," in *Women, Science, and Technology* (first edition). New York: Routledge, pp. 117–31. Copyright © 2001 by Mary Barbercheck. Reprinted by permission of the author.

Subramaniam, Banu (2001), "The Aliens Have Landed! Reflections on the Rhetoric of Biological Invasions," in *Making Threats: Biofears and Environmental Anxieties.* New York: Rowman & Littlefield, pp. 135–48. Reprinted by permission of Indiana University Press.

Faulkner, Wendy, "The Power *and* the Pleasure? A Research Agenda for 'Making Gender Stick' to Engineers," *Science, Technology, & Human Values* 25, no. 1 (2000): 87–119. Copyright © 2000 by Sage Publications. Reprinted by Permission of Sage Publications, Inc.

Section III. Technologies Born of Difference

Tone, Andrea, "A Medical Fit for Contraceptives," in *Devices and Desires: A History of Contraceptives in America,* pp. 117–49. New York: Hill and Wang. Copyright © 2001 by Andrea Tone. Reprinted by permission of Hill and Wang, a division of Farrar, Straus and Giroux, LLC.

Bleier, Ruth, "Sociobiology, Biological Determinism, and Human Behavior," in *Science and Gender: A Critique of Biology and Its Theories on Women,* pp. 15–48. New York: Teachers College Press. Copyright © 1984 by Teachers College, Columbia University. All rights reserved. Reprinted by permission of the publisher.

Kessler, Suzanne, "The Medical Construction of Gender: Case Management of Intersexed Infants," *Signs: Journal of Women in Culture and Society* 16, no. 1 (1990): 3–26. Copyright © 1990 by the University of Chicago Press. Reprinted with permission from the University of Chicago Press.

Fausto-Sterling, Anne, "The Bare Bones of Sex: Sex and Gender," *Signs: Journal of Women in Culture & Society* 30, no. 2 (2005): 1491–1527. Copyright © 2005 by the University of Chicago Press. Reprinted with permission from the University of Chicago Press.

Section IV. The Next Generation

Keller, Evelyn Fox, "Gender and Science: An Update," in *Secrets of Life, Secrets of Death: Essays on Language, Gender and Science,* pp. 15–36. New York: Routledge. Copyright © 1992 by Taylor & Francis/Routledge, Inc. Reprinted by permission of Taylor & Francis/Routledge, Inc.

Fedigan, Linda Marie, "The Paradox of Feminist Primatology," in *Feminism in the 20th*

Century, ed. Angela Kreiger, Elizabeth Lunbeck, and Londa Schiebinger, pp. 46–72. Reprinted by permission of the University of Chicago Press

Maines, Rachel, "Socially Camouflaged Technologies: The Case of the Electromechanical Vibrator," *IEEE Technology & Society Magazine* 8, no. 2 (June 1989): 3–11. Copyright © 1989 by IEEE. Reprinted with permission from IEEE.

Balsamo, Anne, "On the Cutting Edge: Cosmetic Surgery and New Imaging Technologies," in *Technologies of the Gendered Body: Reading Cyborg Women*, pp. 56–79. Copyright © 1995 by Duke University Press. All rights reserved. Used by permission of the publisher.

Section V. Reproducible Insights

Perry, Ruth, "Engendering Environmental Thinking: A Feminist Analysis of the Present Crisis," *The Yale Journal of Criticism* 6 (1993): 1–17. Copyright © 2000 by Yale University and The Johns Hopkins University Press. Reprinted with permission of The Johns Hopkins University Press.

Rana, Shruti, "Fulfilling Technology's Promise: Enforcing the Rights of Women Caught in the Global High-Tech Underclass," *Berkeley Women's Law Journal* 15 (2000): 272–311. Copyright © 2000 by the Regents of the University of California. Reprinted from *Berkeley Women's Law Journal* by permission of the Regents of the University of California.

Martin, Emily, "Premenstrual Syndrome, Work Discipline, and Anger," in *The Woman in the Body*, pp. 113–38. Boston: Beacon Press. Copyright © 1987, 1992 by Emily Martin. Reprinted by permission of Beacon Press, Boston.

Woliver, Laura, "Reproductive Technologies, Surrogacy Arrangements, and the Politics of Motherhood," in *Mothers in Law: Feminist Theory and the Legal Regulation of Motherhood*, ed. M. A. Fineman and I. Karpin, pp. 346–59. New York: Columbia University Press. Copyright © 1995 by Columbia University Press. Reprinted by permission of Columbia University Press.

Index

Page references to non-textual information such as Figures or Tables are in *italic* print, while references to notes have the letter 'n' following the relevant note.